W0106571

Vibrations at Surfaces

Vibrations at Surfaces

Edited by

R. Caudano
J.-M. Gilles
and
A. A. Lucas

Facultés Universitaires
Notre-Dame de la Paix
Namur, Belgium

PLENUM PRESS • NEW YORK AND LONDON

Library of Congress Cataloging in Publication Data

Main entry under title:

Vibrations at Surfaces.

"Proceedings of the Second International Conference on Vibrations at Surfaces,
held September 10-12, 1980, at the Facultes Notre-Dame de la Paix, Namur,
Belgium" — Verso of t.p.
 Bibliography: p.
 Includes index.
 1. Vibrational spectra — Congresses. I. Caudano, R. II. Gilles, J.-M. III. Lucas, A.
A. IV. International Conference on Vibrations at Surfaces (2nd : 1980 : Facultes
universitaires Notre Dame de la Paix, Namur) V. Facultes universitaires Notre-Dame
de la Paix, Namur.
QD96.V53V53   541.3'453   81-15830
ISBN-13: 978-1-4684-4060-7   e-ISBN-13: 978-1-4684-4058-4   AACR2
DOI: 10.1007/978-1-4684-4058-4

Proceedings of the Second International Conference on
Vibrations at Surfaces, held September 10-12, 1980, at
the Facultés Notre-Dame de la Paix, Namur, Belgium

© 1982 Plenum Press, New York
Softcover reprint of the hardcover 1st edition 1982
A Division of Plenum Publishing Corporation
233 Spring Street, New York, N.Y. 10013

All rights reserved

No part of this book may be reproduced, stored in a retrieval system, or transmitted
in any form or by any means, electronic, mechanical, photocopying, microfilming,
recording, or otherwise, without written permission from the Publisher

PREFACE

    This volume contains most of the invited and contributed papers presented at the second international conference devoted to the general topic "Vibrations at Surfaces" and which took place from 10 to 12 September 1980 at the Facultés Notre-Dame de la Paix in Namur, Belgium.

    The conference was organized to review the large amount of information gathered in this field over the late seventies as a result of the rapid improvements and dissemination of surface spectroscopic technique such a electron energy loss, infrared and Raman surface spectroscopies.

    Much time was devoted to Raman spectroscopy of adsorbed molecules. After several years of vivid debate over the causes of the observed large enhancement of Raman cross section, a clearer picture emerges from the papers presented here: the actual value of the enhancement factor does depend in a complicated manner on long range surface roughness, atomic-scale roughness and the dielectric properties of the substrate as well as on the electronic structure of the molecule in its adsorbed state.

    Less controversial are the results obtained with electron energy loss spectroscopy (EELS) and several sessions of the conference were devoted to the approach. As witnessed by the growing number of laboratories using the technique, EELS is now a mature spectroscopic tool for the characterization and analysis of the chemisorption bond.

    One of the ultimate goals of surface vibrational spectroscopies being a better understanding of the energy transfer and dissipation processes and of chemical reaction paths at solid surfaces, a few papers were included to deal in the advances in this area.

    Inelastic atomic and molecular beam scattering as a way to investigate surface vibrations was not developed in the conference except for one paper. At the writing of this preface, important new advances are being made with this technique which, in the next few years, may be expected to join force with photon and electron spectroscopies for the experimental studies of dynamical surface properties.

We have great pleasure in acknowledging the enthusiastic response and participation of our invited speakers and we thank all authors for their diligent preparation of the manuscripts.

We are particularly indebted to Mr. J. Ghijsen, our Conference Secretary, for his outstanding job in meeting deadlines before and after the conference and in keeping people happy during the meeting. Much of the conference success is to be credited to him.

We are grateful to our younger colleagues, Drs. Darville, Delrue, Pireaux, Riga and Mr. Thiry for their very valuable help with the practical organization of the meeting and the editing of the manuscripts.

Nearly all of the papers were typed by Mme. Coulonval-Polet, whom we thank most sincerely for her usual efficiency and care. We are also thankful to Mr. Lotin of the Audio-Visual Center for taking care of the proper reproduction of the figures and tables in the manuscripts.

The accommodation and social programs were organized by Mr. G. Kelner and his staff at the Public Relation Office with their customary superior style. Many thanks to them for their outstanding service.

Finally we would like to express our deepest gratitude to all of our sponsors for providing the credit, financial and otherwise, which helped us ensure a participation at the highest level:

> Facultés Notre-Dame de la Paix
> IBM Belgium and IBM Europe
> Belgian National Science Foundation (FNRS)
> Belgian Ministry for Science Policy
> State University of Mons
> Shell
> Esso Belgium
> Labofina
> ICI
> Belgian Physical Society
> European Physical Society

> > R. Caudano
> > J.M. Gilles
> > A. Lucas

CONTENTS

1. Clean and Adsorbed Surfaces

5. Other Spectroscopies

6. <u>Surface Molecular Dynamics and Reactivity</u>

THEORY OF PHONONS ON CLEAN AND ADSORBED SURFACES

L. Dobrzynski and B. Djafari-Rouhani

Laboratoire des Surfaces et Interfaces, associé au
C.N.R.S.
I.S.E.N. 3, rue François Baës, 59046 LILLE CEDEX FRANCE

ABSTRACT

We review the main aspects of vibrations on clean and adsorbed surfaces. Surface acoustical modes (Rayleigh waves) are discussed, in particular their dispersion and the modifications due to adsorption. The main physical effects associated with surface optical phonons are also examined.

The main results for the vibrational quantities like the specific heat, the entropies and the atomic mean square displacements are also studied.

The elastic interaction energies of one or two defects near clean and adsorbed surfaces are given.

The stability of surfaces with and without adsorbate (relaxation , thermal expansion,...) are analyzed in relation with vibrational properties.

INTRODUCTION

The study of vibrations at surfaces started with Lord Rayleigh[1] in 1887. He predicted the existence of acoustical surface waves, now well known as Rayleigh waves. The first experiments to excite surface waves with comb-shaped electrode transducers were realized[2] only in 1965. One knows[3] how many applications followed since, in particular in electronics.

Subsequently, the vibrations of semi-infinite or finite crystals began to be studied from a lattice dynamical, as opposed to a continuum, point of view[4] . It was discovered that surface modes of optical character can also exist in crystals with more than one atom in a primitive unit cell. It was suggested[4] that such optical surface modes may lead to additional lines in the infrared absorption and Raman spectra of crystals.

The introduction of a crystal surface also modifies the frequencies and displacement fields of the bulk wavelike modes. It follows that the presence of surfaces will alter the bulk frequency distribution functions, of the order of the ratio of the number of atoms in the surface to the number in the interior. Consequently the presence of surfaces will alter the temperature dependences of the vibrational thermodynamic functions of a crystal from their bulk forms. One expects also that dynamical properties of individual atoms, such as their mean square displacements are different for atoms in the surface layers of a crystal from what they are for atoms in the bulk of the crystal. It is important to take these differences into account in the interpretation of experimental data obtained by the techniques of low-energy electron diffraction from crystal surfaces, or the Mössbauer effect.

The existence of adsorbate induced optical phonons was first shown on simple two-dimensional[5] and three dimensional[6] models. These high-frequency modes are measured through the inelastic scattering of low energy electrons, atoms or neutrons, as well as by infrared and Raman techniques and most of this conference is devoted to the studies[7] . The adsorbed atoms modify also all the vibrational surface properties discussed above.

Another question of interest is to study the energy of interaction of defects near a surface with or without an adsorbed layer of atoms and try to understand when defects are attracted or repeled by a surface. These studies began recently and we will just mention what's known for the elastic energy of interaction .

The stability of surfaces and monolayers, i.e. relaxation, thermal expansion, phase transitions within adsorbed monolayers,.... attracts a lot of interest. In the context of this review on vibrational properties, we will try to underline some simple ideas which may help to solve these open questions.

The next section of this paper will be devoted to surface vibration modes. Section III will deal with thermodynamic vibrational properties, Section IV with the interaction of defects with surfaces and the last Section with the stability of surfaces. This review is not intended to be exhaustive and the reader may find more detailed discussions and references in a recent textbook[8].

SURFACE MODES

Surface vibration waves are modes in which the displacement amplitudes are relatively large for atoms in the vicinity of the surface and decay with increasing distance into the crystal in an essentially exponential fashion. Because the crystal retains translational periodicity in the directions parallel to the free surface the atomic displacement amplitudes are wavelike in directions parallel to the surface, and the two-dimensional continuum of surface modes is characterized by a two-dimensional continuum wave vector $\vec{k}_{/\!/}$, whose components are also parallel to the surface. The frequencies of the localized surface modes lie in ranges (gaps) forbidden to the normal modes of the infinitely extended crystal.

Surface modes can be characterized as either acoustic or optical, depending on whether or not their frequencies vanish with the components of $\vec{k}_{/\!/}$. In the former case, the surface modes split off from the bottom of the lowest band of acoustic bulk mode frequencies having the same wave vector $\vec{k}_{/\!/}$.

However such a classification is too crude. Sometimes surface modes may also split off from the top of the bulk bands, if the surface layers are contracted and hence the surface force constants increased. It is also often the case that "window gaps" are observed in the spectrum of acoustic and optical vibration modes of a bulk crystal. Localized surface modes can also exist in these "window gaps" and cannot be classified simply as either acoustic or optical surface modes.

The earliest surface waves studied are those which propagate along the stress-free, planar surface of an isotropic elastic medium. The frequency ω of these Rayleigh[1] waves was found to be given by

$$\omega = C_R \, k_{/\!/} \tag{1}$$

where C_R, the speed of these surface waves, is the only solution of the equation

$$\left(\frac{C_R^2}{C_t^2}\right)^3 - 8\left(\frac{C_R^2}{C_t^2}\right)^2 + 8\left(3 - 2\frac{C_t^2}{C_1^2}\right)\frac{C_R^2}{C_t^2} - 16\left(1 - \frac{C_t^2}{C_1^2}\right) = 0 \tag{2}$$

which satisfies the condition $(C_R/C_t) < 1$. In this equation, C_t and C_1 stand respectively for the bulk transverse and longitudinal speeds of sound. Physical values of the ratio (C_R/C_t) range from 0.96 for $(C_1^2/C_t^2) = \infty$, corresponding to an incompressible solid, to 0.69 for $(C_1^2/C_t^2) = 4/3$, the smallest value consistent with the stability of the elastic medium. For example, in the typical case in which $C_\ell^2 = 3\, C_t^2$, we find that $(C_R/C_t) = 0.9194$.

The particle displacements in a Rayleigh wave execute ellipses in the sagittal plane, i.e., the plane containing both the normal to the surface and the direction of propagation of the wave. The attenuation lengths of the wave into the elastic medium are of the order of the wavelength $\lambda = (2\pi/k_\text{//})$ of the wave along the surface.

For larger values of $k_\text{//}$, the Rayleigh waves show a dispersion and depart from the law given by Eq. (1). This dispersion was computed for different atomic models[8]. Let us give here only the first correction to this law obtained[9] in closed form for a simple cubic model[4] for which $C_1^2 = 3\, C_t^2$

$$\omega^2 = 0.846\ C_t^2 k_\text{//}^2\ (1 - 0.057\ |k_\text{//}|a + ...) \tag{3}$$

where a is the lattice parameter.

When an adsorbed layer of thickness h is deposited on the surface of an elastic substrate, one expects the frequency of the Rayleigh waves to be changed. Let $\{C'_{11},\ C'_{12},\ C'_{44}\}$ be the elastic constants of the adsorbed film and ρ' its mass density. The eigenmodes of thin deposited films were studied[10] extensively. Let us give here in closed form the first correction to the frequency of the Rayleigh wave for a thin adsorbed layer[9] $(hk_\text{//} \ll 1)$

$$\omega^2 = C_R^2\ k_\text{//}^2\ (1 - \gamma\ |k_\text{//}|\ h + ...) \tag{4a}$$

where

$$\gamma = \frac{4C'_{44}}{C'_{11}}(C'_{12}+C'_{44})\beta_t - \rho'C_R^2(\beta_t+\beta_1)/2C_{44}\ \{1+\beta_t^2-\beta_t\beta_1(\frac{1}{\beta_t^2} + \frac{1}{\frac{C_{11}}{C_{44}} - 1+\beta_t^2})\} \tag{4b}$$

with

$$\beta_t = (1 - \frac{C_R^2}{C_t^2})^{1/2}\ ;\ \beta_1 = (1 - \frac{C_R^2}{C_1^2})^{1/2} \tag{4c}$$

Remember that for the substrate $C_1^2 = C_{11}/\rho$ and $C_t^2 = C_{44}/\rho$.

The total dispersion in presence of a thin layer is obtained when adding in the frame of the above models the corrections in $|k_\text{//}|^3$ given by Eqs (3) and (4).

These results are of particular interest in connection with experimental studies of adsorption with the help of Rayleigh waves. Preliminary results[11] show that this technique is sensitive

to about one hundredth of an adsorbed monolayer. It looks very
promising for the study of the changes of long range order within
adsorbed monolayers.

Other types of surface acoustical waves are also of great
interest. We do not describe them here and send the interested
readers to other more detailed reviews[3,8].

A great deal of theoretical results appeared for the different
surface modes of ionic, nonionic and covalent crystals. These works
were reviewed recently[8]. Let us just insist here on the fact that
surface localized modes may split off the bulk optical continuum
for a given value of $\vec{k}_{/\!/}$ and appear in gaps between optical and
acoustical bands. These surface optical modes as well as the adsorbate induced optical vibrations, can be measured by different surface
spectroscopic methods[7]. A great deal of experimental and theoretical papers appeared on this subject and were also recently
reviewed[7,8,12]. Let us therefore underline here only a few main
qualitative effects.

Adsorbed atoms lighter than the substrate atoms increase the
frequencies of the surface modes and new modes may also appear
above the bulk phonon bands for a given value of $\vec{k}_{/\!/}$. An increase in
the surface force constants has a similar effect as a light monolayer. Due to this effect, it was observed on surface steps that
surface modes may split off from the top of the bulk acoustical
bands[13].

It is possible for surface modes to exist even within some
bulk subbands, when their polarizations are orthogonal to those of
the bulk phonons having the same frequency.

A branch of surface waves may correspond to localized modes
in one part of the surface two dimensional Brillouin zone and to
resonant modes in another part of this zone. The resonant modes
may produce sharp features in the phonon density of states. Their
frequencies fall within the bulk bands, at the difference of true
localized modes which exist only in gaps.

In many instances, the atoms within an adsorbed monolayer
assume a configuration with symmetry lower than that associated
with the surface of the substrate. The effect of the superstructure
is to induce new zone boundaries in the surface Brillouin zone. One
has to fold back all the phonon spectra within a smaller Brillouin
zone. As a consequence, gaps in the surface mode dispersion relations open up at the zone boundaries and one is getting more distinct surface phonon branches. In particular the surface modes
which were at the edges if the Brillouin zone for an unreconstruc t-
ed surface are now at $k_{/\!/} = 0$ and can be detected by the usual
spectroscopic methods[14]. Some parts of the folded localized sur-

face branches may fall within bulk bands and be transformed in surface resonant modes.

Before any quantitative fit of the experimental results for the optical surface phonons can be made, one needs to know the surface adatom distance as well as the adsorption geometry. This distance is for some systems known from experimental determinations, otherwise one may approximate it from molecular interatomic distances. Therefore models[13] which give the vibration frequencies as well as the interatomic distances are of great interest.

Specific examples of the effects above described appear in different papers at this conference. Therefore we feel that it is unnecessary to illustrate here these fundamental considerations.

THERMODYNAMIC VIBRATIONAL PROPERTIES

The frequency distribution functions for infinite and finite crystals differ by an amount proportional to the ratio of the surface area of the finite crystal to its volume. This effect was observed for small particle powders by neutron scattering[15]. As a consequence, every extensive vibrational property of a finite crystal, including the thermodynamic functions, should have a contribution which is proportional to the volume of the crystal.

Surface Specific Heat

The low temperature surface contribution to the vibrational specific heat was found[16,8] for an isotropic elastic solid to be

$$\Delta C_v(T) = 3\pi \frac{k_B^3}{h^2} \xi(3) \frac{2C_t^4 - 3C_t^2 C_1^2 + 3C_1^2}{C_t^2 C_1^2 (C_1^2 - C_t^2)} ST^2 + O(T^2) \tag{5}$$

where S is the surface area, T the temperature and $\xi(3)$ the Riemann zeta function of argument 3.

At temperatures well above the Debye temperature, the total vibrational specific heat of the crystal must be $3N k_B$, where N is the number of atoms; hence, the surface specific heat must be zero in this limit.

The surface specific heat, therefore, starts at zero for T = 0, rises to a maximum, and then approaches zero at high temperatures. Experimentally, the surface specific heat can be observed[17] on fine powders (large S) at low temperatures. The presence of an adsorbed monolayer of atoms contributes[18] an ST^3 term to the above law[5].

Surface Entropy

The vibrational contribution is also important for the evaluation of the surface entropies[8],[19]. The surface entropies are of the order of k_B at high temperatures and increase near surface defects, steps, kinks... It is easy to obtain an order of magnitude estimation in an Einstein approximation. Let ω_B be the bulk Einstein frequency and ω_S the Einstein frequency of a surface atom. The squares of these two frequencies are roughly proportional to the number of nearest neighbours N_B and N_S of a bulk and a surface atom.

$$\frac{\omega_S^2}{\omega_B^2} = \frac{N_S}{N_B} \tag{6}$$

Note that N_B/N_S is usually of the order of 1.2 to 2.
The vibrational entropy of an harmonic oscillator is given at high temperature by[8]

$$S \simeq k_B \left(\frac{1}{2} - \ln \frac{\hbar\,\omega}{k_B T} \right) \tag{7}$$

The variation of entropy due to a surface is then per surface atom

$$\Delta S = k_B \ln \frac{\omega_B}{\omega_S} = \frac{1}{2} k_B \ln \frac{N_B}{N_S} . \tag{8}$$

The measure of the surface free energies

$$\Delta F = \Delta E - T\Delta S \tag{9}$$

as a function of T, provides estimations of ΔS[8].

The variations of entropies due to adsorption can be evaluated in the same manner[8],[20].

Mean Square Displacements of Surface Atoms

Low energy électron diffraction (LEED) provides a means of investigating the mean square displacements of surface atoms through measurements of the Debye-Waller factors[21]. For a single simple harmonic oscillator, the Boltzmann equipartition of energy E provides us with

$$M\omega^2 \langle u^2 \rangle = \langle E \rangle = 3\,k_B T \tag{10}$$

where M is the mass of the oscillator and $<u^2>$ the mean square displacement of an atom, assumed to be the same in'the three directions of space. Then with the help of the Einstein approximation (Eq. 6), one obtains easily an estimate of the ratio between the mean square displacements of surface and bulk atoms

$$\frac{<u^2>}{<u^2>_B} = \frac{\omega_B^2}{\omega_S^2} = \frac{N_B}{N_S} \tag{11}$$

This shows that the mean square displacements are about two times larger for surface atoms than for bulk atoms.

More detailed calculations[8] show that surface atom mean square displacements perpendicular to the surface are usually bigger than the parallel ones.

With the same simple approach, one understands at once that the mean square displacements of adsorbed atoms are also different of the surface ones[22] . Let ω_A be the Einstein frequency of an adsorbed atom of mass M'. Then with the help of Eq. (10), one obtains for the mean square displacement $<u^2>_A$ of the adsorbed atom

$$\frac{<u^2>_A}{<u^2>_B} = \frac{M}{M'} \frac{\omega_B^2}{\omega_A^2} \tag{12}$$

The difference $<u^2>_S$ between the surface and bulk mean square displacements is inversely proportionnal to the distance x_3 from the surface, already at a few interatomic distances away from the surface[23] let us define

$$\nu = C_{44}/C_{11} \tag{13}$$

Then for an isotropic solid one obtains [23-24], respectively, for the components perpendicular and parallel to the surface

$$<u_\perp^2(x_3)>_S = \frac{k_B T}{16\pi C_{44}} \frac{(3-2\nu + \nu^2)}{(1 - \nu)} \frac{1}{x_3} + \ldots \tag{14}$$

and

$$<u_\parallel^2(x_3)>_S = \frac{k_B T}{32\pi C_{44}} \frac{(3-4\nu + 3\nu^2)}{(1 - \nu)} \frac{1}{x_3} + \ldots \tag{15}$$

When an isotropic adsorbed layer of thickness h and elastic constants
$\{C'_{44}, C'_{11}\}$ is deposited on the surface the total atom mean square
displacements can be put in the following form

$$\langle u^2_\alpha(x_3)\rangle = \langle u^2_\alpha\rangle_B + \langle u^2_\alpha(x_3)\rangle_S + \langle u^2_\alpha(x_3)\rangle_A \qquad (16)$$

When defining

$$\nu' = \frac{C'_{44}}{C'_{11}} \quad \text{and} \quad \gamma = \frac{C'_{44}}{C_{44}} \qquad (17)$$

one obtains[9] for $h/x_3 \ll 1$, the corrections to the laws (14) and
(15) valid for atoms within the substrate

$$\langle u^2_\perp(x_3)\rangle_A = \frac{-k_B T}{16\pi C_{44}} \frac{h}{x_3^2} \gamma (1-\nu') \frac{(3-2\nu+\nu^2)}{(1-\nu)^2} \qquad (18)$$

$$\langle u^2_{/\!/}(x_3)\rangle_A = -\frac{k_B T}{16\pi C_{44}} \frac{h}{x_3^2} \gamma [\, 1+(1-\nu') \frac{(1-2\nu+3\nu^2)}{2(1-\nu)^2} \,] \qquad (19)$$

These asymptotic laws are important because in L.E.E.D. the
electrons penetrate a few layers inside the solid. Then due to
multiple scattering of the electrons, one is not able to analyse
the temperature dependence of the LEED intensities by a simple
Debye-Waller factor as for X rays diffraction by bulk atoms[25].

INTERACTION OF DEFECTS WITH SURFACES

There has been considerable interest recently in the interaction of adatoms on a solid surface. This interaction can be
direct and it can also be indirect, through the substrate. The
direct, dipole-dipole, interaction between two adatoms on a metal
surface has been determined[26]. The indirect interaction between
two adatoms on a metal substrate, mediated by the conduction
electrons, first discussed by Grimley[27] is still under study. The
indirect interaction of two adatoms through the phonon field of
the substrate was also investigated[28-29]. The interaction between
two adatoms mediated by the elastic distorsion of the substrate to
which they give rise, was also calculated[30-31] and shown to be
three or four orders of magnitude larger than that arising from the
phonon field which is its leading quantum correction.

So the interaction between two adatoms is still an open problem
although one knows which are the three more important physical

effects, namely the interaction mediated by the elastic distorsion of the substrate, the direct dipole-dipole interaction and for metal substrates the indirect interaction by the conduction electrons.

More work will be necessary before being able to make these theoretical results quantitative and to compare them to the experimental results obtained by field emission microscopy[32] . A recent review[33] appeared recently on this subject.

Therefore we will just underline here somewhat different, but related results; namely the elastic interaction of one and then two point defects with the stress-free, planar boundary (situated at $x_3 = 0$) of the semi-infinite isotropic elastic medium in which the point defect is situated.

We denote by $\vec{F}$ the body force per unit volume due to the point defect. The defect is represented by the superposition of three mutually perpendicular double forces without moment centered at the point x_o. That is, we express $F_\alpha(\vec{x})$ as

$$F_\alpha(\vec{x}) = -A_\alpha \frac{\partial}{\partial x_\alpha} \delta(\vec{x} - \vec{x}_o), \alpha = 1,2,3 \tag{20}$$

where A_α is a constant with the dimensions of force times lenght. In what follows we suppose the defect isotropic ($A_1 = A_2 = A_3 = A$) The energy of interaction of this defect with the surface is defined as the difference between the strain energy when the defect is at a distance x from the surface and the strain energy when the defect has been removed to infinity.

$$U(x_3) = U_S(x_3) - U_S(\infty) \tag{21}$$

In the limit $x_3 \to 0$, the atomic character of the surface is important and such effects as dislocations, surface roughness, can give important contributions to the strain energy of interaction of the defect with the surface. Therefore we give in what follows only the asymptotic form of the interaction energies $U(x_3)$ expected to be accurate within a few percents already at a few atomic distances from the free surface[34].

$$U_S(x_3) = -\frac{A^2}{8\pi C_{11}} \frac{\nu}{(1-\nu)} \frac{1}{x_{03}^3} + \cdots \tag{22}$$

ν was defined by Eq. (12).

One sees that the defect is always attracted by the free surface as $\nu < 1$. The presence of the adsorbed layer defined above contributes to this interaction energy a first correction term[9] , valid for (h/x_{03}) small

$$U_A(x_3) = \frac{3h}{8\pi} A^2 \frac{\gamma}{C_{44}} \frac{\nu^2}{(1-\nu)^2} (1-\nu') \frac{1}{x_{03}^4} + \ldots \tag{23}$$

So the total energy of interaction of the defect with the surface covered by the thin layer is

$$U(x_3) = -\frac{A^2}{8\pi C_{11}} \frac{\nu}{(1-\nu)} \frac{1}{x_{o3}^3} [1-3\gamma \frac{(1-\nu')}{(1-\nu)} \frac{h}{x_{o3}} + \ldots] \tag{24}$$

In the case of two point defects, the density of body force characterizing the defects is now the sum of two contributions

$$\vec{F}(\vec{x}) = \vec{F}^{(1)} (\vec{x}|\vec{x}^{(1)}) + \vec{F}^{(2)}(\vec{x}|\vec{x}^{(2)}) \tag{25}$$

each of which is associated with one of the point defects. In the above equation $\vec{x}^{(1)}$ and $\vec{x}^{(2)}$ are the positions of the two defects which in general, need not be identical. The interaction strain energy is defined as the difference between the total strain energy and the sum of the self-strain energies of the individual defects. It is possible[34] to separate this interaction strain energy into a part which corresponds to an infinite medium and a part which reflects the presence of a stree-free surface

$$U(\vec{x}^{(1)}; \vec{x}^{(2)}) = U^{(\infty)} (\vec{x}^{(1)}; \vec{x}^{(2)}) + U^{(i)}(\vec{x}^{(1)}; \vec{x}^{(2)}) \tag{26}$$

Note that the first term in the above equation is non zero only for anisotropic defects. It vanishes then here.

The interaction strain energy between two defects near a free surface was obtained[34] as

$$U_S^{(i)}(x^{(1)}; x^{(2)}) = -\frac{\nu A^2}{\pi C_{11}(1-\nu)} \frac{2}{(R_{/\!/}^2 + \overline{R}_3^2)^{3/2}} P_2(\frac{\overline{R}_3}{(R_{/\!/}^2+\overline{R}_3^2)^{1/2}}) \tag{27}$$

where

$$\vec{R}_{/\!/} = \vec{x}_{/\!/}^{(1)} - \vec{x}_{/\!/}^{(2)} \; ; \; \overline{R}_3 = |x_3^{(1)}| + |x_3^{(2)}| \tag{28}$$

and

$$P_2(x) = \frac{1}{2} (3 x^2 - 1) \tag{29}$$

This interaction energy is positive in regions where $\overline{R}_3{}^2 > \frac{1}{2} R_{/\!/}^2$ negative in regions where $\overline{R}_3{}^2 < \frac{1}{2} R_{/\!/}{}^2$ and varies essentially as the inverse cube of the distance between the two defects.

As above, an adsorbed layer contributes to the interaction energy a first correction term[9] , valid for $[h/(R_{/\!/}^2 + \overline{R}_3{}^2)^{1/2}]$ small

$$U_A^{(i)}(\vec{x}^{(1)}; \vec{x}^{(2)}) = \frac{2h}{\pi} \gamma(1-\nu') \frac{\nu A^2}{C_{/\!/} (1-\nu)^2} \frac{6}{(R_{/\!/}^2 + \overline{R}_3{}^2)^2} P_3\left(\frac{\overline{R}_3}{(R_{/\!/}^2 + \overline{R}_3{}^2)^{1/2}}\right)$$

(30)

where

$$P_3(x) = \frac{1}{2} (5 x^3 - 3x)$$

(31)

The presence of a stress on the surface modifies also the above results and may even in some cases tend to repel the defect away from the surface[35].

SURFACE INSTABILITIES AND SUPERSTRUCTURES

When a crystal is cut by a plane, the atoms may keep the same positions as in the infinite crystal. But different other possibilities may occur to move the atoms out of these normal positions[8]:

a) The atoms on the first plane or planes near the surface may experience non zero forces. In this non equilibrium situation, the distance between the first planes will change in order to have equilibrium. This will lead to surface *relaxation* and a thermal expansion different than in a bulk surface.

b) The atoms near the surface may be in equilibrium, but with unstable positions. This instability may be described by soft surface phonons. The displacements of the atoms towards stable equilibrium will be controlled by anharmonic forces. In this situation, each atom can be labelled by its original position and the new surface structure is specified by finite (and usually small) displacements from the normal positions. In this case, one can speak of *superstructures*.

c) The atoms in normal positions are in a stable equilibrium configuration. But, there exists other situations of lower energy for the first plane (or planes).Simple cases may occur when the first plane keeps the same symmetry as in the bulk but with a different lattice parameter or when the first plane has a different symmetry. In these last cases, one speaks of *reconstruction* or incommensurate surface structures.

Other even more interesting instabilities and phase transitions are known in particular within physisorbed monolayers. It would be really too ambitious to review here all the works done on these still wide open problems and try to give definite explanations to these phenomena. Let us rather send to a recent review paper[36].

However within the context of this review on vibrations of surfaces and adsorbed monolayers, we may try to relate some of these phase transitions to the atom mean square displacements and the Lindeman criterium of melting[37].

Let us consider an adsorbed monolayer. The adsorbed atoms have a mass M' and interact between themselves by force constants β'' and with the substrate atoms with a force constant β'. A straightforward generalisation of Eq. (10) gives us the components of the mean square displacements of these atoms at high temperatures

$$<u^2_\alpha>_A = k_BT \sum_{k_\parallel j} \frac{1}{M'\omega^2_j(\vec{k}_\parallel)} \qquad \alpha = 1,2,3 \qquad (32)$$

In a frozen substrate apprixation and at the limit of long wavelengths, in a simple model[37] one obtains easily a three time degenerate branch of vibration frequencies

$$\omega^2_j = \beta' + \beta''k_\parallel^2 a^2 \qquad j = 1,2,3 \qquad (33\bar{a})$$

a being the lattice parameter.

Then

$$<u^2_\alpha>_A = \frac{k_BT}{2\pi} \int_0^{k_D} \frac{a^2 k_\parallel dk_\parallel}{\beta'+\beta''k_\parallel^2 a^2} \qquad (34)$$

Where k_D is a Debye type cutt-off. Finally,

$$<u^2_\alpha>_A = \frac{k_BT}{4\pi\beta''} \ln (1 + \frac{\beta''}{\beta'} k_\parallel^2 a^2) \qquad (35)$$

This very simple model shows that if $\beta' \to 0$, $<u^2_\alpha>_A \to \infty$. One cannot have long range order in two dimensions. Another limit of interest is $\beta''/\beta' \gg 1$. Because of the logarithm in the result (35) one sees that even a very week interaction with the substrate may stabilize the adsorbed monolayer.

In a more realistic model, one should distinguish between the components parallel $<u_{/\!/}^2>_A$ and perpendicular $<u_\perp^2>_A$ to the surface of the atom mean square displacements. The interesting physical point is to know if the $<u_{/\!/}^2>_A$ can be bigger or smaller than the $<u_\perp^2>_A$, as these two situations have clearly different physical meanings. In the first case one could expect the physisorbed layer to slide on the surface after losing long-range order.

In the second case desorption would be the leading effect.

6. CONCLUSION

This very short review of the theory of phonons on clean and adsorbed surfaces is of course incomplete. We deliberately by lack of space,did not mention several important aspects and cited at most only the pioneering works for each subject treated above. Our only excuse is that we refer the readers to more complete book[8] and review papers[3,7,10,12,13,33,36]

Let us finally insist on two questions which seem to be very important and still wide open for future studies, although already reviewed : namely surface roughness[38] and interface vibrations[39].

REFERENCES

1. Lord Rayleigh, Proc. London Math. Soc., 17, 4 (1887).
2. R.M. White and F.W. Voltmer, Appl. Phys. Lett., 7, 314 (1965).
3. See for example, L. Dieulesaint and D. Royer, Handbook of Surfaces and Interfaces, Ed. L. Dobrzynski, Garland STPM Press, New-York, Vol. II, p. 65 (1978).
4. I.M. Lifshitz and L.N. Rosenzweig, Zh. Eksp. Teor. Fiz., 18, 1012 (1948) and L.N. Rosenzweig, Tr. Fiz. Otdel. Fiz. Mat. Fakul'teta Khark Gos. Univ., 2, 19 (1950).
5. H. Kaplan, Phys. Rev., 125, 1271 (1962).
6. L. Dobrzynski and D.L. Mills, J. Phys. and Chem. Solids, 30, 1043 (1969).
7. See for references the other review papers at this Conference.
8. See for references and a more detailed discussion : A.A. Maradudin, R.F. Wallis and L. Dobrzynski, Surface Phonons and Polaritons, Handbook of Surfaces and Interfaces, Ed. L. Dobrzynski, Garland STPM Press, New-York, Vol. III (1980).
9. B. Djafari-Rouhani, L. Dobrzynski and V. Velasco, to be published.
10. G.W. Farnell and E.L. Adler, Physical Acoustics, Vol. IX, p. 35.
11. J. Pouliquen, M. Depoorter and A. Defebvre, Proc. of the Int. Conf. on Solid Surfaces, Cannes, 1980.
12. See for references and a more detailed discussion : L. Dobrzynski, G. Allan, B. Djafari-Rouhani, B.K. Agrawal and J. Lopez, Proc. of the Symposium on Vibrations at Surfaces, Cambridge 1979, to be published.
13. G. Allan, in the same Conference.

14. H. Ibach, M. Bruchmann, Phys. Rev. Lett., 44, 36 (1980).
15. K.H. Rieder and E.M. Hortl, Phys. Rev. Letters, 20, 209 (1968).
16. M. Dupuis, R. Mazo and L. Onsager, J. Chem. Phys., 33, 1452 (1960)
17. J.H. Barkmann, R.L. Anderson and T.E. Brackett, J. Chem. Phys.,
    42, 1112 (1965).
18. B. Djafari-Rouhani and L. Dobrzynski, Journal de Physique, 37,
    L-213 (1976).
19. L. Dobrzynski and J. Friedel, Surf. Sci., 12, 469 (1968).
20. L. Dobrzynski, Ann. Phys., 4, 637 (1969).
21. A.U. Macrae and L.H. Germer, Phys. Rev. Letters, 8, 489 (1962).
22. J.B. Theeten, L. Dobrzynski and J.L. Domange, Surf. Sci. 34,
    145 (1973).
23. J. Lajzerowicz and L. Dobrzynski, Phys. Rev. B 14, 2695 (1976).
24. R.F. Wallis, A.A. Maradudin and L. Dobrzynski, Phys. rev. B 15,
    5681 (1977).
25. C.B. Duke and G.E. Laramore, Phys. Rev. B 2, 4765 (1970).
26. W. Kohn and K.H. Lau, Solid State Comm., 18, 553 (1976).
27. T.B. Grimley, Proc. Phys. Soc., 90, 751 (1967).
28. M. Schick and C.E. Campbell, Phys. Rev. A2, 1591 (1970).
29. S.L. Cunningham, L. Dobrzynski and A.A. Maradudin, Phys. Rev.
    B 7, 4643 (1973).
30. L.H. Lau and W. Kohn, Surf. Sci., 65, 607 (1977).
31. A.M. Stoneham, Solid State Comm., 24, 425 (1977).
32. T.T. Tsang, Phys. Rev. B 6, 417 (1972).
33. For references, see M.C. Desjonqueres, Journal de Physique, 41,
    C3-243 (1980).
34. A.A. Maradudin and R.F. Wallis, Surf. Sci. 91, 423 (1980).
35. R.F. Wallis, A.A. Maradudin, L. Dobrzynski and B. Djafari-
    Rouhani, Proc. of the Int. Conf. on Solid Surfaces, Cannes 1980.
36. J. Villain, in "Ordering in strongly fluctuating condensed
    matter systems". Edited by Tormod Riste (Plenum Publishing
    Corporation), 1980, p. 221.
37. L. Dobrzynski and J. Lajzerowicz, Phys. Rev. B 12, 1358 (1975).
38. D. Castiel, A. Eguiluz, A.A. Maradudin, D.L. Mills and R.F.
    Wallis, Proceedings of the International Conf. on Lattice Dyna-
    mics (Paris 1977) Ed. M. Balkanski.
39. B. Djafari-Rouhani, L. Dobrzynski and P. Masri, to be published.

DESORPTION BY RESONANT MULTIPHOTON EXCITATION OF INTERNAL ADSORBATE VIBRATION[*]

J. Heidberg, H. Stein and E. Riehl

Institut für Physikalische und Theoretische Chemie der
Universität Erlangen-Nürnberg
Erlangen, FR Germany

ABSTRACT

Infrared laser induced desorption by resonant excitation of internal adsorbate vibration is demonstrated. Fast desorption takes place at low fluences D < 0.1 J cm^{-2} by multiphoton absorption from TEA CO_2 laser pulses in CH_3F-NaCl. Desorption yield Φ versus excitation frequency and Φ versus fluence plots as well as linear absorption spectra are presented. Considering localized and nonlocalized excitation a model of unimolecular desorption induced by monochromatic infrared is presented and found to be in concert with first experiments. The apparatus with the ultra high vacuum cryostat developed is described briefly.

Vibrational spectra exhibit, distinct from electronic spectra, a rich structure with sharp lines rendering high specificity in chemical analysis of molecular species. This holds especially for surfaces. The possibility of using this high specificity in selective excitation of chemical reactions was enhanced by the discovery that monochromatic intense infrared radiation can deposit tens of photons in isolated molecules and induce efficient and isotope-selective processes[1]. To elucidate whether a certain process may be promoted over others by stimulating a particular molecular vibration which is related to the motion along the reaction coordinate would be of interest indeed.

From picosecond relaxation measurements on large molecules in the gas phase, however, ultrashort life times of vibrational excita-

[*]Dedicated to Professor Klaus Schäfer on the Occasion of his 70th Birthday.

tion in the ground electronic state and very fast energy randomiza-
tion ($\gtrsim 10^{12}$ s^{-1}) have been inferred.[2] Moreover, the view has been
adopted that in solid matrices at low temperature rapid relaxation
prevents multiphoton excitation and vibrational ladder climbing in
matrix isolated molecules.[3] On the other hand, Legay and coworkers
have demonstrated that many small molecules in matrix and solid
state exhibit long vibrational life times not far from the radiative
life time. It was shown especially that the life time of the ν_3
vibration of fluoromethane isolated in krypton matrix is relatively
long, $\gtrsim$ 1 µs, a rotational mode not a translational being the
main relaxation channel in vibrational deactivation.[4] Only a
few pioneering investigations have been devoted to infrared induced
processes on solid surfaces by vibrational excitation.[5] Infrared
induced surface desorption by resonant vibrational excitation was
first shown only recently.[6]

In the present work, an ultrahigh vacuum cryostat developed
for spectroscopy and infrared laser chemistry on surfaces, could
be used to observe frequency selective desorption of fluoromethane
from alkali halide surfaces by resonant excitation of an adsorbate
vibration. Desorption yield versus excitation frequency spectra
were measured and found to be in agreement with transmission infrared spectra of the adsorbate. By resonance characteristics it
could be clearly distinguished between desorption and evaporation
from multi-layer adsorbates.[7] At surprisingly low infrared
fluence values, $\lesssim$ 0.1 J cm^{-2}, fast desorption occurred.

Types of Interaction between Infrared and Adsorption Systems

In Fig. 1. types of interaction between infrared radiation
and adsorption systems are proposed as suitable for this work.
All types of interaction considered have been observed, emphasis
being put on the substance- and site-selective resonant localized
interaction of monochromatic infrared radiation with surfaces.
Not included is the non-resonant localized interaction which may
nevertheless exist employing high intensity monochromatic radiation
interacting with adsorbate only.

An example of a non-resonant, non-localized process in general
is the infrared heating of metal surfaces. All the surface area
struck by the radiation is heated. The light is adsorbed essentially by the conduction electrons and the time for electronic
relaxation i.e. transfer of energy to lattice vibrations is much
shorter, 10^{-13}s, than ordinary chemical reaction times.

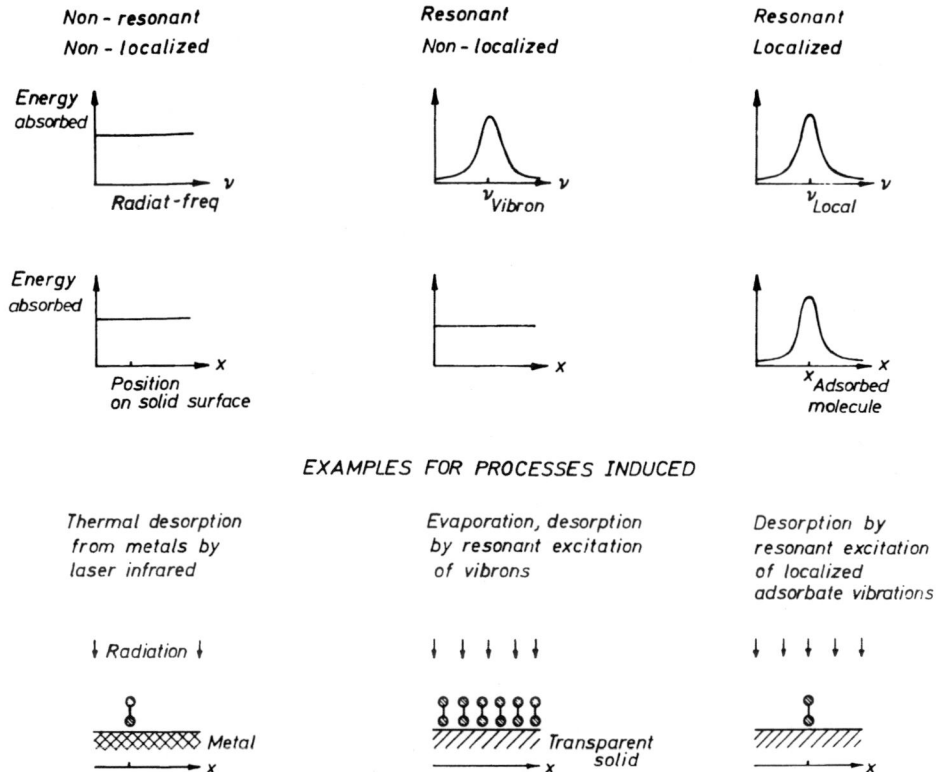

Fig. 1. : Types of interaction between infrared
and adsorption systems.

In general the normal thermal equilibrium laws e.g. for thermal
desorption are adequate : [8]

$$\tau_A^{-1} = A \exp \left[- E / k T \right] \quad , \qquad (1)$$

where τ_A is the mean sojourn time, τ_A^{-1} the unimolecular rate
coefficient, A the pre-exponenctial factor, being of the order
$10^{13} s^{-1}$, E the activation energy of desorption, k the Boltzmann
constant and T the temperature of the surface. An appreciable

fraction of molecules will desorb, when their mean sojourn time τ_A
on the surface becomes shorter than the high temperature period
which, using laser pulses, is roughly equal to the pulse width.
Extreme short pulses can be used ($< 10^{-10}$ s), where particular
caution must be taken in applying equilibrium kinetics. In any
case, under the defined conditions of ultrahigh vacuum, temperatures,
surface concentrations and rates can be attained which approach
values encountered in technical processes.

Absorption of light takes place not on the surface but
in the metal up to a depth of several hundred Ångstrom, the skin
depth; heat moves by conduction to greater depth,

$$d \simeq (t_p \kappa / c \rho)^{1/2} , \tag{2}$$

t_p being the heating time, κ the thermal conductivity, c the heat
capacity and ρ the density of the metal. The order of magnitude
of the maximum temperature increase for a heat pulse of Gaussian or
rectangular shape can be estimated from

$$\Delta T \simeq \epsilon D . (\kappa \rho c t_p)^{-1/2} \tag{3}$$

$\epsilon(\lambda)$ being the optical adsorptivity at wavelength λ, t_p the
halfwidth (FWHM) and pulse duration for Gaussian and rectangular
pulses, respectively.

Most interesting, if applicable, is resonant localized absorption in view of specificity and efficiency of substance purification,
separation, surface conversion, implantation annealing and catalysis. Adsorbate normal vibrations can be excited, which in general
are associated with spectral lines and appear to be highly localized.
Neighboring adsorbent atoms may have some amplitude modulated at
the adsorbate mode frequency, but this amplitude should fall off
with the distance from the adsorbate very quickly. By resonant
interaction with infrared radiation, pumping of selected localized
modes in the adsorbate is possible. Even with moderately intense
monochromatic infrared laser pulses, multiphoton absorption may
occur, inducing subsequent processes such as surface migration,
phase transition, reaction and desorption.

Sharp resonant localized interactions between infrared and
adsorbates can occur, of course, only if there is no strong light
absorption by the adsorbent. From the relatively little reliable
experimental data on the absorption by highly transparent solids,

it may be gathered that the most transparent solids in the mid infrared are alkali halides.   The absorption coefficient α of NaCl at
1000 cm^{-1}, 100 K is 1 x 10^{-7} μm^{-1}, the depth from which heat travels
by conduction according to (2) being much smaller than 1 μm,
so that thermal heating is negligible. α (NaCl) exhibits no
structure and drops exponentially with frequency in the short
wavelength side of the low phonon region, the onset frequency of
transparency may be taken as 3 ν_{LO}, ν_{LO} being the frequency of the
longitudinal optical fundamental.   Damage threshold for NaCl is
reported to be 17 J cm^{2}.   In this work no damage was observed
on the polished single NaCl crystals but on the NaCl films, bowl
shaped cavities of 0.2 mm diameter appeared, which might result from
single, otherwise undetected, giant pulses of G W cm^{-2} intensity.

Resonant excitation of an adsorbate vibration may not only be
achieved by transmission employing highly transparent adsorbents,
but also with surface electromagnetic waves.

At high surface coverage the internal vibrations of the
molecules constituting the adsorbate may be coupled into vibrons.
In this case the excitation may be resonant but non-localized.
Intermolecular vibration-vibration transfer becomes important as
soon as optical pumping and the relaxation to the phonon heat
bath is slower than V-V transfer.

Usually resonant V-V transfer is extremely fast.   Two types
of mechanisms may be distinguished : 1. V-V exchange between two
individual molecules. 2. Collective vibrational energy transfer
through the adsorbate.   For dipole-dipole interaction the energy
transfer between two molecules has been studied by Theodor Förster,
the probability $W_{D \to A}$ of energy transfer from an excited molecule
(donor) to a molecule in the ground state (acceptor) for random
orientation of the molecules and sharp lines, as usual for vibrational adsorbate lines, being

$$W_{D \to A} = \text{const} \ \frac{1}{(n \ R_{DA})^6} \ \frac{1}{\tau_D} \ \frac{1}{\tau_A} \ \int f_D(\nu) \ . \ f_A(\nu) \ d\nu, \qquad (4)$$

where n is the refractive index of the medium, R_{DA} the distance
between donor and acceptor molecules, τ_A and τ_D their radiative
life times, f_D and f_A their normalized line shapes, ν the vibrational frequency, hν the transferred quantum. Also collisions between
the molecules may provide a significant transfer mechanism.   Within
the Landau-Teller model for resonant VV-transfer in binary collisions
the process

$|V> + |V'> \rightarrow |V+1> + |V'-1>$ has a probability per unit time $\alpha(V+1) V'$. V-V transfer has been studied experimentally in the solid state only very recently. No experimental work is know concerned with V-V transfer in adsorbates. To give the order of magnitude of quasi-resonant V-V transfer rates, the value of the pseudo first order rate coefficient k' for vibrational hopping in solid CO at 30 K is cited :[9]

$$^{12}C^{16}O(v=1) + {}^{13}C^{16}O(v=0) \longrightarrow {}^{12}C^{16}O(v=0) + {}^{13}C^{16}O(v=1),$$

$k' = k\left[^{13}C^{16}O\right] = 4 \times 10^9 \text{ s}^{-1}$ (natural abundance, $\left[^{13}C^{16}O\right]$ mole fraction of $^{13}C^{16}O$). Considerations of this kind become especially important in the discussion of desorption from multi-layer adsorbates where molecules of the first layer can selectively be excited and molecular cluster and / or molecules from higher layers may desorb.

EXPERIMENTAL

APPARATUS. A block diagram of the apparatus is shown in Fig. 2. The main components are an ultrahigh vacuum cryostat with a quadrupole mass spectrometer QM used in time-of-flight mode, time resolution being limited by the time constant of the electro-meter pre-amplifier, a frequency tunable TEA CO_2 laser furnished with an intracavity aperture to generate infrared pulses of 200 ns (FWHM) duration and approximate TEM_{00} spatial distribution. Oscillation occurred simultaneously on several longitudinal modes at a single rotational vibrational line. The radiation beam was narrowed by focusing with two mirrors M_1 and M_2 in confocal arrangement. Ultrahigh vacuum cryogenic components and broad band high radiation intensity UHV tight windows will be described elsewhere.

Linear infrared absorption was measured in transmission under UHV using a double beam grating spectrophotometer (Perkin Elmer-Bodenseewerk 225).

SUBSTANCES. Fluoromethane (Matheson > 99%, < 1% SiF_4, $(CH_3)_2O$) was transferred into a cold trap at 77 K, degassed for 15 min at 10^{-5} mbar and fractionally distilled into glass bulbs, the final CH_3F pressure being 200 mbar. The mass spectrum is given in Fig. 3.

Sodium chloride films were prepared by evaporation at pressures < 5 x 10^{-9} mbar on (100) NaCl, single crystals grown from p.a. purity material (Dr. Korth, Kiel).

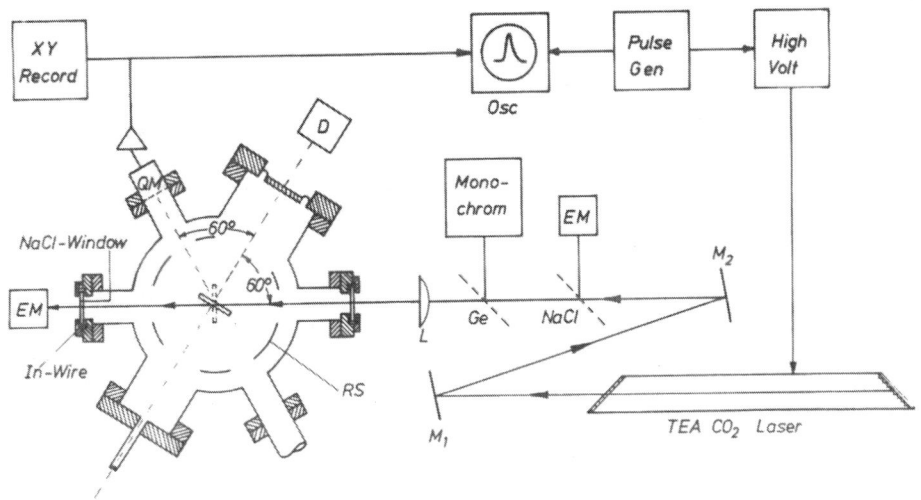

Fig. 2. : Blockdiagram of ultrahigh vacuum cryostat
and apparatus for laser induced processes.

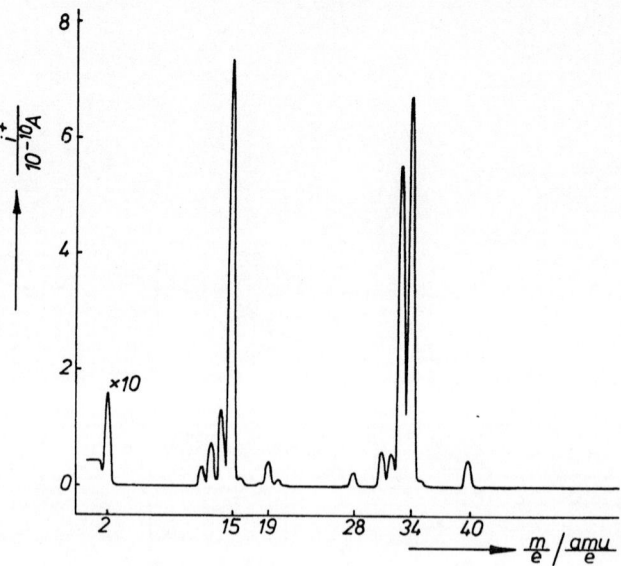

Fig. 3. : Mass spectrum of CH_3F used. Total pressure
7 x 10^{-10} mbar.   Substrate temperature 64 K.

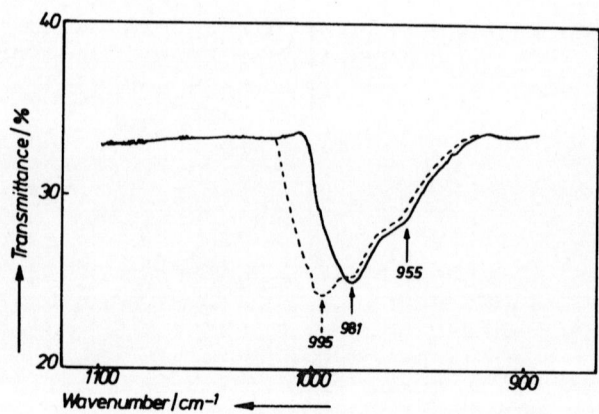

Fig. 4. : Linear absorption spectrum. Adsorption system CH_3F-NaCl (film) : full curve.
Adsorption system CH_3F-NaCl (film) at higher
coverage (multiple layer) : dashed curve.

RESULTS AND DISCUSSION

LINEAR INFRARED ABSORPTION

By linear infrared absorption spectroscopy it could quantitatively be distinguished between monolayer and multilayer adsorption, (Fig. 4., further spectra to be published), the latter having vibrational frequencies equal to those in the solid (ν_3 (solid) = 995 cm^{-1}), the former showing a strong shift to lower wavenumber with decreasing coverage[10].

INFRARED-INDUCED DESORPTION
TIME OF FLIGHT DISTRIBUTION

The desorption pulse after a resonant infrared pulse is shown in Fig. 5, where CH$_3$F ions current (mass 34 see Fig. 3) versus time is displayed. The frequency of the exciting laser pulse was 975 cm^{-1} , the fluence D = 0.05 J cm^{-2} . 90 Langmuir CH$_3$F had been admitted to the cryostat, the adsorbent being at a temperature of 77 K. The total pressure was 5 x 10^{-9}Torr (CH$_3$F, for residual gases see Fig. 3.). The distribution of times of flight of the desorbed molecules, is given under certain restricting conditions by

$$g(\tau_f) = \frac{d\,N\,(\tau_f)}{d\tau_f} = a\,\tau_f^{-4}\;\exp\left[-\left(\frac{\tau_{f0}}{\tau_f}\right)^2\right] \qquad (4)$$

where $\tau_{f0} = 1 / (2\,kT\,/\,m)^{1/2}$, l being the distance between adsorbate struck and ionization chamber, and $N(\tau_f)$ the number of molecules striking the unit area of the detector in unit time having flight times between τ_f and $\tau_f + d\tau_f$.

$g(\tau_f)$ has its maximum at τ_{fmax}, which is related to the most frequent molecular velocity v_0 by

$$\tau_{fmax} = 1 / (\sqrt{2}\,v_0) , \qquad v_0 = (2kT/m)^{1/2}. \qquad (5)$$

The halfwidth (FWHM) of g is approximately equal to τ_{fmax} so that v_0 may be determined without knowing the exact starting time of the molecules. Valuable information, such as surface temperature, velocity distribution, can be obtained from a careful analysis of the time of flight distribution. Here we give only a rough estimate of the lower bound of the substrate temperature, which turned out to be $\geqslant$ 20K.

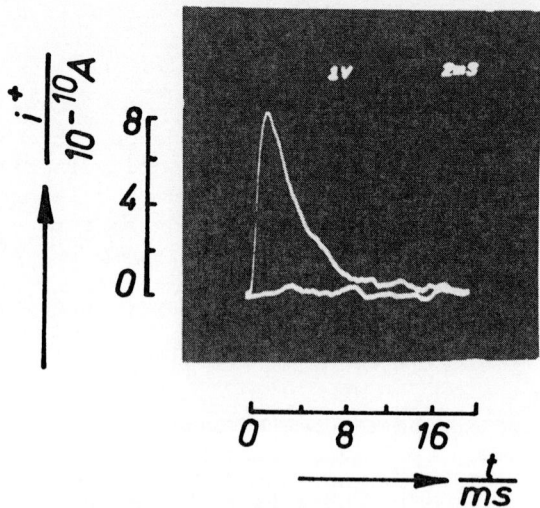

Fig. 5. : Desorption after resonant infrared pulse.
CH$_3$F$^+$ current versus time. Laser frequency
975.9 cm^{-1}. 0.05 J cm^{-2}. 90 L. 5 x 10^{-9} Torr.
CH$_3$F - NaCl (film) 77 K.

QUALITATIVE CHEMICAL ANALYSIS OF DESORBATE.   BACKGROUND -
DESORPTION

A very large number of desorption pulses systematically
varying mass, excitation frequency, fluence, coverage in the
ranges cited was determined.   From these data it was concluded
that neutral CH$_3$F molecules are desorbed under present conditions,
including not excluded is desorption of dimers and vibrationally
excited molecules.

The thermal desorption from metal surfaces induced by
scattered light was determined, this background being a serious
problem in laser induced desorption.   Shooting pulses of compa-
ratively high fluences D $\lesssim$ 0.12 J cm^{-2}at an angle of incidence
of 60° on the sample (77 K) even at very low pressures (3 x 10^{-10}
mbar) desorption occurred at the following m/e values :

$\frac{m}{e}$ (amu)   2, 12, 13, 14, 15, 16, 18, 28, 32, 44

indicating H$_2$, CH$_4$, H$_2$O, CO, N$_2$, O$_2$, CO$_2$.

The ion current desorption pulses were small, in the 10^{-12}A range, except that of H_2 which was 10^{-10} A.   No molecules with m/e ratios 4 (He), 19(F), 33, 34(CH_3F), 35(Cl), 40(Ar) were desorbed, though present in the residual gas, except CH_3F.   No background desorption was observed at normal indidence even under higher fluence $D \stackrel{<}{\sim} 0.2$ J cm^{-2}

DESORPTION SPECTRA RESONANCES

In Fig. 6. the dependence of the desorption yield (CH_3F ion current) on the frequency of the exciting infrared radiation is presented.

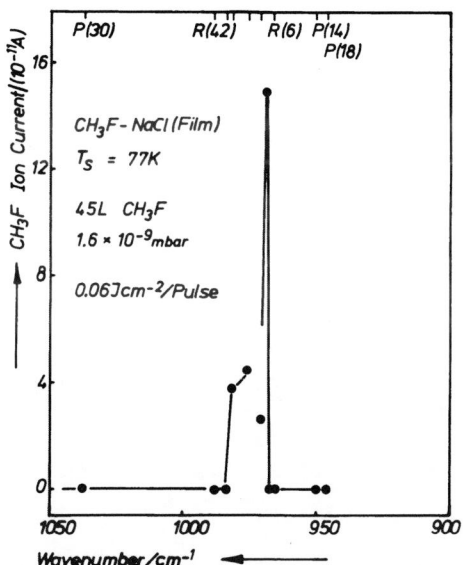

Fig. 6. :   Resonance characteristics of desorption.
             Desorption yield (CH_3F ion current) versus
             excitation frequency;

The resonance characteristics of molecular emission (CH_3F ion current) for multi-layer adsorption attained after larger gas admission is displayed in Fig. 7.   Apparently the width of the yield vs. frequency line became broader with increasing fluence. Comparison of spectra 5,6 and 7 shows that the primary process in this desorption induced by monochromatic infrared radiation (DIMIR) is the resonant excitation of the ν_3 internal vibration in the adsorbate CH_3F-NaCl.

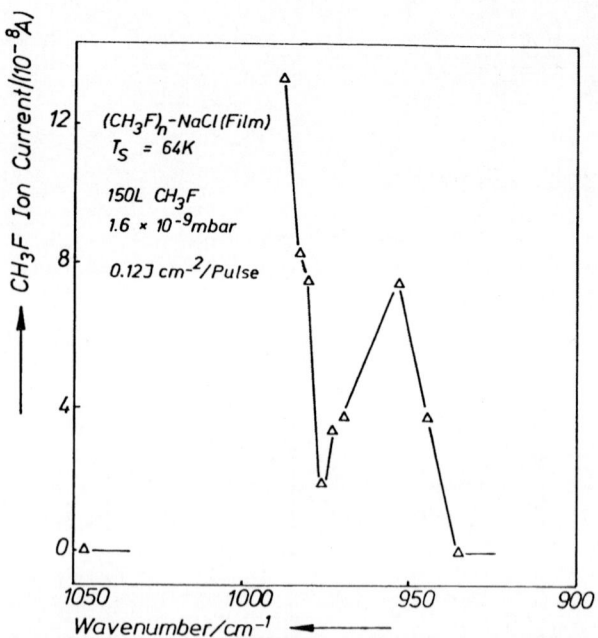

Fig. 7. : Resonance characteristics of emission.
Emission yield (CH₃F ion current) versus
excitation frequency of higher coverage (mul-
ti layer) CH₃F - NaCl (film).

RESONANT LOCALIZED EXCITATION

       At sufficiently high infrared intensity and low surface
coverage, when infrared pumping rate is large compared to intermolecular vibration-vibration (VV) transfer rate, infrared pumping
will be dominant in carrying energy up the vibrational ladder.
The vibrations and energy levels involved in the excitation are
shown schematically in Fig. 8.   Step-by-step one photon transitions
due to coherent IR radiation between levels with many closely spaced
sublevels will take place, the excitation then appearing incoherent
with smoothly changing coarse grained populations of the levels.
For the adsorption system, the levels proposed involve a mixture of
the internal molecular vibration ν_3 and external modes including
T_z.   The proposed mechanism is, moreover, in concert with the

observed dependence of desorption yield on fluence as discussed
below. Anharmonic compensation can be achieved by this mechanism.
The activation energy of desorption is estimated to be equal to
2 or 3 photons ($\nu_3 \leftarrow 0$) from vapor pressure values at different
temperatures), the frequency for the normal translational vibration
T_z is not know unfortunately.   The fundamental frequency of T_Z in
CO_2 - NaCl was calculated to be 78 cm^{-1}, for SO_2 - CsCl   measured
to be 220 cm^{-1} and 300 cm^{-1} (to be published).   Coherent pumping
of the adsorbate modes, conceived to be a single anharmonic mode,
via direct N photon transition to the required vibrational excitation
level probably demands higher intensities than in fact needed for
the desorption. [1c].

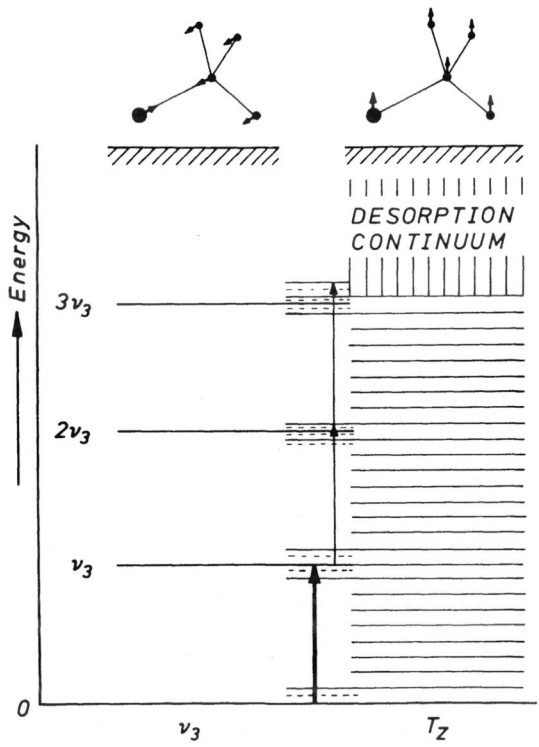

Fig. 8. : Energy level scheme for ν_3 and T_Z vibration
          of CH_3F - NaCl.   Desorption energy is
          equal to 2 to 3 photons (ν_3 $\leftarrow$ 0).

RESONANT NON-LOCALIZED EXCITATION

If on the other hand, vibrational energy conserving intermolecular VV- transfer is faster than pumping by infrared, only the
fundamental vibrational transition will be induced and subsequent
VV-processes transport the energy up the vibrational ladder according
to Fig. 9. and

$$|v> + \quad |v'> \rightarrow |v+1> + |v'-1>, \tag{6}$$

neglecting multiquantum processes.

When the pumped spectral transition is inhomogeneously broadened, as is expected for the adsorbed molecules, Fig. 4., at the
applied infrared intensity of $\gtrsim$ 300 kW cm^{-2} sufficient to saturate
the fundamental, but smaller than needed to induce appreciably higher
transitions (in the gas)[11], then the following distinct ranges of
infrared pulse duration t_p may be distinguished :

1. If t_p is short compared to the VV-transfer times then the fractional population of adsorbed molecules exactly at resonance can
be excited and only the corresponding fractional absorption
energy takes place.

2. If t_p is of the order of the characteristic time of VV-transfer
between the nearly resonant fundamentals (probably mixed with
external modes) within the inhomogeneous line width of Fig. 4, the
shape and structure of the spectral line becomes irrelevant and in
case of saturation the total population of the upper and ground
state can be equalized, the average energy absorbed by a molecule
being $1/2\ \hbar\omega$.

3. If t_p is large compared to τ_{vv}, the characteristic time for
VV-exchange which carries energy up the vibrational ladder
according to (6), then additional energy can be absorbed (which is
also the case, when relaxation is fast).

4. If t_p is larger than τ_{vv} but shorter than τ_{vp}, the characteristic
time of vibrational relaxation into the phonon bath of the adsorbent,
then many vibrational quanta can be stored by the adsorbed molecule
and therefore are available for subsequent processes of interest
such as desorption.

Though long pulses were produced by adjusting the laser gas
mixture in some cases, the fluence variation was generally accomplished by a change of the average pulse intensity at nearly constant
pulse shape and length.   Therefore both localized and non-localized
excitation will occur.   Selectivity is expected to be enhanced

in localized excitation at higher infrared intensities. Desirable
are experiments varying systematically infrared intensity, pulse
duration and surface coverage.

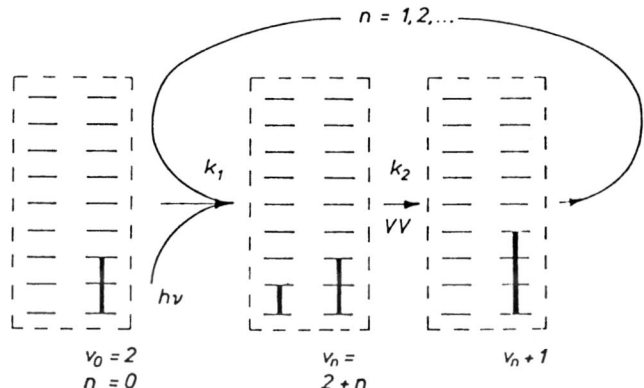

Fig. 9. : Non-localized excitation : Infrared adsorption
         and vibrational excitation induced by VV
         transfer.

DEPENDENCE OF DESORPTION YIELD ON FLUENCE

     The effect of the infrared pulse fluence on desorption yield
is demonstrated in Fig. 10. and 11.   In Fig. 10. the desorption
yield dependence upon fluence at constant coverage is presented.
Constant coverage was accomplished by defined gas admission and
striking at 60° incidence a different spot of the sample at each
laser pulse, having certain fluence.   Desorption (Θ_0 - Θ) was
extraordinarily high and fluence dependence unusually weak in this
case. Θ_0 and Θ are the number of adsorbed CH_3F molecules in the
light path before and after the pulse.   Apparently desorption
of weakly bound molecules is induced by the first infrared pulse
after gas admission.   Remarkably, the slope in Fig. 10. is considerably larger than in the corresponding plot of the relation between
multi-layer evaporation and fluence.[7] The results shown in Fig. 11.
were obtained by firing a sequence of pulses at the same spot,

thereby depleting the spot of CH$_3$F molecules until it is practically vacated.   Considering DIMIR a unimolecular process

$$-\frac{d\ \ln\theta/\theta_o}{dt} = k_u(I,\ D,\ t)\ ,\tag{7}$$

fluence D, intensity I and time t being connected by

$$D(t) = \int_o^t I(t')dt'\ ,\tag{8}$$

one obtains under certain conditions, see below, setting

$$k_u = const.\ I^m\tag{9}$$

for the logarithm of the desorption yield

$$\ln\frac{\theta_o - \theta}{\theta_o} = m\ \ln D + const.,\qquad (\theta_o-\theta)/\theta_o \ll 1\tag{10}$$

Surface migration of CH$_3$F even at 77 K must be considered.

The fluence required for desorption is extraordinarily low The dependence of desorption on fluence, being high and also remarkably different from that of multi-layer evaporation, deserves further investigation.

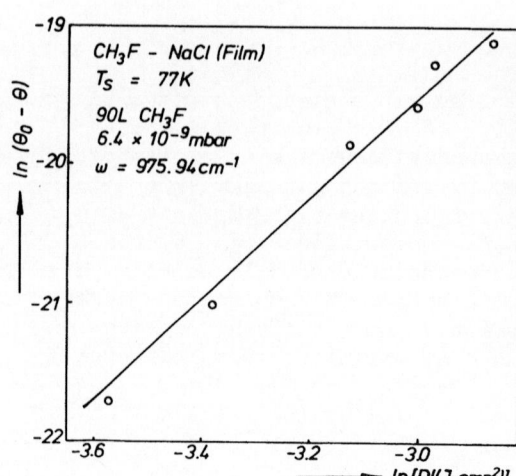

Fig. 10. : Dependence of desorption $(\theta_o-\theta)$ on fluence D of exciting infrared pulse at constant coverage after CH$_3$F admission.

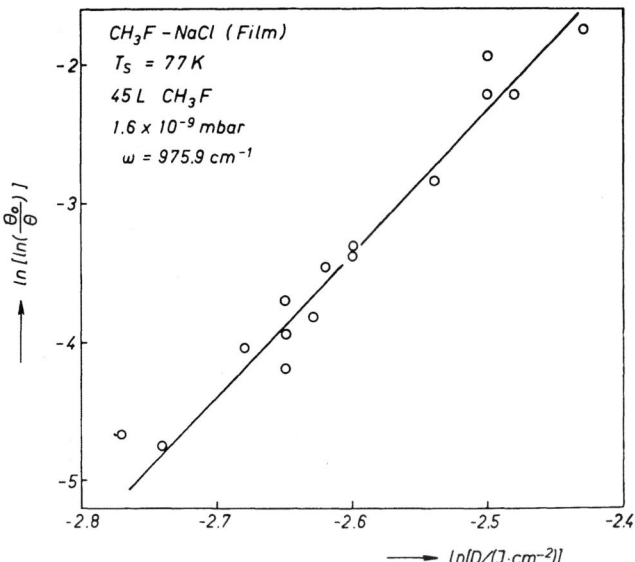

Fig. 11. : Dependence of desorption yield $\phi = (\theta_o - \theta)/\theta_o$
on fluence D in a sequence of pulses
at the same spot with depletion of coverage.
Double logarithmic plot ln ln (1 - ϕ) versus
ln D.

MODEL OF UNIMOLECULAR DESORPTION INDUCED BY MONOCHROMATIC INFRARED. RELATION BETWEEN DESORPTION YIELD AND FLUENCE

Equation (10) may be considered as an empirical representation
of the relationship between desorption yield and fluence.   The
observed power law (Fig. 10 and 11), with m ≫ 1, (m=10) in eq. (9),
though also consistent with coherant multiphoton excitation, is
interpreted as a non-steady state effect within the model of unimolecular desorption.   For steady state this model predicts a
roughly linear relation between rate coefficient and intensity.
The model of unimolecular processes[12] may be applied to desorption
induced by monochromatic laser infrared under a set of conditions :

a) If the intensity I is constant during the irradiation,

b) steady state is reached after sufficiently long irradiation
   times and

c) linear rate equations are adequate to describe the changes of the
   coarse grained population p_v of the vibrational levels of the

adsorbed molecule :

$$\frac{d\ p_v}{dt} = \sum_r K_{vr}\ \dot{p}_r - p_v \sum_r K_{rv} - k_v p_v \tag{11}$$

or in matrix notation

$$\dot{p} = Kp. \tag{12}$$

As an initial approximation we suppose that the populations p_v change only by radiant energy transfer and desorption. Then the coefficients K_{vr} for step by step transitions between neighboring levels are proportional to the radiation intensity I, $K_{vr} = k_{vr} I$, whereas the specific rate constants for desorption k_v do not depend on intensity. The simple linear rate equations (11) can be used to describe the time evolution of the populations under the following conditions which appear to be relevant to the present problem :

d) if the density of vibrational states is high in certain ranges, so that the states with similar values of energy and other significant observables can be collected into groups or "levels", each having a sufficiently large number N_v of states. Adjacent levels are separated by the frequency of the pumped fundamental or the IR frequency (Fig. 8.).

e) the states of the lower level are initially randomly populated (diagonal density matrix) e.g. by thermal motion.

f) the Rabi frequencies of the adsorbed molecule in the infrared field are sufficiently high, i.e. larger than the frequency separation of states within each level. Deducing eq. (12) form the Schrödinger eq. for the amplitudes of the molecular states is a general problem of statistical mechanics and will not be repeated at least in certain cases.

For K independent of time (12) may be written

$$p(t) = \exp\left[K.t\right] p(0) \tag{13}$$

The explicit solution is obtained by diagonalizing K, using the detailed balance condition $K_{vr}p_v^e = K_{rv}p_v^e$, $p_v^e \sim N_v$ (for thermal desorption $p_v^e \sim N_v \exp(-E/kT)$, p_v^e denoting the equilibrium population of level v,

$$p(t) = U \exp(\Lambda t)U^{-1}.p(0). \tag{14}$$

The eigenvalues λ_1, λ_2, ... of K, being the elements of Λ, are ≤ 0. Supposing the largest eigenvalue λ_1 to be the only one close to zero (or zero) then after sufficiently long time, the contributions from all terms $\exp(\lambda_k t)$, except $\exp(\lambda_1 t)$, may be neglected, defining a "steady state". Then

$$P_v = \exp(\lambda_1 t) \sum_r u_{vr}P_r(0) \tag{15}$$

and

$$-\frac{d \ln \sum_v P_v}{dt} = -\lambda_1 \equiv k_u \tag{16}$$

or

$$-\frac{d \ln(\theta/\theta_o)}{dt} = k_u \tag{17}$$

which may be written (for constant intensity during irradiation)

$$-\frac{d \ln(\theta/\theta_o)}{I \, dt} = \frac{k_u}{I} \tag{18}$$

or

$$-\frac{d \ln(\theta/\theta_o)}{dD} = \frac{k_u}{I} = k_u^{(st)} \tag{19}$$

Numerical evaluation of (16) shows that k_u/I is only slightly dependent upon intensity within a range of intermediate intensities in concert with general considerations : At low intensity, the first transitions in the excitation will be rate determining in the overall desorption. In the region of small vibrational quantum number the density of states is in general low and excitation will be described as coherent resonant transitions. A more than linear

increase of the rate coefficient with intensity in multiphoton
processes is usual in this case.   In the limit of high intensity
all IR induced up and down transitions on the vibrational ladder
are faster than the unimolecular dissociation step, described by
k_v, which is of the order of 10^{12} to $10^{13} s^{-1}$.
The unimolecular rate coefficient is then independent of intensity.

In describing infrared induced processes on surfaces, pumping
by intermolecular VV-transfer and vibrational relaxation into the
phonon bath of the adsorbent have generally to be taken into account
e.g. by adding terms of the form

$$\sum_r F_{vr} P_r \; - \; P_v \sum_r F_{rv} \; + \; \sum_r R_{vr} P_r \; - \; P_v \sum_r R_{rv} \tag{20}$$

to the right hand side of eq. (11), both processes being independent
of infrared intensity, but dependent upon coverage, particularly the
VV-transfer.

g) If VV-transfer is very fast compared to optical pumping,
$\tau_{vv} \ll \tau_{op}$, and relaxation slower than excitation and the unimolecular dissociation step, $\tau_{vp} \gg \tau_{vv}$, τ_{op}, k_v^{-1} , then eq. (11)
holds approximately.   The form of the rate law for the case
$\tau_{vp} \lesssim \tau_{vv}$, τ_{op}, k_v^{-1} is being studied.   Provided that the conditions a) to g) are fulfilled, and in addition that steady state is
reached after long irradiation times or at high yields, eq. (19)
is valid independent of the intensity profile of the laser pulse.
In this case fluence alone determines yield, a linear relationship
between the steady state unimolecular rate coefficient k_u and
intensity I, as well as a linear dependence of desorption yield
upon fluence are predicted.   First experiments appear to be in
concert with the prediction.

ACKNOWLEDGEMENTS

We deeply thank Professor G. Wedler and Professor W. Jaenicke
for support of this work, Dipl. -Chem. I. Hussla for efficient and
friendly help, and E. Wallner for outstanding work in building the
main apparatus, Priv. -Doz. Dr. M. Quack for illuminating discussions
and sending us preprints before publication.

Grants from Deutsche Forschungsgemeinschaft, Zerweck Fonds des
Universitätsbundes Erlangen-Nürnberg and Fonds der chemischen Industrie are gratefully acknowledged.

REFERENCES.

1. R.V. Ambartzumian, V.S. Letokhov in "Chemical and Biochemical
   Applications of Lasers", Vol. 3, Ed. C.B. Moore, Acad. Press,
   New York 1977.
   N. Bloembergen, E. Yablonivitch, Physics Today 31 (1978) 23.
   C.D. Cantrell, S.M. Freund, J.L. Lyman in "Laser Handbook",
   Vol. 3, Ed. M.L. Stitch, North-Holland, Amsterdam 1979.
   P.A. Schulz, A.S. Sudbø, D.J. Krajnovich, H.S. Kwok, Y.R. Shen,
   Y.T. Lee, Ann. Rev. Phys. Chem. 30 379 (1979).
   J.L. Lyman, G.P. Quigley, O.P. Judd in "Multiple Photon Excitation
   and Dissociation of Polyatomic Molecules" ed. C. Cantrell, Sprin-
   ger, Heidelberg 1980, and LA-UR 79-2605 (Report of the Los
   Alamos Scientific Lab. 1979).
   M. Kneba and J. Wolfrum, Max-Planck-Institut für Strömungsfor-
   schung, Göttingen, Bericht 102, 1980, to be publ. in Ann. Rev.
   Phys. Chem.
   W. Fuß and K.L. Kompa, Projektgruppe für Laserforschung, Max-
   Planck-Gesellschaft, Garching bei München, Bericht 30, 1980,
   Submitt. Ann. Rev. Phys. Chem.
2. J.P. Maier, A. Seilmeier, A. Laubereau and W. Kaiser, Chem.
   Phys. Letters 46, 527 (1977).
3. B. Davies, M. Poliakoff, K.P. Smith and J.J. Turner, Chem.
   Phys. Letters 58, 28 (1978), M. Poliakoff in Laser-induced
   Processes in Molecules P. 304, Eds. K.L. Kompa and S.D. Smith,
   Springer Verlag Berlin 1979.
4. F. Legay, in Chemical and Biochemical Applications of Lasers,
   Vol. II, C. Bradley Moore, Ed. Academic Press, New York 1977.
   L. Abouf-Marguin, B. Gauthier-Roy and F. Legay, Chem. Phys.
   23, 443 (1977).
5. M.S. Djidjoev, R.V. Khokhlov, A.V. Kiselev, V.I. Lygin, V.A.
   Namiot, A.I. Osipov, V.I. Panchenko and B.I. Provotorov, in Tuna-
   ble Lasers and Applications, p. 100, Eds. A. Mooradian, T. Jaeger,
   P. Stokseth, Springer Verlag, Berlin 1976.
   M.S. Slutsky and T.F. George, Chem. Phys.Letters 57, 474 (1978).
   J. Heidberg, H. Stein, A. Nestmann, E. Hoefs, I. Hussla, in sym-
   phosium Laser-Solid Interactions and Laser Processing, Materials
   Research Soc., Boston 1978.
   Laser-Solid Interactions and Laser Processing, p. 49 Eds. S.D.
   Ferris, H.J. Leamy, J.M. Poate, American Institute of Physics,
   New York 1979.
6. J. Heidberg, H. Stein und E. Riehl, Abstracts, Hauptyersamml.
   Deutsche Bunsen-Gesellschaft, München, May 15-17, 1980.
7. J. Heidberg, H. Stein, E. Riehl, A. Nestmann und E. Hoefs,
   Z. Phys. Chem. N.F. in press.
8. L.P. Levine, J.F. Ready, E. Bernal, J. Appl. Phys. 38, 331
   (1967).
   R. Gauthier, M. Babout, C. Guittard, Proc. 7th Intern. Vac.
   Congr. and 3rd Int. Conf. Solid. Surfaces, Wien, 1977.

G. Ertl, M. Neumann, Z. Naturforschg. 27a, 1607 (1972).
H. Hartwig, P. Mioduszewski, A. Popiescyk, 3rd International
Conf. on Plasma Surface Interactions in Controlled Fusion Devices,
Culham 1978, Leybold-Heraeus Inform. 1978.
J.P. Cowin, D.J. Auerbach, C. Becker, L. Wharton, Surface Sci.
78, 545 (1978).
G. Wedler, H. Ruhmann, private communication.
D. Menzel, in Interactions on Metal Surfaces,
R. Gomer, Ed. Springer, Berlin 1975.
9. N. Legay-Sommaire and F. Legay, IEEE J. Quantum Electron. Vol.
QE-16, 308 (1980). Chem. Phys. Letters 52, 213 (1977).
10. J. Heidberg und A. Nestmann, unpublished.
A. Nestmann, Dissertation, Erlangen 1978.
11. R.A. Forber, R.E. McNair, S.F. Fulghum, M.S. Feld,
B.J. Feldman, J.Chem. Phys. 72, 4693 (1980).
12. M. Quack, J. Chem. Phys. 69, 1282 (1978; 70, (1979).
M. Quack, P. Humbert, and H. van der Bergh, J. Chem. Phys., in
press.

FACE CENTERED CUBIC TRANSITION METAL SURFACE VIBRATIONS

G. Allan * and J. Lopez **

* Laboratoire de Physique des Solides
  Institut Supérieur d'Electronique du Nord
  3, rue F. Baës, 59046 Lille Cédex, France

** Laboratoire de Physico-Chimie
   Université Claude Bernard
   43, Bd du 11 Novembre 1918, 69622 Villeurbanne Cédex,
   France

ABSTRACT

We calculate the surface atom and adatom positions or vibration spectra from the electronic structure. The valence bands are described in the tight-binding approximation. We fit the d band parameters to bulk properties. The clean surface atomic positions are obtained by a direct minimization of the crystal energy. The agreement with the experimental results is quite good. All the clean surfaces are found to be contracted. The effect of this contraction on the surface vibration spectra is studied. The contraction increases with the surface roughness. It enhances the force constants between the atoms close to the surface leading to localized step vibrations above the bulk frequency spectrum. The model is also extended to chemisorbed chalcogenide atoms. The experimental positions and vibration frequencies are fitted. Then we can study the influence of chemisorption on the surface relaxation or vibrations and also the elastic indirect interaction between adatoms.

INTRODUCTION

During the last few years, our knowledge of surface atom positions or vibration frequencies has been strongly improved. Ion scattering spectroscopy seems to be a more direct way to determine clean surface atom or adatom positions than the Low Energy Electron

Diffraction whose interpretation requires a large amount of computation. Surface Excited X-Ray Attenuation Fine Structure seems also
very promising. The resolution of the Characteristic Energy Loss
Spectrometry is now sufficient to observe vibration spectra in the
range of bulk phonon frequencies. Less improvements have been made
in "a priori" calculations of all these quantities which must be
obtained from the electronic structure. Elaborate models accurately
determine bulk properties like the cohesive energy, the interatomic
distances, the bulk modulus[1] . Their extension to bulk phonon
calculations is not straightforward. We also know that a surface
problem is still more difficult due to the lack of periodicity in
the direction perpendicular to the surface. In the semiconductor[2,3]
and transition metal cases[4-7], the tight-binding approximation seems
well suited to calculate such bulk and surface properties.

In the next section, we shall recall the model we have developed to calculate the interatomic distances and the vibration spectra
in a simple tight-binding approximation. The model is not precisely
"a priori" since we shall fit some parameters to bulk properties like
the cohesive energy or the phonon dispersion curves. Nevertheless it
gives a coherent way to describe lattice properties from the
electronic structure. In the third section, the bulk parameters are
used to study the atomic relaxation or the vibration spectra near
face centered cubic transition metal defects (vacancy[5] or clean
surfaces[5,8]). The results we get agree quite well with the experimental measurements. The procedure is the same for adatoms. New
parameters are needed. They are fitted to experimental measurements
and then used to calculate the influence of the chemisorption on
the surface atom relaxation, on the surface phonons or on the
indirect elastic interaction between chemisorbed atoms.

SIMPLE TIGHT-BINDING EXPRESSION OF THE ENERGY AS A FUNCTION OF
ATOMIC DISPLACEMENTS

Most of the transition metal properties only depend on the d
electrons as the d band density of states is larger[10] than the s
one. The cohesive energy and other quantities which are related to
the integrated density of states does not depend very much on
details of the density such as peaks in this density. The most
important parameter is the d band width[10]. Such a rough description
of the d band may be easily obtained in the tight-binding approximation. The p states of the chalcogenide adatom will be also studied
within this approximation. A rough density of states may also be
obtained from the knowledge of only its first few moments. The
moment μ_n of the density of states is defined as :

$$\mu_n = \int E^n \, n(E) \, dE \qquad (1)$$

They are easily calculated from the trace of the n-th power of the hamiltonian in terms of the usual resonance integrals. For example the first moments $\mu_n^{i\nu}$ of the local density of states $n_{i\nu}(E)$ corresponding to the ν orbital on atom i are :

$$\mu_o^{i\nu} = 1 \tag{2}$$

$$\mu_1^{i\nu} = <i\nu|H|i\nu>$$

$$= E_{i\nu} + V_i \tag{3}$$

$$\mu_2^{i\nu} = \sum_{\nu'} <i\nu|H|i\nu'> <i\nu'|H|i\nu>$$

$$+ \sum_{j\nu''} <i\nu|H|j\nu''> <j\nu''|H|i\nu>$$

$$= (E_{i\nu} + V_i)^2 + \sum_{j,\nu''} (H_{ij}^{\nu\nu''})^2 \tag{4}$$

where $E_{i\nu}$ is the ν orbital atomic level and V_i the intraatomic matrix element of the selfconsistent potential due to charge transfer between the atoms close to a defect (surface, vacancy, adatom). This potential must be self-consistently determined. However the electron-electron correlations reduce the Coulomb integrals such as an exact calculation remains difficult. Nevertheless we cannot neglect the potential in the energy calculation for a non self-consistent model allows unphysical and too large charge transfers and even leads to crystal instabilities[5]. Near a vacancy[11] the self consistent potential reduces the charge transfers one gets in a non self consistent model. It is still more efficient near a surface due to a very large surface dipole layer Coulomb integral. For defects like atomic displacements, we may consider two cases. For large wave vector vibrations (close to the Brillouin zone boundaries) the atomic planes normal to the wave vector are alternatively positive and negative like a dipole layer near a surface. So we expect a small charge transfer between planes. From phonon dispersion curves at small wavevector, one can deduce the lattice elastic constants. They may also be obtained from crystal deformations which keep each atom neutral. So also in this case the charge transfers must be small. We shall see that an easy way in our model to determine the self-consistent potential is to assume in any case that each atom remains neutral. This approximation allows to accurately determine the potential since the error we made is of the order of the true charge transfer (which would be small) divided by the density of states at the Fermi level (which we expect to be large for a d band transition metal).

In the second moment expression (4) the sum over ν'' is extended to the N_j orbitals located on the j-th nearest neighbour of the atom i. We then assume that $H_{ij}^{\nu\nu''}$ does not depend too much upon ν and ν'' and take for these hopping integral a mean value β_{ij} independent of the direction $\vec{R}_{ij}$

$$\beta_{ij} = \beta_{ij} \exp(-q_{ij} R_{ij}) \tag{5}$$

We have already noticed that in our calculation, the exact shape of the density of states is not very important provided we fit the exact first moments. Rectangular or Gaussian bands have been used. We have taken a Gaussian local density of states on atom i :

$$n_i(E) = \frac{2N_i}{\sqrt{2\pi}\mu_i} \exp - \frac{(E-E_i-V_i)^2}{2\mu_i} \tag{6}$$

where $\mu_i = \sum_j N_j \beta_{ij}^2$ \qquad (7)

The neutral approximation shows that $(E_F-E_i-V_i)/(2\mu_i)^{1/2}$ does not depend upon atomic displacements. The local contribution of the valence bands to the crystal cohesive energy is :

$$E_{A_i} = \int^{E_F} E\, n_i(E) dE - N_d E_i - N_d V_i \tag{8}$$

It is the one-electron energy differences between the crystal and the free atom. The last term is due to electron Coulomb inter-action counted twice in the one-electron energies. An additional term $1/2 \,\delta\rho_i V_i$ generally occurs but is equal to zero as the atoms remain neutral in our model. We get :

$$E_{A_i} = -2N_i \left[\frac{\sum_j N_j \beta_{ij}^2}{2\pi} \right]^{1/2} \exp - \frac{(E_F-E_i-V_i)^2}{(2\sum_j N_j \beta_{ij}^2)} \tag{9}$$

where E_F is the bulk Fermi level.

The crystal is stabilized at short distances by repulsive Born-Mayer potentials

$$C_{ij} = C_o \exp (-p_{ij} R_{ij}) \tag{10}$$

They are essentially due to s electron repulsions.

To simplify the expression, we shall only consider the case of transition metals. The extension to oxygen or sulphur adatoms is straightforward. The transition metal cohesive energy for the atom i is then :

$$E_{C_i} = - 10 \sqrt{\frac{\mu_i}{2\pi}} \exp (- \frac{E_F^2}{2\mu}) + \sum_j C_{ij} \tag{11}$$

where μ is the second moment for the bulk perfect infinite crystal. If we use the bulk equilibrium condition , we get :

$$E_{C_i} = \frac{E_c}{(1 - \frac{q}{p})} \left[\sqrt{\frac{\mu_i}{\mu}} - \frac{q}{p} \frac{1}{N} \sum_j \exp (-p R_{ij}) \right] \tag{12}$$

where E_c is the bulk cohesive energy and N the bulk coordination number, p and q are transition metal values for p_{ij} and q_{ij}. We call $\vec{u}_i$ the displacement of the atom i from the relaxed equilibrium position $\vec{R}_i$ which is different from the unrelaxed one (which is a lattice site $\vec{R}_i^0$). If we put :

$$d_{ij} = (\vec{u}_i - \vec{u}_j) . \frac{\vec{R}_i - \vec{R}_j}{R_{ij}}$$

$$u_{ij} = |\vec{u}_i - \vec{u}_j| \tag{13}$$

$$\delta R_{ij} = |\vec{R}_{ij} - \vec{R}_{ij}^0|$$

a second order expansion of the energy (12) is straightforward

$$E_{C_i} = E_{C_i}^0 + \frac{pqE_c}{2N(1-q/p)} \times \left\{ \sum_j \frac{2}{p} \left[\frac{N^{1/2} \exp(-2q\delta R_{ij})}{[\sum_\ell \exp(-2q\delta R_{i\ell})]^{1/2}} - \exp(-q\delta R_{ij}) \right] \right.$$

$$\times \left[d_{ij} + \frac{u^2_{ij} - d^2_{ij}}{2R_{ij}} \right]$$

$$+ \sum_j d^2_{ij} \left[\exp - (p\delta R_{ij}) - \frac{2q}{p} \frac{N^{1/2} \exp(-2q\delta R_{ij})}{[\sum_\ell \exp(-2q\delta R_{i\ell})]^{1/2}} \right]$$

$$\left. + \sum_{jj'} d_{ij} d_{ij'} \frac{q}{p} \frac{N^{1/2} \exp(-2q\delta R_{ij}) \exp(-2q R_{ij'})}{[\sum_\ell \exp(-2q\delta R_{i\ell})]^{3/2}} \right\} \quad (14)$$

where $E_{C_i}^0$ is the energy of the atom i in the unrelaxed position

$$E_{C_i}^0 = \frac{E_c}{1 - \frac{q}{p}} \left[\sqrt{\frac{N_{ci}}{N}} - \frac{q}{p} \frac{N_{ci}}{N} \right] \quad (15)$$

where N_{ci} is the atom i coordination number. The expression (15) also allows to calculate the monovacancy formation energy[5,13], the surface tension or the adatom binding energy.

APPLICATION TO THE SURFACE ATOMS AND TO ADATOMS

We first use (14) to calculate the bulk phonon spectrum and to determine the parameters p and q. Then we shall apply the fitted parameters to calculate the surface relaxation.

Bulk Phonon Spectrum

For the bulk perfect crystal, ($\delta R_{ij} = 0$ and $N_{ci} = N$) we get from (14) :

$$E_{C_i} = E^o_{C_i} + \frac{pqE_c}{2N\,(1-p/q)} \left\{ (1 - \frac{2q}{p}) \sum_j d^2_{ij} + \frac{q}{pN} \sum_{jj'} d_{ij}\, d_{ij'} \right\} \quad (16)$$

The bulk maximum frequency ν_M is :

$$m(2\pi\nu_M)^2 = 16 \frac{(pR_o)\,(qR_o)E_c\,(1 - \frac{2q}{p})}{N(1 - q/p)\,R^2_o} \quad (17)$$

where m is the atomic mass. We remark that the displacement energy
coefficients only depend on ν_M and q/p. Moreover for any phonon
frequency, the ratio ν/ν_M is only function of q/p. Figure 1 gives
the Nickel phonon spectrum for different values of q/p. The case
q/p = 0 is identical to a force constant between nearest neighbours model. If q/p is different from zero, we introduce an angular
force constant. The agreement with the experimental values is
quite good for q/p = 1/5. Let us remark this value gives a transition
metal monovacancy formation energy equals to $3/8|E_c|$ which agrees
very well with the experimental value[5,13]. The monovacancy formation
volume also agrees with the experimental value[5]. This relaxation
volume depends on the forces acting on the vacancy neighbours. This
calculation checks the first order terms of (14). We now use the same
parameters for clean f.c.c. transition metal surfaces.

Clean f.c.c. Transition Metal Surfaces

For the (100) and (110) surfaces, the minimization of (14) is
analytical[5] We get a steep exponential decrease of the atom displacements near the surface. For other surfaces, we use a numerical
energy minimization. As the displacements also rapidly decrease as
one goes into the bulk we have to only consider a few layers close
to the surface plane.

For an unrelaxed lattice, let us first remark that the forces
acting on atoms close to the surface plane are proportional to
$(\sqrt{N_i/N} - 1)$. They give rise to a surface contraction increasing with
the number of dangling bonds, i.e. with the surface roughness.
Table 1 gives the result we get for Nickel surfaces (pR_o = 11.0,
qR_o = 2.2). The (001) surface value disagrees with the LEED result.
Such a discrepancy may be due to adatoms which can reverse the sign
of the surface relaxation. This was the case of the Pt(111)
surface[8,14]. Recent measurements[15] have shown such a surface is
expanded with a small hydrogen coverage.

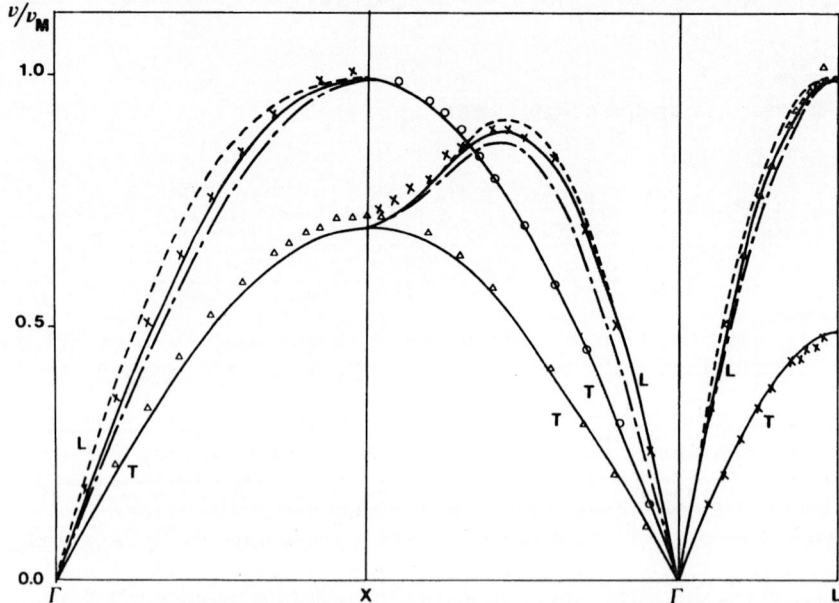

Fig. 1 - Phonon dispersion curves for nickel. The transverse modes
do not depend on q/p. The longitudinal modes are plotted
for three values of q/p. q/p = 0 (dot-dashed lines), q/p =
1/3 (dotted lines) and q/p = 1/5 (full line) compared with
the experimental values (+,0,△).

Table I - Nickel Surface Contraction.

Surface	Coordination number	Contraction in %	
		Theory	Exp.
111	9	1.5	1
001	8	2.7	2.5
110	7	6.5	4.5

The second order terms of (14) increase when the crystal is contracted ($\delta R_{ij} < 0$). Then the surface force constants are larger than the bulk ones. Near a step edge, the contraction is large (about $0.08 R_0$[6,8]) and the force constants between the atoms close to the step are strongly enhanced[6,8]. A step localized frequency appears above the bulk phonon spectrum. Such a step localized vibration has been observed for a 6(111) x ($\bar{1}$11) Platinum surface[12]. But it seems to also occur for other steps[6] . It is rather difficult to observe such a vibration which rapidly disappears as atoms are adsorbed on the step sites. This preferential impurity adsorption cancels the step edge contraction and then the force constant enhancement.

Chalcogenide Adatom Vibrations

The adsorption of atoms or molecules on a clean surface modifies the surface relaxation. One may observe by ELS the own molecule vibration frequencies which are slightly modified by the adsorption but also new frequencies corresponding to a surface metal atom-adatom vibration. We shall only consider the coupling of the incoming electrons with the vibration dipole perpendicular to the surface. The corresponding surface phonon wave vector is equal to zero. For a clean surface, no localized state appears at the center of the surface Brillouin zone excepted for stepped surfaces. But recently, Nickel surface phonons near the clean surface Brillouin zone boundaries have been observed[16]. Oxygen and sulphur ($\sqrt{3}$ x $\sqrt{3}$)R30° and p(2x2) structures are observed on the Nickel (111) surface. Figure 2 shows that the structure reciprocal lattice vectors correspond to symmetry points on the boundaries of the free surface Brillouin zone (labelled K and M). For these points, localized surface vibrations exist in triangular gaps[17]. If the adsorption does not perturb the clean surface vibrations, the surface localized phonons may induce a vibration of the adatom and create an efficient dipole perpendicular to the surface.

We may simply extend our model to study the chalcogenide adsorption. But we must use four new parameters corresponding to the metal-adatom bond, respectively β_{Ni-O}', c_{Ni-O}', P_{Ni-O} and q_{Ni-O}. To simplify further, we have taken :

$$P_{Ni-O} = P_{Ni-Ni}$$

$$(18)$$

$$q_{Ni-O} = q_{Ni-Ni}$$

and we have tried to fit all the measured positions and vibration frequencies of the chalcogenide adatom with the two remaining

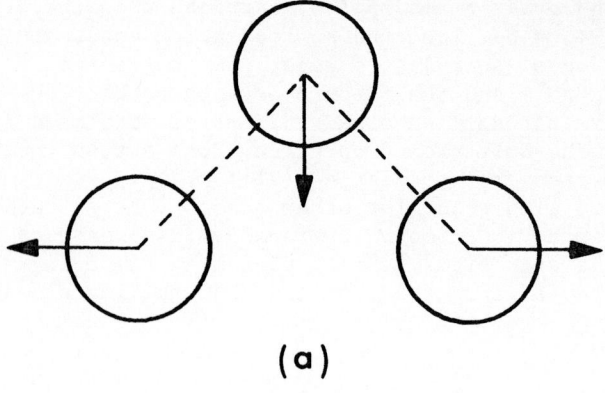

(a)

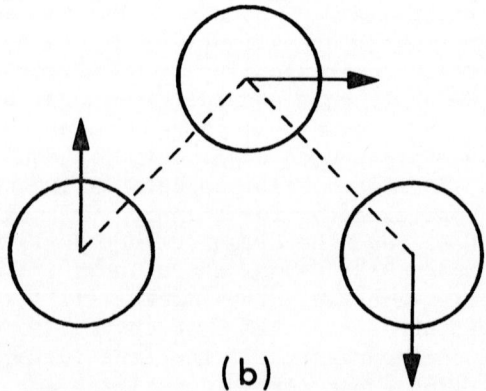

(b)

Fig. 2 – Brillouin zones of the (111) f.c.c. free surface (full
         line) of the ($\sqrt{3}$ x $\sqrt{3}$) R30° structure (a) (dotted line) and
         of the p(2x2) structure (b) (dotted line).

parameters. Table II shows that the fit we get is generally quite
good for

$$\beta^{o}_{Ni-O} = 1.29 \; \beta^{o}_{Ni-Ni} \quad , \quad \beta^{o}_{Ni-S} = 0.95 \; \beta^{o}_{Ni-Ni}$$

$$\hspace{10cm} (19)$$

$$c^{o}_{Ni-O} = 0.32 \; c^{o}_{Ni-Ni} \quad , \quad c^{o}_{Ni-S} = 0.66 \; c^{o}_{Ni-Ni}$$

The calculated c(2x2) oxygen frequency is much larger than the
experimental one whereas the agreement between the sulphur ones is

Table II - Calculated and experimental distances between the adatom and the nickel surface plane and of the adatom vibration frequency for zero wave vector parallel to the surface. ω_M is the nickel maximum frequency.

Structure	Oxygen				Sulphur			
	$d(\mathring{A})$	$d_{exp}(\mathring{A})$	ω/ω_M	ω_{exp}/ω_M	$d(\mathring{A})$	$d_{exp}(\mathring{A})$	ω/ω_M	ω_{exp}/ω_M
p(2 x 2) Ni (001)	0.80	0.90 ± 0.1	1.48	1.45	1.28	1.30 ± 0.1	1.26	1.27
c(2 x 2) Ni (001)	0.85	0.90 ± 0.1	1.30	1.08	1.29	1.28 ± 0.1	1.17	1.20
(√3x√3)R(30°)Ni(111)	1.26		1.93	1.97	1.60		1.40	
p(2 x 2) Ni(111)	1.26	1.20 ± 0.1	1.93	1.97	1.60	1.40 ± 0.1	1.42	
c(2 x 2) Ni(110)					0.84	0.87 ±0.03	1.33	

good. We must remark that in this structure, a Nickel surface atom has two oxygen neighbours so we may expect a strong indirect electronic interaction between the adatoms via the substrate. In the calculation of the second moments of the surface Nickel atom density of states, we partly take this effect into account. But the fourth order moment is also strongly concerned by the indirect interaction. But β^O_{Ni-O} is larger than β^O_{Ni-S} and the distance Ni-O is smaller than the Ni-S one (1.98 Å instead of 2.18 Å). So the contribution to the fourth moments of jumps between the oxygen adatoms via a Nickel substrate atom is 6 times larger than the sulphur one. In one case, it is the main part of the fourth moments whereas in the other it is almost negligible compared to the adatom-Nickel-Nickel walks.

Let us now examine the phonon spectra. They have been calculated at the center of the structure Brillouin zone using a slab of 6 to 12 planes deposited on a frozen Nickel substrate. This is quite sufficient to describe surface localized vibrations whose amplitude exponentially decreases as one goes into the bulk. The discrete phonon frequencies then are gaussian broadened to get a continuous spectrum. The local phonon density $n_i(\omega)$ on atom i is then :

$$n_i(\omega) = \sum_k |u_i(\omega_k)|^2 \exp - [(\omega - \omega_k)^2/2\sigma]$$ (20)

where $u_i(\omega_k)$ is the vibration amplitude on atom i for the k-th eigenfrequency ω_k. σ is the gussian mean-standard deviation.

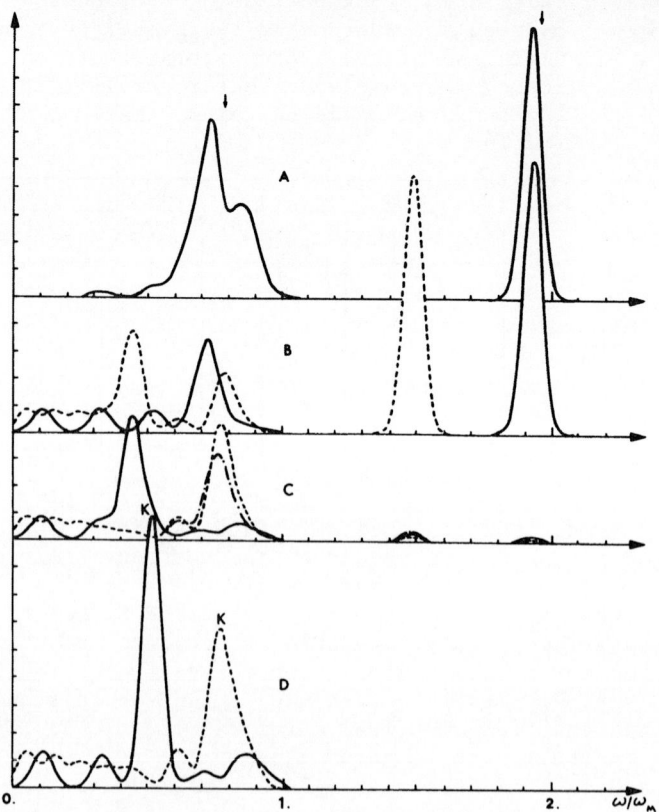

Fig. 3 - Oxygen ($\sqrt{3}$ x $\sqrt{3}$)R30° structure local phonon spectra on the
oxygen adatom (b), on one of its nickel neighbours (c)
and on a clean nickel (111) surface (d). The curve (a)
shows the vibration dipole. The arrows indicate the
positions of the experimental ELS peaks. The full line is
the polarization normal to the surface and the dotted
lines the polarizations on the surface plane.

If we assume that a charge transfer occurs between the adatom
and its N_A neighbours, we can put :

   $q_i = 1$ for the oxygen adatom

   $q_i = - 1/N_A$ for one of its N_A neighbours.

so that we can calculate the dipole $D(\omega)$.

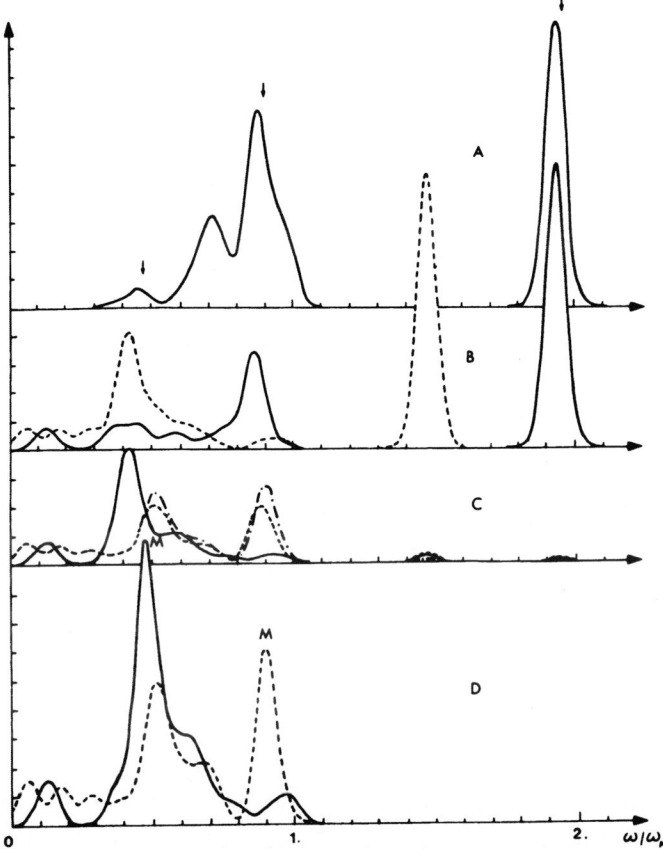

Fig. 4 - Oxygen p(2x2) structure on nickel (111) surface. The
curves are labelled as on figure 2.

$$D(\omega) = \sum_{i,k} q_i \, u_i \, (\omega_k) \, \exp - [\, (\omega - \omega_k)^2 / 2\sigma] \qquad (21)$$

This method eliminates out of phase nickel surface atoms vibrations.
The peaks agree quite well with the experimental results for the
(111) surface[16]. For the (100) surface, the surface doublet phonon
is also observed for the p(2x2) structure[18] whereas it seems difficult to observe a complete c(2x2) structure[18] without oxygen atoms
penetrating in the first Nickel (100) layers (Figs. 3-6).

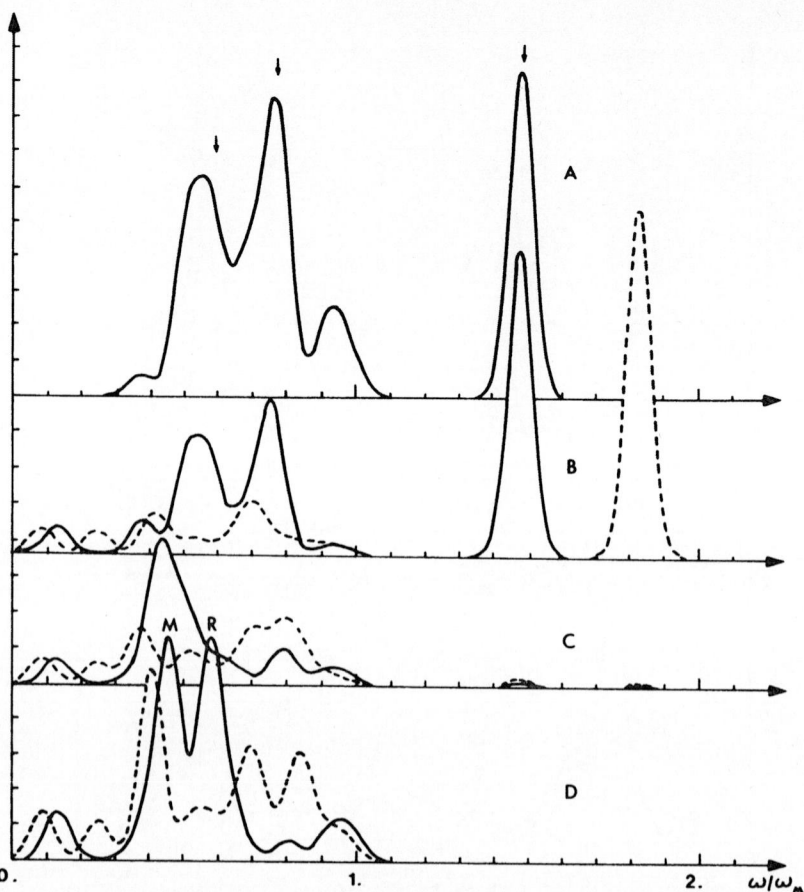

Fig. 5 - Oxygen p(2x2) structure on nickel (001) surface. The
         curves are labelled as on Figure 2.

One must also remark that the main experimental peaks
correspond to clean surface atom vibrations parallel to the surface. The Rayleigh wave which is mainly perpendicular to the surface
would induce a parallel oxygen adatom vibration rather than a
perpendicular one (Fig. 7). Due to symmetry, it also seems difficult
to get an efficient dipole for the c(2x2) structure. Figures 3 to 6
also show that the surface modes are slightly shifted by the oxygen
adsorption.

Now, we can use the fitted parameters to study for example the
adatom frequency dispersion as a function of the wave vector parallel
to the surface. This dispersion is small excepted when we get a force
constant between the adatoms. This is the case of the c(2x2) (100)
surface structure. But even in this case, for sulphur, we only get

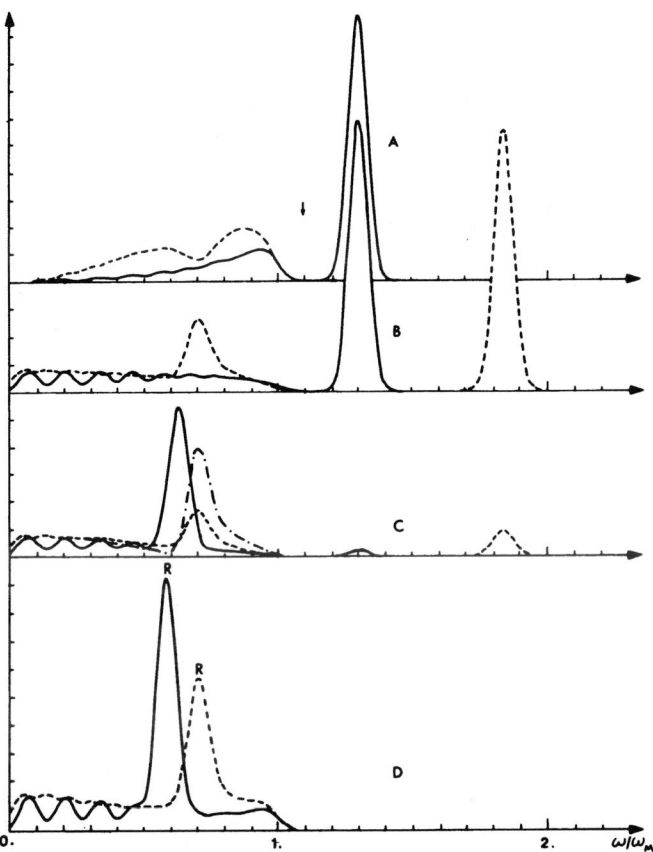

Fig. 6 - Oxygen c(2x2) structure on nickel (001) surface. The
curves are labelled as on Figure 2.

a relative frequency shift equal to ten per cent [9]. We can also
study the elastic indirect interaction between the adatoms[19]. The
interaction anisotropy increases with the surface roughness. It is
very small for a dense (111) plane and quite large for a stepped
surface. The interaction rapidly decreases with the adatom distance.
Finally let us also notice that the adsorption cancels the surface
contraction as it has been observed by LEED[20] or by ISS.

CONCLUSION

We have calculated the surface atom and adatoms positions and
the vibrations from a very simple electronic structure. In spite of
the simplicity of the band model, the agreement with the experimental results is not too bad. With a few parameters we can fit

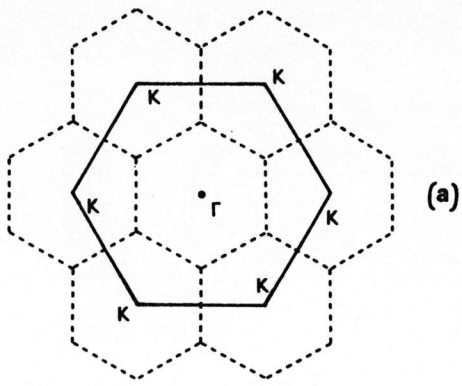

(a)

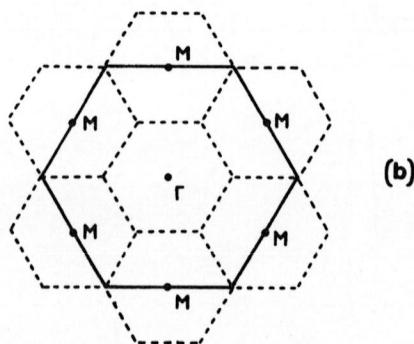

(b)

Fig. 7 - Relative displacements of the surface atom and the adatom.
The vibration (a) gives a dipole perpendicular to the
surface and corresponds to a surface mode polarization in
the surface plane.

almost all the measured positions and frequencies. Improvements
of the model can be made (higher order moments d and p function
anisotropy). But it is not sure that the model would remain
tractable for a rather small improvement of the results.

REFERENCES

1. V.L. Moruzzi, A.R. Williams, J.F. Janak, Phys. Rev. B15, 2854
   (1977).
2. M. Lannoo, J. Phys. (Paris),40, 461 (1979).

3. J. Chadi, CECAM Conf. on Atomic Positions at Solid Surfaces (Paris 1980) (unpublished).
4. F. Ducastelle, J. Phys. (Paris),31, 1055 (1970).
5. G. Allan, M. Lannoo, Surf. Sci., 40, 375 (1973); Phys. Stat. Sol. (b) 74, 403 (1976).
6. G. Allan, Surf. Sci., 89, 142 (1979).
7. B. Stupfel, Thesis (Strasbourg 1980) (unpublished).
8. G. Allan, Surf. Sci., 85, 37 (1979).
9. G. Allan, J. Lopez, Surf. Sci., 95, 214 (1980); 3rd Intern. Conf. Surf. Sci. (Cannes 1980) (to be published).
10. J. Friedel, The Physics of Metals, Ed. J.M. Ziman (Cambridge University Press, 1969).
11. G. Allan, Ann. Phys. (Paris), 5, 169 (1970).
12. H. Ibach, D. Bruchmann, Phys. Rev. Letters, 41, 958 (1978).
13. G. Allan, J. Wach, 3rd Intern. Conf. Surf. Sci. (Cannes 1980) (to be published).
14. J.F. Van der Veen, R.G. Smeenk, R.M. Tromp, F.A. Saris Surf. Sci. 79, 219 (1979); J.A. Davies, D.P. Jackson, N. Matsunami, P.R. Norton, J.V. Andersen, Surf. Sci., 78, 274 (1978).
15. J.A. Davies, D.P. Jackson, P.R. Norton, D.E. Posner W.N. Unertl, Sol. Stat. Comm., 34, 41 (1980).
16. H. Ibach, D. Bruchmann, Phys. Rev. Letters, 44, 36 (1980).
17. V.R. Velasco, F. Yndurain, Surf. Sci., 85, 107 (1979).
18. H. Ibach, D. Bruchmann (private communication).
19. J. Lopez, G. Allan (to be published).
20. K.O. Legg, F. Jona, D.N. Jepsen, P.M. Marcus, J. Phys. C (Solid State Phys.) 10, 937 (1977); Phys. Rev. B16, 5271 (1977).

ATOM MOTION AT MODEL CRYSTAL SURFACES

John Earle Black ★

Physics Department - University of California

Irvine, California 92717

ABSTRACT.

We have studied the motion of atoms in and near the surfaces of model crystals using an expectation value of the form $<u_\alpha(\vec{\ell})u_\beta(\vec{\ell}')>$, where $u_\alpha(\vec{\ell})$ is the αth Cartesian component of displacement of atom ℓ from its equilibrium position. The method of continued fractions has been used to construct the spectral densities $\rho_{\alpha\beta}(\vec{\ell}\vec{\ell}';\omega)$. Analysis of these densities provides insights into which phonons contribute to atomic displacements, parallel and perpendicular to the surface, and which phonons contribute to the correlated motion of neighbouring atoms. We have compared the spectral densities with phonon dispersion curves obtained from slab calculations, and have examined the usefulness of spectral density in interpreting the results of electron energy loss spectroscopy experiments. In this paper we present a brief description of the method, and the results for tungsten, rhodium, nickel, and for oxygen adatoms on nickel.

Recently there have been several papers dealing with atom dynamics at metal surfaces in which a continued fraction technique is employed. This technique allows one to calculate spectral densities and mean square displacements. It has been used by Black, Laks and Mills[1] to study atom motion at the W(100) surface, by Mosteller and Landman[2] to study atom motions at a step on the Pt(111) surface, and by Black[3] to study motion of adatoms on the Ni(111) surface.

★ On leave of absence from Physics Department, Brock University, St. Catharines, Ontario L2S 3A1, CANADA.
UCI Technical Report #80-56.

It is the intention of this paper to concentrate on what can. be achieved with the application of the continued fraction method. We begin with a discussion of the theory of the method, and the accuracy to be expected. This is followed by an examination of spectral density and mean square displacement obtained for a number of metals, and for various models of the interaction between atoms. In the third section of the paper prominent peaks in the spectral density are interpreted using dispersion curve data obtained by means of slab calculations.[4]

In the fourth section of the paper we examine the application of the method to the study of atom correlations in the surface. Then in the fifth section we apply the method to the motion of adatoms at the nickel (111) surface. We conclude with a comparison of our data with the electron energy loss spectra obtained by Ibach and Bruchmann[5] and Ibach[6], and with a few remarks about future applications of the continued fraction method.

THE CONTINUED FRACTION TECHNIQUE

It can be shown[1] that the continued fraction technique used by Haydock, Heine and Kelly[7] in the study of electronic states may also be used in the study of atom dynamics. In particular it may be used to calculate the Green's function defined as

$$U_{\alpha\beta}(\vec{\ell}\vec{\ell}';z) = \sum_s \frac{e_\alpha^{(s)}(\vec{\ell}) \, e_\beta^{(s)}(\vec{\ell}')}{\omega_s^2 - z^2} \quad \ldots \tag{1}$$

Here $e_\alpha^{(s)}(\vec{\ell})$ is the αth component of the eigenvector at site $\vec{\ell}$ for the sth mode, and ω_s is the eigenfrequency of the sth mode.

A spectral density is then obtained from the Green's function using the relation

$$\rho_{\alpha\beta}(\vec{\ell}\vec{\ell}';\omega) = \frac{\omega}{i\pi} (U_{\alpha\beta}(\vec{\ell}\vec{\ell}';\omega + i\in) - U_{\alpha\beta}(\vec{\ell}\vec{\ell}';\omega - i\in))\ldots \tag{2}$$

in the limit as $\in$ goes to zero. The density can be used to determine the correlation between the displacement $u_\alpha(\vec{\ell})$ of the atom at $\vec{\ell}$ in the α direction and the displacement $u_\beta(\vec{\ell}')$ of the atom at $\vec{\ell}'$ in the β direction. We find

$$<u_\alpha(\vec{\ell})u_\beta(\vec{\ell}')> = \frac{\hbar}{2M} \int_0^\infty \frac{(1 + 2\bar{n}_\omega)}{\omega} \rho_{\alpha\beta}(\vec{\ell}\vec{\ell}';\omega)d\omega \quad \ldots \tag{3}$$

where $\bar{n}_\omega$ is the Bose-Einstein function, $\hbar$ is the Planck's constant, and M is the mass of the atom. We see that $\rho_{\alpha\beta}(\vec{\ell}\vec{\ell}';\omega)$ is a measure of the frequencies which contribute most to the atomic displacements. It may be called an effective local phonon density of states.

The Green's function $U_{\alpha\alpha}(\vec{\ell}\vec{\ell}';z)$ is the easiest to determine. Moreover, for $\beta = \alpha$, $\vec{\ell}' = \vec{\ell}$ in Equation (3) we obtain the well known mean square displacement of the atom at $\vec{\ell}$ in the α direction. For this case we have to evaluate the continued fraction

$$U_{\alpha\alpha}(\vec{\ell}\vec{\ell}';z) = \cfrac{1}{z^2 - A_1 - \cfrac{B_2}{z^2 - A_2 - \cfrac{B_3}{z^2 - A_3 - \cdots}}} \cdots \qquad (4)$$

The coefficients A_n, B_n in the above expression are obtained by an iterative procedure. In our model of the crystal we specify the number of neighbours with which each atom interacts through central forces. In the iteration the atom of interest is first allowed to interact with only its nearest neighbours. This yields (A_1, B_2). It is then allowed to interact with the neigbours of its nearest neigbours to obtain (A_2, B_3) and so on. In our studies we have used clusters of about 1500 atoms. This has permitted determination of the (A_n, B_{n+1}) out to n = 8. The procedure developed by Haydock, Heine and Kelly then allows one to replace the remainder of the continued fraction with a simple function of z^2, provided A_n, B_{n+1} has converged to a constant value.

In our studies the coefficients have not quite reached their constante values. As a consequence the spectral densities, and mean square displacements, depend on the details of our termination of the continued fraction. This provides a measure of the accuracy of our calculation. We find the mean square displacements are accurate to about $\pm$ 5 %, while peaks in the spectral density are accurate to $\pm$ 0.1 THz.

We conclude this section with a few remarks about the model crystals. For tungsten a three neigbour model of atomic forces was used based on fits to phonon dispersion data in the bulk by Chen and Brockhouse[8]. For nickel a nearest neighbour model based on a fit to phonon dispersion data by Clark, Herman and Wallis[9] was used. We also employed, for nickel and rhodium, a two neighbour model with angle bending forces described by Black, Campbell and Wallis[10]. Finally we used data calculated by Upton and Goddard[11] to calculate nearest neighbour force constants for adatoms of H, O and S at a site of three fold symmetry above the nickel (111) surface. In the various models no attempt has been made to allow for surface relaxation.

SPECTRAL DENSITIES OF SURFACE ATOMS

In Figure 1 we show spectral densities for vibrations of surface atoms perpendicular to the surface. We find one or two peaks at about 3 THz for a number of model crystals and surfaces studied. Thus for the rhodium (100) surface there are two peaks, but only one peak is seen on the rhodium (111) surface when a two neighbour model with angle bending forces is used. We also note that tungsten, when a three neighbour model is used, and nickel, when a nearest neighbour model is used, have single peaks in the (100) spectra.

Not shown in the figure are results for a two neighbour model of Ni(100). In that case the peak at 3.95 THz is just starting to split. Apparently, with our 1500 atom cluster, the peak separation of about 0.5 THz is just at the limits of resolution of the continued fraction method. Accordingly the absence of two peaks in the other cases studied may be simply due to the fact that they are not resolved.

The bottom curve is for bulk nickel with a single neighbour model. Note that it bears little resemblence to the surface density. In all models studied we have found this to be the case. We also find that the density one layer below the surface is quite different qualitatively from both its surface and bulk values. Two layers below the surface the density resembles its bulk value.

In Figure 2 we show spectral densities for atom vibrations parallel to the surface. All models give a sharp peak at the high end of the spectrum, and only tungsten has several strong peaks below this. It is not known if this character of tungsten occurs because it has BCC structure, or because a three neighbour model was used. Typically the densities parallel to the surface resemble their bulk forms one layer below the surface. We also find an upward shift of the high frequency peak as we move further into the bulk.

It is a straightforward calculation to obtain mean square displacements from the densities. Results of these calculations are described in the various papers of the references.

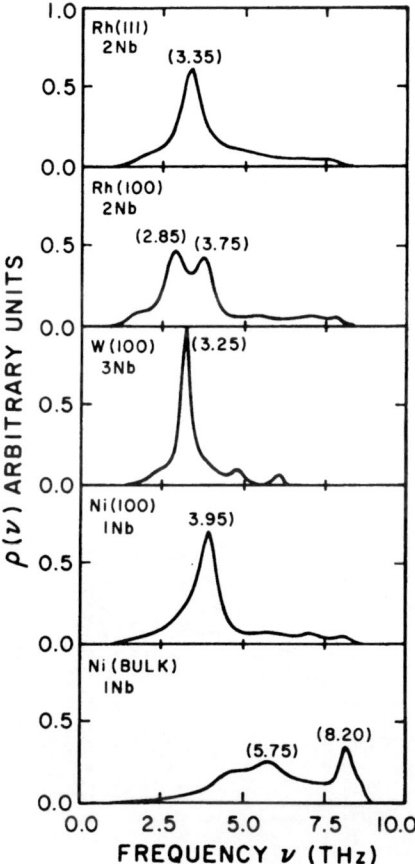

Figure 1 – Spectral density for vibrations perpendicular to the
surface of an atom at the surface.

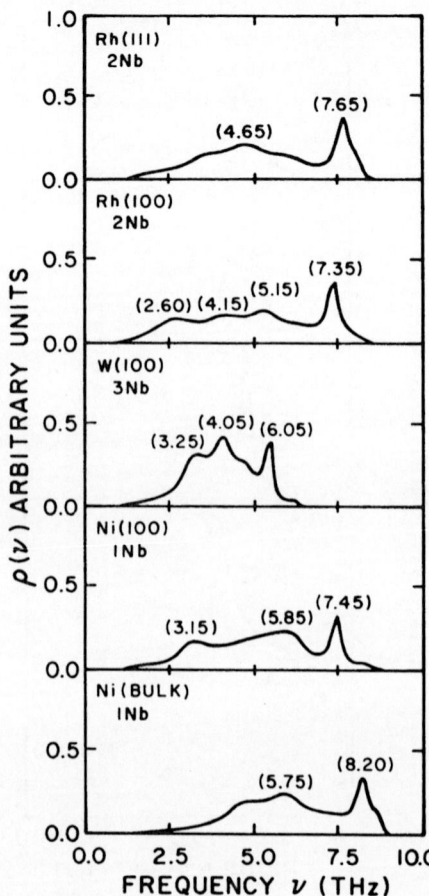

Figure 2 - Spectral density for vibrations parallel to the surface
    of an atom at the surface.

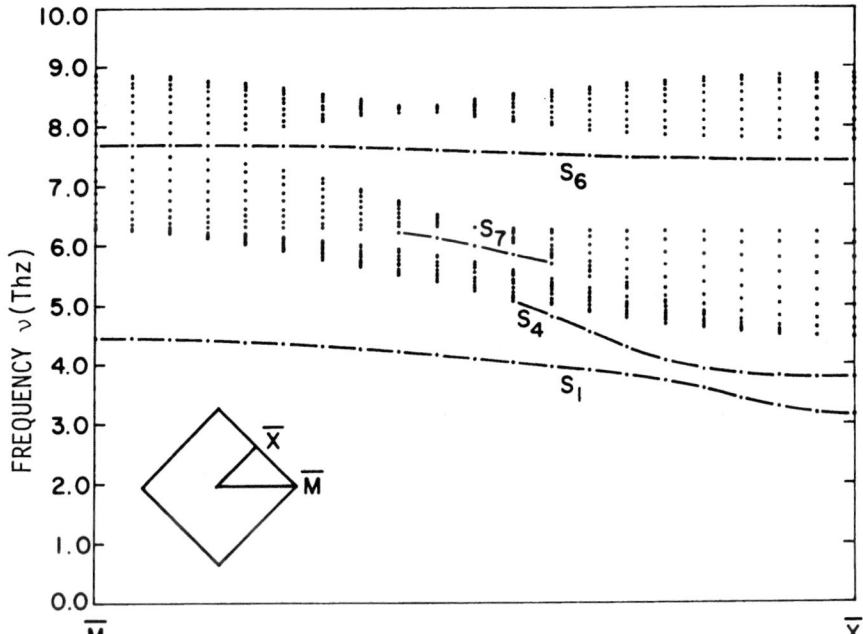

Figure 3 – Dispersion data between the $\bar{M}$ and $\bar{X}$ points on the
surface Brillouin zone boundary of nickel (100). Nearest
neighbour forces only

THE RELATION BETWEEN SPECTRAL DENSITY AND DISPERSION CURVES OBTAINED BY SLAB CALCULATIONS

In Figure 3 we show the results of a slab calculation of the
Ni(100) dispersion curves with 20 atomic layers. The calculation
parallels that of Allen Alldredge and de Wette[12], and is described
in Black et al[10]. Only the wavevectors between $\bar{X}$ and $\bar{M}$ on the zone
boundary are shown. We can, it would seem, find the origin of
features on the spectral density curves by examining the flat
regions of these dispersion curves (where we expect the highest
density of states). Thus the peak at 3.95 THz in Figure 1 seems
to originate in the surface modes S_1 at and near $\bar{M}$ (4.5 THz) and
S_4 at and near $\bar{X}$ (3.75 THz). A study of rhodium[10] shows these two
modes are much more widely separated in frequency, which explains
why the two peaks in rhodium of Figure 1 are well separated.

The feature at 7.45 THz in Figure 2 for nickel is apparently
due to the mode S_6 in the vicinity of $\bar{X}$. We also note that the
bulk peak at 8.20 THz is at the frequency of the bottleneck region
between $\bar{X}$ and $\bar{M}$.

In reference 10 we present a detailed study of the slab modes
and densities for a two neighbour model of rhodium with angle bending
constants. There the identification of peaks with surface modes is
improved by examining the eigenvectors at various depths in the
crystal.

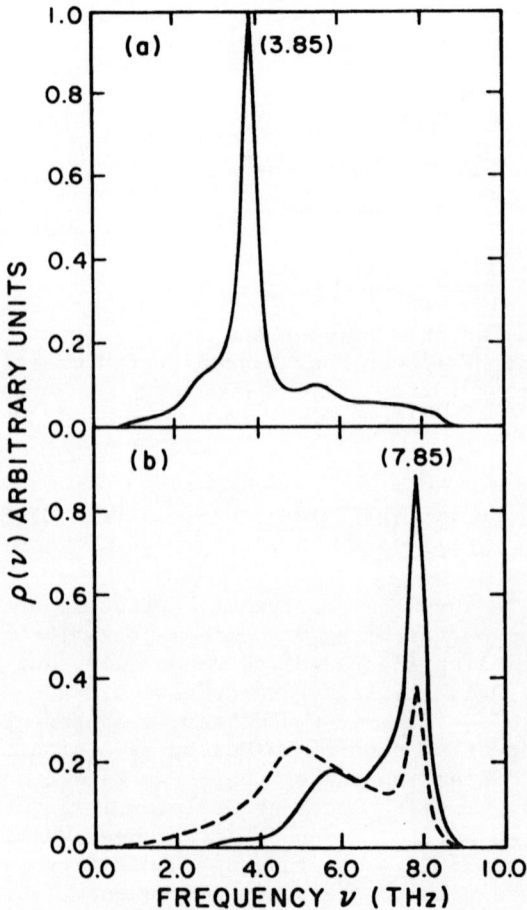

Figure 4 - Spectral density for vibrations at the nickel (111)
          surface. a) motion perpendicular to the surface
                  b) motion parallel to the surface (dashed) and
          a breathing mode with three nickel atoms (continuous).
          A nearest neighbour force constant model is used.

SPECTRAL DENSITIES INVOLVING MORE THAN ONE ATOM

The continued fraction technique can be applied to more than
one atom. In Figure 4 we show densities for motion parallel and
perpendicular to the nickel (111) surface obtained with a nearest
neighbour model. As an aid to understanding what happens to an
adatom placed above the surface at a site of three-fold symmetry
we have also examined the density for a coordinate which consists
of the sum of the three surface atom displacements away from their
common center. As you can see this has a strong feature at 7.85 THz.
Thus we expect to see an adatom above the surface moving up and
down as the three atoms below it "breathe".

We have examined a number of correlations in W(100)[1] . They
are somewhat difficult to calculate if a number of atoms are involved, but reveal dynamical features which are difficult to get at
by other means.

NICKEL IN THE PRESENCE OF A SINGLE ADATOM

We have examined[3] the dynamics of H, O, S and Ni adatoms
which are located above the Ni(111) surface. In Figure 5 the results
are shown for an oxygen adatom. Here we compare the motion parallel
and perpendicular to the surface of the adatom with that of the
neighbouring nickel atoms. The peaks are at the same locations
(withing the accuracy of $\pm$ 0.1 THz). There is evidence that the
oxygen adatom is "riding" on the nickel substrate. This is supported by the mean square displacements which, as shown in Equation 3,
depend inversely on the atom mass. We find, at 400K, the values
of 12.3×10^{-19} cm^2 and 7.9×10^{-19} cm^2 for neighbouring nickel
atom motion normal and parallel to the surface, whereas the oxygen
adatom values are 8.1×10^{-19} cm^2 and 12.6×10^{-19} cm^2 for normal
and parallel motion respectively.

The above remarks apply to the frequency range below 10 THz.
As can be seen from the scale factors of Figure 5 the oxygen spectral density is reduced considerably below that of the nickel atoms.
In reference 1 it was shown that the total integral under the density curves should be unity. The remaining oxygen density lies at
17.2 THz in the perpendicular case and 15.4 THz in the parallel
case, under delta functions. These modes are the frequency shifted
modes of 15.4 THz and 13.3 THz introduced as input data for the
force constants of the oxygen nickel interaction. The input data is
in the form of a perpendicular frequency (with the oxygen adatom
coupled to an infinitely massive substrate) and the distance of
the adatom above the substrate. It was obtained from the work of
Upton and Goddard[11]. The application of the continued fraction
technique then allows a comparison of the Upton and Goddard calculation with the experimental frequency obtained by Ibach and

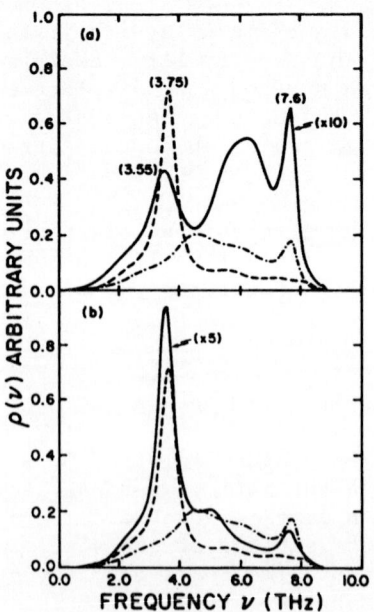

Figure 5 - Spectral density for an oxygen adatom at the Ni(111)
          surface (continuous) and the normal (dashed) and
          parallel (dot-dash) densities for adjacent nickel atoms.
          a) Normal to the surface
          b) Parallel to the surface.
          A nearest neighbour force constant model is used.

Bruchmann[5] for oxygen. They agree to within computational error.

    We have found, for motion perpendicular to the surface, that
the frequency shifts of oxygen sulfur, and hydrogen adatoms at the
nickel (111) surface can be estimated to within 20 % from the
simple formula

$$\omega = \omega_o (1 + \frac{M_A}{M_s \times 3 \sin^2 \theta})^{1/2} \ldots \tag{5}$$

where ω is the shifted frequency, ω_o the unshifted frequency, M_A the adatom mass, M_s the substrate mass, and θ the angle between the surface and a nickel oxygen bond.

A similar formula can be derived for the Nickel (100) surface (replace 3 by 4 in the Equation for ω) for an adatom at a site of four-fold symmetry. A comparison of the work of Upton and Goddard with the experimental data of Andersson[12] indicates very good agreement for oxygen and sulfur adatoms on the nickel (100) surface.

INTERPRETATION OF EELS DATA

Ibach and Bruchmann[5] examined electron energy loss spectra for oxygen adsorbed on the Ni(111) surface. They observed peaks at the locations shown in Table 1. Note the two peaks for a p(2x2) oxygen overlayer and a single peak for the $\sqrt{3}$ x $\sqrt{3}$ R 30° overlayer. They interpreted the peaks as due to dipole excitation normal to the surface of the nickel phonon modes in the substrate, the dipole excitation at the $\bar{\Gamma}$ point of the overlayer of oxygen exciting surface phonons at the $\bar{M}$ and $\bar{K}$ points of the substrate lattice.

Allan and Lopez[13] have examined dipole excitation at the Ni(111) and Ni(100) surfaces. They have used detailed models of the Ni-Ni and O-Ni interaction at the surface, but have restricted their study to the wavevectors $\bar{K}$ and $\bar{M}$. Their findings support the Ibach and Bruchmann interpretation of their Ni(111) data.

Our approach is to examine the total spectral density in order to see what it tells us about the atoms involved in the dipole scattering. In Table 1 we show the location of our calculated peaks. In Figure 5 we saw how the perpendicular and parallel motion of the nickel atoms produced normal motion of the adatom. Note also that the oxygen atom reduces the frequencies of these modes from their bare nickel values of Figure 4. The final calculated entry in Table 1 for Ni(111) allows for further frequency reduction due to a p(2x2) overlayer. It is an estimate based on a study in which oxygen atoms were located at a fcc-like site above the Ni(111) surface. We see the periodic estimate lies below the experimental p(2x2) values. This may be due to the fact that we have not allowed for the relaxation of the nickel (111) surface which is believed to contract by about 1 % according to LEED data.

Table 1.

A comparison of EELS and one atom spectral density peaks.

	Adatom Structure	Spectral Peaks THz (cm^{-1}) (± 0.1 THz; ± 4 cm^{-1})			
		Parallel Motion		Perpendicular	
(111) Surface					
theory	No Adatom	7.85 (263)	-	-	3.95 (129)
	Single Adatom	7.6 (254)	-	-	3.55 (119)
	Periodic Estimate	7.3 (244)	-	-	3.25 (109)
experiment	p(2 × 2)	7.9 (265)	-	-	4.0 (135)
	√3 × √3 R 30°	7.2 (240)	-	-	-
(100) Surface					
theory	No Adatom	7.5 (250)	5.85 (196)	3.15 (105)	3.95 (132)
	Periodic Estimate	6.9 (230)	5.25 (176)	2.55 (85)	3.35 (112)
experiment	p(2 × 2)	6.9 (230)	5.35 (180)	-	-
	c(2 × 2)	6.9 (230)	5.35 (180)	2.7 (90)	-

Recently Ibach[6] has obtained data for p(2x2) and c(2x2) oxygen overlayers on Ni(100). We have used the downward shift of 20 cm^{-1} (0.6 THz) obtained in our Ni(111) calculations to estimate the shifted location of peaks in Figure 1[d] and 2[d]. In Table I the agreement with experiment is surprisingly good. We note that the very strong feature of the normal Rayleigh wave is not present in the EELS data, somehow it is not dipole active for the overlayers studied, or the Ni(100) surface. We also note that the feature at 3.15 THz of Figure 2[d] is a surface mode which has a strong perpendicular component 1 layer below the surface[10]. It is the mode S_1 at $\bar{X}$ of Figure 3 (it should not be excited with a c(2x2) overlayer). The feature at 5.85 THz lies in the vicinity of S_4 between $\bar{X}$ and $\bar{M}$. It may also be the bottleneck region which contributes.

Our spectral density method does not allow us to determine which peaks are dipole active. We are presently studying the Ni(100) surface using a spectral density which is the Fourier transformed version of that described in Equation 2. This will enable us to determine actual dipole moments at selected points in reciprocal space, from which a more precise comparison of theory and experiment can be made.

DISCUSSION

We have used the continued fraction technique to determine spectral density and mean square displacement for a number of different problems. The spectral densities thus obtained are useful in interpreting EELS experiments. In particular they provide indications of the shifts in frequency to be expected due to the presence of a single isolated adatom.

The results obtained for oxygen adatoms on Ni(111) and Ni(100) surfaces are in excellent qualitative agreement with the experimental data of Ibach and the theoretical data of Upton and Goddard. A modified continued fraction technique, one in which employs a Fourier transformed density of states, is presently being used to study dipole scattering. It will, we believe, allow a more precise contact with theory and experiment. In particular it may allow EELS data to be used in determination of the substrate interatomic force constant at various surfaces.

AKNOWLEDGEMENTS

I wish to thank Professors D.L. Mills and R.F. Wallis for many helpful comments during the course of the spectral density investigations. This research has been supported by the Department of Energy through contract No. DE-ATO379-ER10432 and by a grant from the National Science and Engineering Research Council of Canada.

REFERENCES

1. J.E. Black, B. Laks and D.L. Mills, Phys. Rev. B 15 August 1980.
2. M. Mostoller and U. Landman, Phys. Rev. B20, 1755, 1979.
3. J.E. Black, Surface Sci. (to be publ.)
   J.E. Black (to be publ.)
4. R.F. Wallis, Progress Surface Science $\underline{4}$, 233, 1973.
5. H. Ibach and D. Bruchmann, Phys Rev. Letters $\underline{44}$, 36, 1980.
6. H. Ibach - private communication.
7. R. Haydock, V. Heine and M.J. Kelly, J. Phys. C. Solid St. Phys.
   $\underline{8}$, 2591, 1975.
8. S.H. Chen and B.N. Brockhouse, Solid State Communication $\underline{2}$, 73,
   1964.
9. B.C. Clark, R. Herman and R.F. Wallis, Phys. Rev. $\underline{139}$, 860, 1965.
10. J.E. Black, D. Campbell and R.F. Wallis (to be publ.)
11. T.H. Upton and W.A. Goddard III, ISISS 1979, Surface Science,
    Recent Progress and Perspectives (to be publ. by the Chemical
    Rubber Company).
12. S. Andersson, Surface Science $\underline{79}$, 385, 1979.
13. G. Allan and J. Lopez (to be publ. in Surf. Sci.)

INFRARED REFLECTION ABSORPTION AND EMISSION STUDY OF METHYL ALCOHOL ADSORPTION ON MoO$_3$ FILMS

M. Ito

Faculty of Engineering, Keio University, 3-14-1, Hiyoshi

Kohoku-ku, Yokohama, Japan 223

ABSTRACT

The adsorption of methyl alcohol on well oriented MoO$_3$ films has been investigated with infrared reflection absorption and emission spectroscopies. Methyl alcohol adsorbs dissociatively to produce methoxy species at 100 K.

The band at 1045 cm^{-1} was attributed to the C-O stretching vibration of chemisorbed methoxy species. A similar band was dectected at higher frequency, 1065 cm^{-1} , above room temperatures and under high exposure of gaseous methanol. The C-O bond in the latter species has some of its double bond character.

INTRODUCTION

The knowledge about surface vibrations of adsorbed species on top layers on metal catalysts is very important in order to clarify the chemical reaction mechanisms. Recent infrared reflection absorption studies (IRRAS) have made it possible to observe the chemisorbed species on single crystal surfaces. Considering that IRRAS can be applied in the presence of large quantities of gas phase adsorbate molecules, we have concentrated on the development of the infrared emission spectroscopy as a technique for the study of adsorbed species present on real catalytic surfaces under high temperature and pressure.

The spectra in infrared emission spectroscopy are observed as the spontaneous radiative decay from higher energy levels of surface emitting centers excited by sample heating, to lower levels.

Therefore, in principle, the same information is obtainable from emission and transmission spectra. Infrared emission studies for adsorbed species so far reported have been limited on powder poly-crystalline surfaces[1] and no results on well oriented films or single crystal surfaces has been published.

Bradshaw has developed a rotating polarizer infrared modulation technique[2] which has some advantages for studying surface vibration ; it is capable of high spectral resolution and does not damage or disturb the adsorbed layer. The great advantage can be seen in its application to real surfaces under high gas pressure. Thus, the use of a rotating polarizer method has enabled to obtain the weak emittance spectra from a monolayer of chemisorbed species on a small area surface. The example of methyl alcohol adsorption on MoO_3 films will be presented in which both modulation infrared reflection absorption and emission spectroscopies were used.

EXPERIMENTAL

Sample preparation

Well oriented films of MoO_3 could be obtained by evaporating MoO_3 at 500 K on a silver plate, 40 x 30 mm in size and 200 - 500 Å in thickness. Metal surfaces are generally highly reflecting and therefore at the same time they have a low emissivity in the infra-red region. Thus, the silver plate was chosen as a substrate. After evaporation the sample was annealed at 700 - 750 K since, above these temperatures, the film melts and volatilizes. The sample was then remounted in a UHV cell and degassed at 450 K for one hour. MoO_3 powder (99.99 %, Mitsuwa Chem. Co. Ltd.) and spectroscopy grade methyl alcohol (99.5 %) were used. The purity of the alcohol was checked by mass spectrometry.

Polarization modulation measurements

The diagram of the polarization modulation spectrometer used for the experiment is shown schematically in Fig. 1. The UHV IR cell was basically the same as that used in earlier reflection absorption work[3] but with a modified re-entrant. The pressure in the cell was 2×10^{-10} Torr... The sample was placed in the cell at the entrance focus of the spectrometer and was held at temperatures ranging between 330 and 450 K by heating from behind.

In polarization modulation, the IR beam is alternately polarized perpendicular(s) and parallel(p) to the plane of incidence by rotating the polarizer. Only the parallel component(p) is absorbed by the oriented molecule adsorbed on the metal surface (surface selection rule), and the perpendicular component(s) acts in this tech-

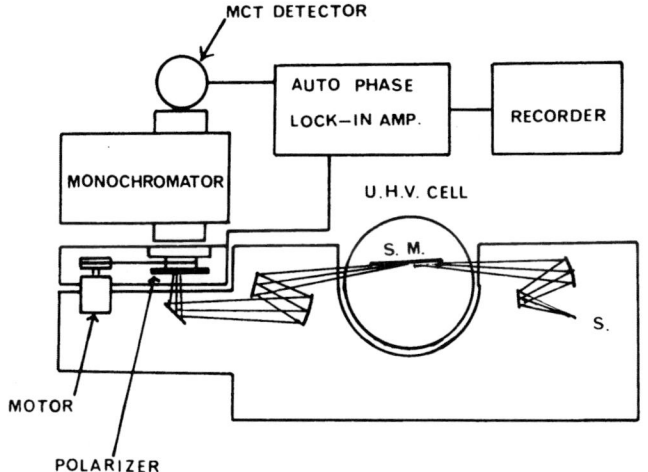

Fig. 1. The arrangement of the polarization modulation spectrometer used for infrared reflection absorption and emission experiments. In emission study, light source(s) is switched off.

nique as a reference. Both p and s polarized components of the radiation are absorbed to exactly the same extent by the randomly oriented gas-phase molecules. The angle between the normal to the emitting surface and the optical axis of the monochromator was varied by rotating the sample plate in order to increase the sensitivity and to test the orientation of the MoO_3 films[4]. A polarizer located in front of the monochromator is rotated at 600 rpm to give a signal of 20 Hz at the liquid nitrogen cooled HgCdTe detector. The signal from the detector is input to a low noise preamplifier and then ot an auto-phase lock-in amplifier.

RESULTS

Reflection absorption and emission spectra of MoO_3 films

The relfection absorption spectra of well oriented MoO_3 films exhibited a strong band at 996 cm^{-1} and a very weak broad band around 910 cm^{-1}. No other band was found in the 4000 - 400 cm^{-1} region. Fig. 2-a (dotted line) represents the reflection absorption spectrum of MoO_3 film at room temperature. The spectrum is in contrast with the transmission results, where in addition to these bands, more intense bands were observed in the lower frequency region. The full lines show the emission spectra at 363, 423 and 483 K (Fig. 2-b ; c; d, respectively). The ripple around 800 cm^{-1} in emission spectra is due to the change of grating of the monochromator. The intensity of the 996 cm^{-1} band in reflection absorption spectra increased in intensity with incident angle and reached its maximum at

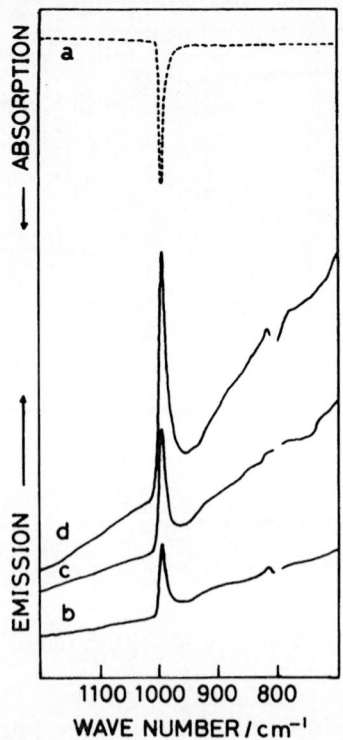

Fig. 2. Infrared spectra of molybdenum trioxide.
       a. Reflection absorption spectrum at room temperature, the
          angle of incidence is 80°
       b. Emission spectrum at 360 K
       c. Emission spectrum at 420 K
       d. Emission spectrum at 480 K

a large angle value of 83° as shown in Fig. 3. Hence, the angle of
the sample plate was fixed at this value (These spectra were obtained by using a conventional Jasco. IR-G double beam spectrometer).
The background level in emission spectra around the 10μ region.
(Fig. 2-b, c, d) increased toward the low frequency side. However,
in polarization modulation emission measurements, a relatively flat
base line could be obtained, though differences in reflectivity for
perpendicular and parallel light normally occur and this causes the
slope in the background curve observed in reflection absorption measurements.

Adsorption of methyl alcohol on MoO_3 film

    Fig. 4 shows the reflection absorption spectra for methanol

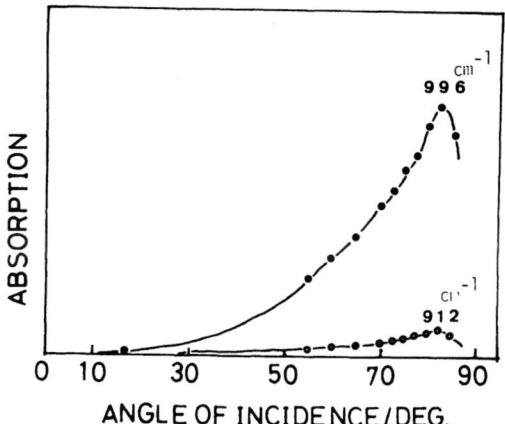

Fig. 3. Intensities for 996 and 912 cm^{-1} bands for infrared reflection absorption spectra at various incident angles.

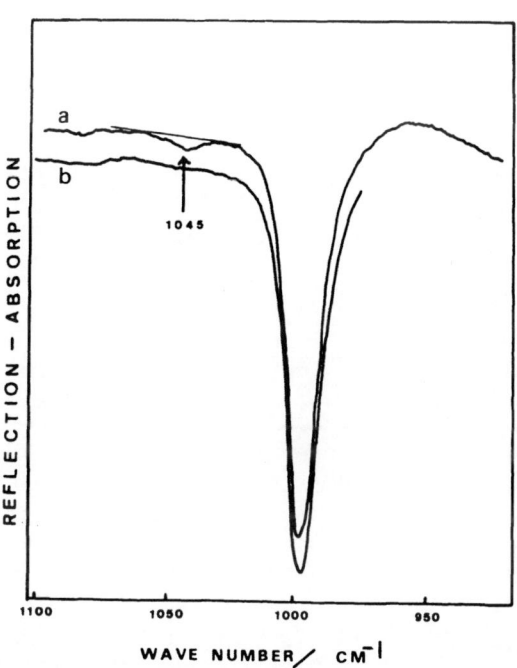

Fig. 4. Polarization modulation infrared reflection absorption spectra.
a. 10 L. methanol at 235 K
b. Evacuated at room temperature.

adsorption on MoO$_3$ film. Alcohol adsorption was done at about 150 K.
Fig. 4-b is the spectrum before exposing to alcohol. After 10 L
(1L : 1 x 10^{-6} Torr. sec.) exposure of methanol to the surface (Fig.
3-a), a peak appeared around 1045 cm^{-1} and could easily be removed
by warming the substrate above room temperature. The band can be
assigned to a C-O stretching vibration, since the corresponding absorption in gas phase alcohol appears at 1033 cm^{-1}. Since films were prepared in another vacuum cell, they were contaminated during the
transfer into the UHV cell. This may be the reason for which the
intensity of the band was weak.

Fig. 5 represents the emission spectra obtained by polarization
modulation technique. Fig. 5-a is the spectrum before exposing to
alcohol at a pressure of 10^{-9} Torr. The 996 cm^{-1} band is the only
band observed in the region. After admitting 30 Torr of alcohol at
the same temperature (Fig.5 -b) a band at 1065 cm^{-1} appeared. The
intensity of this band increased by raising the temperature up to
458 K (Fig. 5-c). By evacuating the cell to less than 10^{-3} Torr
the band at 1065 cm^{-1} disappeared. The band begun to appear at a
pressure of 0.5 Torr of alcohol and the emissivity increased with
temperature and with the pressure of gaseous alcohol. This intensity variation was reversible. The gas phase alcohol was necessary for
the occurrence of the band. The band at 996 cm^{-1} decreased slightly
in intensity by adsorption of methanol. When MoO$_3$ films were exposed to a large amount of alcohol above 475 K, the strong 996 cm^{-1}
band disappeared, and no film was left on the sample plate owing to
sublimation of MoO$_3$ film.

DISCUSSION

Well oriented MoO$_3$ film

MoO has been described as a layer structure based on two dimensional networks of MoO$_3$ octahedra. The Mo-O distance for one terminal oxygen is shorter than for the other bridging oxygens, showing
an analogy with a double bond Mo = O. In this structure the Mo = O
group projects perpendicularly to the plane of the bridging Mo   Mo
group. The intensity of 996 cm^{-1} band showed a remarkable angle dependence with respect to the plane of incidence. Strong intensity
of the band and also angle dependence can be expected since infrared
reflection absorption and also emission spectroscopies have only large sensitivity for vibrations having a dipole moment change perpendicular to the metal surface (surface selection rule). Transmission
spectra of the MoO films on KBr and reflection spectra evaporated
at room temperature (poorly oriented films) showed extra intense
bands which can be assignable to Mo-O-Mo stretching vibration band.
Thus, the band appearing at 996 cm^{-1} can be ascribed to the Mo = O
stretching vibration band, which implies that the MoO$_3$ films evaporated on a silver substrate can grow up with Mo = O groups projecting

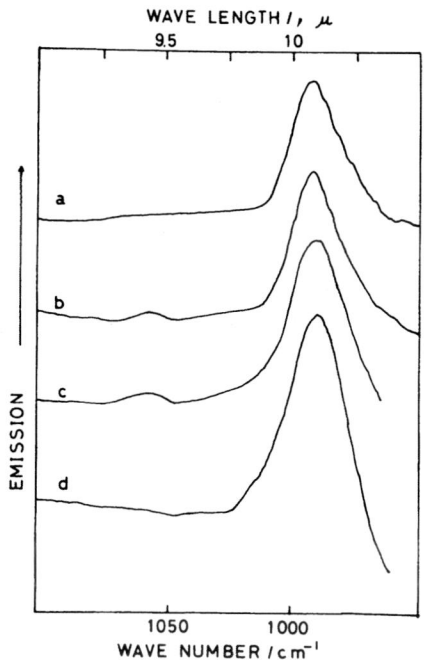

Fig. 5. Polarization modulation infrared emission spectra.
   a. in UHV at 413 K
   b. 30 Torr addition of methanol at 413 K
   c. heating at 458 K
   d. after evacuation at 458 K.

from the surface. Volta, Desquesnes and Moraweck recently reported that selective sites for formaldehyde formation from methyl alcohol are located on (020) MoO₃ planes (Volta, Desquesnes and Moraweck, 1980).

Adsorption spectra of methyl alcohol on a MoO₃ film

   The frequencies of the C-O stretching band on several surfaces are indicated in Table 1 where the vibrations of gaseous and liquid methanol are also included. Demuth and Ibach identified a quasi-stable intermediate species from methanol on Ni(111) as a methoxy species. They showed that the methoxy species decomposes near room temperature leaving carbon monoxide and chemisorbed oxygen. Sexton reported methyl alcohol adsorption on Cu(100) (Sexton, 1979).

   The stable methoxy species exists on Cu(100) even at room temperature. The observed C-O stretching frequency, 1060 cm⁻¹ in this study, is substantially higher than those reported on Ni(111)

Table 1.

Comparison of the C-O stretching vibration of gaseous and liquid methanol to the observed bands. The units are cm^{-1}.

gaseous CH_3OH	liquid CH_3OH	methoxy CH_3O on $Cu(100)$	methoxy CH_3O on $Ni(111)$	this work
1033	1030	1010	1040	1045 at 150 K 1065 at 458 K

(1040 cm^{-1}) or Cu(100)(1010 cm^{-1}) surfaces. This implies that significant changes in the hybridization state of the carbon atoms in methoxide have occurred. The intermediate species was only formed following high exposures of alcohol at elevated temperatures. At low exposures the adsorbed methoxide was rapidly converted to aldehyde. Thus, the formation of the intermediate species required the presence of undissociated methyl alcohol on the surface. The above observations suggest that the oriented MoO_3 film was more reactive than the Ni(111) or Cu(100) surfaces for decomposition of methoxides to aldehyde.

We hope to be able in the future to detect the radiation emitted by species adsorbed on a metal surface at the sub-monolayer level.

Part of the cost of this research was met by the Scientific Research Grant of the Ministry of Education (n° 484023), to which the author's thanks are due.

REFERENCES

1. M. Primet, P. Fouilloux and B. Imelik, Surface Sci. 85 (1979) 457-470.
2. A.M. Bradshaw and F. Hoffmann, Surface Sci. 52 (1975) 449-454.
3. M. Ito and W. Suetaka, J. Catal. 54 (1978) 13-23.
4. R.G. Greenler, Surface Sci. 69 (1977) 647-651.
5. J.C. Volta, W. Desquesnes and B. Moraweck, The Seventh International Congress on Catalysis, Preprints of Communications, C4, Tokyo (1980).
6. T. Shimanouchi, Table of Molecular Vibrational Frequencies, Part I, NSRDS-NBS6 (1967) 63-64.
7. J.E. Demuth and H. Ibach, Chem. Phys. Lett. 60 (1979) 395-399.
8. B.A. Sexton, Surface Sci. 88 (1979) 299-318.

ROTATIONAL ACCOMMODATION AT SURFACES VIA RAYLEIGH PHONONS

B. Feuerbacher

Astronomy Division, Space Science Department of ESA
ESTEC
2200 AG Noordwijk, The Netherlands

ABSTRACT

Time-of-flight measurements have been performed on the inelastic scattering of room-temperature beams of H_2 and D_2 molecules from a LiF(100) surface. The results show a discrete loss peak near the molecular rotational excitation energy that disperses with scattering angle. This indicates that rotationally inelastic scattering is dissipative rather than diffractive, i.e. it includes phonon excitations in the solid. The phonons involved can be identified as single Rayleigh surface phonons.

INTRODUCTION

The scattering of low-energy neutral particles from solid surfaces has recently become the subject of increasing interest[1,2]. This is because such experiments can give information on the atom-solid interaction potential[3], the geometrical structure of the surface[4], and the two-dimensional elementary excitations of the solid surface[5,6]. The latter has led to a direct observation of the Rayleigh phonon dispersion relation on a lithium fluoride crystal[7]. The present paper is concerned with a particular aspect of this scattering process, namely the interaction of a molecule having internal degrees of freedom with a surface, and the coupling of the corresponding quantum states to the elementary excitations of the solid. Such problems are relevant not only for the fundamental energy transfer processes concerned, but also for the energetic accommodation of molecules having internal degrees of freedom on solid surfaces in adsorption-desorption processes, and for the energy balance between dust and gas in interstellar clouds.

While an increasing amount of data is becoming available on atom-surface scattering both from the experimental and theoretical side, little is known on the scattering of molecules from solid surfaces. Following the early experiments of Estermann and Stern[8], Boato et al.[9] have performed elastic measurements of H_2 scattering from a LiF single crystal surface. Their results show additional diffraction maxima due to rotationally inelastic scattering. The analysis of these maxima relates them to processes where no momentum or energy is transferred between molecule and solid, i.e. the surface has merely acted to transform molecular translational energy into rotational energy in the scattering process[9,10]. This leads to scattering into discrete directions, given by the conservation of the parallel momentum component to within surface reciprocal lattice vectors, and should result in a single, discrete energy loss peak at the difference between the initial molecular translation energy E_i and the rotational excitation energy ΔE_{rot}[11].

The present study is the first to include energy resolution in a molecular-surface scattering experiment. Using the H_2-LiF system, it is shown that indeed an inelastic peak is found in the vicinity of the rotational excitation energy, however varying in energy position with scattering angle. This indicates the predominance of a double excitation process. The transformation of translation energy into a rotational excitation occurs with a simultaneous creation or annihilation of a phonon, which serves to ensure energy and momentum conservation over a range of scattering parameters. The phonon involved in this process can be identified as a Rayleigh surface phonon, and the data allow to observe part of its dispersion relation in a direct manner.

EXPERIMENTAL

The experimental setup consists of a beam source, in this case a pulsed free jet source, a UHV scattering chamber, and a flight tube with detectors providing time-of-flight analysis (Fig. 1). The system is pumped by a 140 1/sec ion pump combined with a 200 1/sec turbomolecular pump and a titanium sublimator. A second turbomolecular pump evacuates the flight tube.

The beam source operates on the principle of a fast switching valve combined with a free jet nozzle[12]. It is schematically shown in Fig. 2. Fast current pulses applied to two coils open or close the valve by means of induced current repulsion in an aluminium disk. The time delay Δt and the pulse current can be varied to optimize valve performance for various gases. Pulse lengths of 10 to 100 μsec can be achieved with about 10^{23} mol/sec. sterad intensity. Speed ratios from about 15 for Hydrogen to 25 for Neon have been obtained with a 0.5 mm nozzle orifice. The pulse repetition rate can be adapted to the pumping capacity of the

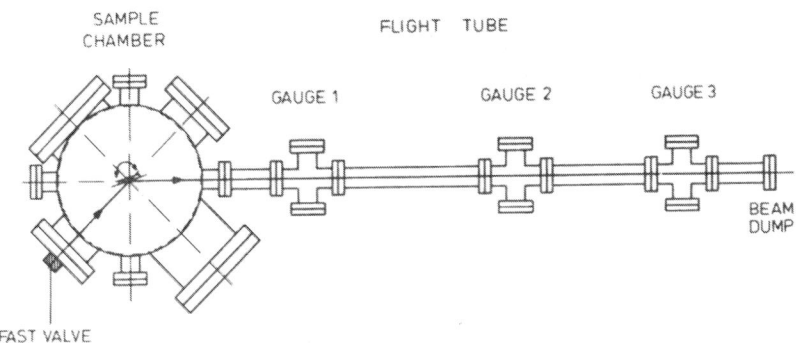

Fig. 1 - Schematic of the experimental setup showing beam source
(fast valve), UHV chamber with rotable LiF sample crystal,
and the flight tube with three gauge positions.

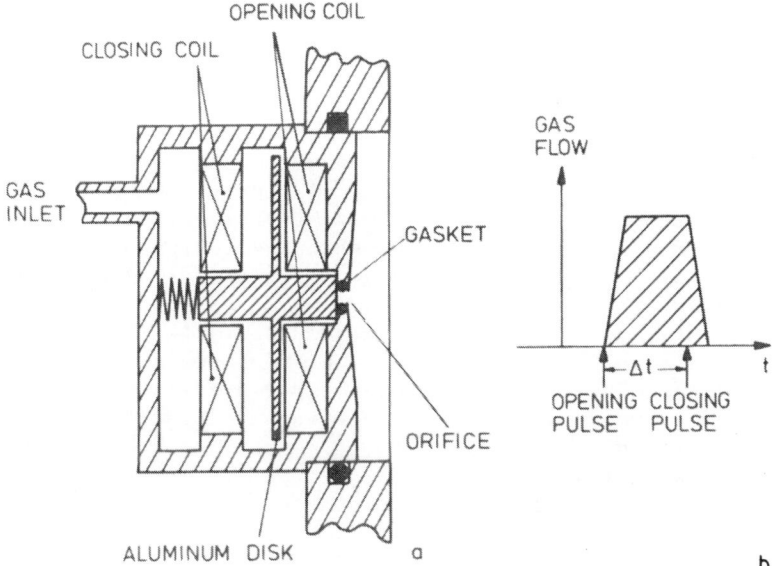

Fig. 2 - a) Schematic view of the fast valve used for pulsed beam
generation. Fast current pulses are applied to the
coils which move the aluminium disk by induced current
repulsion.
    b) The gas pulse duration is determined by the time delay
Δt between opening and closing pulse.

system and was set to about 1 pulse per 10 sec in the present
measurements. Spectra were compiled by adding 100 to 200 pulses
in a multichannel analyzer with a resolution of 10 μsec per channel.

The pulses were detected using commercial UHV vacuum gauges
in the Bayard-Alpert configuration, provided with a fast electro-
meter amplifier. A voltage-to-frequency converter forms the pulses
required for the multichannel analyzer. Gauge 2 (Fig. 2), at a
distance of 905 mm from the sample, was used to obtain the spectra.
The other two gauges served to give an absolute measurement of the
molecular velocity.

The scattering surface was a (100) plane of a LiF single
crystal, annealed for several ten hours in vacuum at about 700 K
after cleavage in air. Standard cleaning procedures ensured an
atomically clean surface. Background pressures in the low 10^{-10}
Torr range raised to about 10^{-8} Torr during a pulse. The scattering
configuration had a fixed angle of 135° between source and detector.
Angular variation was achieved by rotating the crystal through
an angle $\Delta\Theta$, with positive values referring to a decreased angle of
incidence.

RESULTS

Measurements were performed with H_2 and D_2 molecules in order
to provide internal consistency using two similar molecules with
different rotational excitation energy. Typical time-of-flight
spectra for scattering of a H_2 beam having a translation energy of
79 meV are shown in Fig. 3a. The lowest curve represents specular
scattering at 67.5° angle of incidence along a <10> azimuth. Only
the elastic peak is seen, at the position corresponding to the
initial molecular velocity E_0. In the angular range $-7° < \Delta\Theta < -2°$
a secondary structure is observed that changes position with
scattering angle. The flight time of the secondary structure is
longer, so it corresponds to a loss of translational energy of the
molecule. It should be noted that the secondary peak has approximately the same half width as the elastic structure. A slight shift
to shorter flight times is also observed for the primary peak at
increasing negative angles.

A similar set of spectra for D_2 scattering is shown in Fig.3b.
The initial translation energy was 88 meV in this case. The lowest
curve again corresponds to specular scattering. A secondary structure
is observed for a quite different angular region than for H_2 , and
it is much closer to the position of the specular structure. In
contrast to H_2 scattering, the inelastic structure dominates the
peak near the elastic position. The intensity factors in Fig. 3 are
given relative to the specular intensity in each case. There is a
factor 30 less intensity for deuterium scattering on an absolute

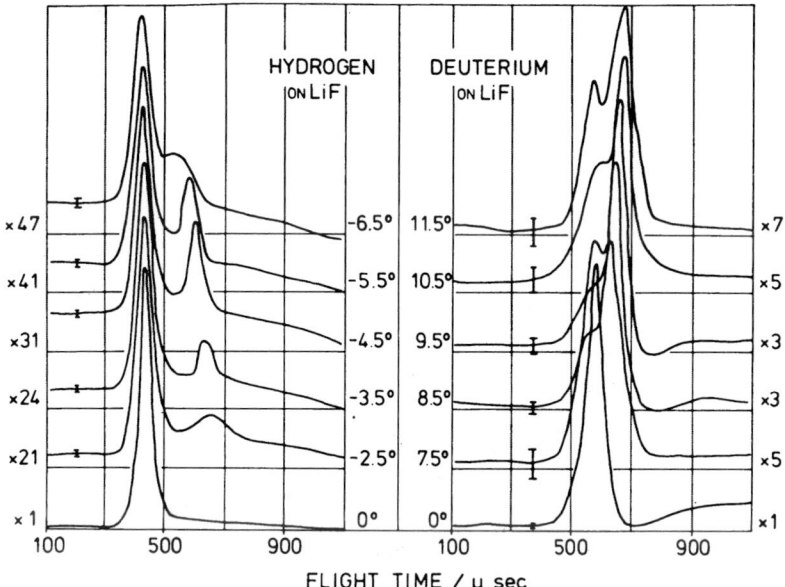

Fig. 3 – Time-of-flight spectra of Hydrogen (a) and Deuterium (b) scattered from a (100) LiF surface along a < 10 > azimuth for various sample angles (as given along the centerline), relative to the specular condition. Fixed source–detector angle is 135°, flight distance from the surface is 905 mm. All curves are normalized to the specular intensities (0°) by factors as indicated. Beam energies are 79 meV for H_2 and 88 meV for D_2.

scale. Therefore the error bars on the D_2 curves are considerably larger than on the H_2 curves.

EVALUATION

The spectra shown in Fig. 3 are presented as measured in terms of flight time, which depends on the beam energy, molecule mass and flight distance. The data can be converted into energy spectra using the independently measured beam velocity. Such a conversion for the case of H_2 scattering is presented in Fig. 4 . The energy scale is referred to the initial molecular beam energy and therefore represents energy gain and loss during scattering as indicated at the top. The data are given in histogram form, the width of each step corresponding to a single channel of the multi-channel spectra. It is apparent that the constant flight time intervals provide an energy resolution varying over the measurement range with a poor resolution in the energy gain region (corresponding to short flight times).

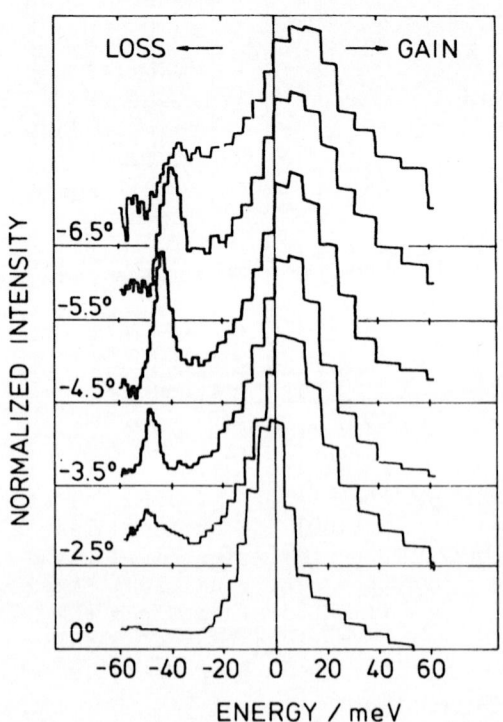

Fig. 4 - Energy spectra derived from time-of-flight measurements for H_2 scattering from a LiF(100) surface. Energies are given relative to the incident beam energy, 79 meV. The steps arise from a constant 10 µsec sampling width in the TOF spectra. Intensities are normalized to the specular peak.

The spectra show a clear loss peak at energies between -35 meV and -55 meV. Note that the measured energy width of the loss peak at e.g. -4.5° is about 6 meV at half height, equivalent to the energy spread of the incident beam. A badly resolved shift is also noticeable for the main peak near the elastic position.

In molecular beam scattering from surfaces a unique relation exists between energy and momentum transfer in a scattering event, depending on the geometric scattering conditions. This relation is

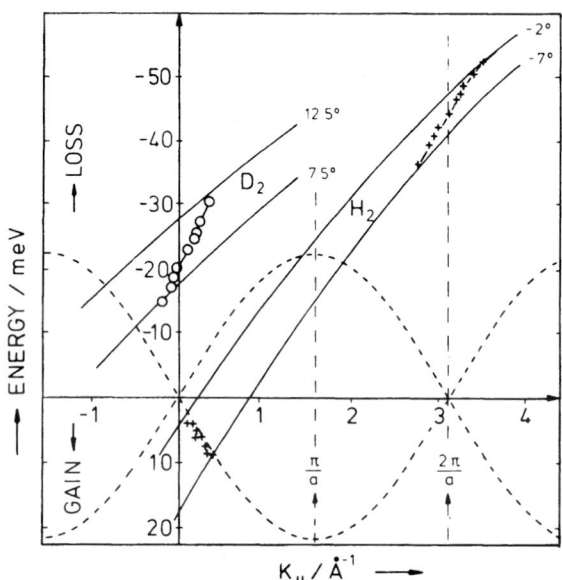

Fig. 5 - Plot of energy vs parallel momentum transfer for H_2 and
D_2 scattering from a LiF(100) surface. The solid lines
represent coupling parabolas derived from Eq. (1) for
sample angles as indicated. Crosses and circles are peak
positions for H_2 and D_2 ,respectively. The broken line
gives the Rayleigh phonon dispersion (Ref. 14) in an extended zone scheme for phonon creation (upper trace) and
annihilation (lower trace).

derived based on the conservation laws of energy and the momentum
component parallel to the surface,K, and can be written[13]

$$\frac{\Delta E}{E_i} = 1 - (1 - \frac{\Delta K}{K_i})^2 \frac{\sin^2 (\theta_s - \Delta\theta)}{\sin^2 (\theta_s + \Delta\theta)} \qquad (1)$$

Here ΔE and ΔK are energy and momentum transfer respectively, E_i and K_i are initial energy and parallel momentum component, Θ_s is the fixed scattering angle given by source and detector positions, and $\Delta\Theta$ the sample rotation relative to the specular position. In a plot of energy vs. parallel momentum, relation (1) gives parabolas for each sample angle $\Delta\Theta$, which allow to relate the energy values derived in Fig. 4 to their corresponding momentum values. Such a plot is presented in Fig. 5. For H_2 , two parabolas derived from equation (1) are shown for the extreme angles $\Delta\Theta = -2°$ and $-7°$. The crosses indicate the measured energy position of the loss peak (upper group near -40 meV) and the shift of the main peak (lower group between 0 and 10 meV), and the momentum transfer corresponding to these losses. In the same diagram, the broken sinusoidal curve represents the dispersion of the Rayleigh phonon in an extended zone scheme according to the calculations of Benedek and Garcia[14]. The dispersion relation is shown twice corresponding to phonon creation (upper trace) and annihilation (lower trace).

The reduced energy-momentum values for D_2 scattering are also plotted in Fig. 5 as circles. Due to the higher noise component in those spectra, the near-specular peak could not be evaluated. All observed points are found to correspond to small positive or negative momentum transfers.

DISCUSSION

The energy gain structure for H_2 scattering, found at 0 to 10 meV in Fig. 5, can clearly be associated with the absorption of single Rayleigh phonons. A similar effect has been demonstrated earlier in the scattering of He atoms from LiF[6] . As apparent from Fig. 4, this feature is not well resolved and the cross positions have been obtained from peak centroids.

As far as the energy loss structure for H_2 and D_2 are concerned, both appear to be centered to either zero momentum transfer (D_2) or transfer of a full reciprocal lattice vector (H_2). This is indicative of rotationally inelastic scattering with vanishing energy and momentum transfer (the latter to within reciprocal lattice vectors) between molecule and solid, as observed earlier in elastic scattering studies[9] . Indeed, the loss energies at these momentum values correspond closely to the $0 \rightarrow 2$ rotational transition, which is 44 meV for H_2 and 22 meV for D_2 . Deviating from this point the peak does not just decrease in intensity while oc-curring at constant energy, as would be expected for a rotational-ly inelastic event without interaction with phonons. In contrast, the structure stays discrete and shows a clear dispersion, which is approximately parallel for both molecules. This is indicative of a dissipative scattering process, involving discrete phonon excitations in the solid.

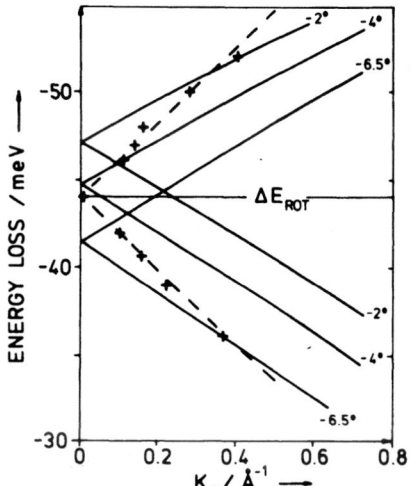

Fig. 6 – Energy vs momentum transfer for the rotationally inelastic
peak in H_2 scattering from LiF(100) in a reduced zone
presentation. Solid lines give coupling parabolas for
sample angles $\Delta\Theta$ as indicated. The broken line shows a
Rayleigh phonon dispersion (Ref. 14) shifted vertically by
the rotational excitation energy ΔE_{rot}.

        A closer inspection of the dissipative scattering is presen-
ted in Fig. 6 for the H_2 results, which have been obtained with a
superior signal-to-noise ratio. Here the data are shown in a reduced
zone scheme. It is seen that the rotationally inelastic loss of
$\Delta E_{rot} = 44$ meV occurs at the diffractive condition for vanishing
molecule-surface momentum transfer. In the same figure the broken
line indicates the dispersion of Rayleigh phonons, shifted vertically in energy by ΔE_{rot}. The close coincidence of the dispersion
curve with the position of the loss peaks, together with the
discrete nature of the losses, suggests that dissipative scattering
takes place via creation or annihilation of single Rayleigh
phonons in the surface. The scattering process can consequently
be regarded as a two step process, where translational energy of
the molecule is converted into a rotational excitation, with

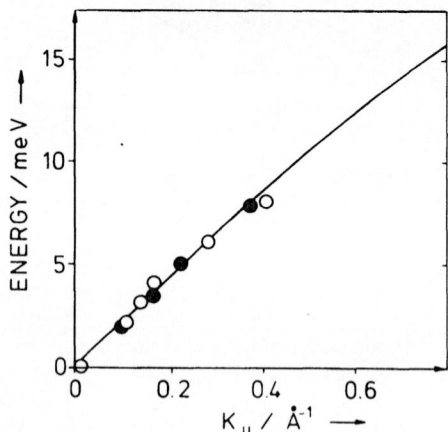

Fig. 7 - Experimental phonon dispersion relation derived from
rotationally inelastic H_2 scattering on a LiF(100) surface. Open and closed points refer to phonon creation and
annihilation, respectively. The full line represents a
calculated dispersion curve along the (10) azimuth including relaxation (Ref. 14).

simultaneous emission of absorption of a Rayleigh phonon to satisfy the scattering conditions away from the diffractive case. A
remarkable feature is the fact that this process seems to occur
with a strong selection rule on the direction of the phonon momentum, namely for antiparallel K for phonon creation and parallel K
for phonon annihilation. The background of this selection rule is
yet unclear.

Under the above assumption, the present data can be used to
obtain an experimental plot of the Rayleigh phonon dispersion
relation, as shown in Fig. 7. This has been obtained by sub-

tracting the rotational excitation energy for H_2 from the measured
loss energies, and plotting the absolute value of the difference as
a function of the reduced momentum transfer. A good agreement with
the theoretical dispersion relation of Benedek and Garcia[14] is
found for their calculation on a relaxed LiF surface, which has
been experimentally confirmed by Doak et al. [7]. An attractive
feature of this method is the fact that the rotationally inelastic
event shifts the phonon losses into a region where a higher resolution can be obtained than in the vicinity of the specular peak
(see Fig. 4).

CONCLUSIONS

The results of the present study on inelastic molecular
beam scattering on surfaces demonstrate that rotational excitation
of a molecule occurs in a dissipative rather than a diffractive
way. This implies that a direct coupling exists between the internal
excitations of both the molecule and the solid. Scattering conditions away from the diffractive case are fulfilled by the creation
or annihilation of phonons. The phonons involved are shown to be
single Rayleigh phonons in the crystal surface. The data allow to
plot a dispersion relation for Rayleigh phonons from the rotationally inelastic structure in good agreement with theoretical predictions and previous experimental results.

REFERENCES

 1. J.P. Toennies, Appl. Phys. 3, 91 (1974).
 2. B. Feuerbacher in "Adsorbate Vibrations at Surfaces", ed. by
    R.F. Willis, Springer, Heidelberg 1980.
 3. H. Wilsch in "Topics in Surface Chemistry", ed. by E. Kay and
    P.S. Bagus, Plenum Press, 1978, p. 135.
 4. K.H. Rieder and T. Engel, Phys. Rev. Letters 43, 373 (1979).
 5. F. Brusdeylins, R.B. Doak and J.P. Toennies, Phys. Rev. Letters
    44, 1417 (1980).
 6. B. Feuerbacher, M.R. Adriaens, and H. Thuis, Surface Sci. 94,
    L 171 (1980).
 7. G. Brusdeylins, R.B. Doak and J.P. Toennies, Proc. 4th Int.
    Conf. Solid Surfaces, Cannes 1980.
 8. I. Estermann and O. Stern, Z. Physik 61, 95 (1930).
 9. G. Boato, P. Cantini and L. Mattera, Japan. J. Appl. Phys.
    Suppl. 2, 553 (1974); J. Chem. Phys. 65, 544 (1976).
10. G. Wolken, J. Chem. Phys. 59, 1159 (1973); ibid. 62, 2730 (1975).
11. R.M. Logan, Mol. Phys. 17, 147 (1969).
12. W. Allison, M.R. Adriaens and B. Feuerbacher, to be published.
13. G. Benedek, Phys. Rev. Letters 35, 234 (1975).
14. G. Benedek and N. Garcia, to be published.

ARRANGEMENT OF CO MOLECULES ADSORBED ON LOW INDEX METAL SURFACES :

A COMPARISON BETWEEN LEED AND HREELS OR IRS

J.P. Biberian ★ and M.A. Van Hove ★★

★ Laboratoire Surfaces-Interfaces, ERA CNRS 070899
Faculté des Sciences de Luminy - 13288 Marseille
Cédex 2 - France

★★ Materials and Molecular Research Division, Lawrence
Berkeley Laboratory, and Department of Chemistry,
University of California, Berkeley, California 94720
USA

ABSTRACT

The adsorption of carbon monoxide on low index metal surfaces has been the object of intense research, presenting some apparently conflicting results. Thus, for three systems, CO on Cu(100), on Cu(111) and on Ru(0001), HREELS or IRS show only one C-O stretching frequency while an interpretation of the LEED diagram with a compact monolayer of CO gives at least two adsorption sites. A re-examination of these systems with a model based on a finite coincidence unit cell and one adsorption site, allows a better interpretation of all the data. Laser simulation of LEED is used to test this interpretation and to determine molecular positions within the larger coincidence unit cells.

INTRODUCTION

In the past few years Infra Red Spectroscopy (IRS) and High Resolution Energy Electron Loss Spectroscopy (HREELS) have brought new data on the adsorption of CO molecules on low index metal surfaces. By observing the position of the C-O stretching frequency, the type of adsorption site can, in principle, be determined. Following a frequently used assignment, if the C-O stretching frequency is above 2000 cm^{-1}, the CO molecules are adsorbed on top sites, between 1700 cm^{-1} and 2000 cm^{-1} they are adsorbed on bridge sites, and below 1700 cm^{-1} on three fold sites. These values are determined by IRS from metal carbonyls and can vary from one metal to another.

On the other hand, Low Energy Electron Diffraction (LEED) has been used extensively to study the adsorption of CO on low index metal surfaces. A simple analysis of the LEED patterns gives only the unit cell of the adsorbed layer. In order to determine the actual position of the CO molecules on the surface, an I-V profile analysis is necessary.

At low coverage and for the few simple structures studied so far by IRS, HREELS and LEED I-V profile analysis, there is perfect agreement between both types of experiments. For the Cu(100)-c(2x2) CO structures, it has been shown that the CO molecules sit on top sites by IRS[1-2], HREELS[3-4] and LEED[5]. The same results have been found for the Ni(100)-c(2x2) CO structure where top site adsorption is found by HREELS[6], IRS[7] and LEED[5,8,9] . For the Pd(100)-(2$\sqrt{2}$ x $\sqrt{2}$)R45° 2CO structure, IRS[10,11] indicates bridge sites as does a LEED I-V profile analysis[12]; in this particular case even more simply, symmetry arguments of the LEED pattern[13] already imply bridge sites. It has been shown that for the Rh(111)-($\sqrt{3}$ x $\sqrt{3}$)R30° CO structure, the CO molecules are on top sites by HREELS[14] and LEED[15]

At high coverages, the LEED patterns become more complex, and there are ambiguities in the determination of the adsorbate surface unit cells. Nevertheless,there is no contradiction between IRS or HREELS and LEED for CO adsorbed on the following surfaces : Pt(100), Pt(111), Pd(100), Pd(111), Ni(100), Ni(111) and Rh(111). However, IRS and HREELS show only top site CO on Cu(100), Cu(111) and Ru(0001), whilst LEED analysis indicates a variety of adsorption sites if a compressed hexagonal structure is assumed.

The purpose of this paper is to show that an alternative model for interpreting the LEED patterns can reconcile both types of experiments.

ANALYSIS OF THE LEED STRUCTURES

The Double Diffraction Model

In this model, it is assumed that the LEED diagrams are composed of three types of spots : the substrate spots, the adsorbate spots, and the multiple diffraction spots created by the combination of the substrate and adsorbate lattices. This analysis leads one to assume that some of the spots exist only by virtue of multiple diffraction. Yet it is known that multiple diffraction spots are about an order of magnitude weaker than single diffraction spots. Therefore such an interpretation is quite questionable since most of the structures observed exhibit intense "multiple diffraction spots". This model leads to the "compact model" for the adsorbed layer where the adsorbate forms a pseudohexagonal (i.e. compressed hexagonal) layer on top of the substrate.

The Coincidence Lattice Model

Another interpretation of the LEED structures is to assume that there is a coincidence unit cell between the adsorbate and the substrate, and that all (or almost all) the spots are diffraction spots of this coincidence unit cell. Multiple diffraction contributes little to the actual intensity of the spots. Since the coincidence unit cell contains more than one molecule per cell, some of the spots can have nearly zero intensity because of the structure factor. The interpretation of the LEED pattern with this model gives easily the coincidence unit cell, but not its contents. In order to determine the actual position of the molecules in the cell, an I-V profile analysis is necessary. However, we shall confine our attention to the gross two-dimensional structure (i.e. the adsorption sites) and ignore the dimension perpendicular to the surface (including bond lengths). Optical simulation with laser diffraction has been used before for this more limited purpose and we find it very suitable in our context as well. We use a He-Ne laser as light source and a grid computer-drawn on a slide as the diffraction object, the distance between the dots representing the CO molecules being ~ 30 μm.

CO ON Cu (100)

The adsorption of CO on Cu(100) has been studied by IRS[1-2] as well as by HREELS[3-4]. Both types of experiments show that there is only one peak at all coverages for the C-O stretching frequency 2079 cm^{-1} as determined by IRS or 2089 cm^{-1} by HREELS. This is characteristic of top site adsorption for the CO molecules at all coverages.

LEED observations for CO adsorbed on Cu(100) have been reported by various authors[16-19]. As the coverage increases, first a c(2x2) structure is observed, then a c($7\sqrt{2}$ x $\sqrt{2}$)R45° structure. Figure 1 shows a schematic of these LEED diagrams. There is no ambiguity in interpreting the c(2x2) structure with the CO molecules sitting on top sites, as reported in section 1.

The c($7\sqrt{2}$ x $\sqrt{2}$)R45° structure has been interpreted first with the "compact model" as a pseudohexagonal structure[17]. The reciprocal unit cell of the CO overlayer is drawn in figure 1b (the two

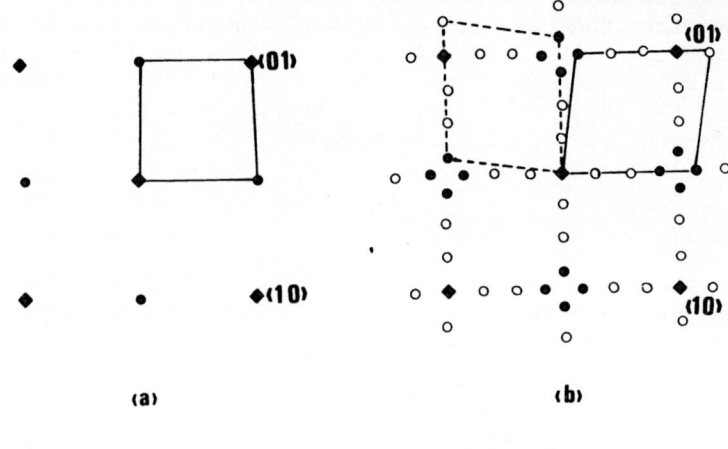

♦ Substrate spots   • Intense extra spots
                             ○ Weak extra spots

Fig. 1 – LEED patterns observed when CO is adsorbed on Cu(100).
    a) the c(2x2) structure.
    b) the c(7$\sqrt{2}$ x $\sqrt{2}$)R45° structure.

orthogonal domains are equivalent). Figure 2a shows the position
of the CO molecules with respect to the copper substrate. In this
model, the molecules are on top, bridge and intermediate sites in
disagreement with the IRS and HREELS results. In order to explain
the difference between the LEED interpretation and the Work
Function measurements, Pritchard has proposed another model with
two types of sites : top and bridge (fig. 2b). But there is no
evidence by IRS or HREELS of the existence of bridge CO molecules.
So again, this model cannot be accepted. The only model that can
explain both the LEED observations and the IRS or HREELS spectra
is shown in figure 2c, where all the CO molecules are on top sites.
A similar model has been proposed for the adsorption of sulfur on
Fe(100)[20] and lead on gold(100)[21]. This model has the c2 mm symmetry, the highest compatible with the substrate. It can be
analysed as domains of c(2x2) (single rows) separated by antiphases
(double rows). A laser simulation of diffraction from one domain
of the c(7$\sqrt{2}$ x $\sqrt{2}$)R45° structure with all molecules adsorbed on
top sites is shown in fig. 3. It should be noticed that the spots
around the (1/2,1/2) position are intense as observed by LEED
(figure 1b). The main argument against this model is that some of
the CO molecules are at a distance of 2.56 Å (diameter of the
copper atoms) much smaller than the Van der Waals radius of the CO
molecules : 3.3 Å. This point will be discussed further.

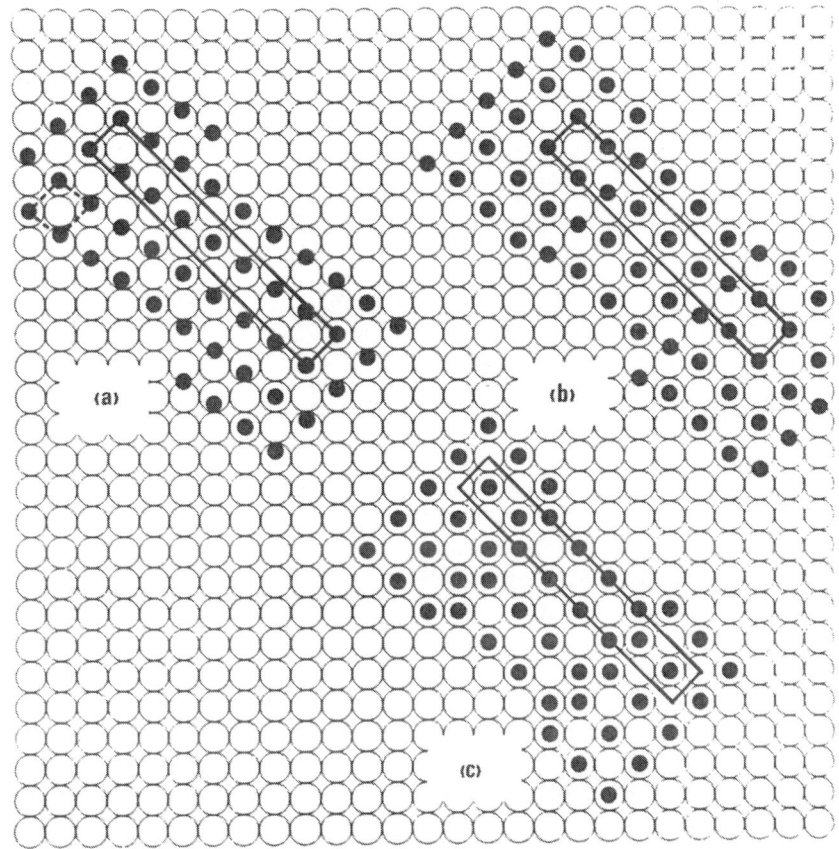

$\bigcirc$ Copper surface atoms $\bullet$ Adsorbed CO molecules

Fig. 2 - The Cu(100)-c(7$\sqrt{2}$ x $\sqrt{2}$)R45° CO structure
   a) "compact model", the CO molecules form a pseudo-
      hexagonal arrangement.
   b) model with top and bridge sites only.
   c) model with only top sites.

CO ON Cu(111)

   The adsorption of CO on Cu(111) has been studied by IRS[1,18,22,23]
only one peak is present at all coverages, which starts around
2076 cm[-1] but shifts slightly with increasing coverage. This
indicates the presence of only top site CO molecules at all
coverages as for CO on Cu(100).

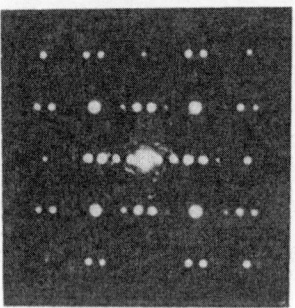

Fig. 3 - A laser simulation of one domain of the Cu(100)-
c(7√2 x √2)R45° CO structure with all molecules on top.

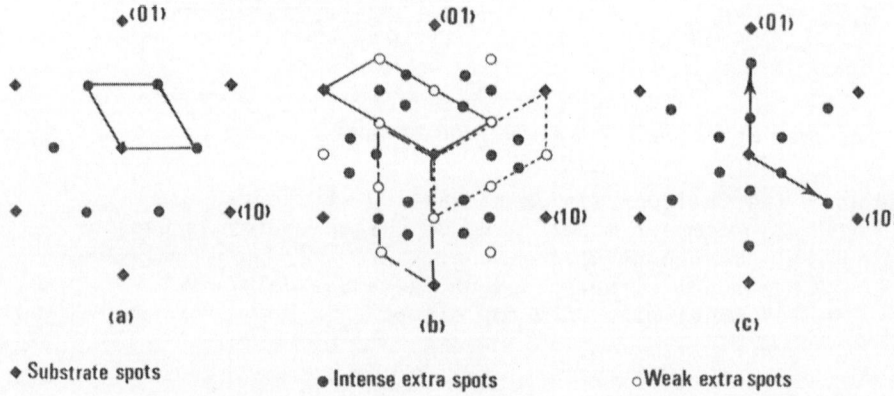

◆ Substrate spots   ● Intense extra spots   ○ Weak extra spots

Fig. 4 - LEED patterns observed when CO is adsorbed on Cu(111).
          a) the (√3 x √3)R30° structure,
          b) the c(4x2) structure,
          c) the "hexagonal" structure.

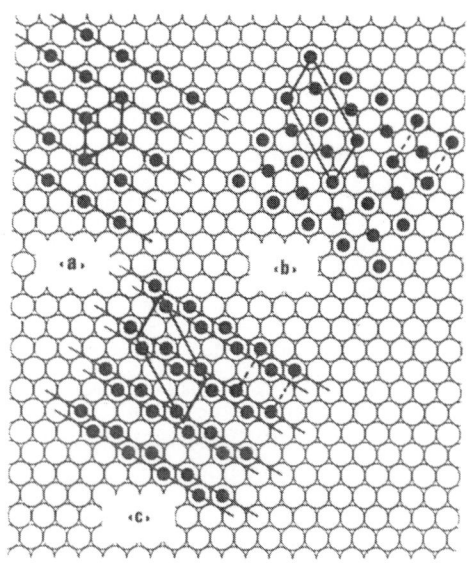

○ Copper surface atoms

● Adsorbed CO molecules

Fig. 5 – Models for CO adsorbed on Cu(111),
    a) the $(\sqrt{3} \times \sqrt{3})R30°$ structure, all the CO molecules are
       on top.
    b) the c(4x2) structure with on top and bridge molecules.
    c) the c(4x2) structure with only on top molecules.

As the coverage of CO increases, the following sequence of
LEED structures is observed : $(\sqrt{3} \times \sqrt{3})R30°$, c(4x2), "hexagonal
structure" as represented in Figure 4[1,18,24] . The interpretation
of the $(\sqrt{3} \times \sqrt{3})R30°$ structure is straightforward with one molecule
per unit cell and on top sites (Fig. 5a). The c(4x2) structure
which has been incorrectly identified as $(\sqrt{7} \times \sqrt{7})R49°$ by Kessler
and Thieme[24] can be interpreted using the "compact model" as a
pseudohexagonal layer of CO molecules sitting on top and bridge
sites (Fig. 5b). This model does not match the IRS data showing

only top sites. So the only possible model with CO molecules adsorbed on top sites is shown in Figure 5c. The CO molecules form
double rows separated by a $\sqrt{3}$ distance, the same as the interrow
distance of the $(\sqrt{3} \times \sqrt{3})R30°$ structure. Again as for the Cu(100)-
$(7\sqrt{2} \times \sqrt{2})R45°$ c(2x2) structure the CO molecules are too close together, and a relaxation is possible, see section 6 for details. A
laser simulation of one domain of the c(4x2) structure of figure 5c
is shown in Figure 6, and it can be noticed that only the extra
spots visible in the LEED diagram are present in the laser simulation. The high coverage "hexagonal structure" will be analysed
in another paper[25].

CO ON Ru(0001)

     The adsorption of CO on Ru(0001) has been studied by HREELS[26]
and IRS[27]; they show the presence of only one adsorption peak that
shifts continuously from 1984 cm^{-1} to 2061 cm^{-1} with increasing
coverage. This is interpreted as CO molecules sitting on top
sites at all coverages.

     LEED shows the following sequence of structures[28] : $(\sqrt{3} \times \sqrt{3})$
R30°, $(2\sqrt{3} \times 2\sqrt{3})R30°$ and an "hexagonal structure" as shown in
Figure 7. Again the $(\sqrt{3} \times \sqrt{3})R30°$ structure is easily interpreted
with one molecule per unit cell sitting on top sites (Figure 8a).
The two other structures have been interpreted by Williams and
Weinberg[28] as rotated compact hexagonal planes of CO (Figures 8b
and c). However in these models, the molecules are adsorbed on a
variety of sites, and not only on top sites as deduced from the
HREELS and IRS data. We are going to reanalyse these two structures
using the coïncidence lattice model starting with the "hexagonal
structure". Figure 9 shows how this structure can be decomposed
into three domains of $(\sqrt{3}x4)$ rectangular structures (one domain
is shown), where we use a shorthand notation for rectangular unit

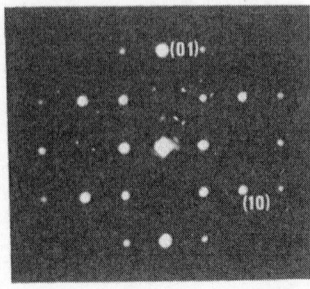

Fig. 6 - A laser simulation of the Cu(111)-c(4x2) CO structure
          with all molecules on top.

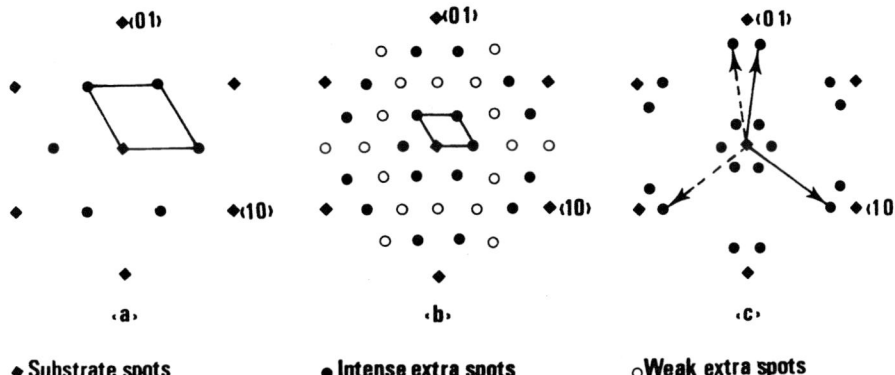

Fig. 7 - LEED patterns observed when CO is adsorbed on Ru(0001),
a) the $(\sqrt{3} \times \sqrt{3})R30°$ structure,
b) the $(2\sqrt{3} \times 2\sqrt{3})R30°$ structure,
c) the rotated "hexagonal" structure.

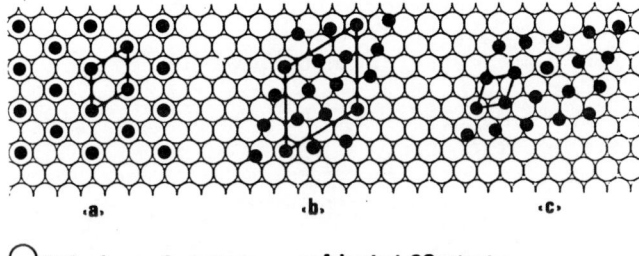

Fig. 8 - "Compact" models for CO on Ru(0001)
a) the $(\sqrt{3} \times \sqrt{3})R30°$ structure, with all the CO molecules
on top,
b) the $(2\sqrt{3} \times 2\sqrt{3})R30°$ structure with molecules on
various sites,
c) the "hexagonal" structure, again with molecules on
various sites.

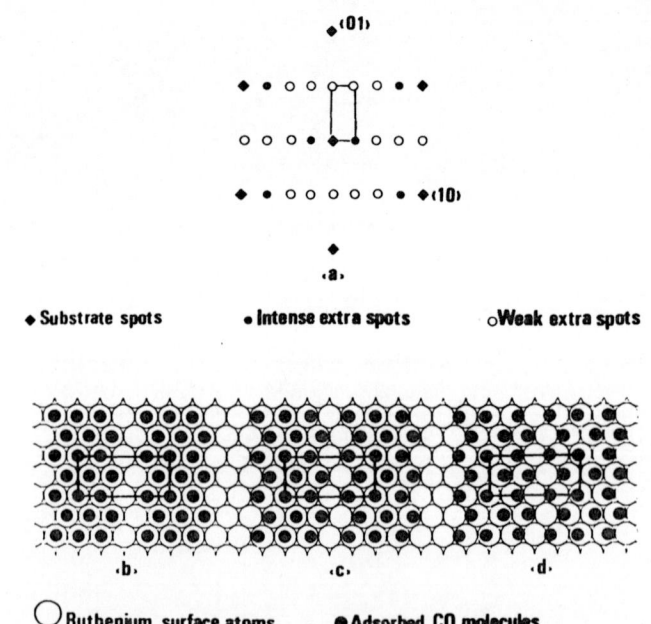

Fig. 9 - The Ru(0001)-hexagonal CO structure
   a) one domain of the ($\sqrt{3}$x4) Rectangular structure showing that the "hexagonal" structure can be decomposed into three of these structures,
   b) the ($\sqrt{3}$x4) Rect. structure with only on top CO molecules,
   c) same structure as b), but there is a parallel relaxation of the rows,
   d) same structure as b), but with a more hexagonal relaxation without rotation.

cells. We have performed laser simulations on all the possible models with CO molecules adsorbed on top sites with five and six molecules per cell, i.e. coverages 5/8 and 3/4[25]. The only satisfactory model is the one shown in Fig. 9b. Laser simulations of this model as well as models obtained by relaxation of this model in two different ways (Fig. 9c and d) are shown in Figure 10.

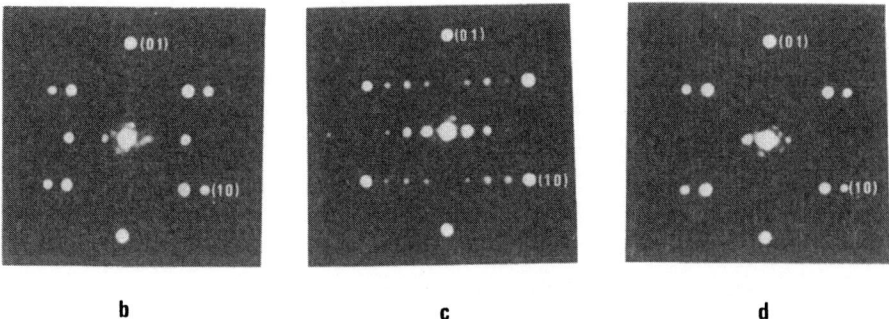

Fig. 10 - Laser simulations of one domain of the Ru(0001)-($\sqrt{3}$x4)
          Rect. CO structures of the models of figures 9b, c and d.

The laser simulation experiment of Figure 10 d is in good fit
with the LEED pattern (Figure 9a). In this model, all the CO molecules are on top sites, or slightly shifted off the site.

A similar interpretation of the ($2\sqrt{3}$ x $2\sqrt{3}$)R30° structure can
be made by noting that it can be decomposed into 3 domains of
($2\sqrt{3}$ x $\sqrt{3}$)R30° (or ($\sqrt{3}$x3)Rectangular) structures, Figure 11 a.
Since this structure is intermediate between the ($\sqrt{3}$ x $\sqrt{3}$)R30° and
($\sqrt{3}$x4)Rect structures, its coverage which is of the n/6 type must
satisfy the following inequality $1/3 < n/6 < 3/4$. The only possibilities are n = 3 or 4. The only model that is satisfactory from
the laser simulation point of view is shown in Figure 11 b. The
laser diffraction patterns of this structure as well as of two
relaxations of this structure are shown in Figure 12. The laser
simulation experiment of Figure 12 d resembles the LEED diagram
of Figure 11 a. Again in the corresponding model (Figure 11 d),
all the CO molecules are on top sites, or slightly shifted.

DISCUSSION - CONCLUSION

In this paper, we have shown that an alternative interpretation of the LEED patterns reconciles IRS or HREELS results with
the structures deduced from LEED. The assumptions made in this
model are :

 i - the CO molecules are adsorbed on specific sites (top, bridge,
     three-fold) at all coverages.
ii - the LEED structures are due to coïncidence lattices.

However, the main argument against this model is the occasional
close packing of the CO molecules, some of them having diameters
comparable to metal atoms. But as noticed by Pfnür et al.[27], the
CO molecules might tilt and make an angle with the surface, so
that whilst staying on top sites, they are no longer perpendicular

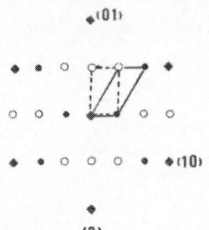

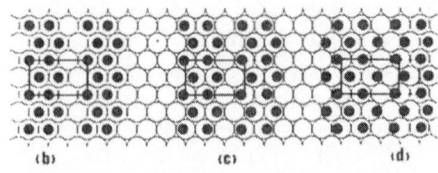

Fig. 11 – The Ru(0001)–$(2\sqrt{3}x\sqrt{3})$R30° CO structure
   a) one domain of the $(2\sqrt{3}x\sqrt{3})$R30° structure (unit cell
      shown in full line) showing that the $(2\sqrt{3}x2\sqrt{3})$R30°
      structure can be composed of three of those domains.
      In dashed line, the $(\sqrt{3}x3)$Rect. unit cell equivalent
      to the one in full line.
   b) the $(\sqrt{3}x3)$Rect. structure with only on top CO mole-
      cules.
   c) same as b), but there is a parallel relaxation of
      the rows.
   d) same as b), but a more hexagonal relaxation without
      rotation.

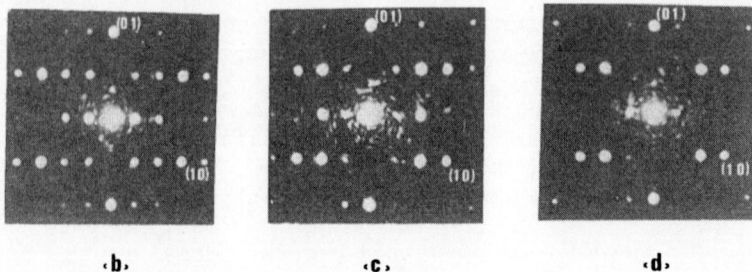

Fig. 12 – Laser simulations of one domain of the Ru(0001)–$(\sqrt{3}x4)$
          Rect. CO structures of the models of figures 11b, c and
          d.

to the surface, the term "linear" being preferred[27]. This is
favoured by the laser simulations showing that shifted models give
better fits with the LEED patterns. However, for the Cu(100)-
c($7\sqrt{2}$ x $\sqrt{2}$)R45° CO structure, the model without shift gives a
good fit for the LEED pattern so it might be of interest to consider
in more detail the problem of the CO "diameter". Most of the authors
have compared CO-CO distances to the Van der Waals radius of CO,
but this is not a correct comparison because when adsorption occurs,
there is chemisorption and electron transfer that drastically
changes the "diameter" of the molecule. It is preferable to compare
the CO-CO distance in chemisorbed CO on metal surfaces to the same
distance in metal carbonyls. It has been shown that in some cases,
like vanadium carbonyls[25], the linearly bound CO molecules are at
distances of about 2.7 Å, much smaller than the Van der Waals
radius. In Biscarbonyl annulene ($C_{16}H_{10}O_2$), the bridged CO molecules are even at shorter distances of about 1.5 Å[29].

Finally, we should note that the accuracy of our comparison
between LEED patterns and laser simulations is questionable
because the LEED patterns are taken from published photographs
which give imprecise relative intensities of the LEED diffraction
spots. It would be desirable to observe the diffraction patterns
at all energies in order to determine if the spots are systematically intense or weak or even missing.

REFERENCES

1. J. Pritchard, J. Vac. Sci. Techn. 9 (1972) 895.
2. K. Horn and J. Prichard, Surface Sci. 55 (1976) 701.
3. S. Andersson, Surface Sci. 89 (1979) 477.
4. B.A. Sexton, Surface Sci. 88 (1979) 299.
5. S. Andersson and J.B. Pendry, Phys. Rev. Lett. 43 (1979) 363.
6. S. Andersson, Solid State Comm. 21 (1977) 75.
7. J.D. Fedyk, P. Mahaffy and M.J. Dignam, Surface Sci. 89 (1979)
   424.
8. M. Passler, A. Ignatiev, F. Jona, D.W. Jepsen and P.M. Marcus,
   Phys. Rev. Lett. 43 (1979) 360.
9. K. Heinz, E. Lang and K. Müller, Surface Sci. 87 (1979) 595.
10. F.M. Hoffmann and A.M. Bradshaw, Proc. 7th Intern. Vac. Congr.
    & 3rd Intern. Conf. Solid Surfaces (Vienna 1977), p. 1167.
11. A.M. Bradshaw and F.M. Hoffmann, Surface Sci. 72 (1978) 513.
12. R.J. Behm, K. Christmann, G. Ertl, M.A. Van Hove, P.A. Thiel
    and W.H. Weinberg, Surface Sci. 88 (1979) L 59.
13. R.L. Park and H.H. Maden, Surface Sci. 11 (1968) 158.
14. L.H. Dubois and G.A. Somorjai, to be published in Surface Science.
15. R.J. Koestner, M.A. Van Hove and G.A. Somorjai, to be published.
16. M.A. Chesters and J. Pritchard, Surface Sci. 28 (1971) 460.
17. J.C. Tracy, J. Chem. Phys. 56 (1972) 2748.
18. J. Pritchard, Surface Sci. 79 (1979) 231.

19. C.R. Brundle and K. Wandelt, Proc. 7th Intern. Vac. Congr. & 3rd Intern. Conf. Solid Surfaces (Vienna 1977) p. 1171.
20. M. Huber and J. Oudar, Surface Sci. 47 (1975) 605.
21. J.P. Biberian and M. Huber, Surface Sci. 55 (1976) 259.
22. J. Pritchard, T. Catterick and R.K. Gupta, Surface Sci. 53 (1975) 1.
23. P. Hollins and J. Pritchard, Surface Sci. 89 (1979) 486.
24. J. Kessler and F. Thieme, Surface Sci. 67 (1977) 405.
25. J.P. Biberian and M.A. Van Hove, to be published.
26. G.E. Thomas and W.H. Weinberg, J. Chem. Phys. 70 (1979) 1437.
27. H. Pfnür, D. Menzel, F.M. Hoffmann, A. Ortega and A.M. Bradshaw, Surface Sci. 93 (1980) 431.
28. E.D. Williams and W.H. Weinberg, Surface Sci. 82 (1979) 93.
29. R. Destro and M. Simonetta, Acta Cryst. B 33 (1977) 3219.

EFFECTS OF MANY BODY FORCES ON THE ADSORBATE INDUCED PHONON

STRUCTURE

Gerold Doyen

Institut für Physikalische Chemie der Universität
Sophiënstrasse 11
D8000 München 2, West-Germany

ABSTRACT

A model is constructed, which permits to study many body forces by gradually deviating from the standard pairwise additive central forces. It is found that an arbitrarily small deviation opens the possibility of adsorbate induced soft phonon modes. Numerical examples are given with a rough parameterization to describe adsorbed hydrogen on a Pt(111) – surface.

INTRODUCTION

Localized modes and the modification of the phonon spectrum associated with adsorbed particles have been studied theoretically in several publications[1] . In all these investigations the adparticle has been coupled to the substrate atoms by springs, i.e., by pairwise additive central forces. Chemisorption forces are, however, known to possess strong many body character. The assumption of pairwise additive forces would predict wrong adsorption sites and would largely overestimate the variation of the adsorption energy parallel to the surface[2] . For this reason it appears desirable to check, if new qualitative effects in the adsorbate induced phonon structure can be expected in the case of many body forces. This paper is meant to be a first step in that direction. It was initiated by an attempt to understand the abnormal frequency factor for hydrogen desorbing from a Pt(111) – surface[3] . Evidence will be given that this unusual behaviour might in fact be correlated with the many body character of the adsorption force, which could induce a soft phonon mode.

THE MODEL

For the lattice vibrations the simplest possible model will be adopted. This is the Rosenstock-Newell model[4], which assumes mutual independence of the displacements in the x, y and z directions. In the following only one direction will be considered, which might be perpendicular or parallel to the surface. The harmonic approximation is used throughout. The Hamiltonian is then written as a sum of two terms :

$$H = H_o + V \tag{1}$$

H_o consists of two independent parts, the vibration of the ad-particle with the lattice fixed, and the motion of the lattice with a clean surface :

$$H_o = \frac{1}{2} \left(p_A^2 + \omega_A^2 q_A^2 + \sum_k (p_k^2 + \omega_k^2 q_k^2) \right) \tag{2}$$

ω_A and ω_k are the adsorbate-metal frequency and the phonon frequencies, respectively. p_A, q_A and p_k, q_k are the generalized momenta and coordinates. The interaction between the adsorbate motion and the phonons has the form :

$$V = -q_A \sum_k V_k q_k - \sum_{\substack{kl \\ k \neq l}} V_{kl} q_k q_l - \frac{1}{2} \sum_k V_{kk} q_k^2 \tag{3}$$

This arises from an expansion of the interaction potential in the displacements of the phonons and keeping only terms up to second order. A reasonable restriction on the force constants V_k and V_{kl} is that the total momentum should be conserved, if the interaction between the adsorbate vibration and the phonons is switched on. Caroli et al.[5] pointed out that spurious effects for the desorption dynamics could arise, if the condition of momentum conservation is not fulfilled. This restriction still leaves a large class of potentials open for consideration. Here a special choice is made :

$$V = - \gamma\sqrt{(M/m)} \omega_S^2 q_A q_x + \frac{n}{2}\gamma\omega_S^2 q_x^2 + \frac{1-n}{N} \gamma q_x \sum_n \omega_n^2 q_n \tag{4}$$

$$- \frac{1-n}{2N} \gamma \sum_n \omega_n^2 q_n^2$$

q_x is the generalized coordinate for the surface atom nearest to the adsorbed particle. The q_n characterize the normal modes for the indented substrate, i.e., without the surface atom "x".

γ is the ratio of the force constants :

$$\gamma = \frac{m\omega_A^2}{M\omega_S^2} \tag{5}$$

The parameter η varies between 0 and 1. It is easily verified that
the total momentum is conserved, i.e. :

$$[-i\vec{\nabla}, \frac{1}{2} \omega_A^2 q_A^2 + V] = 0 \quad \text{for } N\omega_S^2 = \sum_n \omega_n^2$$

where N is the number of substrate atoms and differentiation is
with respect to the real coordinates.
V is a very special potential with the following properties :
i) if η decreases from 1 to 0, the potential transforms smoothly
   from a pairwise additive central potential to one with extreme
   many body character;
ii) direct adsorbate induced phonon – phonon scattering is missing.
Point i) is highly desirable, because it is possible to see what
happens, if one gradually deviates from the spring model, which
has successfully described many properties of the adsorbate –
metal vibration. Point ii), which was introduced for mathematical
simplicity, has the disadvantage that the adsorbate – metal interaction is of infinite range. This becomes obvious if one specializes to an Einstein solid.'Therefore below the case of finite
range will be studied for a cluster model.

   The interaction potential eq. (4) couples the adparticle to
a surface atom and modifies the interaction of this surface atom
with the indented solid. The latter point constitutes the many
body character of the potential V. In order to proceed one needs
to know the coupling of the surface atom "x" to the other metal
atoms. The Green function for the surface atom neglecting the
potential V is :

$$G_x^o(\omega^2) = (\omega^2 - \omega_S^2 - \Sigma_x(\omega^2))^{-1} \tag{8}$$

where Σ_x is the embedding self-energy defined by :

$$\Sigma_x(\omega^2) = \sum_n \frac{|W_n|^2}{\omega^2 - \omega_n^2} \tag{9}$$

W_m are the force constants, which connect the surface atom movement
to the rest of the solid.The resolution of this movement in the
spectrum of squared frequencies is :

$$\rho_x(\omega^2) = -\frac{1}{\pi} \frac{\mathrm{Im}\Sigma_x}{(\omega^2 - \omega_s^2 - \mathrm{Re}\Sigma_x)^2 + (\mathrm{Im}\Sigma_x)^2} \tag{10}$$

The philosophy is to choose the W_n in such a way that $\rho_x(\omega^2)$ has the desired properties. For the numerical example discussed below it will be assumed that $\mathrm{Im}\Sigma_x(\omega^2)$ is proportionnal to a Debye phonon density of states (in squared frenquencies) and that it can be integrated to yield $\omega_S^4/(\pi K)$, where K is the coordination number of the surface atom. This means that the atom "x" is coupled by springs to K neighbouring metal atoms.

ADSORBATE INDUCED PHONON STRUCTURE

For the described model the exact surface atom Green function $G_x(\omega^2)$ can be evaluated analytically. From a numerical point of view it is, however, more convenient to directly solve the equations of motion by discretizing the phonon spectrum. This method is in close analogy to that one used in chemisorption calculations and therefore will not be described here[2] . It is rapidly converging and allows a solution to arbitrary accuracy.

Numerical calculations have been performed for parameters describing a hydrogen atom on a Pt(111)-surface : m = 1; M = 198; ω_A = 150 meV; ω_S = 14 meV (surface Debye energy); K = 9. This yields for the coupling parameter : γ = 0.568. With these parameters the model has been investigated for all physically possible values of η. It turns out that the adsorbate - metal vibration is hardly affected by the many body character of the forces. It changes by less than 0.4 meV from 150.8 meV for η= 1 (spring model) to 150.4 meV for η = 0.1. The change of the spectral resolution of the surface atom motion is shown in Fig. 1. For every value of η there are always two peaks in the spectrum, one at the bulk Debye energy of 20 meV resulting from the high phonon density of states and another one at lower energies. The latter peak results from the interaction of the surface atom with the indented solid. It should be noted that ω_s is the surface atom frequency only, if the other metal atoms are fixed at their equilibrium positions. This frequency has been chosen smaller than the bulk Debye frequency in order to account for the reduced coordination of the surface atom. If the indented lattice moves, it will push down the surface frequency and will broaden it into a resonance, because it is embedded into a continuum of phonon frequencies.

The spectrum for η = 1 (spring model) is nearly identical to that for a clean surface, because the adsorbate-metal frequency lies far above the phonon band and the interaction can therefore

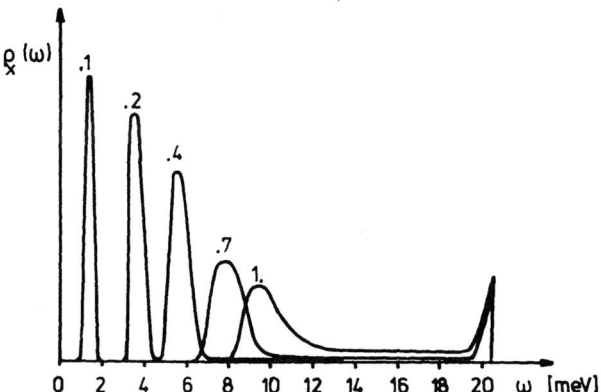

Figure 1 - Spectral resolution $\rho_x(\omega)$ of the surface atom vibration
in the presence of an adsorbed particle. The numbers
ascribed to the different curves are the values of the
parameter η. The values of the other parameters are
given in the text.

be considered as small. The more surprising is, what happens for
decreasing values of η : the resonance moves to lower energies and
sharpens. The decrease of width is due to the Debye phonon density
of states, which tends to zero proportional to ω^2. The energy shift
results from the decrease of the q_x^2 - term in V (cf. eq. (4)).
This means that the distorting force resulting from the adsorbate
motion is not restricted to one metal atom but acts on a cluster
of substrate atoms and makes these atoms move coherently. This
leads to a long wavelength vibration with a low frequency. For
values of η slightly below 0.1 this will induce an instability with
a subsequent re-construction of the metal surface.

The treated model includes the phonon dispersion and gives an
estimate of the width and the resonance character of the adsorbate
induced (soft) phonon mode. A considerable simplification is
achieved, if one neglects dispersion and specializes to an Einstein
solid, i.e., $\omega_n = \omega_S$ for all n. In this case the metal atoms of
the indented solid vibrate independently and the local phonon mode

$$q_o = \Lambda^{-1/2} \sum_n (W_n - \frac{1-\eta}{N} \gamma \omega_S^2) q_n \tag{11}$$

with

$$\Lambda = \sum_n (W_n - \frac{1-\eta}{N} \gamma \omega_S^2)^2 \tag{12}$$

is the only mode with linear coupling to the surface atom. All the orthogonalized modes decouple. There remain, however, adsorbate induced quadratic terms, which renormalize the vibrational frequency of each metal atom. This demonstrates the infinite range of the potential V. Abolishing this infinite range a cluster model is now discussed, where the surface atom couples to K independent harmonic oscillators (metal atoms) which are somehow anchored in space. The adparticle – cluster potential is then :

$$V_{cluster} = -\gamma\sqrt{(M/m)}\omega_S^2 q_A q_x + \frac{\eta}{2}\gamma\omega_S^2 q_x^2 + \frac{1-\eta}{K}\gamma\omega_S^2 q_x \sum_n q_n - \frac{1-\eta}{2K}\gamma\omega_S^2 \sum_n q_n^2 \tag{13}$$

The normal modes of the clean cluster are : K-1 modes with frequency ω_S, one mode where the surface atom moves against the cluster with frequency $\omega_S(1+K^{-1/2})$, and one mode where the whole cluster moves in the same direction as the surface atom with frequency $\omega_S(1-K^{-1/2})$. The latter two modes are the cluster version of the two peaks in $\rho_x(\omega)$ discussed above. Their modification by the adsorbate interaction is depicted in Fig. 2. It can easily be shown that an instability will occur, if

$$\gamma = \frac{K-1}{(K+3)(1-\eta)} \tag{14}$$

This means that the possibility of an adsorbate induced soft phonon mode exists , if $\eta \neq 1$. The pairwise additive potentials are very special in that they would never permit an instability. An arbitrarily small deviation would be sufficient to destroy this (probably unphysical) property. Inserting in eq. (14) the critical value of η from the example of the dispersion model above one finds that a cluster model with the same η would not be unstable. This is a consequence of the reduced range of the interaction. A longer range facilitates instabilities. For $\eta = 0$ and $K \gg 1$ the cluster model reduces to the example treated in[3] . If applied to hydrogen

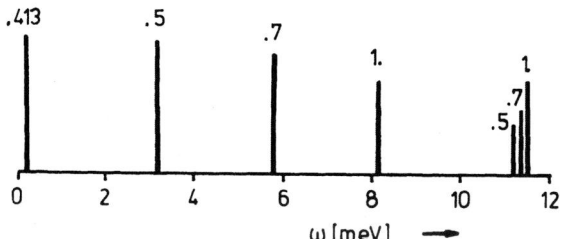

Figure 2 - Spectral resolution of the surface atom vibration for
the cluster model, which consists of discrete levels.
The height of the vertical bars is proportional to the
weight of the corresponding state in the spectral
resolution. The numbers on top give the value of η, i.e.
for each value of η the spectrum consists only of two
peaks. The bar around 11 meV labelled ".5" is also
valid for $\eta = 0.413$.

on Pt(111) this limiting model requires a value $\gamma = 1$ at the instability. This might appear too large in view of the measured hydrogen
frequency of 150 meV[6] . It has been demonstrated here that for
finite values of K the instability might occur at lower values of
γ . The experimental evidence that this adsorbate induced instability occurs in fact for hydrogen on Pt has been discussed in
detail in [3] and will not be repeated here. Another hint in this
direction is the abnormal width of the 150 meV - excitation in
electron energy loss spectroscopy. This is theoretically expected
for an unstable surface, although - for lack of space - this
cannot be outlined here.

REFERENCES

1. T.B. Grimley, Proc. Phys. Soc. (London) 79, 1203 (1962);
   M. Ashkin, Phys. Rev. 136, A821 (1964);
   F.O. Goodman, Surf. Sci. 5, 283 (1966);
   L. Dobrzynski, Surf. Sci. 20, 99 (1970);
   S.L. Cunningham, L. Dobrzynski and A.A. Maradudin, Phys. Rev.
   B7, 4643 (1973);
   J. Budimir, M.S. Jhon and J.S. Dahler, Surf. Sci. 80, 175 (1980).
2. G. Doyen and G. Ertl, Surf. Sci. 65, 641 (1977).
3. R.J. Madix, K. Christmann, G. Doyen and G. Ertl, preprint (1978).
4. H.B. Rosenstock and G.F. Newell, J. Chem. Phys. 21, 1607 (1953).
5. C. Caroli, B. Roulet and D. Saint-James, Phys. Rev. B18, 545
   (1978).
6. A.M. Baro, H. Ibach and H.D. Bruchmann, Surf. Sci. 88, 384 (1979).

VIBRATIONAL LIFETIMES FOR MOLECULES ADSORBED ON METAL SURFACES

M. Persson and B.N.J. Persson *

Institute of Theoretical Physics
Chalmers University of Technology
S-412 96 Göteborg, Sweden

ABSTRACT

Theoretical calculations of the decay of vibrations of molecules adsorbed on a metal surface due to excitation of electron-
hole pairs in the metal are described and discussed. Both a
dipole- and a resonance-coupling model are considered. Calculated
lifetimes are compared with the measured value for the system
CO/Cu(100).

INTRODUCTION

The prime experimental tools for studying molecular vibrations
at metal surfaces are infrared reflection absorption spectroscopy
(IRS) and high resolution electron energy loss spectroscopy (EELS).
The resolution of these methods is sometimes high enough to allow a
determination of the inherent width of a vibrational resonance. Thus
there is a need to understand the nature of these vibrational linewidths. This paper treats the deexcitation of the vibration through
excitation of electron-hole pairs in the metal, which causes homogeneous broadening.

In Sections 2 and 3 two models for the electron-vibration
coupling are presented for the cases with the molecule physisorbed
and chemisorbed respectively. For a more detailed discussion of the
models the reader is referred to Refs. 1 and 2. Each model is
applied to the system CO/Cu(100) and compared with experimental
results as described in Sections 2 and 3.

---

* Present address : IFF der KFA, Postfach 1919, D-5170 Jülich , BRD.

DIPOLE COUPLING - PHYSISORPTION

When the molecule is physisorbed, the electron states of the metal and the molecule are separated in space. For such a case we have proposed the following model.[1]:

(i) The molecular vibration is described by a vibrating dipole moment $\mu = |\langle B | \hat{\vec{\mu}} | A \rangle |$, where $|A\rangle$ is the ground state and $|B\rangle$ the first vibrationally excited state of the molecule, $\hat{\vec{\mu}}$ is the dipole-moment operator, and Ω is the vibration frequency.

(ii) The metal is described as an electron gas, confined to one half-space by an infinite barrier, and treated fully in the random phase approximation (IBM, Infinite Barrier Model, see Fig. 1).

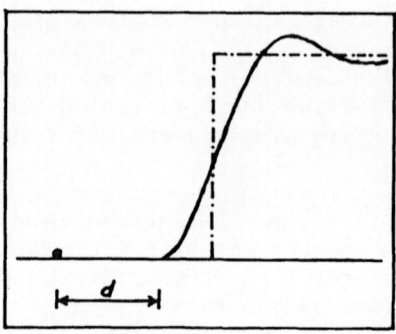

Figure 1 - The electron density in the surface region of a metal treated within the Infinite Barrier Model (IBM). The black dot indicates the position of the vibrating dipole located at distance d outside the infinite barrier.

According to the golden rule the lifetime τ of a vibration state is given by[1] :

$$\frac{1}{\tau} = 2\pi\Omega\rho^2(\varepsilon_F) \left| \int d^3X\, \Psi_{\vec{k}_1}(\vec{X})\, \hat{\phi}_{mol}(\vec{X})\, \Psi_{\vec{k}_o}(\vec{X}) \right|^2 \qquad (1)$$

average over
Fermi surface

where $\Psi_{\vec{k}}(x)$ is the electronic wavefunction in IBM $\rho(\varepsilon_F)$ the density of states at the Fermi level, $\hat{\phi}_{mol}(\vec{x})$ the dipole potential from the molecule, screened by the semi-infinite electron gas ($\hat{\phi}_{mol}$ is proportional to μ).

The actual calculation of the lifetime is not performed by direct evaluation of Eq. (1) but in an equivalent way. The energy loss rate is first evaluated by integrating Poynting's vector over the metal surface. The Poynting vector is given by the solution of an integral equation with the dielectric function for IBM as the kernel. This integral equation is solved numerically for different parameters. We have chosen the parameters for the C-O stretch vibration $\mu = 0.04$ ea$_o$ (τ is proportional to μ^{-2}) and $\hbar\Omega = 0.25$ eV.

In Fig. 2 the lifetime τ is shown as a function of the distance d between the dipole and the infinite barrier with the electron gas parameter $r_s = 2$. The curve labelled SCIB (semiclassical infinite barrier model) is obtained from a simplified description of the metal[3] . In Fig. 3 the lifetime τ is shown as a function of the electron gas parameter r_s for a fixed distance d = 0. The lifetime decreases with increasing density of the electron gas.

In Fig. 4 the lifetime τ is shown as a function of the vibration frequency Ω for d = 0 and $r_s = 2$. The linear dependence is expected (see Eq. (1)) and is simply a consequence of the restrictions in phase space available for the electrons.

With this theory one can make comparison with experiment for e.g. the system CO/Cu(100). The Cu metal is treated as an electron gas with $r_s = 2.7$. This immediately determines the position of the infinite barrier[1]. From LEED measurements one knows the adsorption site and the distance between the lattice plane and the center of mass of the CO molecule[4] (see Fig. 5), which gives d $\approx$ 0.73 Å. The dynamic dipole moment μ and the vibration frequency Ω are determined from IRS data, $\mu = 0.04$ ea$_o$ and $\hbar\Omega = 0.25$ eV. With these parameters one obtains the lifetime

$$\tau(\text{theory}) \approx 10^{-10} \text{ s}.$$

which should be compared with the measured value [9]

$$\tau(\text{exp}) \approx 1.3 \times 10^{-12} \text{ s}.$$

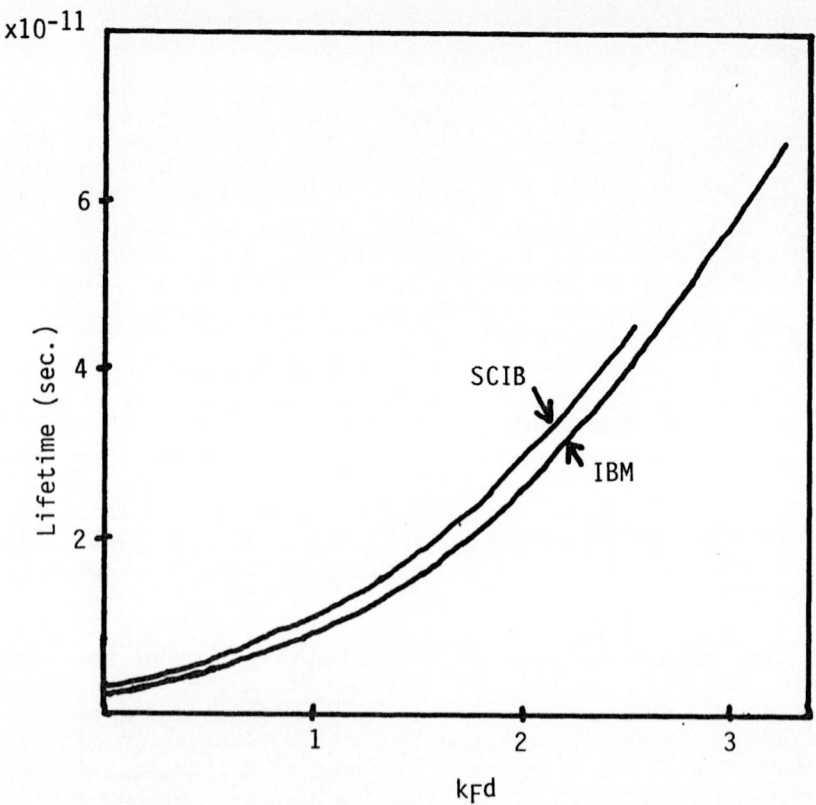

Figure 2 –   The calculated lifetime of a vibrationally excited molecule as a function of the distance d outside the metal
surface in two models, with parameters $\hbar\Omega$ = 0.25 eV,
μ = 0.04 ea$_o$ and r$_s$ = 2.

This is a large deviation, which we don't think is due to our
simplified description of the metal. Instead we argue that CO is
chemisorbed rather than physisorbed on Cu(100), which requires
another model for the electron vibration coupling.

RESONANCE COUPLING – CHEMISORPTION

For a chemisorbed molecule the electrons are shared between
the metal and the molecule. This gives rise to a coupling between
the electrons and the vibration, which cannot be described by a
long range dipole coupling. Upon chemisorption of CO on Cu the
molecular orbitals turn into orbital resonances. The electronic
structure that emerges from experiments is shown in Fig. 6. The

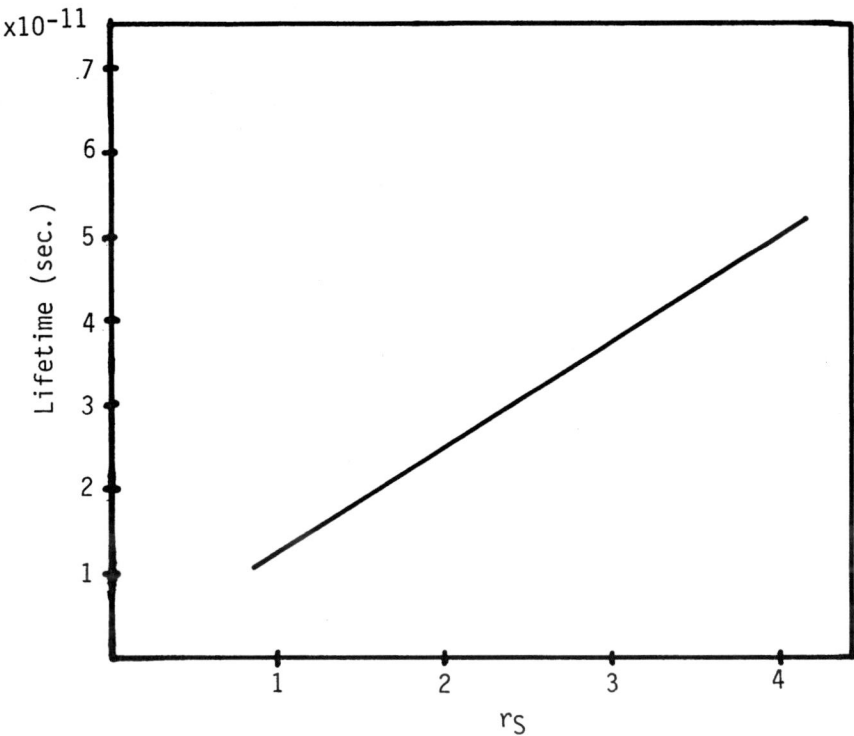

Fig. 3 – The calculated lifetime of a vibrationally excited
         molecule as a function of the electron density parameter
         r_s ($\hbar\Omega$ = 0.25 eV, μ = 0.04 ea_o and k_Fd = 0).

low-lying CO-induced levels $5\tilde{\sigma}$, $1\tilde{\pi}$, $4\tilde{\sigma}$ have been identified in
photoemission spectra[10] , while the partial occupancy of the $2\tilde{\pi}\star$
resonance can only be inferred from the shift in frequency of the
C–O stretch mode, as observed in EELS. The occupancy of the $2\tilde{\pi}\star$
resonance and its variation during the vibration should cause
large fluctuations between the metal and the molecule, accompanied
with dissipation of vibrational energy into electron hole pairs in
the metal.

     We shall use a description in terms of a model Hamiltonian
of Anderson-Newns' type[2]

$$\mathcal{H} = \varepsilon_a + \Delta_{aa}(b+b^+)a^+a + \sum_k E_k a_k^+ a_k + \hbar\Omega b^+ b + \sum_k (V_{ak} a^+ a_k + h.c.)$$

where $|k\rangle$ is a metal orbital with energy ε_k, $|a\rangle$ a $2\pi\star$ molecular
orbital with energy ε_a, and

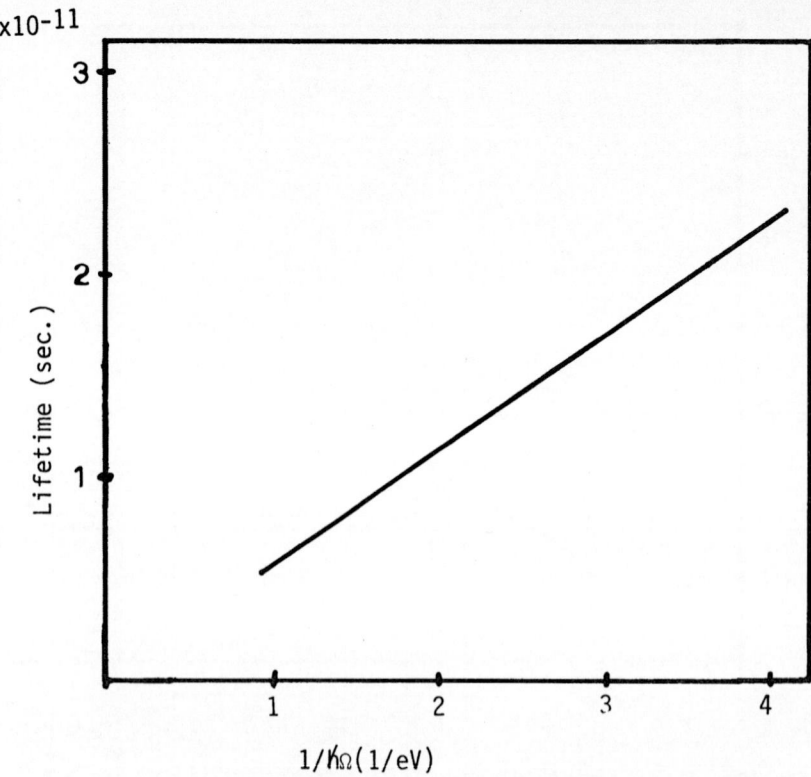

Figure 4 – The calculated lifetime of a vibrationally excited
          molecule as a function of the excitation energy $\hbar\Omega$ of
          the molecule ($\mu = 0.04$ ea$_o$, $k_F d = 0$ and $r_s = 2$).

$$Q = (\hbar / 2m^\star \; \Omega)^{1/2} \; (b^+ + b) \equiv Q_o (b^+ + b)$$

is the displacement coordinate for the harmonic C-O stretch
vibration mode, $m^\star$ being the reduced mass. The electron vibration
coupling is given by

$$\Delta_{aa} = (\hbar / 2m^\star \; \Omega)^{1/2} \; \frac{\partial \varepsilon_a}{\partial Q}(o)$$

The golden rule of perturbation theory with $\Delta_{aa} (b^+ + b)$ as the
perturbation gives after a few algebraic manipulations the lifetime for the first excited vibrational state as[2],[8]

$$\frac{1}{\tau} = 2\pi\Omega \; (\partial n_a)^2.$$

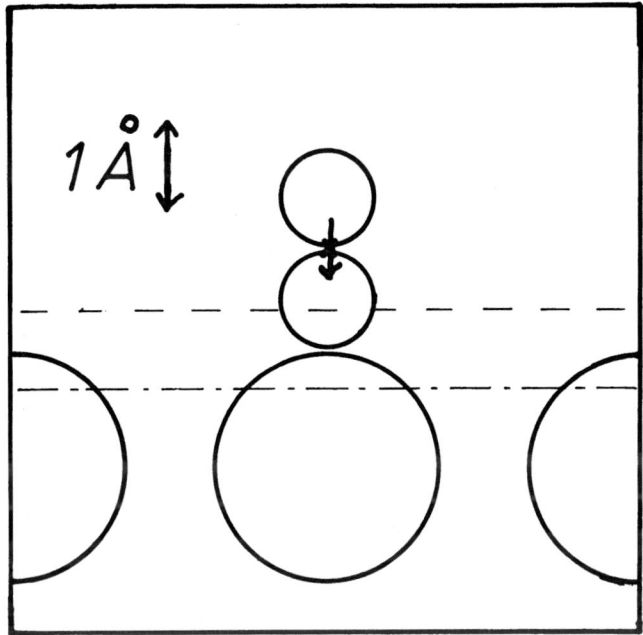

Figure 5 - The adsorbed CO molecule on a Cu(100) surface with hard
           sphere representation of the atoms. Dashed line gives the
           location of the infinite barrier; the dash-dotted line the
           jellium edge, and the arrow the vibrating dipole.

Here δn_a is the fluctuation in the occupancy of orbital $|a\rangle$ during
one vibration. It is related to the change in dynamic dipole moment
upon chemisorption. For an adsorbed monolayer one has earlier interpreted experimental results in terms of no change in the dynamic
dipole moment from the gas phase value. Recently it has been discovered, however, certain insufficiencies of the theories, on which
such interpretations of EELS and IRS were based.

     In the earlier theories one neglected the screening of the
dipole moment due to electronic polarizabilities of the adsorbed
CO molecules. If this is taken into account, one obtains a larger
dynamic dipole moment, e.g. $\mu = 0.09$ ea_o for CO adsorbed on Cu(100)[5].

     We make the reasonable assumption that the difference 0.05 ea_o
in dynamic dipole moment between a chemisorbed and a free CO
molecule is due to charge transfer between the metal and the $2\pi*$
orbitals. This charge will give a contribution to the dynamic
dipole moment with the true charge described by point charges, and

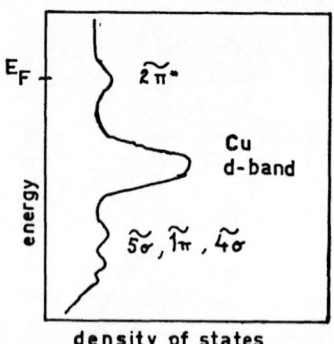

Figure 6 – Schematic characterization of the local one electron
density of states for CO adsorbed on Cu(100).

$$\Delta\mu \simeq -q_c d_c - q_o d_o$$

where d_o and d_c are the equilibrium distances between the centres
of the oxygen and carbon atoms respectively, and the "image plane"
(see Fig. 7). These distances can be determined by LEED measurements[4] and from the theory of the image plane[6] . From the electron
density of the $2\pi*$ orbital[7] , we deduce $q_c \approx q_o$. Since $\Delta\mu = 0.05ea_o$,
one gets $q_c \approx - 0.016$ e and so $q = -e\delta n \approx - 0.03$ e. Now there are
four $2\pi*$ orbitals (spin up, spin down, px and py) so the lifetime
is

$$\frac{1}{\tau} = 2\pi\Omega \times 4 \times \left(\frac{\delta n}{4}\right)^2 = \frac{\pi\Omega(\delta n)^2}{2}$$

Using the known vibration frequency, one gets

$$\tau(\text{theory}) \approx 2 \times 10^{-12} \text{ s.}$$

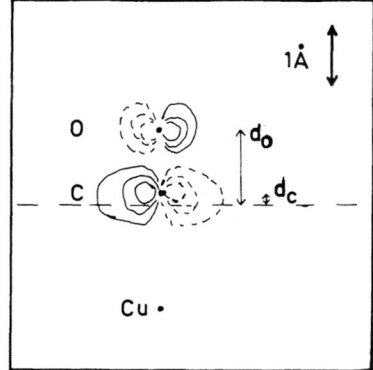

Figure 7 - The adsorbed CO molecule on a Cu(100) surface with the
          position of the atomic nuclei indicated. The wave
          function contour of the $2\pi\star$ molecular orbital[7] are
          drawn. The dashed line indicates the image plane.

This value lies close to the experimental lifetime [9]

$$\tau(\text{exp.}) \approx 1.3 \times 10^{-12} \text{ s.}$$

CONCLUDING REMARKS

     According to estimates in section 3, the coupling of the
vibration to electron hole pairs in the metal could cause an
appreciable part of the decay of the vibrations. Another mechanism
for homogeneous broadening could be deexcitation into phonons
in the metal through anharmonic coupling, but it seems unlikely
to be efficient, as the vibration frequency is 8 times greater than
the highest phonon frequency in the metal. We think that the results
of the present paper clearly indicate that electronic excitations

are important for the rate of energy transfer between the metal and adsorbed species. This should have significance also for other non-adiabatic effects appearing in e.g. reaction rates for catalytic processes on metal surfaces.

ACKNOWLEDGEMENTS

The authors are grateful to B.I. Lundqvist for reading the manuscript and giving useful comments. We would also like to thank D.M. Newns, W.L. Schaich and G. Wendin for useful discussions and S. Andersson and R. Ryberg for providing unpublished results.

REFERENCES

1. B.N.J. Persson, M. Persson, Surface Science (in print).
2. B.N.J. Persson, M. Persson, Solid State Communications (in print).
3. B.N.J. Persson, J. Phys. C 11, 4251 (1978).
4. S. Andersson, J.B. Pendry, J. Phys. C 13, 3547 (1980).
5. B.N.J. Persson, R. Ryberg, submitted to Phys. Rev. B.
   S. Andersson, B.N.J. Persson, to be published.
6. N.D. Lang, W. Kohn, Phys. Rev. B17, 3541 (1973).
7. J.B. Johnson, W.G. Klemperer, J. Am. Chem. Soc. 99, 7132 (1977).
8. D.M. Newns, private communication.
9. R. Ryberg, submitted to Phys. Rev. B.
10. H. Conrad, G. Ertl, J. Küppers and E.E. Latta. Solid St. Commun. 17, 613 (1975).

SURFACE MODES OF A SPHERICAL VOID AT THE PROXIMITY OF A SURFACE [*]

A. Ronveaux and A. Magnus †

Département de Physique
Facultés Universitaires N.D. de la Paix - B-5000 NAMUR

† Analyse Numérique
   Université Catholique de Louvain
   2, Chemin du Cyclotron - B-1348 LOUVAIN-LA-NEUVE

ABSTRACT

The surface modes of a spherical void situated inside a semi-infinite dielectric are computed in the long-wavelength approximation.

Numerical eigenvalues for the dielectric constant $\in_{\ell m}$ are given as function of R/D where D is the distance between the void of radius R and the surface, for a large number of values of the quantum number m.

The eigenfrequencies in the I.R. band are also computed for a typical ionic crystal using the known $\in(\omega)$ for LiF.

INTRODUCTION

The knowledge of the electromagnetic modes of a void situated inside a semi-infinite medium is of obvious interest in solid state physics. In the quasi-static approximation, the solution of the Laplace equation for the potential and the usual boundary conditions of the Electromagnetism quantizes the dielectric constant $\in$. It is well known for example that for an isolated void the dielectric constant is given by $-\frac{\ell}{\ell+1}$ and for an isolated semi-infinite region

[*] Work performed in the framework of the IRIS program (Institute for Research in Interface Sciences) of the Belgian Ministry for Science Policy.

the dielectric constant $\in$ equals -1.

A general formulation and the corresponding matrix equation were already given[1,2] in the context of surface plasmons.

The two equivalent formulations [1] and [2] use matrix elements which were not known in analytical form at that time. Explicitly the following integral appears naturally in this half plane problem :

$$I_{\ell,n}^{m} = \int_{0}^{1} t^{\ell+n} P_{\ell+1}^{m}(t) P_{n}^{m}(t) dt \tag{1}$$

It was only recently that one of us[3] gave the value of that integral. The perturbed eigenvalues were given numerically by successive truncation of the infinite matrix and are expressed in power series of the dimensionless parameter $\xi = \frac{R}{z_O}$ where R is the radius of the spherical void, and z_O is the distance from the center of the sphere to the plane.

In the search of the surface plasmon modes in [1] and [2] , the "state equation" for the dielectric constant $\in(\omega)$ is the following :

$$\in = 1 - \frac{\omega_p^2}{\omega^2} \tag{2}$$

where ω_p is the plasma electronic frequency. ($\omega_p^2 = \frac{4\pi n e^2}{m}$, n electronic density).

In order to take into account of a damping, the following complex $\in(\omega)$ law was already used[4] :

$$\in(\omega) = 1 - \frac{\omega_p^2}{\omega(\omega+i\gamma)} \tag{3}$$

In this note we compute the phonon modes of an ionic crystal using the well known dielectric law :

$$\in(\omega) = \in_{\infty} \frac{\omega^2 - \omega_L^2}{\omega^2 - \omega_T^2} \tag{4}$$

The constants $\in_{\infty}, \omega_L$ and ω_T correspond to Lithium Fluoride.

BASIC EQUATION AND NUMERICAL RESULTS

The coupled system of integral equation for the potential $V(\vec{r})$ reduces to[4] :

$$2 \pi \lambda V(P) = \int_{sphere} V(M) \frac{\partial \frac{1}{r_{PM}}}{\partial r} d S_M + \int_{plane} V(M) \frac{\partial \frac{1}{r_{PM}}}{\partial z} d S_M \quad (5)$$

where r_{PM} is the euclidian distance between the 2 points P and M both belonging to the two surfaces.

$\lambda = \frac{\in + 1}{\in - 1}$ where $\in$ is the unknown dielectric constant of the medium inside the plane and outside the void and the normal derivative points towards the medium.

The boundary conditions of electromagnetism are included of course in the integral equations; continuity of the potential relates univocally, for each mode, the potential on the plane to the potential on the sphere.

If we keep as unknowns the Fourier components $V_{\ell,m}$ of the potential on the sphere :

$$V(P) = \sum_{m = -\infty}^{+\infty} \sum_{\ell = |m|}^{\infty} V_{\ell,m} P_{\ell}^{|m|} (\cos \theta) e^{im\phi} \quad (6)$$

The integral equations reduce for each value of m to the following infinite system of linear equation for $V_{\ell,m}$:

$$\lambda (\lambda + \frac{1}{2\ell + 1}) V_{\ell,m}$$

$$= \frac{(\ell - m + 1) !}{(\ell + m) !} \sum_{n = m}^{\infty} (-1)^{n-\ell} V_{n,m} \frac{n}{n + \frac{1}{2}} \xi^{n+\ell+1} I_{\ell,n}^{m} \quad (7)$$

with $I_{\ell,n}^{m}$ defined as before is given by :

$$I_{\ell,n}^{m} = \frac{(\ell + n) !}{2^{\ell + n + 1} (\ell + 1 - m) ! (n - m) !} \quad (8)$$

These equations are solved by truncation and give the perturbed eigenmode $\in_{\ell,m}$ and the eigenvector $V_{\ell,m}$ as a function of , using the Q.R. algorithm, on the IBM 370/158 of the Centre de Calcul de l'Université Catholique de Louvain in Louvain-La-Neuve.

Tables I and II give for m = 0,1, in the full range : $0 \leqslant \xi \leqslant 95$ (close to contact), the first perturbed eigenmodes $\in_{\ell,m}$ of the void and the plane.

These figures are reported on graph 1 (m = 0) for the dielectric constant. More tables and graphs are available on request : Internal publications of the 'L.P.M.P.S.'- Department of Physics, Facultés Universitaires Notre-Dame de la Paix, 5000 NAMUR - Belgium.

Close to contact, the void's eigenmodes $(\in_{\ell,m})$ behave like $\not{C}_{\ell,m} \sqrt{1-\xi}^{\,3}$, and the plane's eigenvalues like $\in_{\ell,m} D_{\ell,m}/\sqrt{1-\xi}$.

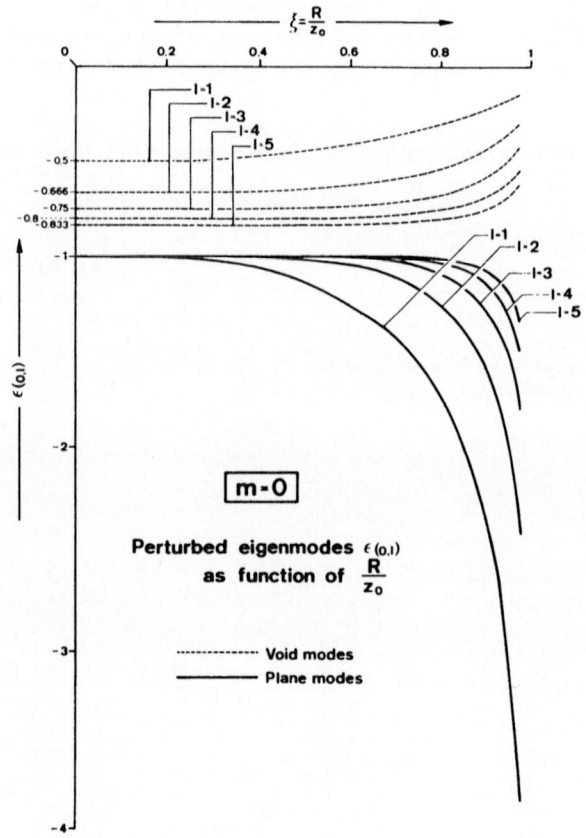

Fig. 1 – Perturbed eigenmodes $\in(0,1)$ as function of $\dfrac{R}{z_o}$.

Table I : Perturbed void's eigenmodes (m = 0,1).
Dielectric constant $\in_{m,\ell}$ (with opposite sign).

m = 0

ℓ \ 1	0	.1	.25	.50	.60	.70	.75	.80	.85	.90	.95
1	.500000	.499439	.491437	.439059	.400767	.350955	.320977	.286801	.247212	.199761	.138089
2	.666666	.666656	.665721	.645809	.623295	.585266	.557006	.519393	.468472	.397416	.290344
3	.750000	.750000	.749917	.742933	.730626	.706039	.686252	.658175	.616030	.547217	.421910
4	.800000	.800000	.799993	.797618	.791203	.776057	.762671	.742237	.709570	.652612	.531300
5	.833333	.833333	.833333	.832560	.829296	.819738	.810548	.796071	.771939	.726054	.618440

m = 1

ℓ \ 1	0	.1	.25	.50	.60	.70	.75	.80	.85	.90	.95
1	.500000	.499919	.495661	.466984	.443778	.410459	.388473	.361393	.326999	.280851	.211360
2	.666666	.666660	.666014	.650049	.630362	.596809	.572285	.540101	.496726	.435135	.336498
3	.750000	.750000	.749936	.744001	.733038	.710603	.692103	.665481	.625690	.562298	.448120
4	.800000	.800000	.799995	.797939	.792121	.777888	.765179	.745854	.714729	.659695	.544984
5	.833333	.833333	.833333	.832657	.829673	.820637	.811854	.797902	.774536	.730423	.625874

Table II : Perturbed plane's eigenmodes (m = 0,1).
Dielectric constant $\in_{m,\ell}$ (with opposite sign).

m = 0

ℓ \ 1	0	.1	.25	.50	.60	.70	.75	.80	.85	.90	.95
1	1.00000	1.00101	1.01651	1.14795	1.27233	1.47605	1.62859	1.84014	2.15561	2.69179	3.91412
2	1.00000	1.00000	1.00040	1.01737	1.05030	1.12656	1.19408	1.29593	1.45816	1.74911	2.44362
3	1.00000	1.00000	1.00000	1.00171	1.00806	1.03243	1.06140	1.11255	1.20405	1.38302	1.84120
4	1.00000	1.00000	1.00000	1.00015	1.00114	1.00734	1.01781	1.04143	1.09246	1.20626	1.52722
5	1.00000	1.00000	1.00000	1.00001	1.00015	1.00151	1.00465	1.01407	1.04067	1/11258	1.34312

m = 1

ℓ \ 1	0	.1	.25	.50	.60	.70	.75	.80	.85	.90	.95
1	1.00000	1.00050	1.00852	1.08627	1.16911	1.31364	1.42548	1.58327	1.82189	2.23220	3.17729
2	1.00000	1.00000	1.00026	1.01170	1.03482	1.09167	1.14419	1.22541	1.35745	1.59829	2.18191
3	1.00000	1.00000	1.00000	1.00127	1.00603	1.02472	1.04770	1.08948	1.16614	1.31923	1.71832
4	1.00000	1.00000	1.00000	1.00012	1.00090	1.00584	1.01428	1.03380	1.07716	1.17624	1.46082
5	1.00000	1.00000	1.00000	1.00001	1.00012	1.00124	1.00384	1.01172	1.03449	1.09777	1.30169

REFERENCES

1. Plasmons de surface autour d'une cavité située dans un métal
   semi-infini.
   A. Ronveaux, A. Moussiaux, Can. J. Phys. $\underline{56}$, 5, 490-496 (1978).
2. The Van der Waals Energy between a void and a metal surface.
   K. Ohtaka, A.A. Lucas, Solid State Commun. $\underline{24}$, 565 (1977).
3. Exact Results in Van der Waals Interaction between Spherical
   Bodies.
   A. Magnus, A. Ronveaux (in preparation).
4. Collective modes of a void-surface coupled system.
   K. Ohtaka, H. Miyazaki, A.A. Lucas, Phys. Rev. B, $\underline{21}$, 2, 1980.
5. Electrostatique
   E. Durand, Tome 1, Masson, 1964.

THEORETICAL STUDY OF THE VIBRATIONAL PROPERTIES OF A (117)F.C.C. SURFACE

G. Armand, P. Masri[*]

Service de Physique Atomique, Section d'Etudes des Int.
Gaz-Solides, CEN. Saclay, n°2,
91190   Gif-sur-Yvette, France

[*] Centre d'Etude Electronique des Solides, Laboratoire
associé au CNRS (LA 21) U.S.T.L. Place Eugène
Bataillon, 34060 Montpellier Cédex, France

ABSTRACT

   The phonon dispersion relations for the (117) face of a F.C.C.
crystal have been calculated by the method of generating coefficients for Green functions. The model takes account only of central forces between nearest neighbour atoms.

   The unit cell contains four atoms, two are located on the
terrace, the other ones being located respectively on the top and
the bottom of the step. The waves propagating in the direction
normal to the step are decoupled into transverse and sagital vibrations. Each vibrational mode gives a surface phonon branch and two
very well defined resonant branches. The different polarization
amplitudes for each atom of the surface cell are determined.

   These results can be qualitatively accounted for by three
times folding the dispersion realtion for the (100) face of F.C.C.
crystal which is the crystallographic configuration of the (117)
terrace.

THE CHEMISORPTION OF CO ON Cu : AN AB INITIO CLUSTER MODEL STUDY

Paul S. Bagus, M. Seel*

IBM Research Laboratory San Jose, CA 95193, USA

* Lehrstuhl für Theoretische Chemie, Universität Erlangen-Nürnberg, 852 Erlangen, FRG

ABSTRACT

We have used ab initio Hartree-Fock wave functions for a
Cu_5CO cluster to model the interaction of CO with a Cu surface at
the head-on absorption site. For a fixed CO distance, 2.173 bohr,
we have obtained an interaction curve for various Cu to C distances.

With this curve, we are able to derive the Cu-C stretching
frequency, the equilibrium bond distance, and the chemisorption
bond strength. The stretching frequency is calculated to be 31
meV which is reasonably close to the 43 meV EELS loss observed by
Andersson. The bond distance and stength are also in reasonable
agreement with observed values. This agreement is strong evidence
that the theoretical approach, including both choice of cluster
and the use of SCF molecular orbital theory, provides a good representation of the interaction of CO with a real Cu surface. The
nature of this interaction will be discussed based on an analysis
of the cluster wave functions.

LIBRATIONAL MODES OF MOLECULES ON METALS

H. Morawitz, T.R. Koehler

IBM Research Laboratory, San Jose, CA 95193, USA

ABSTRACT

The hindered rotational motion of a molecule bound to a metal surface by its image potential is studied. The potential is anharmonic in the tilt angle and leads to low frequency modes with frequencies less than 20 cm^{-1}.

Molecules with permanent dipole moments as well as induced dipole moments are considered and the modification of the image potential due to the finite plasma frequency of the metal included. The distance of the center of mass (charge) of the molecule from the surface, which strongly affects the strength of the image dipole well, is chosen for the case of pyridine and CN- on silver.

The possible connection of the calculated librational modes to recently observed Raman-active modes of pyridine on a silver surface showing strong surface enhancement will be discussed.

VIBRATIONS OF NEUTRALLY ADSORBED ATOMS ON TRANSITION METAL SURFACES

C.M. Sayers

Materials Physics Division, A.E.R.E. Harwell,

Oxfordshire - U.K.

ABSTRACT

An electron band model is used to calculate the vibration frequency of a neutrally adsorbed atom such as hydrogen on a transition metal surface. The bond length increases with increasing adatom co-ordination number, and in the atop position is greater than in the corresponding metal-adatom dimer. For strong bonding there is little difference between the vibration frequency in the atop position and of the dimer, but for weak binding large shifts occur. A simple relation between the adsorption energy and vibration frequency is derived. Bond relaxation is found to be of great importance in determining vibration frequencies.

EELS OF SURFACE PHONONS AND VIBRATIONS OF ADSORBED SPECIES

S. Lehwald and H. Ibach

Institut für Grenzflächenforschung und Vakuumphysik
Kernforschungsanlage Jülich, Postfach 1913
D-5170 Jülich, West Germany

ABSTRACT

By improving the resolution of electron energy loss spectrometers, surface vibrations on platinum and nickel single crystal
surfaces could be observed by EELS. The observed phonons are in
good agreement with recent theoretical results. On a stepped
Pt(111) surface a phonon localized near the step edge and with a
frequency slightly above the maximum bulk frequency has been
found. The phonon is only observed on the clean surface and is
caused by relaxation of the step atoms. The step phonon is excited by dipole scattering and the necessary dynamic dipole moment
is provided by the particular electronic properties of the step
atoms. On the flat Ni(111) and Ni(100) surface, nickel surface
phonons at certain points of the two-dimensional Brillouin zone
have been observed when suitable submonolayer amounts of gases
were adsorbed. The adsorbates provide the appropriate coupling
with the slow electrons. The same phonon frequencies are observed
after adsorbing H_2, O_2, C_2H_2, or NO. Sideband- or multiple excitations including the metal phonons are observed. Examples show
that the Ni-surface phonons can be excited by both dipole and
impact scattering. Implications of these results for the interpretation of vibrational spectra of adsorbed species are outlined.

INTRODUCTION

During recent years electron energy loss spectroscopy (EELS)
has proven to be a powerful technique for studying vibrations of
atoms and molecules adsorbed on single crystal surfaces. From the
vibrational loss spectra identification of adsorbed species is pos-

sible, adsorption sites and –geometries can be deduced. With an
energy resolution of about 10 meV ($\sim$ 80 cm^{-1}) the spectrometers
usually have allowed the measurement of vibrations above 300 cm^{-1}
i.e. the adsorbate metal vibrations and the inner vibrations of
molecular adsorbates could be observed.

By improving the resolution down to 3.5 meV (< 30 cm^{-1}) it
became possible to measure vibrations also in the range of the
metal bulk phonon frequencies and therefore we were able to detect
metal surface phonons.

In the following we review results for surface phonons on a
stepped Pt(111) and on the flat Ni(111) surface. New results for
the flat Ni(100) surface are reported. Comparison is made to recent
theoretical results of Allan and Lopez. Examples are reported which
demonstrate coupling of the phonons to adsorbate vibrations. This
may make the interpretation of vibrational loss spectra of adsorbed species more complicated.

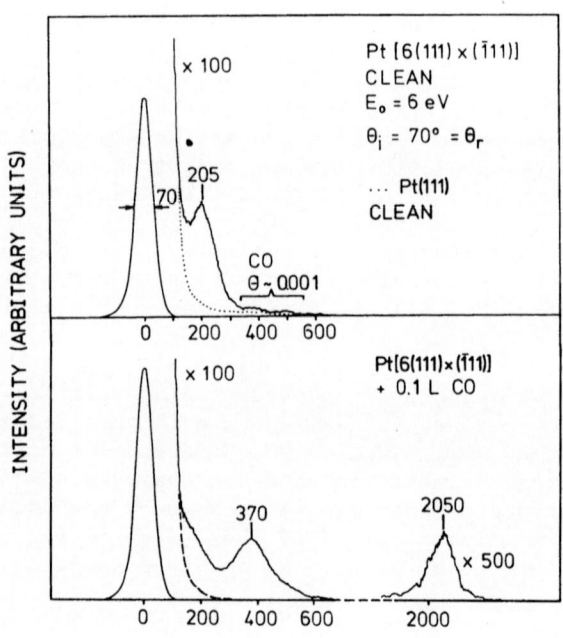

Fig. 1. Upper panel : EELS spectrum of a clean stepped Pt[6(111)x
(Ī11)] surface. The dotted line is the spectrum of a clean
flat Pt(111) surface recorded in situ with the elastic intensity scaled to the stepped surface. Lower panel : Spectrum of the stepped surface after dosing with 0.1 L CO.

STEP LOCALIZED PHONON ON Pt(111)

A direct coupling of a metal surface phonon to the impinging electrons has been observed on a stepped Pt(111) surface[1]. Fig. 1, upper panel, shows electron energy loss spectra of a clean flat Pt(111) surface and of a clean stepped Pt [6(111) x ($\bar{1}$11)] surface. Both crystals were mounted together on a manipulator for an in situ comparison of the flat and stepped surface. The electron energy E_O is 6 eV, the angle of incidence and of detection is 70° off the normal. On the stepped surface a vibration at 205 cm^{-1} is observed, which is slightly above the maximum bulk frequency of 195 cm^{-1} [2]. It is only observed in specular reflection, i.e. the electrons are scattered by the perpendicular dipole moment associated with the vibration, and the wave vector parallel to the surface is equal to zero [18]. The 205 cm^{-1} loss is only observed on the clean surface. It is attributed to a surface phonon localized near the step edge[1], because it is not observed on the flat surface and because small amounts of CO, which has been found to adsorb on top of the step atom[1,3], shift the frequency and reduce the intensity of the loss substantially as seen in Fig. 1, lower panel. The losses at 370 and 2050 cm^{-1} are the carbon-metal and CO-stretching vibration, respectively.

The existence of the phonon and its frequency are in agreement with recent calculations by Allan[4]. He calculated relaxation, force constants, and phonon spectra near the [6(111) x ($\bar{1}$11)] fcc surface taking into account the electronic rearrangement near the step within a tight-binding scheme for the d band electrons. The phonon (with a wave vector parallel to the surface equal to zero) is localized at the step edge atoms and caused by their relaxation. The main component of the vibration is normal to the surface. An inward relaxation of the step atoms provides a higher force constant and hence the frequency above the maximum bulk frequency. The dynamic moment is provided by the particular electronic properties of the step atoms[1,4]. When the relaxation is lifted, the phonon shifts into the bulk frequency range[4]. This is also observed in Fig. 1, lower panel upon adsorption of CO[1].

As the step phonon disappears with adsorption, it gives no implications for the interpretation of vibrational spectra of adsorbates. This is different for the flat surface.

SURFACE PHONONS ON FLAT Ni SURFACES

Electron energy loss spectra of clean flat Pt or Ni surfaces show no excitations of metal surface phonons. But they can be excited by the slow electrons when gases are adsorbed. Nickel surface phonons at certain points of the two-dimensional Brillouin zone have been observed on the flat Ni(111) surface when suitable

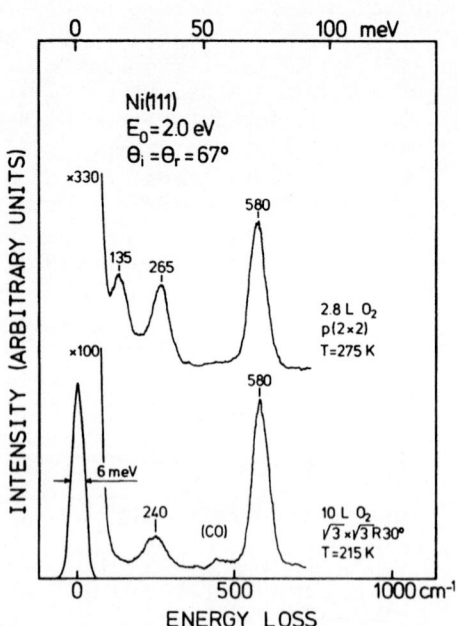

Fig. 2. EELS spectra of Ni(111) with a p(2x2) and $\sqrt{3}$ x $\sqrt{3}$ R 30°
overlayer of oxygen, rspectively.

submonolayer amounts of gases were adsorbed[5]. The adsorbates
provide the appropriate coupling with the slow electrons. Fig. 2
shows electron energy loss spectra of a p(2x2) and a $\sqrt{3}$x $\sqrt{3}$ R 30°
overlayer of oxygen adsorbed on Ni(111). The loss at 580 cm^{-1} is
the perpendicular oxygen-nickel stretching frequency of O atoms
in a threefold hollow site, the losses at 135, 240 and 265 cm^{-1}
are attributed to nickel surface phonons[5]. All losses are observed
only in specular reflection, i.e. they are excited by dipole
scattering. The phonons must be excited via the oxygen dipole.
Because of dipole scattering the oxygen vibrations are observed
with $q_{\parallel} \cong 0$, i.e. in the center of the superstructure Brillouin
zone. The LEED reflexes and hence the centers of the p(2x2) and
$\sqrt{3}$x$\sqrt{3}$ R 30° structure Brillouin zones fall in the $\bar{M}$ and $\bar{K}$ points
of the nickel surface Brillouin zone, respectively, and therefore
nickel surface phonons in these points of the nickel Brillouin zone
should be excited.

Allen et al.[6] calculated phonon spectra of a fcc 21 layer-
slab with (111) surfaces and found five bands of surface modes
below and in the gaps of the bulk bands. Comparison to these
calculations gives the following conclusions[5] : the loss at
135 cm^{-1} is due to excitation of a surface mode in $\bar{M}$, which is
primarily polarized perpendicular to the surface. Using 295 cm^{-1}
for the maximum bulk frequency[7] Allen et al. find this phonon at
110 cm^{-1} in $\bar{M}$. The losses at 265 and 240 cm^{-1} are caused by
excitation of a surface mode in $\bar{M}$ and $\bar{K}$, respectively, which is
primarily polarized parallel to the surface. The predicted
frequencies of Allen et al. are 260 and 245 cm^{-1}, respectively,
in good agreement with the observed values. From the polarization
of the surface phonons and from the observation, that the mode
polarized perpendicular is not excited by a $\sqrt{3} \times \sqrt{3}$ R 30° dipole
overlayer in $\bar{K}$, although it exists there, one can suggest models
for the motion of the nickel atoms[5] , which make the excitation
of the 135 cm^{-1} phonon forbidden by symmetry in $\bar{K}$. The observed
frequencies also agree with those calculated by Velasco et al.[8]
for surface phonons of the free Ni(111) surface in $\bar{M}$ and $\bar{K}$ when
295 cm^{-1} is used for the maximum bulk frequency.

The phonon frequencies should be shifted in the adsorbed case
compared to the free surface. We observe the phonons with the
same frequency for different adsorbates. Fig. 3 shows electron
energy loss spectra of a (2x2) overlayer of H_2 and of a p(2x2) of
C_2H_2 on Ni(111). In the case of hydrogen the excitation of the
260 cm^{-1} phonon in $\bar{M}$ is symmetry forbidden because the (2x2) hydrogen lattice is not primitive[5] . The losses above 300 cm^{-1} in the
case of acetylene are due to acetylene vibrations.

If one adsorbs 2.8 L of oxygen at low temperature it does
not order to a p(2x2) structure but adsorbs in a non ordered overlayer exhibiting a diffuse (1x1) LEED pattern. In that case only
one surface phonon loss at 240 cm^{-1} is observed. The same loss is
observed for non ordered adsorption of C_2H_2 below coverages corresponding to the p(2x2) structure. Probably the loss at 240 cm^{-1}
in these cases belongs to a nickel "breathing mode" coupled to
isolated adsorbates[5] . Black has calculated the breathing mode
on a clean Ni(111) surface to be 260 cm^{-1} [10]. The same phonon
frequency, however, is also observed for a c(4x2) overlayer of NO
on Ni(111) as shown in Fig. 4. The losses at 360 and 1580 cm^{-1}
are attributed to the nitrogen-nickel and N-O stretching vibration,
respectively, of molecularly adsorbed NO in a twofold bridging
site[9]. Therefore the motions of the nickel atoms may be different
from a breathing mode in this case and only the frequencies coincide.

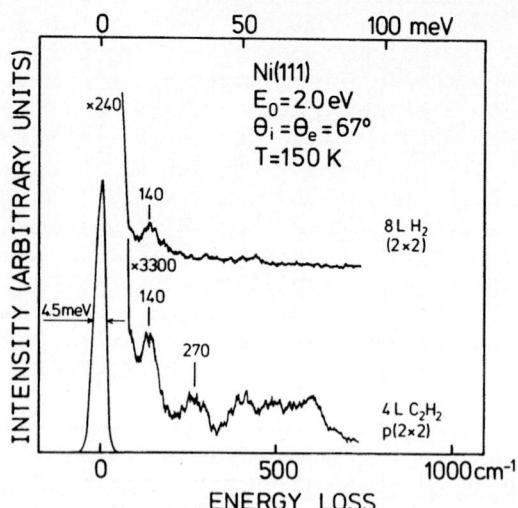

Fig. 3. EELS spectra of Ni(111) with a (2x2) overlayer of hydrogen and a p(2x2) overlayer of acetylene.

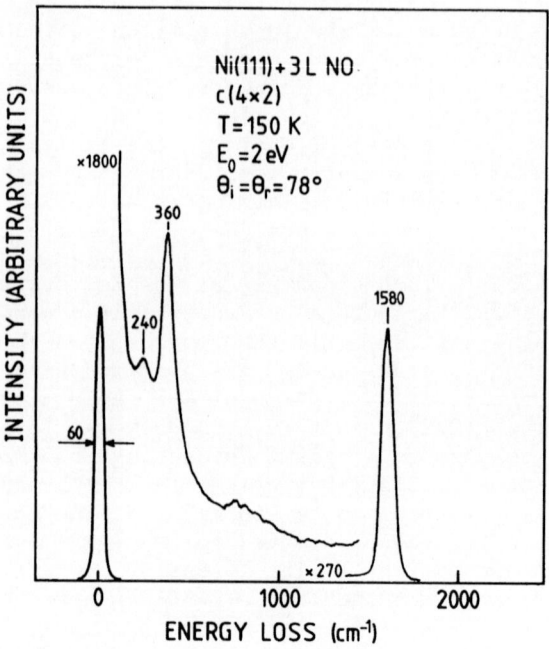

Fig. 4. EELS spectrum of Ni(111) with a c(4x2) overlayer of NO.

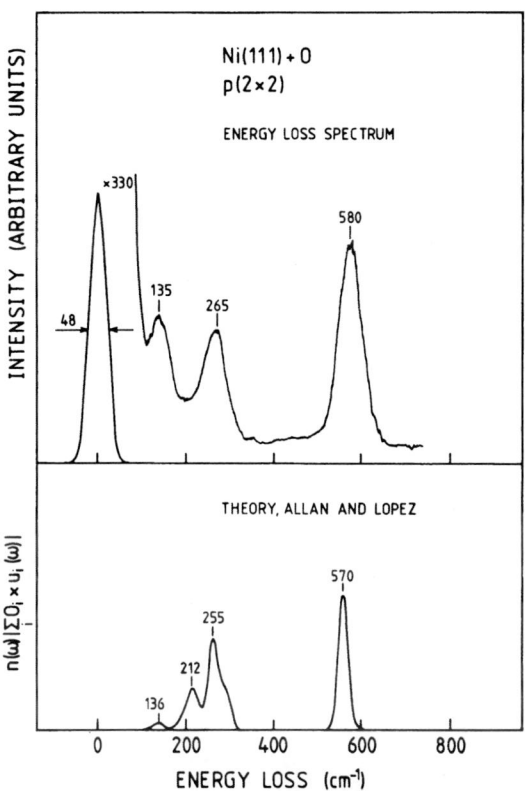

Fig. 5. Comparison of the EELS spectrum of Ni(111) with a p(2x2)
        overlayer of oxygen with phonon frequency calculations
        of Allan and Lopez[11]. Plotted in the lower panel is the
        resulting dipole moment perpendicular to the surface.

Allan and Lopez[11] have calculated the surface phonons of free
and oxygen covered nickel surfaces using the slab method, allowing
for relaxation, and including the nickel d bands and oxygen p
orbitals in a tight-binding approximation. For the (111) surface
they obtained the local density of states in $\bar{M}$ and $\bar{K}$ for vibrations of the Ni and O atoms perpendicular and parallel to the
surface.

Fig. 5 and 6 show a comparison of their results with our
spectra for a p(2x2) and $\sqrt{3}x\sqrt{3}$ R 30° oxygen covered Ni(111) surface, respectively. The resulting vibrating dipole D(ω) perpendicular to the surface[11] :

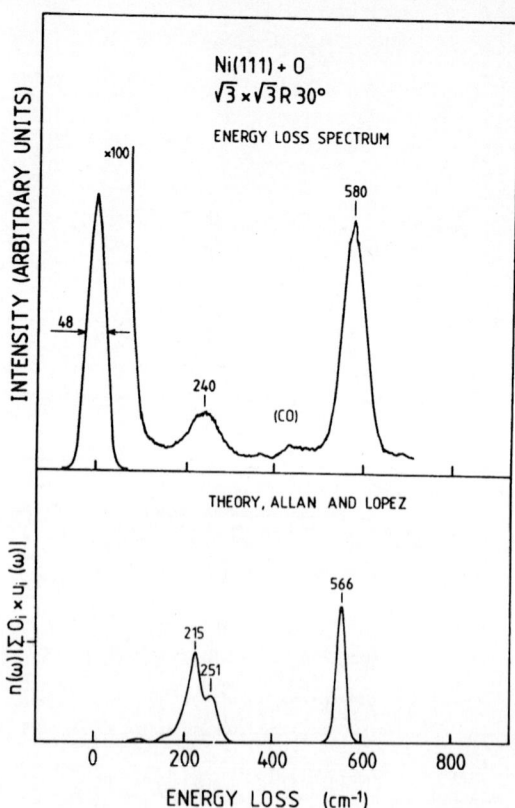

Fig. 6. Comparison of the EELS spectrum of Ni(111) with a $\sqrt{3} \times \sqrt{3}$
R 30° overlayer of oxygen with phonon frequency calculations
of [11].

$$D(\omega) = n(\omega) \left| \sum_i Q_i \times U_i (\omega) \right|$$

is plotted in the lower panel with $n(\omega)$ being the total density
of phonons, $Q_i = 1$ for the oxygen atom, $Q_i = -1/N$ for one of its
N neighbours, and $U_i(\omega)$ the amplitude of vibration on atom i for
the frequency ω. Only the position of the peaks not their intensity is of interest because $D(\omega)$ and not a scattering probability
was calculated. The curves of Allan and Lopez have been brought
to the energy loss scale by using $\omega = 295$ cm^{-1} for the maximum
bulk frequency [7]. The agreement between the observed and calculated phonon frequencies is quite good. Allan and Lopez also find
that in the 135 cm^{-1} phonon the nickel atoms vibrate essentially
perpendicular to the surface and essentially parallel to the surface in the 240 and 260 cm^{-1} phonon. The frequencies of the Ni-
phonons are slightly shifted to lower values with the oxygen on
the surface compared to the free surface[11].

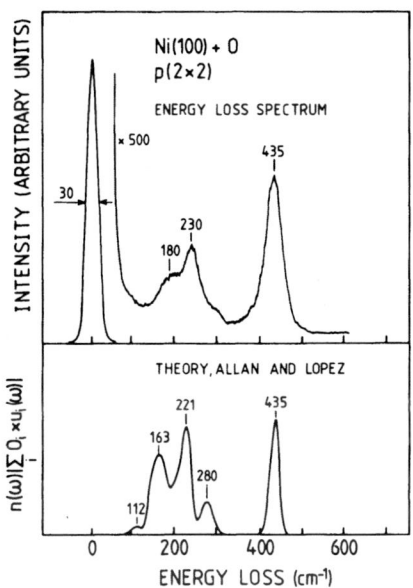

Fig. 7. EELS spectrum of Ni(100) with a p(2x2) overlayer of oxygen
(upper panel) and comparison to phonon frequency calculations of [11] (lower panel). Plotted is the resulting dipole
moment perpendicular to the surface.

With an improved energy resolution below 4 meV we also stu
died surface phonons on the Ni(110) surface covered with oxygen
in a p(2x2) and c(2x2) overlayer. The phonons are therefore excited in the $\bar{M}$ and $\bar{R}$ points of the nickel Brillouin zone in the
case of p(2x2) and in $\bar{R}$ in the case of c(2x2). Simple comparison
of our spectra with the (100) surface phonon calculations of
Allen et al.[6] is not possible. They report a very complex spectrum
for this surface with at least 19 surface bands of changing polarization throughout the Brillouin zone.

The electron energy loss spectra are presented and comparison
is made to results of Allan and Lopez[11] in Fig. 7 and 8. The losses at 435 and 330 cm^{-1}, respectively, are due to the oxygen-
nickel stretching vibrations of O atoms in fourfold hollow sites[12]
the other losses observed are attributed to Ni surface phonons.

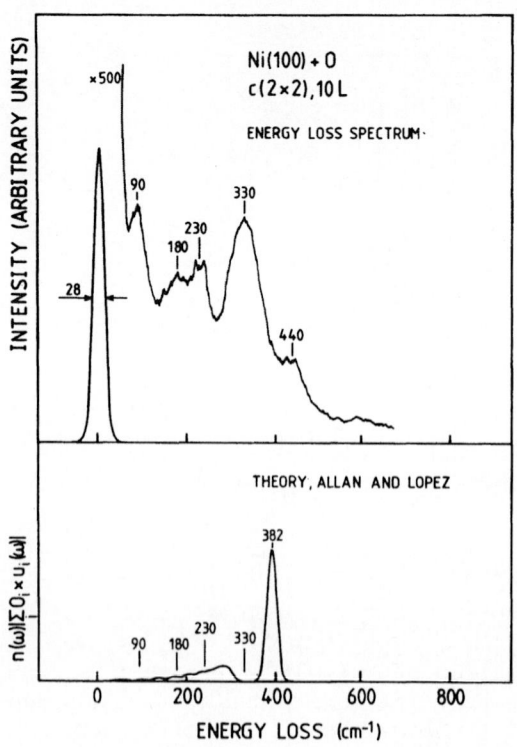

Fig. 8. EELS spectrum of Ni(100) with a c(2x2) overlayer of oxygen
        (upper panel) and comparison to phonon frequency calcula-
        tions of [11] (lower panel).

All losses are observed in specular reflection only. The loss at
230 cm^{-1} in p(2x2)-O has also been observed as a shoulder in the
slope of the elastic beam by Andersson[12] and he also suggested
that it may be due to a nickel vibration. The agreement of the
frequencies observed and those calculated by Allan and Lopez is
reasonable in the case of p(2x2) (Fig. 7), but poor in the case
of c(2x2) (Fig. 8). Probably this may be due to a more complex
interaction between the O and Ni atoms because only in the case
of c(2x2)-O each surface nickel atom has two oxygen neighbours.
A more complex interaction is also reflected in the spectra of
Fig. 9. Here EELS spectra are shown for oxygen on Ni(100) after
increasing exposures. All exposures are made at 150 K. After each
exposure the sample has been warmed to 400 K for five minutes for
developing the LEED pattern. The lowest curve in Fig. 9 is the
spectrum of a p(2x2) overlayer as also shown in Fig. 7. After

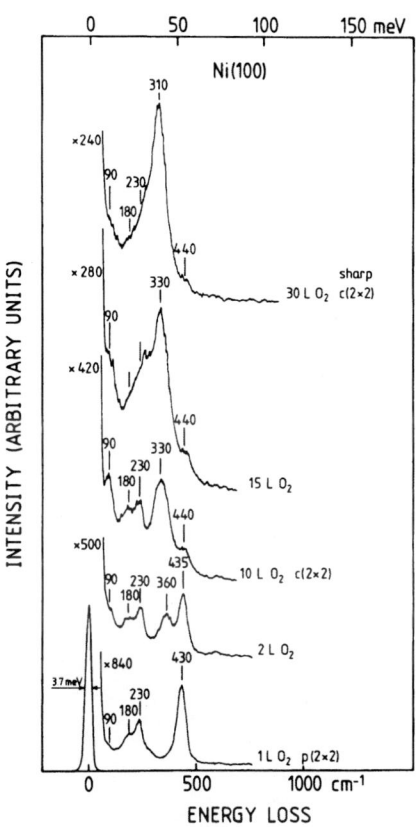

Fig. 9. EELS spectra of Ni(100) covered with oxygen after different
        exposures at 150 K and intermediate warming to evaluate
        the indicated LEED patterns.

higher exposure we get a mixture of the p(2x2) and c(2x2) spectra,
until after an exposure of 10 L the (1/2,0) extra LEED spots are
gone and we observe a c(2x2) structure with the loss spectrum as
also shown in Fig. 8. Between 10 and 30 L the c(2x2) LEED pattern
is observed to be stable; it is sharp and the extra spots only
get more intense compared to the (1x1) spots with increasing
exposure. The EELS spectra, however, change remarkably and the
structure in the spectra is gradually washed out with higher
exposure.

IMPLICATION FOR THE INTERPRETATION OF LOSS SPECTRA.

The result that metal surface phonons can be observed by EELS has to be taken into account for the interpretation of loss spectra of adsorbed species. Firstly, because the surface phonons fall into the same frequency range where bending modes of adsorbates are also expected. Secondly, multiple or sideband excitations can occur including the metal surface phonons. In Fig. 10 the EELS spectrum of Ni(111) with a c(4x2) overlayer of CO is shown. The losses at 400 and 1910 cm^{-1} have been found to be the carbon-nickel and the C-O stretching vibration of CO adsorbed in twofold bridging sites, respectively[13] . All losses are observed in specular reflection only. In Fig. 10 we observe an increase in line-width to 90 cm^{-1} of the carbon-metal stretching vibration and additional losses on both sides of it. This can be explained by multiple excitation including the metal surface phonons or by sideband excitation due to a strong coupling of the carbon-nickel vibration to the nickel surface phonons similar to a recent infrared study of the hydrogen-tungsten vibration on W(100)[14] . Alternatively the losses below and above 400 cm^{-1} could be interpreted as bending modes of the CO adsorbed in a site exhibiting C_S or C_1 symmetry. Because the observed losses agree with the surface phonon frequencies, as indicated in Fig. 10, and because in the c(4x2) overlayer of NO (Fig. 4) the 240 cm^{-1} phonon is excited, we conclude that in Fig. 10 the 240 or 260 cm^{-1} nickel vibration is also excited and in addition multiple and/or sideband excitations are

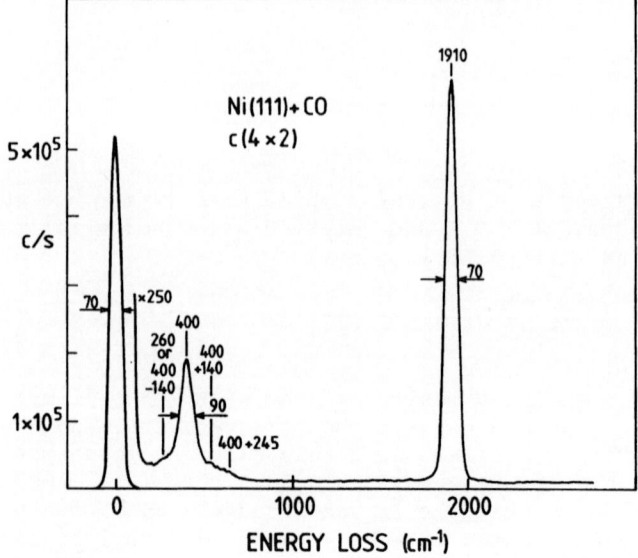

Fig. 10. EELS spectrum of Ni(111) with a c(4x2) overlayer of CO
         at 150 K.

observed. Hence the CO is adsorbed in a symmetry higher than C_s
and is in a twofold bridging site[13].

A second example are the EELS spectra of a p(2x2) overlayer
of acetylene on Ni(111). Spectra for normal and deuterated acetylene are shown in Fig. 11. They are recorded in specular reflection.
The loss at 260 cm^{-1} is due to the surface phonon in $\bar{M}$, as described
before, the surface phonon at 140 cm^{-1} is not resolved here contrary to Fig. 3, because of the lower resolution of ~8 meV
(65 cm^{-1}) compared to 4.5 meV in Fig. 3.

In the Ni(111)-acetylene system we observe dipole plus impact
scattering [15],[16]. Using the deuterated species and from the
angular profiles of the loss intensities we can make an assignment
of all losses to the normal modes [16]. For example, the loss at
1220 cm^{-1} (1190 cm^{-1} for C_2D_2) is due to the C-C stretching vibration. During the course of the assignment we had to decide whether
the losses at 1370 cm^{-1} and the similar one on the high frequency
side of the 1190 cm^{-1} loss, which is around 1330 cm^{-1} , are due
to normal modes or to multiple excitation of the C-C stretch and
of the 140 cm^{-1} surface phonon. The off specular spectra which are

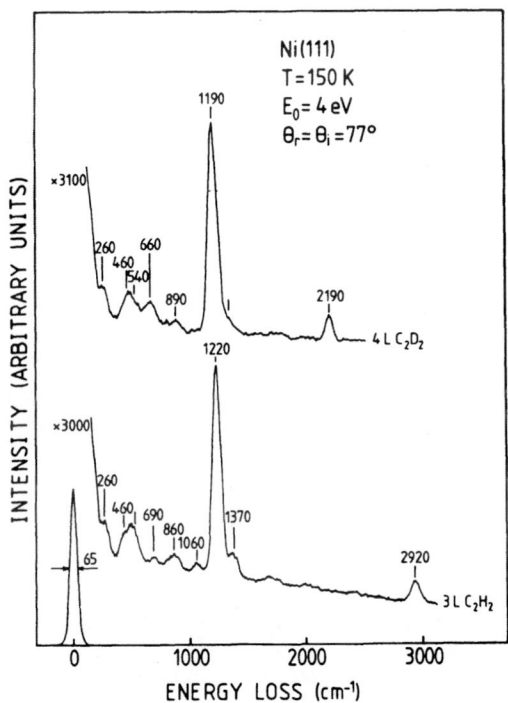

Fig. 11. EELS spectra of Ni(111) with a p(2x2) overlayer of
normal and deuterated acetylene.

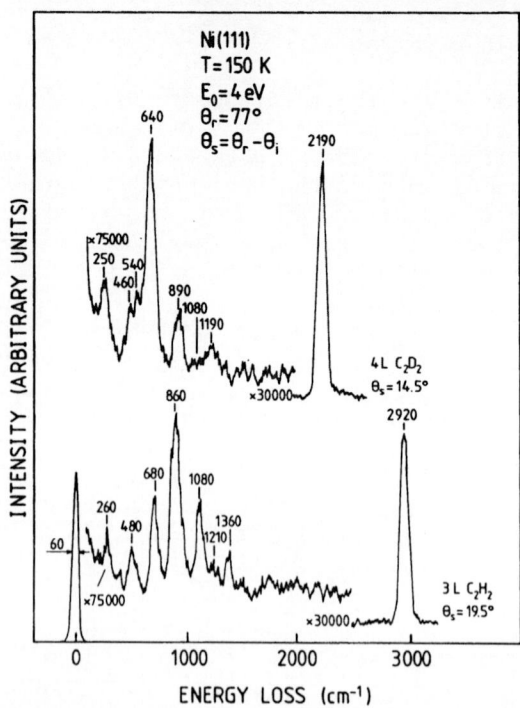

Fig. 12. EELS spectra of Ni(111) with a p(2x2) overlayer of normal
          and deuterated acetylene recorded off-specular.

shown in Fig. 12. could help to answer this question. We find that
the C-C stretch is purely dipole excited and therefore its intensity
drops to the noise level off specular, while the intensity of vibrations which have impact scattering contributions does not.

The intensity of the 1190 cm^{-1} loss in C_2D_2 is a little higher
than that of the 1220 cm^{-1} loss in C_2H_2 because of the different
angles θ_s off specular in Fig. 12. Furtheron we find that in the
case of C_2D_2 the 1330 cm^{-1} loss disappears together with the C-C
stretching loss, whereas the loss at 1360 cm^{-1} in the case of C_2H_2
does not. Therefore we conclude that the observed loss at 1330 cm^{-1}
is due to multiple excitation of the C-C stretching vibration and
of the 140 cm^{-1} nickel surface phonon. In the case of C_2H_2 this
overlaps with a loss due to another acetylene vibration with a frequency of 1360 cm^{-1} associated with a CH-bending mode[16].

Another important result can be drawn from Figs. 11 and 12 : the
C-C stretching vibration, for example, shows no contribution of
impact scattering at the used electron energy of 4 eV, whereas most
of the other losses do. That means we observe selective impact

excitation of vibrational modes. This is in agreement with recent
calculations of the inelastic scattering cross section of electrons
scattered from the vibrations of a c(2x2) overlayer of CO on Ni(100)
at larger angles[17]. The calculations are based on a basic impact
scattering theory[18] and Tong, Li and Mills find that selective
excitation of the C-O and Ni-C stretching vibration occurs at certain electron energies. This turns out to be a normal phenomenon
due to multiple scattering and interference of the scattered electron waves. The spectra of Fig. 12 furtheron show, that also the
loss due to the 260 cm^{-1} nickel surface phonon exhibits a contribution of impact scattering ! In the calculations of Tong, Li and
Mills the substrate has been treated as a rigid lattice[17,18]. The
spectra of Fig. 12, however, show that also the Ni-Ni vibration is
excited by impact scattering and thus also the vibrations of the
substrate atoms have to be included in a complete scattering theory.
Therefore impact scattering can provide substantial information
concerning the bonding and the geometry of the adsorbate complex.

SYNOPSIS

    Surface phonons have been observed on stepped and flat metal
single crystal surfaces by EELS. This fact makes data available
which reflect real microscopic properties of the surface like relaxation, force constants and structures of clean and adsorbed
surfaces. The excitation of surface phonons and their coupling
to other vibrations has to be considered for the interpretation
of vibrational loss spectra of adsorbed species.

ACKNOWLEDGEMENT

    Many results reported here were obtained in collaboration with
D. Bruchmann. We acknowledge the critical reading of the manuscript
by Dr. H. Wagner.

REFERENCES

 1. H. Ibach, D. Bruchmann, Phys. Rev. Lett. <u>41</u>, 958 (1978)
 2. D.H. Dutton, B.N. Brockhouse, A.P. Miller, Can. J. Phys. <u>50</u>,
    2915 (1972).
 3. H. Hopster, H. Ibach, Surf. Sci. <u>77</u>, 109 (1978).
 4. G. Allan, Surf. Sci. 85, 37 (1979).
 5. H. Ibach, D. Bruchmann, Phys. Rev. Lett. <u>44</u>, 36 (1980).
 6. R.E. Allen, G.P. Alldredge, and F.W. de Welte, Phys. Rev. B <u>4</u>,
    1661 (1971).
 7. R.J. Birgeneau, J. Cordes, G. Nolling, H.D.B. Woods, Phys. Rev.
    <u>136</u> A, 1359 (1964).
 8. V.R. Velasco, F. Yudurain, Surf. Sci. <u>85</u>, 107 (1979).

9. S. Lehwald, J.T. Yates, Jr., and H. Ibach, Proceedings of ECOSS 3, Cannes 22.-26.9.1980, p. 221.
10. J.E. Black, private communication.
11. G. Allan, J. Lopez, Surf. Sci. $\underline{95}$, 214 (1980).
12. S. Andersson, Surf. Sci. $\underline{79}$, 385 (1979).
13. W. Erley, H. Wagner, and H. Ibach, Surf. Sci. $\underline{80}$, 612 (1979).
14. Y.J. Chabal, A.J. Sievers, Phys. Rev. Lett. $\underline{44}$, 944 (1980).
15. S. Lehwald, H. Ibach, Nederl. Tijdschrift v. Vakuumtechniek $\underline{18}$, 71 (1980) (abstract).
16. H. Ibach, S. Lehwald, Proceedings of the 27th Nat. Symposium of the American Vacuum Society, Detroit, 14-17.10.1980, to be published.
17. S.Y. Tong, C.H. Li, and D.L. Mills, Phys. Rev. Lett. $\underline{44}$, 407 (1980).
18. C.H. Li, S.Y. Tong, and D.L. Mills, Phys Rev. B $\underline{21}$, 3057 (1980).
19. E. Evans. D.L. Mills, Phys. Rev. B $\underline{5}$, 4126 (1972).

ELECTRONIC SURFACE RESONANCE ENHANCEMENT OF VIBRATIONAL LOSS INTENSITIES

Roy F. Willis

Surface Physics Group /P.C.S., Cavendish Laboratory
Department of Physics, University of Cambridge
Cambridge CB3 OHE, U.K.

ABSTRACT

A high resolution EELS study of the electron reflectivity lineshape and $(E,\vec{k}_{/\!/})$ dispersion relations of surface barrier resonances on W(100) is reported. Their effect on the impact energy dependence of the symmetric stretch-vibration excitation cross-section of chemisorbed hydrogen is to produce resonance enhancement. Results are presented for both the low coverage c(2 x 2)H and high coverage p(1 x 1)H chemisorbed layers. The energies of these resonances would appear to fit a simple Rydberg type series of levels for a smooth monotonic image potential surface barrier. However, the dispersion behaviour of the intense "ground state level" is shown to be anomalous and to deviate strongly from two-dimensional free electron behaviour. Strong resonance intensity is found only for energies corresponding to energy gaps in the surface projection of bulk states. The effect on adsorbate vibrational selection rules is discussed in terms of the degree to which the resonance electron density overlaps the short range scattering potential of the vibrating atoms.

INTRODUCTION

The overall process underlying any observation of inelastic reflection of electrons at surfaces is usually pictured as a sequence of events which include an elastic reflection either preceded or followed by inelastic scattering[1]. Under certain conditions, low energy electrons incident on a crystal surface can become temporarily trapped in electron surface states[2]. The electron-surface interaction produces a surface (electrostatic

image) barrier which refracts the incident electron beam and can
cause total internal reflection of back-scattered diffracted beams.
Non-specular beams emerging from the crystal substrate which have
insufficient kinetic energy normal to the surface to surmount
this surface potential energy barrier are scattered back into the
crystal. If the beam is subsequently strongly reflected by the
substrate atoms, a sustained multiple scattering can occur between
the barrier and the crystal. The interference between the specularly scattered and multiply internally reflected waves produces
narrow fluctuations of elastic (specular) scattering intensity with
respect to variation of the incident beam energy E_0 and incident
direction, defined in terms of the polar angle θ_i. These temporary
or nonstationary states are quantized with respect to the electron's
momentum normal to the surface, forming a discrete Rydberg-like
series of energy levels associated with the Coulombic form of the
surface barrier potential[3]. The lifetime of these states is
relatively long and, since the electrons are confined to a region
close to the surface, we expect that such conditions will lead to
a resonant enhancement of inelastic losses arising from the
vibrational excitation of adsorbate-induced surface modes[4]. This
mechanism of "inelastic surface resonance scattering" is illustrated
in Fig. 1.

An electron can scatter directly from the surface, exciting an
adsorbed molecule X in the process. Alternatively, providing the
$(E,\vec{k}_{/\!/})$ values correspond to an elastic resonance condition, the
electrons can become temporarily trapped in a surface resonance
state, thereby enhancing the vibrational excitation cross-sections.
The inelastic reflection corresponding to excitation energy $E' =$
$E - \Delta E_{loss}$ and parallel momentum transfer $\vec{k}'_{/\!/}$, will exhibit resonance fluctuations for $(E,\vec{k}_{/\!/})$ values displaced by $(E',\vec{k}'_{/\!/})$ from the
elastic resonances. The ratio of the inelastic to elastic intensi-

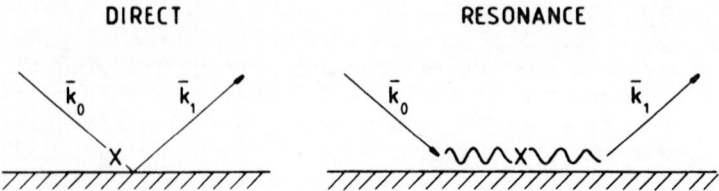

Figure 1 - Schematic drawing illustrating the direct and surface
resonance mechanisms of inelastic interaction with an
adsorbed molecule X : $\vec{k}_0$ and $\vec{k}_1$ refer to the incident
and scattered electron wavevectors respectively.

ties (I_L/I_o) will deviate strongly from that predicted by the dipole scattering theory of vibrational loss intensities observed in the specularly reflected beam direction[4,5]. The extent to which this resonance enhancement can occur will also depend on any modification to the dimensions of the surface potential barrier itself due to the presence of the adsorbate layers[6].

Surface barrier resonances have been observed in specular beam scattering from clean W(100) surfaces[7]. The fine-scale fluctuations in the elastic intensity have been analyzed in terms of a model image potential barrier producing a Rydberg series of trapped states[2,8]. In this paper, we extend these measurements to investigate the associated inelastic surface resonance scattering in the presence of adsorbed hydrogen c(2 x 2) and p(1 x 1) layers, i.e. the so-called β_2 low coverage and β_1 saturation coverage adsorbate phases. In both cases, the hydrogen occupies a bridge site position between two W atoms; it is merely the symmetry of the surface together with the lateral spacing between adjacent W atoms which changes with increasing adsorption[9]. As the inelastic "decay" channel, we choose the vibrational loss corresponding to the symmetric stretching frequency of the hydrogen moving in the Z direction, normal to the surface. Changes occur in both the elastic and inelastic reflectivity line shape and $(E, \vec{k}_{//})$ dispersion behaviour. The results confirm the inelastic resonance scattering mechanism and provide insight into the nature of the overlap between the resonance electron density and the scattering potential of the substrate.

EXPERIMENTAL PROCEDURE

Measurements were carried out on the intensity variation of the (00) beam specularly reflected from a W(100) surface using an electron spectrometer consisting of a monochromator and analyzer of the hemispherical electrostatic deflector type. Both crystal and analyzer were rotable about a central axis such that the polar angle of incidence of the incoming beam could be varied over a range, $23° \lesssim \theta_i \lesssim 80°$, relative to the crystal normal direction. The acceptance angle of the analyzer, as defined by two entrance slits, was $\sim 1.7°$. The impact energy was varied over a range $3 \text{ eV} \lesssim E_o \lesssim 30 \text{ eV}$ by stepping and scanning the spectrometer voltages in a series of measurements corresponding to overlapping scan widths of magnitude 3 to 6 eV and step intervals of 0.01 to 0.02 eV. The overall energy resolution was 25 meV. The absolute number of counts for the elastic I_E and inelastic (energy loss) I_L were recorded simultaneously together with the ratio $R = I_L/I_E$ curve.

After submitting the crystal to a standard oxygen treatment, a clean and reproducible surface was obtained by regular flashing at 2400 K in a UHV system with a base pressure of 1 x 10^{-10} Torr.

The crystal could be cooled to ∼ 125 K by liquid N_2 cooling. Hydrogen gas was introduced into the system via a tube extending to within 2.5 cm of the crystal surface. The c(2 x 2) (low coverage, β_2 phase) hydrogen layer produces a vibrational loss at 155 meV; that associated with the p(1 x 1) (high coverage, β_1 phase) hydrogen layer occurs at 130 meV. Both hydrogen losses have been identified with the symmetric stretching vibration mode associated with bridge-site adsorption and motion normal to the surface[9].

RESULTS

Reflectivity Line Shape

In Figure 2, the specular elastic (E), inelastic (I) and ratio $R = I_L/I_E$ curves for the clean W(100) surface are shown as a function of impact energy and incident polar angle θ_i for $(E, \vec{k}_{/\!/})$ values corresponding to surface resonance conditions associated with the grazing emergence of the $(0\bar{1})$ diffraction beam. The inelastic curve corresponds to the intensity of the smooth background current at 130 meV loss; the curves are very similar at 155 meV so that, within this energy loss range, the inelastic curve is independent of background loss intensity. Its impact energy dependence follows closely the features for the elastic beam, with a slight displacement towards higher energies equivalent to the loss energy. This shift, together with the reflectivity line shape of the curves I and E is responsible for the fluctuations observed in the ratio curves R. These features serve to locate the positions of the singularities, A, B and D observed in both the elastic and inelastic reflectivity curves. The discontinuity D occurs very close in energy to the $(0\bar{1})$ beam grazing emergence condition and, as such, shows angular dispersion with changing incident energy or vice versa. The minima A and B are strongly correlated with this $(0\bar{1})$ beam threshold behaviour. More extensive measurements reveal a sharp third minimum C occurring between B and D, in agreement with the observations of Adnot and Carette[7] . The energies of these features A, B. C in relation to D are indicative of a Rydberg series of resonance states[10]. However, whereas the minimum at C is only resolved over a narrow angular range, $40° \lesssim \theta_i \lesssim 50°$, features A, B and D are observed over the whole range of incident angles shown in Figure 2. This behaviour of the clean crystal will be reported in a separate publication[11].

It is interesting to note that the line shape of these features in both the elastic and inelastic reflectivity curves varies with incident angle over the range $23° < \theta_i < 70°$, becoming particularly sharp for angles $\theta_i \gtrsim 35°$. As we will later show, this behaviour is due to an energy band gap in the bulk crystal's electronic states at these energies and for scattering in the [01] incidence plane. The resonance life-times are increased since current cannot easily leak away into the crystal.

The $(E, \vec{k}_{//})$ values of the surface barrier resonances are linked to the $(0\bar{1})$ beam threshold condition which is independent of the scattering mechanism. However, the line shape of the fluctuation will depend critically on the actual scattering mechanism[2]. Referring to Figure 1, the scattering amplitude is seen to be the superposition of two contributions, namely the direct and the resonance processes. The resultant line shape is the sum of the squared modulus of the resonance amplitude contribution and the cross-term due to the interference between these two processes. This is particularly important in the case of inelastic surface resonance scattering from adsorbate layers.

Figures 3 and 4 illustrate the effect of introducing c(2 x 2) (β_2 phase) and p(1 x 1) (β_1 phase) adsorbed hydrogen layers respectively. In the case of the low coverage β_2 phase (Figure 3), the reflectivity behaviour is changed little save for some slight broadening and a small shift to higher energies due to an increase in the work function ~ 0.25 eV. In this particular case, the curves were recorded along the [11] azimuthal direction ($\phi = 45°$)

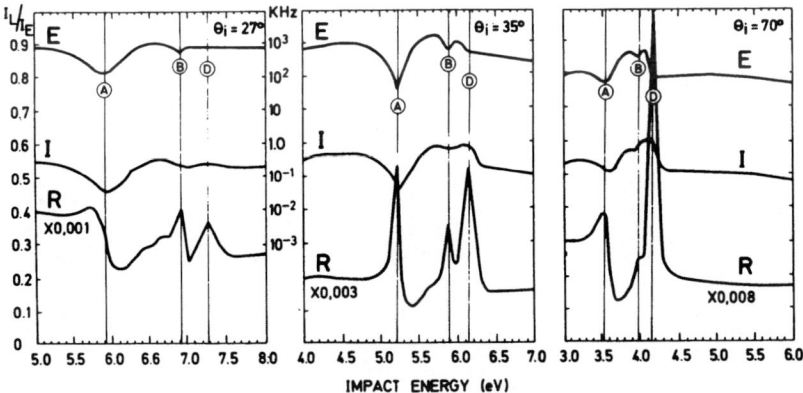

Figure 2 - Absolute counts of the elastic (E) and inelastic (I) intensities plotted on a logarithmic ordinate (right hand scale, KHz) as a function of the electron impact energy in the region of the $(0\bar{1})$ beam emergence threshold for clean W(100). The ratio of the inelastic to elastic intensities, $R = I_L/I_E$ is plotted on the left ordinate for a loss intensity corresponding to the background current at 130 meV loss. Fluctuations in the ratio curve R (with appropriate multiplication factors shown) serve to locate the positions of singularities A,B,D in the elastic reflectivity curve in relation to the inelastic profile, the positions and lineshapes of which vary with incident angle θ_i.

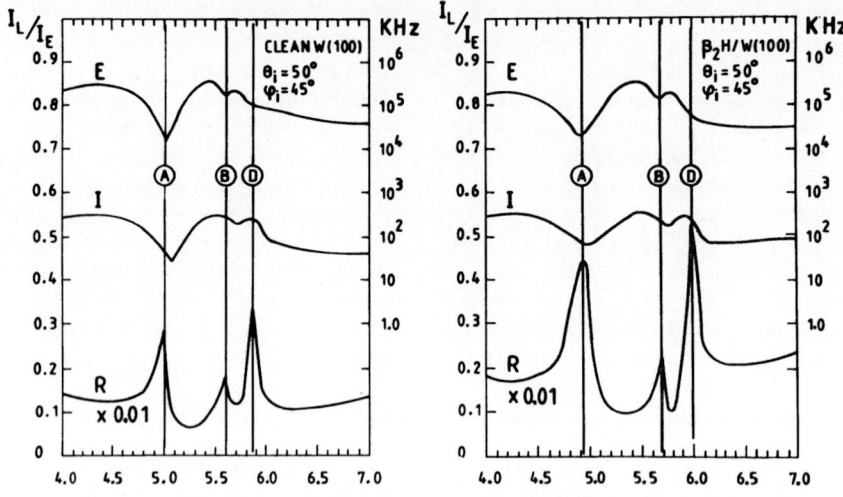

IMPACT ENERGY (eV)

Figure 3 – Comparison of elastic (E), inelastic (I) and ratio
             curves R = I_L/I_E for clean W(100) and for the low
             coverage c(2 x 2)H, β_2-phase. ϕ = 45° refers to an
             angular scan along the [11] direction of the surface
             lattice. The loss intensity corresponds to the 155 meV
             symmetric stretch vibration of hydrogen in a bridge
             site with reduced lateral spacing between the W atoms
             (Ref. 9).

but similar behaviour is observed in the [01] plane of incidence.
(The resonance features also occur at slightly different impact
energies compared with those observed in the [01] incidence plane,
Figures 2 and 4, due to a cos ϕ term in the determination of the
E,$k_\parallel$ values). For the saturated coverage (β_1 phase), significant
changes in the inelastic intensity and impact energy dependence
are observed, Figure 4. The (0$\bar{1}$) beam emergence threshold D is
shifted towards higher energy due to an increased work function
change ~ 0.50 eV. Also, the inelastic reflectivity curve I no
longer closely mirrors the behaviour of the elastic intensity E.
This is particularly true of the deap elastic minimum A which is
more strongly attenuated, and produces large fluctuations in the
ratio curves with changing angle of incidence as a consequence[12].

     Clearly, the p(1 x 1) layer has a stronger influence on the
surface barrier scattering potential than the c(2 x 2)H layers.
Since hydrogen has a very small elastic scattering cross-section
compared with the substrate tungsten atoms, the reasons for this
are not immediately apparent. However, considerable insight into
this behaviour is provided if we plot the (E,$\vec{k}_\parallel$) dispersion

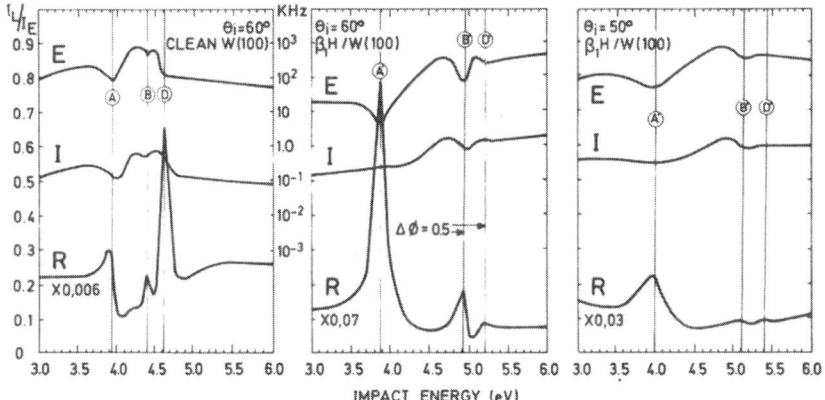

Figure 4 - Comparison of elastic (E), inelastic (I) and ratio
curves R = I_L/I_E for clean W(100) and the saturation
coverages p(1 x 1)H, β_1-phase. Angular scan along the
[10] direction, ϕ = 0°. The loss intensity corresponds to
the 130 meV symmetric stretch vibration of hydrogen in
a bridge site with the lateral spacing between the W
atoms "relaxed" to their bulk lattice value, ~ 3.16 Å.

behaviour of the surface resonance features.

Dispersion Behaviour

The dispersion $E(\vec{k}_\parallel)$ of the surface resonances for clean W(100)
are shown plotted in Figure 5 in relation to the angles of incidence
at which measurements were taken : $(\vec{k}_\parallel + 2\pi\vec{g}) = [(2mE_0)^{1/2}/\hbar]$ sin θ_i.
That is, the experimental points lie on curves approximately parallel
to the free-electron dispersion curve representing the (0$\bar{1}$) beam
threshold condition. In a strictly two-dimensional free-electron
description[2] , the energies of the surface barrier resonances may
be written :

$$E_{ng}(\vec{k}_\parallel) = \frac{1}{2} |\vec{k}_\parallel + 2\pi\vec{g}|^2 + \varepsilon_n \qquad n = 1, 2, 3 \ldots \qquad (1)$$

(i.e. $\hbar$ = m = e = 1 and energy values in Hartree units). The first
term on the right of Eq. (1) is the threshold (grazing emergence)
energy for the diffraction beam indexed by its reciprocal (surface
net) vector $\vec{g}$, and the second term is the energy eigenvalue of the
nth bound-state eigenfunction corresponding to a Coulombic surface
barrier potential. According to the free electron description, the
dispersion behaviour of the surface barrier resonances is the same

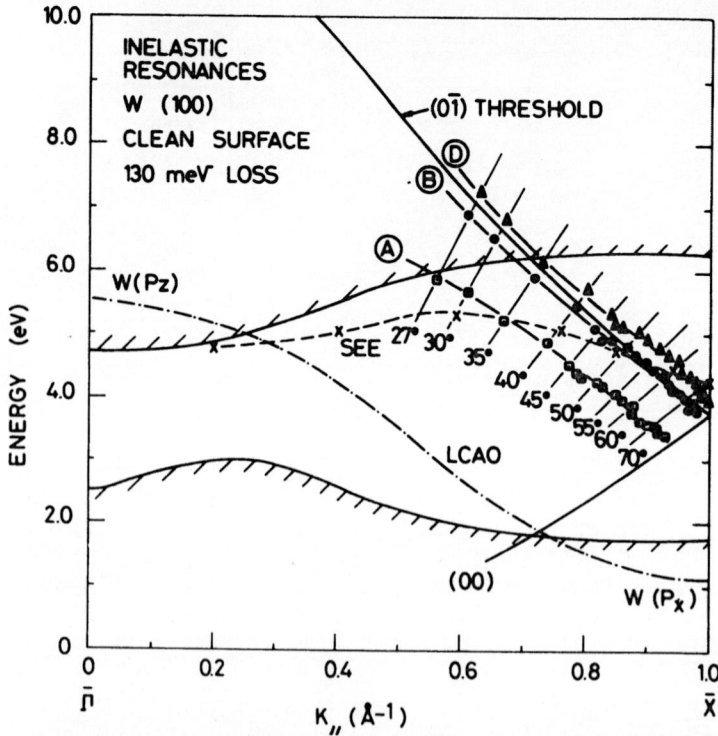

Figure 5 – Dispersion $E(\vec{k}_{//})$ of the surface resonances A,B and D observed for clean W(100) in relation to the (0$\bar{1}$) beam emergence threshold free-electron behaviour. The experimental points are taken from the reflectivity profile curves for different angles of incidence, $27° \lesssim \theta_i \lesssim 80°$ The bulk crystal energy band gap (shaded) together with a surface state resonance feature observed in secondary electron emission (SEE) spectroscopy (Ref. 16) and the LCAO band predicted theoretically (Ref. 17) are also indicated.

as that for the beam threshold except for a downward displacement by an amount equivalent to the binding energy $-\varepsilon_n$. The beam threshold may be regarded as the ionization limit, $n = \infty$, for a Rydberg-like series of such states[13].

In the case of the results shown in Figure 4, feature D in the reflectivity curves lies very close to the beam threshold condition, although some uncertainty exists due to experimental difficulties in establishing more exact values for the impact energies.

Feature B (and C which is not shown and is resolved only over a
limited angular range[7]) parallels this behaviour very closely. On
the other hand, the strong minimum A shows some deviation from the
two-dimensional free-electron behaviour, particularly for angles
of incidence approaching the upper edge of the bulk energy band
gap which extends over several volts in tungsten along the [10] or
$\bar{\Gamma}\bar{X}$ direction (shaded, Figure 5)[14].

There has been much discussion in the literature regarding the
origin of these resonance features on clean W(100) surfaces[10,15].
A much broadened feature observed by the method of angle-resolved
secondary electron emission (SEE, Figure 5)[16] was originally
identified as a surface state resonance (SSR) predicted by a tight-
binding calculation[17] which neglected the long range Coulombic
nature of the surface barrier potential. This LCAO state (Figure 5)
has a dispersion which strongly parallels the behaviour of feature
A, although it is displaced ~2 eV lower in energy. The origin of
this SSR is closely related to the overlap of the tungsten 6p
atomic orbitals. The calculation predicted a rigid shift upwards in
energy with hydrogen adsorption.

In Figure 6 we show the surface resonance dispersion behaviour
in the presence of the p(1 x 1)H layer. Whereas resonance feature B
and the beam threshold D are moved upwards in energy (B' and D'
respectively) due to the increase work function at the surface,
feature A again remains anomalous in that its energy remains relatively unchanged at large angles of incidence, $\theta_i \gtrsim 55°$, and its dispersion behaviour changes at lower angles of incidence, $\theta_i \lesssim 50°$
(curve A'). The inset diagram indicates the H adsorbed in a bridge
site geometry between adjacent W atoms might be expected to have
some influence on the overlap of the tungsten 6p-orbitals, which
are partly responsible for the LCAO "crystal-derived" surface
state[17].

DISCUSSION

Origin of Surface Resonances on W(100)

Considerable progress has recently been made in our understanding of surface resonance band structure and lineshapes from
semi-empirical fits to the experimental reflectivity data based
on a model image potential barrier and the simplest form of the
layer multiple scattering description of resonances[2] . As shown
schematically in Fig. 7a, the surface of the crystal is imagined
as divided in two parts by a plane parallel to the surface (Z_0
in the case of the clean surface and Z_0' in the presence of the
adsorbate), the plane being so positioned that the potential to
one side (say $Z_0 > 0$) is approximately that of the substrate
crystal and that for $Z_0 < 0$ is the surface region. The important

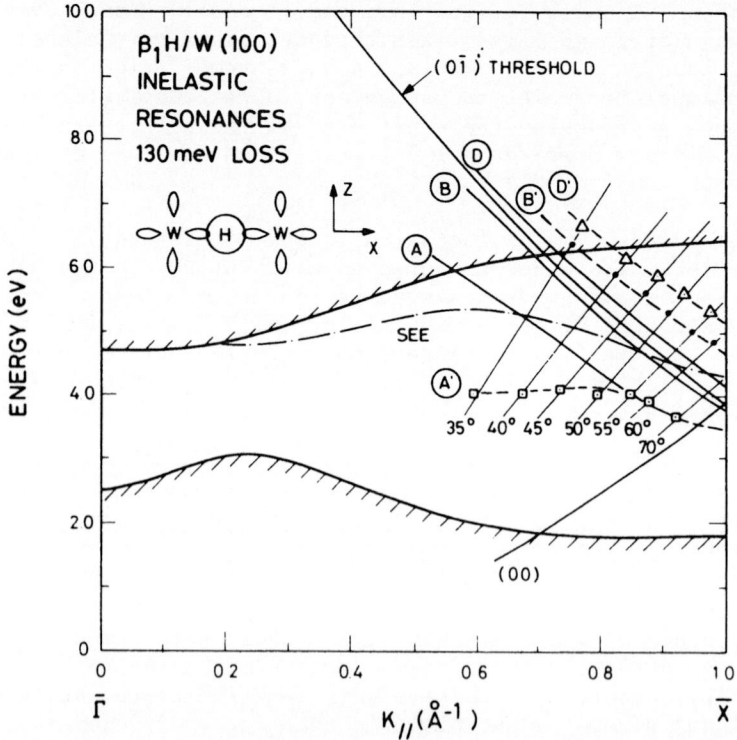

Figure 6 – Surface resonance dispersion behaviour in the presence
of p(l x l)H layer (β_1 –phase). Resonance features A,B
and D on the clean surface are modified, A', B' and D'
in the presence of the hydrogen (inset diagram).

quantities which come into the description of the phase and ampli‐
tude coefficients of those waves multiply reflected within the
surface region are the height (or inner potential) U_O and the
position of the origin of the image potential barrier Z_C relative
to the position of the outermost atomic layer, Z_O or Z_O'. In the
case of W(100), Jennings[8] has derived a value of $Z_C \sim - 3.0$a.u. as
the best fit agreement between theory and experiment (Ref. 7).
However, while this calculation reproduces the positions and
lineshapes of features B and D, the agreement with experiment is
poor for feature A.

Somewhat better agreement has been obtained by McRae[2], also
based on high-resolution intensity data of Adnot and Carette[7].
However, here again problems arise concerning the overall line
shape and, particularly, the location of feature A. For example,

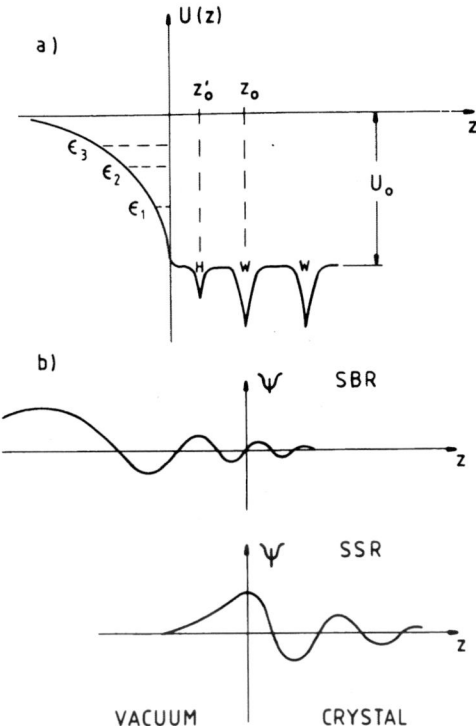

Figure 7 - (a) Model one-dimensional image potential barrier of the
              W(100) crystal surface showing the energetic positions
              of the surface barrier (Rydberg) resonances $\varepsilon_1, \varepsilon_2, \varepsilon_3$..
              in relation to the vacuum potential, $Z = -\infty$. The
              actual position of the bottom of the image potential
              barrier (Z_c) will vary in relation to the centre of
              the topmost atomic layer, depending on whether the
              surface is clean (Z_0) or contains adsorbed hydrogen
              (Z_0'). $U_0(Z)$ is the inner potential which serves to
              locate the depth of barrier in relation to the atomic
              ion-core muffin-tin scattering potentials of the
              substrate.
          (b) Schematic wavefunctions showing the location of re-
              sonance electron density far outside the crystal in
              vacuum for the higher order ($n \gg 1$) surface barrier
              resonance (SBR) and that for a lower order ($n \approx 1$)
              "crystal-induced" surface state resonance (SSR) in
              which the resonance electron density overlaps strong-
              ly into the surface crystal region.

it is not clear from either the experimental data or the semi-
empirical analysis whether feature A can be regarded as the n = 1
ground state or a higher n = 2 state of a Rydberg series. The line
shape analysis places the n = 1 resonance at a peak (rather than a
minimum) in the reflectivity curve at either - 0.33 eV or - 1.73 eV
below the (01) beam threshold. The situation is further confused
by the possible incidence of a Bragg reflection peak which occurs
at an impact energy of ~ 3.5 eV (θ_i ~ 48°) which is close to the
position of the possible n = 1 Rydberg resonance at - 1.73 eV.

This question of the location of the lower order Rydberg states
is important since it relates to the degree of overlap between the
resonance electron density and the scattering potential of the
atoms at the surface[3] . As shown in Fig. 7b, we may distinguish
between two extreme cases : that in which the electronic charge
density remains close to the surface, (surface state resonance,
SSR) and that in which it is concentrated far outside in vacuum
(surface barrier resonance, SBR). The former case is representative
of the lower order "Rydberg states" and corresponds to strong interaction with the scattering potential of the crystal. In this case,
the situation strongly resembles the familiar surface states located near or below the Fermi level, for which the long range part of
the Coulombic potential is not important. In contrast, the weak
interaction case is more representative of states of higher order
in the Rydberg series, which are "true" surface barrier induced
states. Since their wave function amplitude is concentrated far
outside the surface, they are longer lived and not severely lifetime broadened due to inelastic processes associated with scattering within the crystal surface. Also, the two-dimensional free
electron description of electronic motion parallel to the surface
will remain generally true for these higher order states. On the
other hand, the free electron dispersion behaviour of the lower
order states is expected to be modified more strongly by any
deviation from a laterally averaged potential within the surface
arising from, for example, any change in surface structure.

Inelastic Resonance Enhancement of Surface Vibrations

The above distinction is relevant to the question of which
surface resonances are more effective in producing inelastic
resonance scattering enhancement of the vibrational loss intensities ? The behaviour of resonance feature A (Fig. 6) is clearly
seen to follow the "strong overlap" behaviour[2] . We might therefore expect it to be more strongly influenced by the introduction
of the adsorbate layer, and inelastic scattering by the vibrating
surface atoms to be larger than is the case for the weakly overlapping higher order barrier resonances, B, C, etc. This would
appear to be true in the case of the p(1 x 1) layer (Figs. 4 and 6)
but less so in the case of the c(2 x 2) layer, the inelastic

reflectivity profile being little changed (Fig. 3). The reason for
this is not entirely clear at the present time, but it may relate
to the similarity in the symmetry of the c(2 x 2)H layer and the
symmetry of the clean W(100) surface which is known to reconstruct
into a c(2 x 2) surface structure at around 300 K[18]. Any change
in surface symmetry would be expected to have a profound influence
on the actual surface barrier dimensions and shape.

Surface Resonance Band Structure

The diffraction beam threshold condition, $\frac{1}{2} |\vec{k}_{/\!/} + 2\pi\vec{g}|^2$, for
different $\vec{g}$ values collectively make up the surface free-electron
band structure, shown in Fig. 8a. The reciprocal lattice and
surface Brillouin zones appropriate to the p(1 x 1) and c(2 x 2)
surface lattice nets of W(100), Fig. 8b, together with the scattering geometry, fig. 8c, are also indicated for scattering along
the [01] or Δ and [11] or Σ directions of the surface Brillouin
zone. Strong resonance features associated with beam emergence
conditions are represented by filled circles. They correlate closely with gaps in the surface projection of bulk states[11]. Of
particular relevance in the present context, is the dispersion
behaviour of resonance feature A (squares) which is seen to lie on
the curve corresponding to the beam emergence condition for the
1/2 order beams associated with a c(2x2) surface. The fact that the
feature A is strong in the case of the clean reconstructed c(2 x 2)
and low coverage c(2 x 2)H surface structures, but weak in the case
of p(1 x 1)H layer (cf. Figs. 3 and 4) would indicate that its
origin is partly linked to the emergence thresholds of the 1/2 order
beams. This has been established also in the case of simila resonance structures appearing in the reflectivity curves for the
($\bar{3}$/2, 1/2) and ($\bar{1}$/2; 3/2) beams along the Σ symmetry direction. The
results also suggest possible strong interaction between resonances
associated with different emergent beams, as indicated at their
crossing point (open circles, Fig. 8a). This behaviour is indicative
of relatively strong coefficients in the two-dimensional Fourier
expansion of the lateral interaction potential in the crystal surface[19] .

CONCLUDING REMARKS

Some ambiguity therefore remains concerning the actual ordering of the surface resonances on W(100), which may arise as a consequence of the tendency of this surface to reconstruct into a
c(2 x 2) lattice. The increased broadening of the lower order
resonances would suggest shorter lifetimes due to inelastic processes arising from the stronger overlap between the resonance
electron density and the scattering potential of the surface atoms.
This being the case, we can expect that these resonances will have

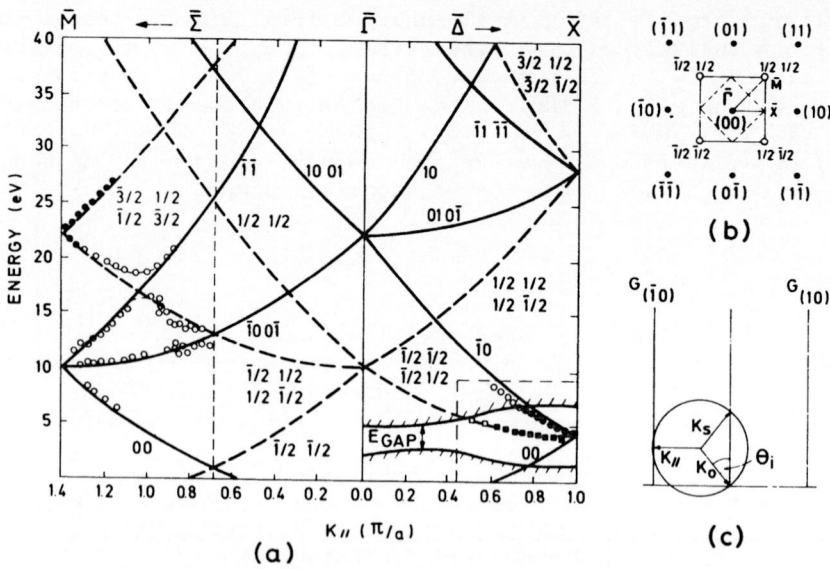

Figure 8 – (a) Two dimensional free-electron band structure
showing the $E(\vec{k}_{\parallel})$ beam threshold behaviour of W(100).
The full lines refer to the p(1 x 1) unreconstructed
surface and the dashed lines show the incidence of
the fractional order beams of the c(2 x 2) reconstructed and β_2-phase hydrogen lattices. The positions
of the threshold surface resonances, which are observed as strong singularities in the reflectivity
profiles at bulk band gap energies, appear as filled
points. Weaker features appear as open points. The
dispersion of the "ground state" feature A in the
reflectivity curves is shown (filled squares) to
correlate with the emergence of 1/2 order beams.
(b) Reciprocal lattice with diffraction beams appropriat-
ely indexed in relation to the p(1 x 1) and c(2 x 2)
surface Brillouin zone dimensions and symmetry
points $\bar{\Gamma}$, $\bar{X}$ and $\bar{M}$.
(c) Scattering geometry in the Ewald sphere construction.

a greater influence on resonant enhancement of vibrational loss
intensities. This will be particularly true for off-specular
electron energy loss spectroscopy (EELS) where we can expect SSR-
type resonances to influence both the angle- and impact-energy
dependence of the vibrational cross-sections due to the short range
nature of the scattering process[20].

    In the case of EELS in the specular beam direction, this
distinction becomes less clear. Although the electron density of

the longer lifetime, higher-order surface barrier resonances (SBR) overlap only weakly with the crystal scattering potential, the electrons are able to excite vibrational modes via the long range dipole interaction (which is also responsible for the Coulombic "image" nature of the surface barrier potential). However, this mechanism is selective in that it favours the excitation of those modes which possess a net dipole moment normal to the surface [1]. To what extent particular modes vibrating parallel to the surface are influenced by the actual symmetries of these resonances propagating along specific directions in the surface has yet to be established. It could be that by tuning the impact energy into one or other of the two kinds of resonance interaction (SBR vs SSR) one could find a "surface-resonance-enhanced vibrational selection rule" sensitive both to the point group symmetry of particular modes at the surface and the exact nature and surface symmetry of the inelastic resonance scattering mechanism.

ACKNOWLEDGEMENTS

I am indebted to M.R. Barnes for invaluable technical assistance with these measurements and to The Royal Society (London) for financial support for this work.

REFERENCES

1. See, for example : Z. Lenac, M. Sunjic, D, Sokcevik and B. Brako, Z. Physik B28, 273 (1977).
2. For a review : E.G. Mc Rae, Rev. Mod. Phys. 51, 541 (1979).
3. P.M. Echenique, J.B. Pendry, J. Phys. C : Sol. State Phys. 11, 2065 (1978).
4. R.F. Willis, W. Ho, E.W. Plummer, Surf. Sci. 80, 543 (1979).
5. A.M. Barò, H. Ibach, H.D. Bruchmann, Surf. Sci. 88, 384 (1979).
6. P.J. Jennings, G.L. Price, Surf. Sci. 95, L205 (1980).
7. A. Adnot, J.D. Carette, Phys. Rev. Lett. 38, 1084 (1977).
8. P.J. Jennings, Surf. Sci. 75, L777 (1978).
9. M.R. Barnes, R.F. Willis, Phys. Rev. Lett. 41, 1779 (1978); R.F. Willis, Surf. Sci. 89, 457 (1979).
10. A. Adnot, J.D. Carette, Phys. Rev. B16, 4703 (1977).
11. R.F. Willis, to be published.
12. W. Ho, R.F. Willis, E.W. Plummer, Phys. Rev. B21, 4202 (1980).
13. G. Malström, J. Rundgren, J. Phys. C : Sol. Stat. Phys. 13, L61 (1980).
14. R.F. Willis, N.E. Christensen, Phys. Rev. B18, 5140 (1978); J. Phys. C : Sol. Stat. Phys. 12, 167 (1979).
15. E.G. McRae, Phys. Rev. B17, 907 (1978); R.F. Willis, ibid B17, 909 (1978).
16. R.F. Willis, B. Feuerbacher, N.E. Christensen, Phys. Rev. Lett. 38, 1087 (1977).

17. N.V. Smith, L.D. Mattheiss, Phys. Rev. Lett. 37, 1494 (1976).
18. T.E. Felter, R.A. Barker, P.J. Estrup, Phys. Rev. Lett. 38, 1138
    (1977); M.K. Debe, D.A. King, Phys. Rev. Lett. 39, 708 (1977).
19. E.G. McRae, J.M. Landwehr, C.W. Caldwell, Phys. Rev. Lett. 38,
    1422 (1977).
20. S.Y. Tong, C.H. Li, D.L. Mills, Phys. Rev. Lett. 44, 407 (1980).

EELS OF COLLECTIVE SURFACE MODES

Stig Andersson

Department of Physics, Chalmers University of Technology
S-412 96 Göteborg, Sweden

ABSTRACT

Vibrational modes of the Cu(100) – CO system have been investigated by inelastic electron scattering. The C–O stretching
vibrational mode of the Cu(100) c(2x2)CO system is found to be
collective. The mode dispersion is dominated by dipole-dipole
interaction. A low-energy mode which is observed to be particularly pronounced for the c(2x2)CO structured situation is interpreted to
be due to the excitation of a surface phonon at the M point of the
substrate surface Brillouin zone.

INTRODUCTION

High resolution electron energy loss spectroscopy (EELS) is
a potential tool to study surface phonon modes and their dispersion
for ordered overlayers on metal substrates. Collective modes of
clean metal surfaces and ordered adsorbate layers have been of
theoretical interest for quite some time and several realistic systems have been studied recently [1] . Spectroscopic studies of the
properties of such modes would provide important insight in the
strength and the nature of the interaction among the surface
species. In practice such experiments are still quite difficult
to perform, resolution and intensity of present instruments being
the major obstacles. Either the frequency shifts are small and
the inelastic intensity weak at large momentum transfer as one
would expect for modes well above the substrate phonon band or
the mode frequency is low as for modes that are strongly coupled to
the substrate surface phonon modes. The latter situation appears,.
however, to be quite favourable and modes in the phonon band have

been observed for adsorbates   on Ni(100)[2]   and Ni(111)[3]

The aim of this paper is to discuss some results concerning the collective behaviour of vibrationally excited CO molecules in the c(2x2)CO structure on Cu(100)[4] and the coupling to substrate surface phonons via the adsorbed CO layer.

EXPERIMENTAL

The experimental data reported in this work refer primarily to the c(2x2) structure of CO on Cu(100).   The structure and stoichiometry of this system is well characterized which makes it suitable for quantitative studies of e.g. vibrational cross sections.   The surface density of CO molecules is n = 0.0767 $\mathring{A}^{-2}$.   The molecule is adsorbed on top of a substrate atom with the carbon end towards the substrate (see Fig. 1.),the bond lengths are 1.90 $\mathring{A}$ and 1.15 $\mathring{A}$ for the Cu-C and C-O bonds respectively[5] .   The molecular axis is essentially normal to the surface plane.[6]

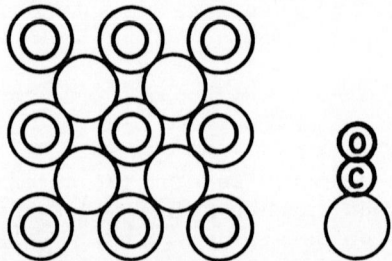

Fig. 1. : Model of the vertical terminal - CO arrangement for the Cu(100) c(2x2)CO structure.

The CO overlayers were formed and·investigated at about 80K substrate temperature.   The adsorption was monitored by LEED and work function measurements.

The spectroscopic measurements presented here were obtained using a new high resolution electron spectrometer.

The spectrometer which is shown schematically in Fig. 2, consists of a selector and an analyzer both of cylindrical mirror design. The specimen and the analyzer can be rotated such that the polar angles of incidence and collection can be varied independently. The work function difference between the spectrometer and the specimen is compensated to within 0.05 eV and the electron energies (E_O) quoted in the spectra refer to the vacuum level. The spectrometer has been operated at an energy resolution of 4-4.5 meV which is close to the designed resolution 3.5-4 meV at the pass energies (0.35 - 0.40 eV) used.

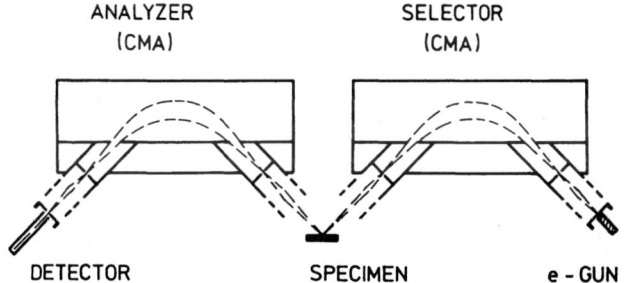

ANALYZER (CMA)   SELECTOR (CMA)

DETECTOR   SPECIMEN   e - GUN

Fig. 2. : Schematic drawing of the new EEL spectrometer arrangement. The analyzer and specimen can be rotated independently around an axis normal to the plane of the paper.

CROSS SECTION AND DISPERSION

Fig. 3. shows a vibrational loss spectrum for the Cu(100) c(2x2)CO system. The spectrum is measured with the new spectrometer described briefly above. The primary energy is $E_0 \stackrel{\scriptscriptstyle >}{\scriptscriptstyle =} 3$ eV, the angle of incidence is $\alpha = 65°$ and the angle of collection is $\Theta = 0°$ i.e. specular condition. The scattering plane containing the incident and the scattered beams is defined by the substrate normal $[100]$ and the $[010]$ direction in the Cu(100) surface plane. The scattering situation and the notation is shown in Fig. 4. The low loss peaks at 42.5 meV and 259 meV correspond to the excitation of the fundamental Cu-CO and C-O stretching vibrations of CO adsorbed in the on-top position[7] as shown in Fig. 1.

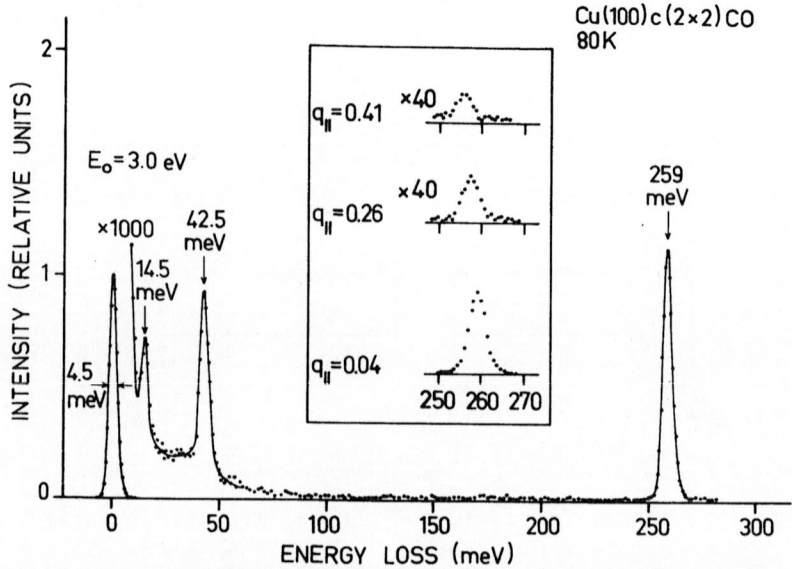

Fig. 3. :   EEL spectrum from the Cu (100) c(2x2)CO structure measured in the specular direction for an energy $E_0 = 3$ eV and angle of incidence $\alpha = 65°$ of the primary electron beam. The C-O stretch loss peak is shown in the inset for three values of the momentum $\underline{q}_{\shortparallel}$ along the $[010]$ surface direction.

The loss peak at low energy, 14.5 meV, is interpreted to be due to the excitation of a surface phonon and wil be discussed in more detail in section 4.   The inset which shows the C-O loss peak at three different scattering conditions will be discussed below.

Dipole scattering, localized excitation

In a previous investigation it was found [8] that the energy
and angular dependence for inelastic electron scattering from the C-O
vibrational mode were in accordance with the dipole excitation
mechanism.   The inelastic scattering cross section was then calculated assuming that the vibrational state was a localized excitation.
The theory for this scattering process has been worked out by
several authors [9,13].   The differential scattering cross section
is at not too large angles from the specular direction to a good
approximation given by [11]

$$\frac{d\sigma}{d\Omega} = n\left(\frac{me}{\pi\varepsilon_o\hbar}\right)^2 \; \mu^2 \; \left|\frac{k'}{k}\right|^2 \; \frac{1}{\cos\alpha} \; \frac{\left|\underline{k}_{||}-\underline{k}_{||}'\right|^2}{\left|\underline{k}-\underline{k}'\right|^4} \tag{1}$$

where n is the number of adsorbed molecules per unit area, μ is
the dipole moment matrix element for the vibrational excitation,
$\underline{k}$, $\underline{k}'$, $\underline{k}_{||}$ and $\underline{k}_{||}'$ are the wave vectors and their surface components
of the incident and scattered electron respectively.

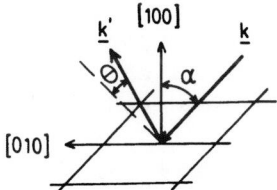

Fig. 4. : Schematic drawing of the scattering situation,
          $\underline{k}$ and $\underline{k}'$ are the wave vectors of the incident
          and scattered electron beam.

In the derivation of (1) it is assumed that the metal substrate acts
as a perfect electron reflector and that the adsorbate is close to
the image plane of a perfect conductor.   This means that only the

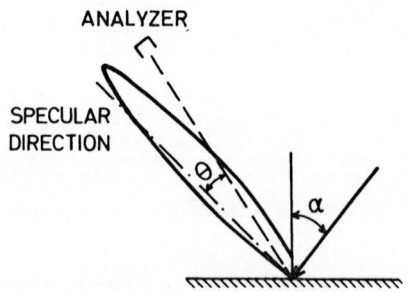

Fig. 5. : Schematic polar plot of the differential cross
         section, $\dfrac{d\sigma}{d\Omega}$, for dipole excitation of the
         C-O stretch vibration for an electron energy
         $E_o \sim 2$ eV.

component of μ along the surface normal contributes to the inelastic electron scattering.   The calculations discussed below will be
related to the vibrational polarizability α_v

$$\alpha_v = \frac{2\,\mu^2}{\hbar\omega} \tag{2}$$

where $\hbar\omega$ is the vibrational excitation energy.   The differential
cross section peaks strongly in a lobe close to the specular direction as shown schematically in Fig. 5.

In order to compare with experiments one has to integrate $d\sigma/d\Omega$ over the solid angle of collection. Fig. 6. and 7. show experimental data for the Cu(100) c(2x2)CO system concerning the energy and angular dependence of the relative loss intensity for the C-O stretching vibration. The spectrometer that was used to obtain these data is different from the one described above. The selector and analyzer have a fix orientation such that the angle between the incident and analyzed electrons is fixed at 95.4° [2]

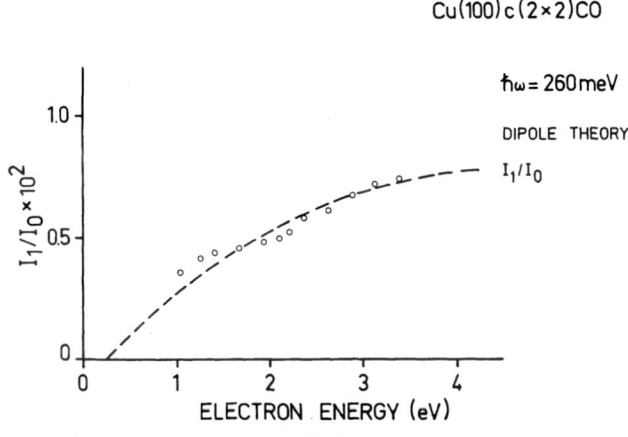

Fig. 6. : Relative intensity of the C-O stretch loss versus primary electron energy, (o) experimental data, (---) dipole theory calculation for $\alpha_V = 0.05$ Å^3. Angle of incidence $\alpha = 47.7°$: specular condition.

The analyzer angular acceptance is a cone of half angle 3°. In the angular distribution measurements the sample was rotated normal to the scattering plane. This obviously changes the angle of incidence as well as the angle of collection. The changes in angle are modest however, ($\Delta\alpha = \pm 5°$) and the reflectivity of the substrate does not change significantly over this range. The data should thus closely resemble those of an experiment with fixed crystal and movable analyzer. The calculated relative loss intensity (dashed curves in Fig. 6. and 7.) has been fitted to the experimental data by adjusting α_V in equation (2).

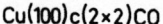

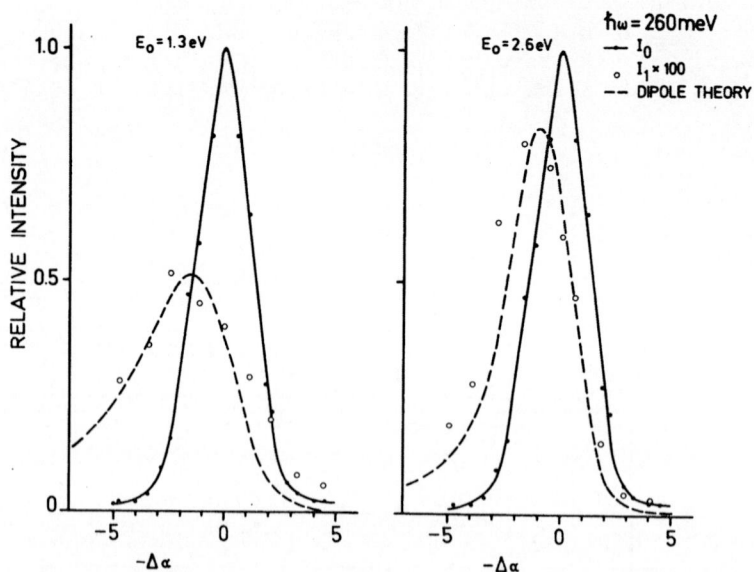

Fig. 7. : The angular dependence of the C-O stretch loss
intensity (o) and the elastic peak intensity
(Θ) versus the change ($-\Delta\alpha$) in the angle
between the analyzer and the specimen surface
normal for a) 1.3. eV and b) 2.6 eV electrons.
The dashed curves are dipole theory calculations for α_v = 0.06 $\mathring{A}^3$.

The energy dependence in Fig. 6. shows a reasonably good fit for
α_v = 0.05 $\mathring{A}$.   The angular dependence data in Fig. 7. were plotted
against $-\Delta\alpha$ i.e. the change in the angle between the analyzer and
the surface normal. The solid curves denote the elastic intensity
distribution. The inelastic intensity is normalized to the elastic at the specular condition.   The experimental data reveal that
the inelastic cross section is sharply peaked in a direction that
does not coincide with the specular as sketched in Fig. 5.   The
maximum moves closer to the specular direction and the peak becomes
narrower as the energy increases i.e. as the momentum transfer

becomes a smaller fraction of the total momentum.   These are also
the characteristic features of the dipole scattering calculations
using equation (1) and α_v = 0.06 Å^3.   It should be noted that
the α_v values found in this treatment is close to the free CO
value 0.05 Å^3.

Dipole scattering, collective excitation

        The inset in Fig. 3. shows the C-O loss peak at three diffe-
rent values of the wave vector $q_{\shortparallel}$ (in units of π/a) along the
[010] direction.

$$|q_{\shortparallel}| = |k_{\shortparallel}| - |k_{\shortparallel}'| \ = |k| \sin\alpha \ - |k'| \ \sin(\alpha-\theta)$$

The loss peak intensity decreases quickly as $q_{\shortparallel}$ is increased (i.e.
θ increases).   More importantly however, is the observation that
the C-O loss peak energy shifts by a few meV toward lower energy.
This means that the vibrating molecules are interacting and hence
the mode is collective.   This situation has recently been taken
into account in the formulation of the dipole scattering theory[4]
and is found to give rise to some significant and interesting
features.   Assuming that the vibrating molecules are interacting
and that the polarizability of the monolayer is $\alpha(q_{\shortparallel},\omega)$ one can
write the transition rate for an electron to be scattered from
state $\psi_{\underline{k}}$ to state $\psi_{\underline{k}'}$

$$w \ \sim \ \text{Im}\ \alpha(\underline{q}_{\shortparallel},\omega) \ |< \ \psi_{\underline{k}'} \ \ |\underline{E}(\underline{0})| \ \psi_{\underline{k}} >|^2 \tag{3}$$

where $<\psi_{\underline{k}'} \ |\underline{E}(\underline{0})\ \psi_{\underline{k}}>$ is the matrix element for

the electron to be scattered from $\psi_{\underline{k}}$ to $\psi_{\underline{k}'}$.   In the case studied
it turns out to be of particular interest   to investigate the
situation where the adsorbed molecules are assumed to interact
with each others via their dipole fields.   From a treatment due
to Mahan and Lucas[14] of such a system one can show that

$$\alpha(q_{\shortparallel},\omega) \ = \ \frac{\alpha_o}{1+\alpha_o U(q_{\shortparallel})} \tag{4}$$

where $u(q_{\shortparallel})$ is the spatial Fourier transform of the dipole field. The polarizability α_o of one single molecule may be written as :

$$\alpha_o = \alpha_e + \frac{\alpha_v}{1 - \frac{\omega}{\omega_o}(\frac{\omega}{\omega_o} + i\,\delta)} \tag{5}$$

where α_e and α_v are the electronic and vibrational polarizabilities respectively, δ is a small number determined by the finite lifetime of the vibrationally excited molecule.   From equation (4) and (5) assuming δ infinitesimal one gets

$$\text{Im } \alpha(q_{\shortparallel}, \omega) = \frac{\pi\,\alpha_v\,\omega_o^2}{2\omega(q_{\shortparallel})\,\left|1 + \alpha_e U(q_{\shortparallel})\right|^2}\,\delta\,(\omega(q_{\shortparallel}) - \omega) \tag{6}$$

$$\left(\frac{\omega(q_{\shortparallel})}{\omega_o}\right)^2 = 1 + \frac{\alpha_v U(q_{\shortparallel})}{1 + \alpha_e U(q_{\shortparallel})} \tag{7}$$

Equation (7) gives the normal mode dispersion relation for the dipole-dipole interacting system.   The differential inelastic scattering cross section is derived from equations (3) and (6)

$$\frac{d\sigma}{d\Omega} = n\,\frac{8m^2 e^2 \omega_o^2}{\hbar^3\,\omega(q_{\shortparallel})}\,\left|\frac{k'}{k}\right|\,\frac{1}{\cos\alpha}\,\frac{\alpha_v}{\left|1 + \alpha_e U(q_{\shortparallel})\right|^2}\,|\langle\psi_{\underline{k}}|\,\underline{E}(0)\,|\psi_{\underline{k}}\rangle|^2 \tag{8}$$

Comparing equations (1) and (8) one finds that the screenning due to the electronic polarizability of the adsorbed molecular layer will reduce $\frac{d\sigma}{d\Omega}$ by a factor $|1 + \alpha_e U(q_{\shortparallel})|^2$.   This turns out to be an appreciable correction for the Cu(100)c(2x2)CO system since typically[15] $\alpha_e \sim$ a few $\mathring{A}^3$ and U(0) $\sim 0.3\,\mathring{A}^3$.

Fig. 8. shows more detailed experimental results for the C-O loss peak intensity as a function of collection angle Θ obtained for the same experimental conditions as for the data of Fig. 1.   The loss intensity has been normalized to the elastic intensity in the specular direction.

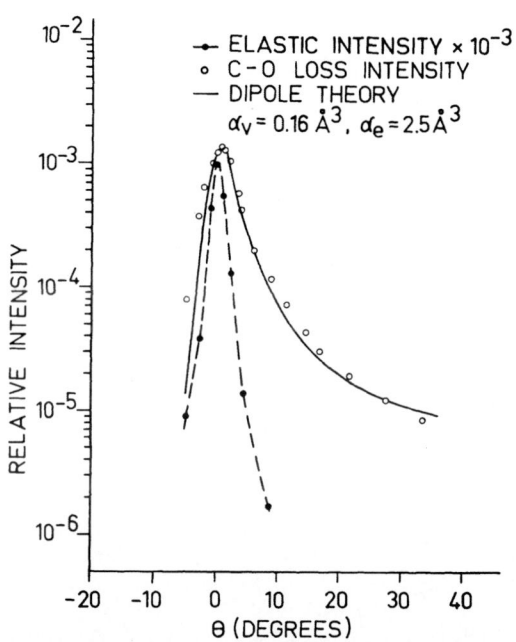

Fig 8. : Experimental C-O loss peak intensity (o) and
         elastic peak intensity (⊖) versus collection
         angle θ for the Cu(100) c(2x2)CO; energy
         E_o = 3 eV and angle α = 65°.   The solid
         curve is a dipole theory calculation.

The calculated relative loss intensity is shown as the solid curve
and was obtained by integrating $\frac{d\sigma}{d\Omega}$ from equation (8) over the
solid angle of collection.   The matrix element $\langle\psi_{k'}|\underline{E}(\underline{0})|\psi_k\rangle$

was evaluated according to a previous treatment[16] which accounts for
the Coulomb interaction between the incident electron and the charge
density it induces in the surface region of the metal. $U(q_{\shortparallel})$ was
obtained for the Cu(100) c(2x2)CO nearest neighbour distance
3.61 Å placing the molecular dipole 0.8 Å above the "image plane"
as defined in the jellium model[17] which gives e.g. $U(0) = 0.30$ Å^{-3}.
The theoretical and experimental data were matched in such a way that
$\alpha_e = 2.5$ Å^3 was found by adjusting the shape of the calculated curve
to the experimental data, $\alpha_v = 0.16$ Å^3 was then obtained by fitting
the magnitude.   Due to inaccuracies in the specimen alignment and
the electron optical adjustment of the spectrometer there is about
30 % scatter in the experimental intensities and hence in α_v.   The
result in Fig. 8 demonstrates that dipole scattering dominates
the inelastic scattering cross-section and confirms the result discussed previously.   It stresses however, the importance on including
the dielectric screening due to the electronic polarizability of the
adsorbed molecules.   Neglecting it would give $\alpha_v = 0.05$ Å^3,
i.e. close to the value found above, and a considerably poorer
fit to the experimental data.

        It is obvious from equations (7) and (8) that both the
dispersion and the cross section depend on α_v and α_e for the
dipole-dipole interaction system.   This gives a test on the validity of the model. Fig. 9 shows the experimental C-O loss energies
as a function of $q_{\shortparallel}$.
The error bars represent the scatter in the experimental data.
The need to work at sufficiently good energy resolution limits
the explored $q_{\shortparallel}$ range to about half the Brillouin zone.   The
solid curve in Fig. 9. is a dispersion relation calculated from
equation (7).   A direct comparison with the cross section data
was facilitated by adopting $\alpha_e = 2.5$ Å^3 which resulted in a
best fit between theory and experiment for $\alpha_v = 0.23$ Å^3.
This figure is consistent with $\alpha_v = 0.16$ Å^3 found above and
the C-O mode dispersion is apparently dominated by dipole-dipole
interactions.   The reliability of the experimental data was tested
for a dilute disordered CO layer ($\sim$1/4 of the c(2x2)CO coverage).
In this case the C-O loss energy was found to remain almost
constant over the range of $q_{\shortparallel}$ explored.   This rules out the
possibility that the measured loss energies could suffer from
any substantial systematic error when the analyzer is rotated.
The α_v values found is 3-4 times larger than for free CO but
are in fact more similar to those found for metal carbonyls[18]

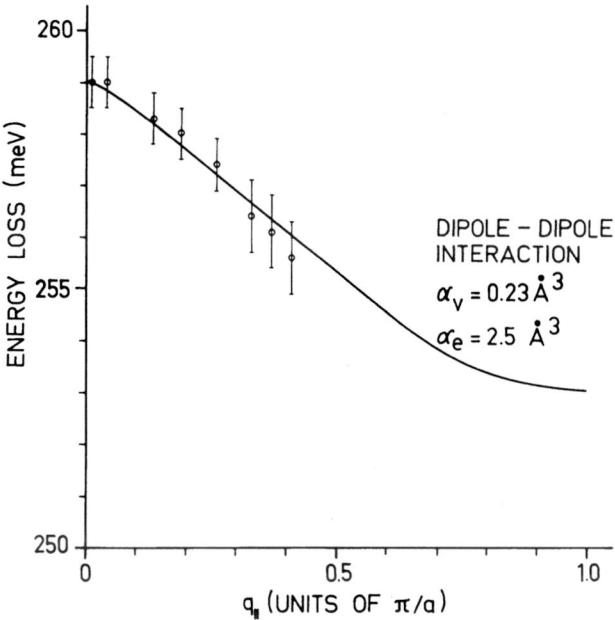

Fig. 9. : Experimental C-O stretch loss energy (o) versus
          the momentum $q_{\shortparallel}$ along the $[010]$ surface direc-
          tion (the same conditions as for Fig. 8.).
          The calculated C-O mode dispersion relation
          (solid curve) is for dipole-dipole interaction.

SURFACE PHONONS

        The c(2x2)CO spectrum shown in Fig. 3. exhibited a loss peak
at 14.5 meV which was suggested to be due to the excitation of a
substrate surface phonon.   In Fig. 10 the low-energy region of the
spectrum is shown in more detail for a slightly different scattering
situation.   The primary energy is E_o = 2 eV, the angle of incidence
is α = 50°, the angle of collection is θ = 0 i.e. specular condition.
This spectrum shows two distinct loss peaks at 14.5 meV and
42.5 meV respectively and two faint peaks at $\sim$ 30 meV and $\sim$ 57 meV.
The 42.5 meV peak is due to the excitation of the Cu-C stretching
mode.   The dispersion of this mode has not yet been investigated,
but it is likely to be rather weak.   The dipole-dipole contribution
to the dispersion will certainly be very weak since the vibrational
polarizability for this mode is more than a magnitude smaller than
that for the C-O stretching mode.

While the 42.5 meV peak grows in intensity approximately proportional to the CO coverage the 14.5 meV peak does not. It appears when the c(2x2)CO structure nucleates and grows quickly in intensity as this structure completes. The peak is apparently associated with the existence of the structure. A consequence of forming this new structure is that the surface Brillouin zone changes as sketched in Fig.11. The substrate surface Brillouin zone (solid) is replaced by the c(2x2)CO surface Brillouin zone (dashed). The $\bar{M}$ point of the substrate zone is now equivalent to the Γ point of the c(2x2) zone. This implies that modes of the substrate surface corresponding to a surface wave vector $q_{\parallel}$, of magnitude $\bar{\Gamma} - \bar{M}$, can now be excited at $q_{\parallel} = 0$, i.e. at about specular condition in the EELS experiment. For Cu(100) there exists a model calculation due to Castiel, Dobrzynski and Spanjaard[1]. If one equates the maximum phonon energy $\hbar\omega_{max}$ of their Cu(100) slab calculation to the maximum energy of bulk copper i.e. 30 meV[19] one observation strongly supports the interpretation of the 14.5 meV peak in terms of the excitation of a surface phonon at the $\bar{M}$ point[20]. The surprisingly good agreement between experiment and theory may be somewhat fortuitous since one expects the adsorbed CO molecules to cause the bare surface mode to shift. The binding energy of CO on Cu(100) is only ~ 0.6 eV [21] so that this effect may be small. The calculation is not selfconsistent what concerns changes in the surface force constants, which gives some unknown uncertainty of the bare surface phonon energy.

The angular dependence (for $E_o = 2$ eV, $\alpha = 50°$) of the 14.5 meV loss intensity peaks strongly around the specular direction which is characteristic of dipole excitation. The dipole moment presumably originates from the adsorbed CO molecules but details of the charge oscillation are yet unknown[22].

The faint peak at 57 meV is obviously due to a coupled mode of the Cu-C stretch mode and the surface phonon modes. The loss around 30 meV can either be a surface mode at the top of the substrate phonon band or an overtone of the 14.5 meV mode. The last suggestion implies an appreciable anharmonicity.

ACKNOWLEDGEMENTS

The author is grateful to his collaborators T. Gustafsson. B.N.J. Persson and E.W. Plummer who have contributed to various parts of the work presented here. Financial support from the Swedish Natural Science Research Council is gratefully acknowledged.

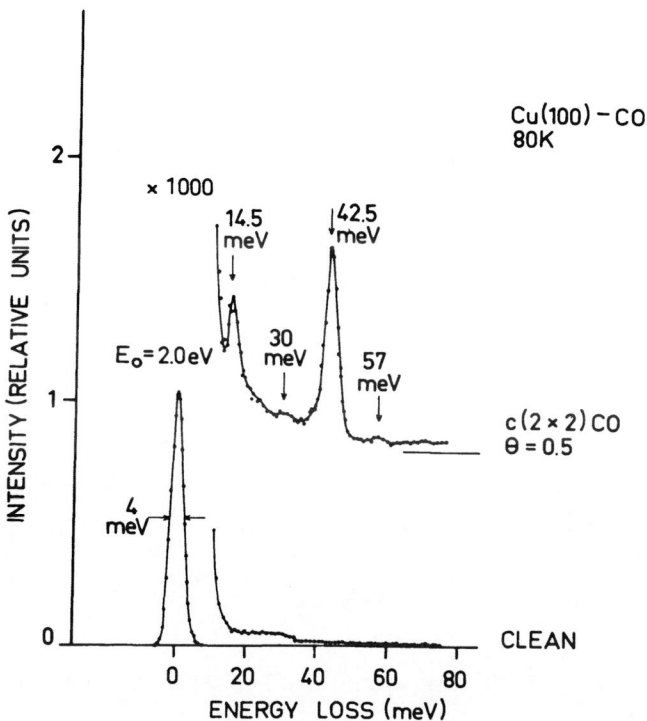

Fig. 10. :   EEL spectrum for the Cu(100) – CO system.
E_o = 2 eV,   α = 50°, specular condition.

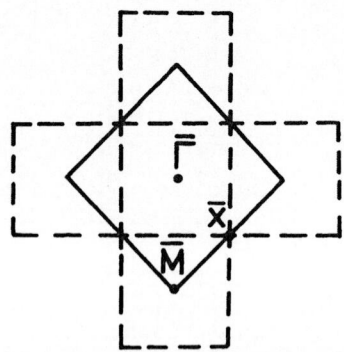

Fig. 11. : Surface Brillouin zone for Cu(100) (solid)
and repeated surface Brillouin zones for
Cu(100) c(2x2)CO (dashed).

REFERENCES

1. See papers by J.E. Black and by G. Allan & J. Lopez in this
   Proceeding.
   S.W. Musser and K.H. Rieder, Phys. Rev. B2, 3034 (1972)
   G. Armand and J.B. Theeten, Phys. Rev. B9, 3969 (1974)
   D.Castiel, L. Dorbrzynski and D. Spanjaard, Surf. Sci. 59, 252
   (1976).
2. S. Andersson, Surf. Sci. 79, 385 (1979).
3. H. Ibach and D. Bruchmann, Phys. Rev. Lett. 44, 36 (1980).
4. S. Andersson and B.N.J. Persson, Phys. Rev. Lett. (1980).
5. S Andersson and J.B. Pendry, Phys. Rev. Lett. 43, 363 (1979).
6. C.L. Allyn, T. Gustafsson and E.W. Plummer, Solid State Comm.
   24 531 (1977).
7. S. Andersson, Surf. Sci. 89, 477 (1979).

8. S. Andersson, B.N.J. Persson, T. Gustafsson and E.W. Plummer, Solid State Commun. $\underline{34}$, 473 (1980).
9. E. Evans and D.L. Mills, Phys. Rev. B$\underline{5}$, 4126 (1972).
10. D.M. Newns, Phys. Lett. 60A, 461 (1977).
11 B.N.J.Persson, Solid State Commun. $\underline{24}$, 573 (1977).
12. D. Sokcevic, Z. Lenac, R. Brako & M. Sunjic, Z. Physik B$\underline{28}$, 273 (1977).
13. F. Delanaye, A. Lucas and G.D. Mahan, Surf. Sci. $\underline{70}$, 629 (1978).
14 G.D.Mahan and A.A. Lucas, J. Chem. Phys. $\underline{68}$, 1344 (1978).
15. The average α_e for a free CO molecule is ~ 1.9 $\AA^3$.
16. B.N.J. Persson, Surf. Sci. $\underline{92}$, 265 (1980).
17. N.D. Lang and W. Kohn, Phys. Rev. B$\underline{7}$, 3541 (1973).
18. M. Birgogne, Spectrochem. Acta 32A, 673 (1976).
19. S.K. Sinha and G.L. Squires, Lattice Dynamics ed. R.F. Wallis, (Pergamon Press, Ltd, London 1965) p. 53.
    E.C.Svensson, B.N. Brockhouse and J.M. Rowe, Phys. Rev. $\underline{155}$, 619 (1971).
20. In their study of surface modes on Ni(111)(Ref. 3) Ibach and Bruchmann suggested the interpretation that surface phonons at certain points of the surface B.Z. were observed.
21. J.C. Tracy, J. Chem. Phys. $\underline{56}$, 2748 (1972)
22. Model calculations for a fcc (100) surface by R.E. Allen, G.P. Alldredge and F.W. de Wette, Phys. Rev. B$\underline{4}$, 1661 (1971) snow that the surface mode along $\bar{\Gamma} - \bar{M}$ is polarized in the sagittal plane and predominantly normal to the surface plane.

THE ADSORPTION AND CO-ADSORPTION WITH CO OF NO ON Cu(110)

STUDIED WITH EELS AND LEED *

J.F. Wendelken

Solid State Division, Oak Ridge National Laboratory

Oak Ridge, TN. 37830, U.S.A.

ABSTRACT

The adsorption of NO on Cu(110) and its interaction with co-adsorbed CO at a crystal temperature of 80 K have been investigated with EELS. Dissociation of NO occurring at higher temperatures or higher exposures has been studied with both EELS and LEED. For exposures of less than 1.2 Langmuir of ^{14}NO at a crystal temperature of 80 K, vibrational energies are initially observed at 106, 195, and 262 meV with EELS. With increasing NO coverage, the first peak reaches a maximum intensity and then diminishes while the second peak grows. However, both peaks vanish for exposures greater than 1.2 Langmuirs. If the crystal is heated above 113 K both peaks also vanish. The third peak, which represents the CO stretching vibration of less than 5 % of a monolayer of CO, shifts upward in energy with increasing NO coverage but does not shift back when the first two peaks vanish. Hence, the NO or its products do not desorb under these conditions. With higher exposures of NO followed by heating to 500 K a distinct loss peak at 49 meV associated with O_{ad} in the long bridge site is observed. No additional vibrational peak is observed for nitrogen. However, a comparison of LEED patterns produced following NO and O adsorption at 80 K followed by annealing to 500 K indicates that N_{ad} is also present. By using the isotopic energy shifts obtained by adsorbing ^{15}NO, the two low exposure peaks are assigned to the bending and stretching modes of adsorbed NO, respectively. Since the intensity of the bending mode peak is greater initially than the stretching

* Research sponsored by the Division of Materials Science, U.S. Department of Energy under contract W-7405-eng-26 with Union Carbide Corporation.

mode, it is concluded that the NO is initially adsorbed in a
strongly bent configuration. Co-adsorption with 0.25 monolayer
of CO results in a substantial suppression of the bending mode and
energy shifts in both the CO and NO stretching vibrations.

INTRODUCTION

The adsorption of NO and other oxides of nitrogen on metal
surfaces is of interest for the wide variety of catalytic reactions
which are known to occur and also for the variety of adsorption
geometries which have been observed with such systems. For
example, NO has been found to adsorb at different sites on Ru(100)[1]
as a function of coverage, it has been found to adsorb in both
linear and bent forms on Ni[2,3] and Pt(100)[4,5], it has been found in
both dissociated and undissociated states on Ni(111)[3,6] and Ni(100)[7]
and it has been found to form new compounds on Cu(100)[8] and Pt(111)[9].
These systems show more complexity than the much studied case of
CO as a result of an unpaired electron in the antibonding $2\pi^*$
orbital of NO. This results in a decreased binding energy for NO
compared to CO[10] and the ability of NO to either donate or accept
an electron in forming a chemical bond[11].

In this paper the adsorption of NO on both clean Cu(110) and
Cu(110) which has been pre-exposed to CO is examined using both
LEED and EELS. The results indicate molecular adsorption on the
clean surface only at low exposures and for temperatures below
113 K. The molecular vibrations observed under these conditions
indicate a strongly bent molecule. At higher exposures, or when
the crystal is heated above 113 K, dissociation is observed with
O_{ad} and probably N_{ad} remaining in the long bridge site.

EXPERIMENTAL APPARATUS AND PROCEDURES

All measurements were made in a single stainless steel UHV
system which contains facilities for high resolution electron
energy loss spectroscopy (EELS), LEED, AES, and quadrupole mass
spectroscopy (QMS). The high resolution spectrometer has been
described before[12] and was operated with an energy resolution of
20 meV for these measurements. The LEED/Auger system is mounted
on a bellows so as to be retractable and possesses a glass LEED
screen which is viewed from behind[13].

The Cu(110) crystal was polished and oriented to within 0.1°
as described previously[14]. Ion bombardment produced a clean surface
as judged by AES and EELS measurements and annealing at 675 K
produced a well ordered surface as judged from the LEED patterns.
NO exposures were made by filling the vacuum chamber with high
purity NO through a leak valve from a metal cylinder in the case

of ^{14}NO and from a glass flask in the case of ^{15}NO. Several NO exposures followed by target cleaning cycles were made for each species of NO to reduce the contamination which results from exchange reactions with the walls. The system was also continuously pumped through a throttle valve during exposures.

EXPERIMENTAL RESULTS AND DISCUSSION

The adsorption of ^{14}NO and ^{15}NO on Cu(110) was examined as a function of exposure and temperature by means of LEED and EELS. As a check on the influence of CO impurities on the observations, the adsorption of NO on Cu(110) with pre-adsorbed CO was also examined.

The LEED observations for NO adsorption on Cu(110) are summarized in Figure 1. When the crystal is exposed to NO at a temperature of 80 K, the only effect is to change the relative intensities, hence the I-V character at the beams. Figure 1(a) shows the 200 eV incident beam LEED pattern obtained after an exposure of 20 L (1 Langmuir = 10^{-6} Torr.s = 1.33 x 10^{-4} Pa.s) at a temperature of 80 K. The pattern observed is identical to that observed after an equivalent oxygen exposure[15] in which the (a,b) beams have near zero intensity and are missing at this energy when a+b = 2n+1. Figure 1(b) shows the (1x1) pattern obtained at this energy after exposure of the clean surface at 225°C to 20 L of NO. The pattern is identical to that obtained for the clean surface with no indication of adsorption at this temperature. If the surface represented by figure 1(a) is heated to 500 K for 5 minutes, the pattern shown in figure 1(c) develops. This pattern is a combination of c(6x2) and (2x1) patterns. With more extended annealing times such as the 30 minutes used for figure 1(d) the pattern shifts to a dominantly (2x1) character, but remnants of the c(6x2) structure are still observed.

It is of interest to compare these LEED results with a study of oxygen adsorption on the same Cu(110) crystal[15]. In this study a 20 L exposure to O_2 at a temperature of 80 K produced the same modified (1x1) LEED pattern as the NO exposure. However, subsequent heating to 400 K produced a sharp c(6x2) LEED pattern and further heating to 500 K produced a sharp (2x1) pattern with no trace of the c(6x2) pattern. EELS measurements indicate that atomic oxygen is responsible for the LEED patterns in the oxygen adsorption case[15]. In the case of NO adsorption, the reluctance of the system to shift from the c(6x2) structure to the (2x1) structure indicates that more than atomic oxygen is present on the surface.

Vibrational spectra obtained with EELS for both ^{14}NO and ^{15}NO on Cu(110) as a function of dosage are shown in Figures 2 and 3 for a crystal temperature of 80 K. Two energy loss peaks are seen

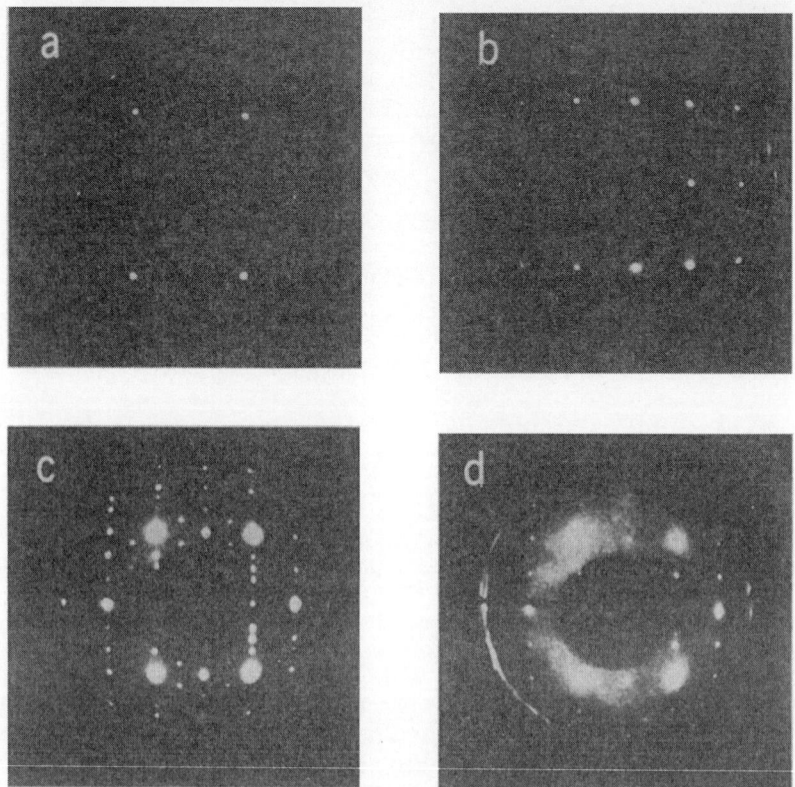

Figure 1 - LEED patterns obtained after exposure of Cu(110) surface
            to a) 20 L of NO at a crystal temperature of T = 80 K,
            b) 20 L of NO at T = 500 K, c) 20 L of NO at T = 80 K
            followed by heating to 500 K for 5 minutes, and d) as
            in (c) but after 30 minutes additional heating at 500 K.

which are associated with NO. For an exposure of 0.5 L the losses
are at 106 meV and 195 meV in the case of ^{14}NO. At 1.2 L, the
losses shift slightly to 104 and 194 meV. Nearly identical results
were obtained for adsorption of ^{15}NO with the first loss peak
shifted downward by 2 meV and the second loss peak by 4 meV for
all exposures. A third loss peak which appears, is the result of
adsorption of CO which results from exchange reactions with the
vacuum chamber walls during NO exposure. Based on an earlier
study of CO on Cu(110)[14] it is estimated that this represents less
than .05 monolayer. As discussed later, pre-adsorption of large
amounts of CO does influence the energies and intensities in the
NO vibrational spectrum, however, at 0.05 monolayer this is not
considered a serious problem. An upward shift in the CO vibrational

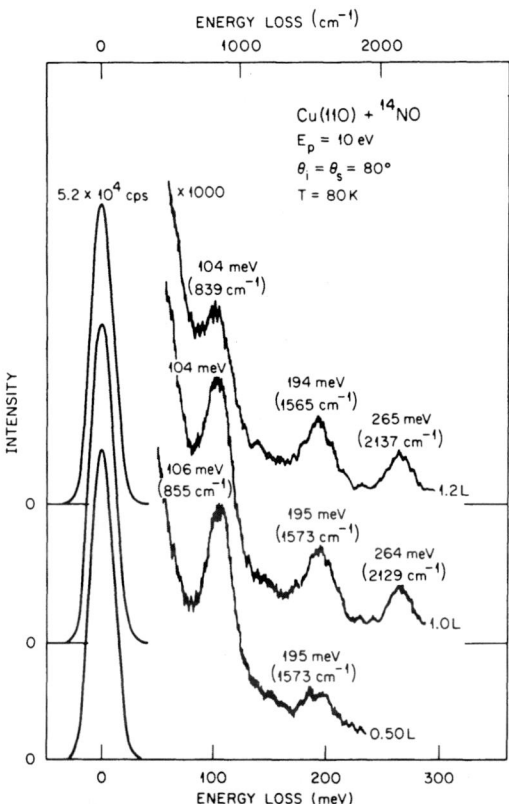

Figure 2 - EEL spectra of the Cu(110) surface at 80 K after several exposures to ^{14}NO.

energy from the 260 meV observed for CO alone[14] is in fact useful to this study as discussed below.

The behaviour of the relative intensities of the loss peaks as a function of coverage is of particular interest. The low energy peak develops first at low coverage. Its maximum intensity occurs at about 0.5 Langmuir and then its intensity decreases. The second NO peak develops more slowly with a maximum at 1.2 Langmuir . For higher exposures, both loss peaks vanish. However, the CO loss peak does not vanish or shift down in energy at this time, so it must be concluded that the NO does not simply vanish from the surface. With an exposure of 20 L as used for the LEED patterns of Figure 1, a shoulder in the loss profile is observed

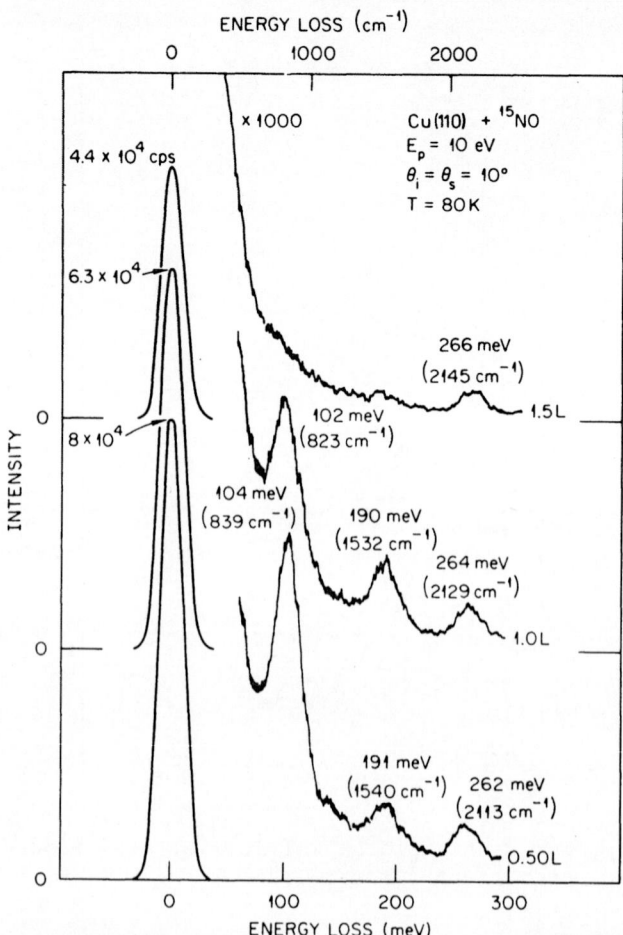

Figure 3 - EEL spectra of the Cu(110) surface at 80 K after
several exposures to [15]NO.

at about 50 meV as shown in Figure 4. Heating to 500 K to produce
the mixed c(6x2) and (2x1) LEED pattern results in a loss peak at
49 meV which is the energy observed for atomic oxygen[15]. With the
much lower exposures which produced the higher energy loss peaks,
no significant effects were observed in the LEED patterns.

By observing the loss profiles as a function of temperature
with an initial exposure of NO of less than 1.0 Langmuir, it was
found as shown in Figure 5 that the NO loss peaks diminish as the

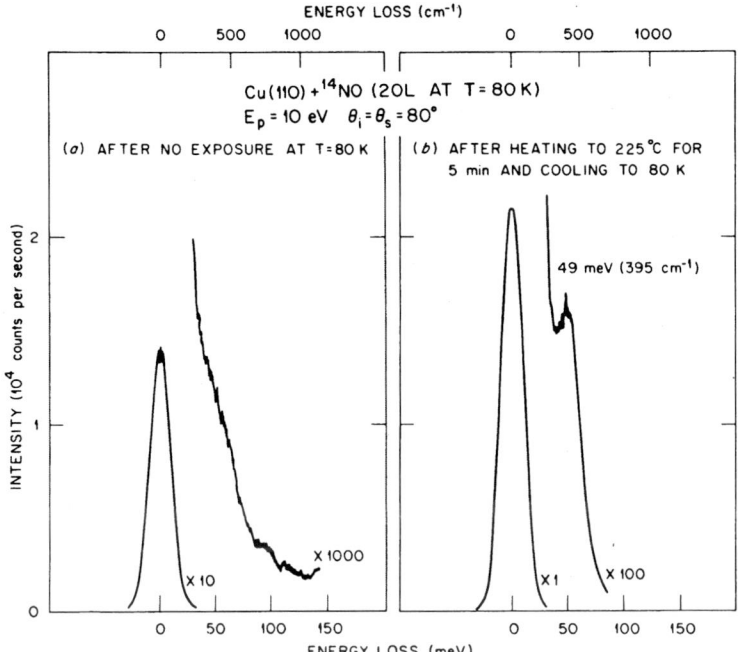

Figure 4 – a) EEL spectra of the Cu(110) after exposure to 20 L
[14]NO at a crystal temperature of 80 K and b) after
heating to 225°C for 5 minutes.

temperature is increased and vanish above the temperature of 113 K.
However, the residual CO loss peak is again unaffected by this
vanishing of the NO peaks.

From the EELS results it is then apparent that the NO adsorbs in a molecular form (not necessarily NO) on Cu(110) for
exposures of less than 1.2 L and for temperatures less than 113 K.
When these conditions are exceeded, it is also apparent from EELS
that atomic oxygen remains on the surface and is located at the
long bridge site as previously observed[15]. The LEED patterns
indicate that not only oxygen is on the surface, so it must be
presumed that atomic nitrogen is also present. The similarity of
the LEED patterns to the oxygen adsorption case suggests that N_{ad}
is in the same site as O_{ad}. The vibrational energy of N_{ad} should
be close to that of O_{ad} and may not have been resolved. A recent
EELS study of NO on Ni(111)[3] found that when the NO dissociated,
both atomic species remain and in equivalent sites. However, the
vibrational loss peak for the O_{ad} dominates the nearby N_{ad} loss

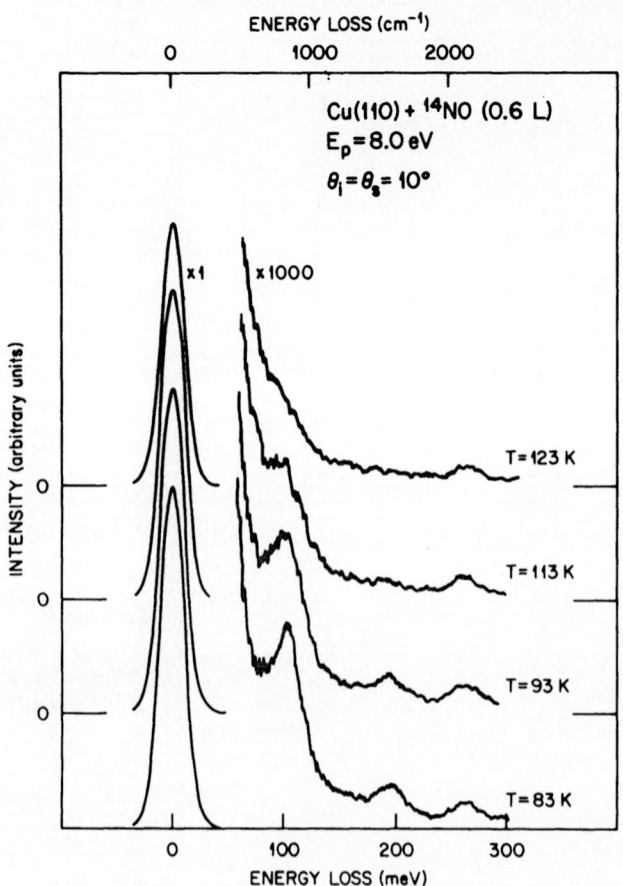

Figure 5 – EEL spectra of the Cu(110) surface as a function of
temperature after exposure to 0.6 L ^{14}NO.

peak which therefore cannot be observed.

    The nature of the molecular adsorption species is less apparent.
In order to interpret the vibrational energies at low coverage and
temperature, it is necessary to consider possibilities beyond
simply NO because of the wide range of behaviour which has been
found for NO on other surfaces. A particularly interesting possibility was given by a recent XPS study of NO on Cu(100)[8] in which
it was concluded that N_2O is formed which desorbs at 110 K leaving

behind only oxygen. However, there are several arguments against
this possibility on Cu(110). First, the gas phase vibrational
energies are 276 meV (N-N stretch), 159 meV (N-O stretch) and 73 meV
(bending)[16]. The N-N stretching mode might be unobservable if the
N-N axis is parallel to the surface as a result of the dipole
selection rule[17]. But, this would require the N-O stretching mode
to shift upward in energy upon adsorption which is not physically
reasonable. Alternatively, all the vibrational modes must undergo
a severe downward energy shift upon adsorption leaving the bending
mode unobservable due to instrumental resolution. The second
argument involves the observed isotopic energy shifts at 4 meV for
the high energy mode and 2 meV for the low energy mode. With gas
phase N_2O, the shifts are 2.5 meV (N-O stretch), 2.1 meV (N-O bend)
and 8.5 meV (N-N stretch)[16]. The ratios of the energy shifts do
not match the present observations for either pair of possibly
observed vibrational modes. Finally, the LEED patterns as already
discussed indicate that more than oxygen remains after heating the
crystal and no N_2O desorption is observed with the quadrupole
spectrometer.

Another possible reaction product is NO_2. This was checked by
adsorption of NO_2 under the same conditions as for the NO studies.
The same loss energies were observed, however, the intensities
were generally weaker. From this it is concluded that NO_2 reacts
to form the same species as results from NO adsorption, but with
less efficiency. Again, the gas phase vibrational energies of
163 meV (symmetric stretch), 93 meV (bending) and 201 meV (asymmetric stretch) with their respective isotopic energy shifts of
1.5 meV, 1.3 meV and 4.7 meV[18] are not consistent with the two
observed vibrational energies. This and more complicated molecular
forms can be ruled out on the basis that there are not enough
observed loss peaks to justify a more complicated structure.

The explanation of the vibrational spectra which appears most
likely is that the NO is adsorbed in a bent configuration. In an
IR study published concerning nitrosyl platinum complexes[19] it was
found that the NO stretching and bending modes at 212 meV and
64 meV shifted down in energy by 3.75 meV and 1.3 meV respectively
when the isotope was changed from ^{14}N to ^{15}N. This corresponds well
to the presently observed shifts of 4 meV and 2 meV. In the above
study a NO-Pt vibration was observed at 36 meV. A loss in this
range might not be resolved with the spectrometer employed in this
experiment. If, then, the vibrational mode near 194 meV is assigned
to be the NO linear stretching mode and the mode near 104 meV is
assigned to be the bending mode, the observed energies and relative
intensities of the two losses can be understood if the molecule is
strongly bent, i.e. the molecular axis has a large component
parallel to the surface. The inelastic intensity is peaked in the
same directions as the elastic intensity, hence the energy loss
process involves dipole scattering in this case and only vibrational

components perpendicular to the surface should be observed[17].
Hence the stretching mode of a strongly bent NO molecule would
appear weak and the bending mode would be strong in the energy
loss spectrum.

In comparing the intensities of two widely separated vibrational energies, it is necessary to consider the relative sensitivity to the two different energies. From the work of Evans and
Mills [20], it may be concluded in the present study that the
measurements are three times more sensitive to the bending mode
than the stretching mode. Even so, the bending mode is seen to have
a large dipole moment perpendicular to the surface. For comparison,
in the EELS study of NO on Ni(111)[3], a bending mode for a bent
NO molecule was observed with an intensity of less than 0.10 of
the stretching mode intensity. It is risky to try to determine a
precise angle for the bending since the relative magnitudes of the
two dipole moments is not known and can be influenced by several
interrelated factors such as site location, orientation, and
coverage.

The low energy of the stretching mode compared to the gas
phase value of 233 meV is an indication either of a multiply coordinated site or a severe distortion of the molecule by strong
bonding forces. The latter is certain due to the observed dissociation at very low temperatures and the former is also likely.
Reference must be made here to the previous study of oxygen
adsorption on Cu(110)[15] in which it was found that atomic oxygen
showed a strong preference for one site independent of coverage,
temperature, or observed LEED pattern. It is probable that the NO
which is on the verge of dissociation would also be drawn to the
same multiple bonding site, i.e. the long bridge position and in
the surface layer rather than above.

The relatively high bending mode energy as compared with
the nitrosyl platinum bending mode at 64 meV[19] and the NO on
Ni(111) bending mode energy of 92 meV[3] is again an indication of
strongly bent orientation. The atom which is "wagging" above the
surface is also likely to be drawn into an adjacent long bridge
position. Since the distance between adjacent long bridge positions
is 2.56 Å and the NO gas phase separation distance is 1.15 Å, the
resulting forces are dissociative. The "wagging" atom of the NO
molecule is not specified here because that is not known from
these measurements. Generally, NO is thought to adsorb with the N
bonding to the metal in analogy with CO. However, calculations in
the case of the nitrosyl platinum complexes[19] indicate very little
difference in observed vibrational energies for the two possible
orientations.

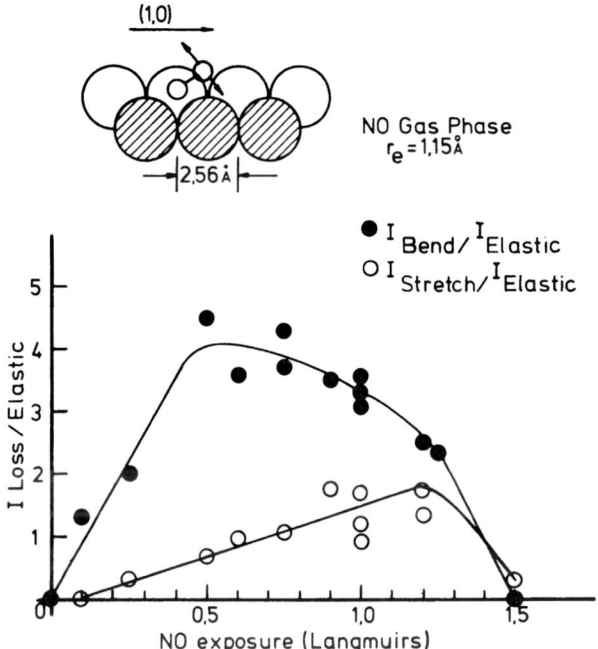

Figure 6 – Plot of ratio $I_{Loss}/I_{Elastic}$ for the two observed vibrational modes associated with NO adsorbed on Cu(110) at T = 80 K. A schematic of the suggested site and orientation for molecular NO adsorption is shown.

In Figure 6 the suggested bonding location and orientation is illustrated along with a plot of the intensities of the two vibrational modes. It might be inferred from the intensity behaviour that the NO molecules are straightening up before dissociating, however, this cannot be correct. If the molecule were to straighten up, the distortion of the N-O bond would be reduced as also the probability of dissociation. This effect has been observed in the case of NO on Ni(111) in which increased coverage results both in a more upright orientation for the molecule and a higher temperature required for dissociation[3]. Further, the bending mode should decrease in energy while the stretching mode should increase. The stretching mode for NO on Ni(111) does in fact increase[3] as coverage increases and the molecules straighten up, while the bending mode remains fixed in energy. These energy shifts could, of course, be offset by other coverage dependent intermolecular interactions with their attendant energy shifts. Coverage dependent intermolecular interactions might also have the effect of causing shifts in the dipole moments which are directly related to the observed intensities. Also, as dissociation occurs and the atomic species are bound in the long bridge sites the bending mode amplitudes of newly adsorbed NO in the remaining empty sites may be reduced. Thus it could be that some more upright molecules are adsorbed at higher exposures. What is clear is that at exposures above 1.2 L, the dissociation process becomes very rapid and goes to completion over a very narrow exposure range. As the adsorption sites become filled with atomic species, further molecular adsorption is prohibited. Molecules which may have been adsorbed in a more upright orientation may also be displaced at this time by more strongly bound atomic species. A better understanding of those possibilities will require measurements with higher resolution particularly to see if the vibrational modes might be split due to the adsorption of NO with two different orientations. If not, then there must be a change in dipole moments vs. coverage.

The effect of co-adsorption of NO with CO is shown in Figure 7. Here EEL spectra are shown for 0.5 L of CO adsorption which corresponds to 0.25 monolayer[14] followed by 0.5 and 1.0 L of ^{14}NO. The obvious effect is the suppression of the bending mode loss. At 0.5 L of NO the NO vibrational energies are shifted from 106 and 195 meV to 103 and 192 meV. The effect of the CO contamination observed in Figures 2, 3 and 5 may then be estimated as producing energy shifts only 20 % as great as here assuming the effects are linear with coverage. From the suppression of the bending mode intensity it also appears that the NO molecule must stand more upright, but this conclusion is not certain since the influence of the CO on the bending mode dipole moment is not known.

CONCLUSIONS

Nitric oxide has been found to adsorb initially in molecular form on Cu(110) at temperatures below 113 K. It is adsorbed in a strongly bent or tilting configuration which represents a pre-dissociation state. The site of the molecule is suggested as the long bridge site. For exposures above 1.2 L at 80 K or at temperatures above 113 K, dissociation is complete. Oxygen atoms and probably nitrogen atoms are then located at the long bridge site in, rather than above, the surface layer. Co-adsorption with carbon monoxide results in shifts in energy of both the NO and CO vibrations and may result in a more upright orientation for the NO.

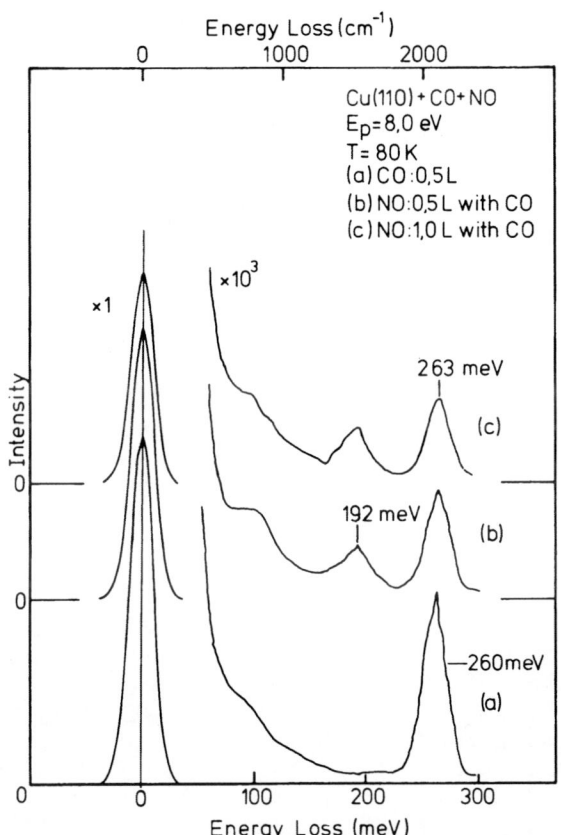

Figure 7 - EEL spectra of the Cu(110) surface at 80 K after
exposure to 0.5 L CO followed by exposures to 0.5 L and
1.0 L of ¹⁴NO.

ACKNOWLEDGEMENTS

Discussions with Drs. H. Ibach, S. Lehwald and J. Yates concerning this work and their recent data for NO on Ni(111) are gratefully acknowledged. The author thanks also the staff of the IGV der KFA Jülich, Jülich, FRG for assistance with preparation of this manuscript.

REFERENCES

1. G.E. Thomas, W.H. Weinberg, Phys. Rev. Letters 41, 1181 (1978).
2. A.F. Carley, S.Rassias, M.W. Roberts, Wang Tang-Han, Surf. Sci. 84, L227 (1979).
3. S. Lehwald, J.T. Yates,Jr., H. Ibach, Proceedings of the International Conference on Solid Surfaces and Third European Conference on Surface Science, Cannes, France September 22-26, 1980, to be published.
4. H.P. Bonzel, G. Pirug, Surf. Sci. 67, 45 (1977).
5. G. Pirug, H.P. Bonzel, H. Hopster, H. Ibach, J. Chem. Phys. 71, 593 (1979).
6. H. Conrad, G. Ertl, J. Küppers, E.E. Latta, Surface Sci. 56, 296 (1975).
7. J.F. Wendelken, S. Lehwald, work in progress.
8. O.W. Johnson, M.H. Matloob, M. Wyn Roberts, J.C.S. Chem. Comm. 40 (1978).
9. H. Ibach, S. Lehwald, Surface Sci. 76, 1 (1978).
10. Handbook of Chemistry and Physics, 46th ed. (Chemical Rubber, Cleveland, (1965)).
11. R. Eisenberg, C.D. Meyer, Acc. Chem. Res. 8, 26 (1975).
12. J.F. Wendelken, F.M. Propst, Rev. Sci. Instrum. 47, 1069 (1976).
13. J.F. Wendelken, S.P. Withrow, P.S. Herrell, Rev. Sci. Instrum. 51, 255 (1980).
14. J.F. Wendelken and M.V.K. Ulehla, J. Vac. Sci. Technol. 16, 441 (1979).
15. J.F. Wendelken, to be published.
16. T. Shimanouchi , Tables of Molecular Vibrational Frequencies, Consolidated Volume I, Nat. Stand. Ref. Data Ser., Nat. Bur. Stand. (U.S.) 39, 1972.
17. H. Froitzheim in Electron Spectroscopy for Surface Analysis, ed. by H. Ibach (Springer-Verlag, New York, 1977).
18. T. Shimanouchi, Tables of Molecular Vibrational Frequencies, Consolidated Volume II, J. Phys. Chem. Ref. Data 6, 993 (1977).
19. E. Miki, K. Mizumachi, T. Ishimori, H. Okuno, Bull. Chem. Soc. Japan 47, 656 (1974).
20. E. Evans, D.L. Mills, Phys. Rev. B5, 4126 (1972).

THE ADSORPTION OF WATER ON Ru(001) : CHEMISORPTION AND HYDROGEN

BONDING *

P.A. Thiel **, F.M. Hoffmann and W.H. Weinberg †

Division of Chemistry and Chemical Engineering
California Institute of Technology
Pasadena, California 91125

ABSTRACT

The adsorption of H_2O on Ru(001) at temperatures of 95 K and
165 K has been investigated using electron energy loss spectroscopy,
low-energy electron diffraction, and ultraviolet photoelectron
spectroscopy. The vibrational data, together with the structural
information available, supports a model in which aggregates of
water molecules form via population of a first (chemisorbed) layer
and subsequent hydrogen bonded layers. At the higher temperature
of adsorption, 165 K, major differences in the vibrational spectra
are apparent. Possible reasons for these differences are discussed.

We have investigated the adsorption of water on the (001) surface of ruthenium using a combination of techniques : electron
energy loss spectroscopy (EELS), low energy electron diffraction
(LEED) and ultra-violet photoemission spectroscopy (UPS). In
addition, we have utilized data previously reported by Madey and
Yates[1] concerning electron-stimulated desorption-ion angular distribution (ESDIAD) patterns of H_2O adsorbed on Ru(001) at 90 K. This
has provided a uniquely detailed description of the properties of
this adsorption system, in which strong intermolecular interactions
(i.e., hydrogen bonds) are reflected in the vibrational and geometric properties of the adsorbate. Previous investigations of the

* Supported by the National Science Foundation under Grant
  n° CHE77-16314.
** IBM Predoctoral Fellow and American Vacuum Society Predoctoral
   Scholar.
† Camille and Henry Dreyfus Foundation Teacher-Scholar.

vibrational properties of water have dealt mainly with dissociative adsorption[2,3]. A study of molecular H_2O adsorbed on Pt(100)[4] concluded that small clusters of H_2O form at submonolayer coverages at 150 K, on the basis of coverage-dependent features in the vibrational spectra. We conclude also that aggregates of H_2O molecules form at low coverage, and we propose a model for their structure based upon the more complete set of information available for this system.

The isolated water molecule is characterized by C_{2v} symmetry. There are three fundamental internal vibrations : the symmetric O-H stretch, the asymmetric O-H stretch and the deformation, or scissoring (ν_s) mode. Adsorption on a metal surface fixes the molecule such that the gas-phase translational and rotational modes are "frustrated" and may also be observed in vibrational spectra. In the infrared spectra of isolated water molecules, crystalline ice, and aquo-metal complexes, the region of the vibrational spectrum between 3200 and 3750 cm^{-1} is associated with the symmetric and asymmetric O-H stretches, and the region between 1570 and 1650 cm^{-1} corresponds to ν_s[5-6]. Absorption in the region 300 to 900 cm^{-1} has been assigned to frustrated rotations, known as "librations" and frustrated translations[5-8]. In our experiments, electron energy loss spectra show that the vibrational modes of water adsorbed on Ru(001) occur in similar energy ranges. Figure 1 shows loss spectra taken following exposure of the Ru surface to various amounts of H_2O at 95 K. Note that the clean surface spectrum is featureless, indicating the absence of contamination. At very low exposures of H_2O, such as 0.03 L, the data indicate that the spectra are particularly sensitive to the "chemisorbed" H_2O molecules, i.e., those directly in contact with the metal surface. Experiments to be discussed later support the assignment of the low-energy feature at 370 cm^{-1} as due to two overlapping modes : a metal-oxygen stretch at 370 to 400 cm^{-1} and a librational mode. A $Ru-OH_2$ vibration at 370 to 400 cm^{-1} is consistent with bonding through the oxygen atom to a single metal atom, by analogy with the infrared spectra of aquometal complexes. In such complexes, this vibration is observed between 310 and 440 cm^{-1} [7-8]. Molecular adsorption is indicated bu observation of the deformation mode at 1510 cm^{-1}. This feature is clearly visible in Fig. 2, where the spectral region above 1000 cm^{-1} is shown on an expanded vertical scale.

At slightly higher coverages of H_2O, there is distinct evidence for strong attractive interactions between the water molecules which lead to aggregation at even submonolayer coverages. One indication of this is the position of the broad and weak O-H stretch at 3320 to 3380 cm^{-1}, which is characteristic of intermolecular hydrogen bonding (cfr. Fig. 2). Non-hydrogen bonded OH groups would be expected to occur at 3750 to 3630 cm^{-1}, from gas-phase and matrix isolation data [5,6,9].

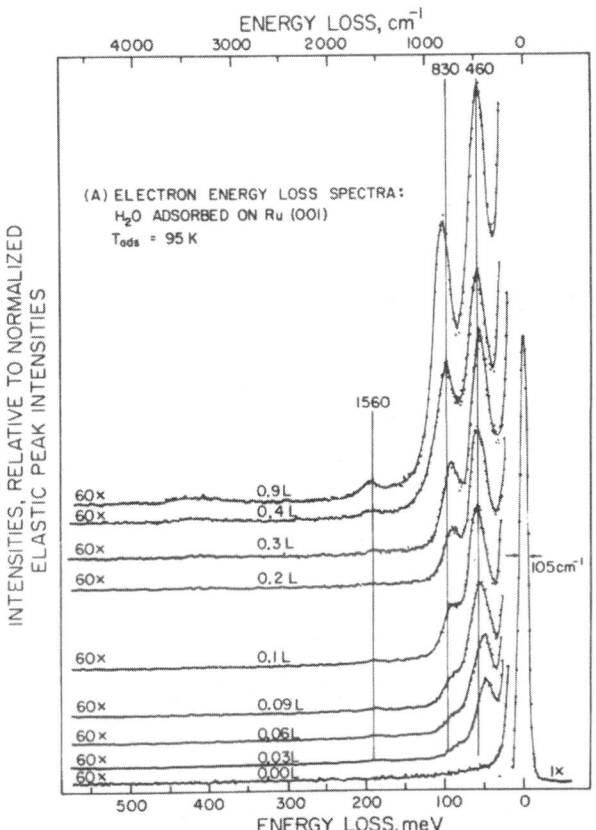

Fig. 1 – Electron energy loss spectra of H_2O adsorbed on Ru(001)
at 95 K. The experimental apparatus is described in 17.
The incident beam energy is 4 eV, and it is 57° from the
surface normal.

The second indication that clusters of water molecules are
forming at submonolayer coverage comes from an examination of the
intense vibrational features between 370 and 900 cm^{-1}. They are
assigned as frustrated rotational and translational modes on the
basis of the following arguments :

1. The frequencies of these modes occur in a range appropriate to
   their assignment as such[5-8]; and
2. The shifts in frequency observed for D_2O relative to H_2O are
   consistent with the assignment of these modes as frustrated
   rotations based upon the ratios of the moments of inertia of
   the isolated H_2O and D_2O molecules, 1.41 to 1.34. The average
   experimental values of the frequency shifts are 1.40 and 1.37

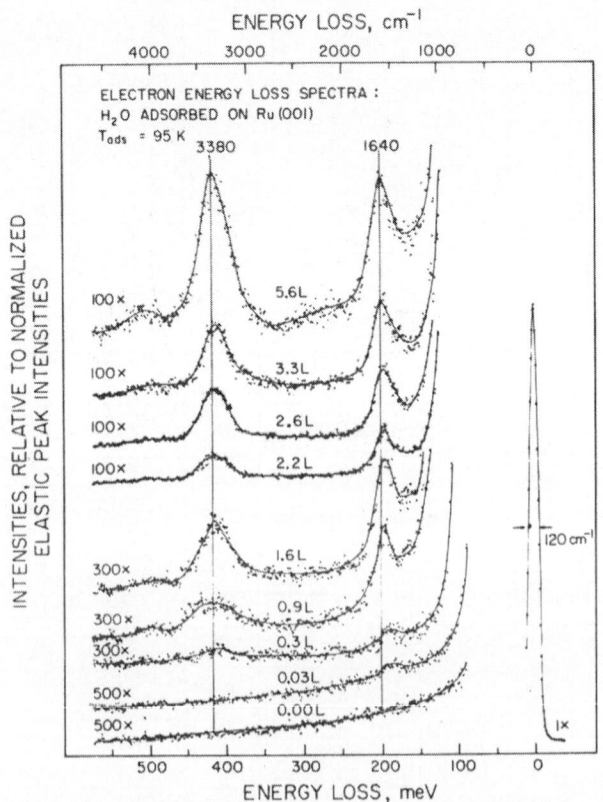

Fig. 2 - Electron energy loss spectra for various exposures of
         H_2O on Ru(001) at 95 K, showing the spectral range
         between 1000 and 4800 cm^{-1} on an expanded vertical scale,
         relative to Fig. 1.

for the modes which occur below 1000 cm^{-1} in Fig. 1, except
at exposures less than 0.08 L. In the latter case deuteration
causes the frequency of the single low energy feature to shift
by an anomalous factor of 1.22. This observation supports the
earlier argument that the low energy mode at very low exposures
is caused by two overlapping modes, one of which (the Ru-OH$_2$
stretch) shifts negligibly upon deuteration.

An examination of the scattering mechanism which governs these
low-frequency modes is further enlightening. For a molecule adsorbed in a site such that local C_{2v} symmetry is preserved, none
of the librational modes are dipole allowed. However, they are the

most intense features observed in the specular scattering experiments of Figure 1. One possible explanation, which we have investigated, is that the scattering mechanism for these modes is non-
dipolar. In Fig. 3 are shown angle-resolved EEL spectra, taken by
moving the energy analyzer from the specular direction along the
polar angle, θ_s. It is assumed that non-dipolar scattering is distinguished in these experiments by a relatively isotropic distribution of scattered electrons away from the specular direction[10].
These data show that all of the modes, except the O-H stretch and
ν_s, decrease in intensity rapidly away from the specular direction,
as does also the elastic peak. This has been illustrated in Fig. 4
by plotting the intensity of the inelastically scattered electrons
as a function of scattering angle. This indicates that the
librational modes are dipole-enhanced. It is interesting to note
that the lowest frequency loss feature changes from 440 to 520 cm^{-1}
as $\Delta\theta_s$ changes from zero; again, this is consistent with the
previous assignment of two overlapping features at this energy,
which may decrease at different relative rates with respect to
$\Delta\theta_s$ and give rise thereby to the apparent change in freqeuncy. The
scattering mechanism is discussed in detail elsewhere[11].

Another possible explanation for the observation of the
librational modes is that the C_{2v} symmetry of the gas phase molecule is reduced to C_s or C_1 as a result of adsorption. Under the
latter two symmetry groups, one or more of the librational modes,
respectively, could become dipole-active. A model in which clusters
of H_2O molecules form, wherein the molecules occupy a distribution
of orientations relative to the surface and are coupled via hydrogen
bonds, provides an adequate and consistent explanation of the observed dipole enhancement of the librational modes.

The third piece of evidence which supports the formation of
clusters comes from a combination of LEED and ESDIAD[1] data, as
shown in Fig. 5. At exposures greater than 0.5 L, we observe a
faint and diffuse ($\sqrt{3}$ x $\sqrt{3}$)R30° LEED pattern, arising from an ordered array if oxygen atoms with respect to the ruthenium substrate,
as shown in Frame (C) of Fig. 5. (Scattering from the hydrogen
atoms is assumed to be insignificant in forming this pattern). This
pattern has been observed also following adsorption of H_2O at 95 K
on the (111) surfaces of Rh[12], Ag[13] and Pt[13]. Further, at submonolayer coverages of water, an hexagonal pattern of H$^+$ emission
has been observed to result from electron bombardment[1]. Under the
assumptions that :

1. the pattern of H$^+$ emission reflects the initial orientation of
   the molecular O-H bond; and
2. the C_2 axis of the gas phase molecule is perpendicular to the
   metal surface in the chemisorbed state, the hydrogen atoms must
   be positioned with respect to the oxygen atoms of the ($\sqrt{3}$ x $\sqrt{3}$)
   R30° lattice as shown in Fig. 5(C).

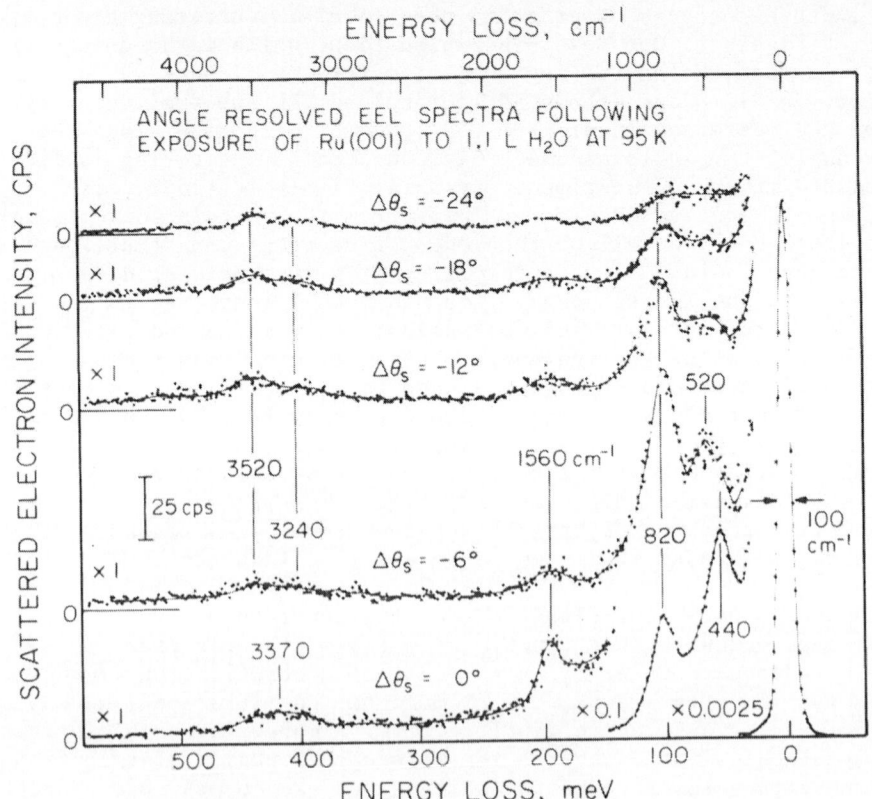

Fig. 3 - Angle-resolved EEL spectra of the Ru(001) surface following exposure to 1.1 L H_2O at 95 K. $\Delta\theta_s$ is the deviation
of the analyzer position (polar angle) from the specular
direction.

An intuitively appealing explanation of the physical forces which
create this combination of orientations is available if a second-
layer H_2O molecule is positioned within the $(\sqrt{3} \times \sqrt{3})R30°$ unit cell,
hydrogen-bonded to at least two of the first-layer H_2O molecules.
Such a structure is suggested from an examination of the crystal
structures of ices Ic and Ih[5,6,13]. Two possible structures of this
type are shown in Fig. 6. Thus, the combination of LEED and ESDIAD
data further support the hypothesis that aggregates of hydrogen-
bonded H_2O molecules form at submonolayer coverages.

Heating a surface covered with H_2O causes irreversible ther-
mally-induced changes which do not simply reflect a change in
coverage, because the vibrational spectra do not merely revert to
lower-coverage forms. This phenomenon was investigated therefore
by adsorbing H_2O at a slightly higher temperature of adsorption,

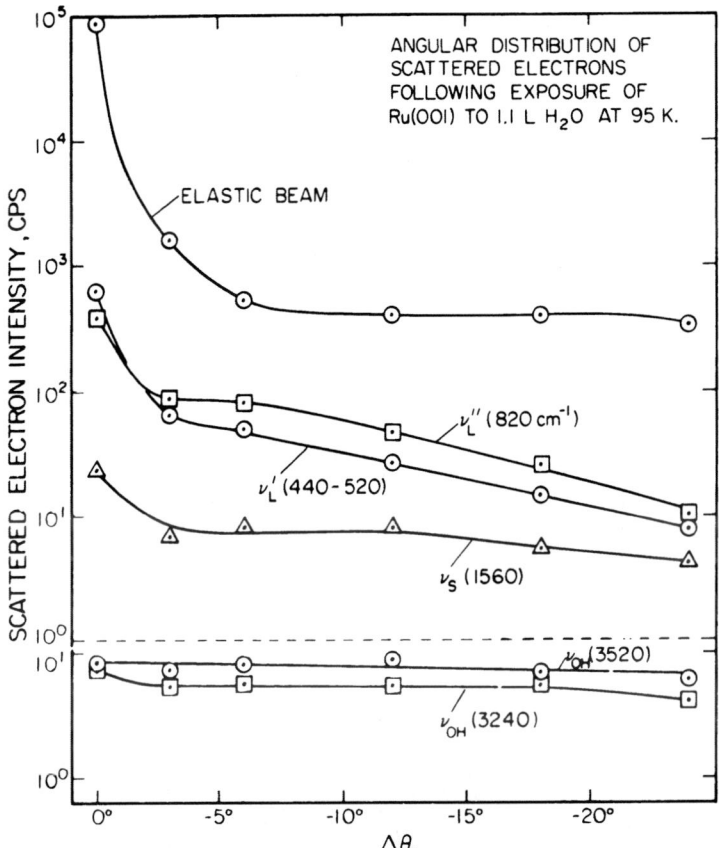

Fig. 4 - Intensities of the loss features of Fig. 3 as functions
of polar scattering angle, θ_s. In the specular direction,
$\Delta\theta_s \equiv 0°$.

165 K. Thermal desorption spectra indicate that at this temperature
and these pressures of H_2O ($\leqslant 1 \times 10^{-7}$ torr), multilayers of ice
do not form on the surface; saturation is achieved with a 3 L
exposure of H_2O. The vibrational spectra obtained under these
conditions differ markedly from those obtained at an adsorption
temperature of 95 K, particularly for exposures of H_2O greater than
1 L. The loss due to the O-H stretching mode remains at a value
typical of intermolecular hydrogen bonding, 3240 cm^{-1}, and the
scissoring mode is somewhat lowered in frequency; the most noticeable differences, however, are in the region below 1000 cm^{-1}
(cf. Fig. 7). At saturation at 165 K, three modes are now resolved.
Isotopic substitution reveals that the two modes at 940 and 710 cm^{-1}
are librational modes with an average isotopic frequency ratio of

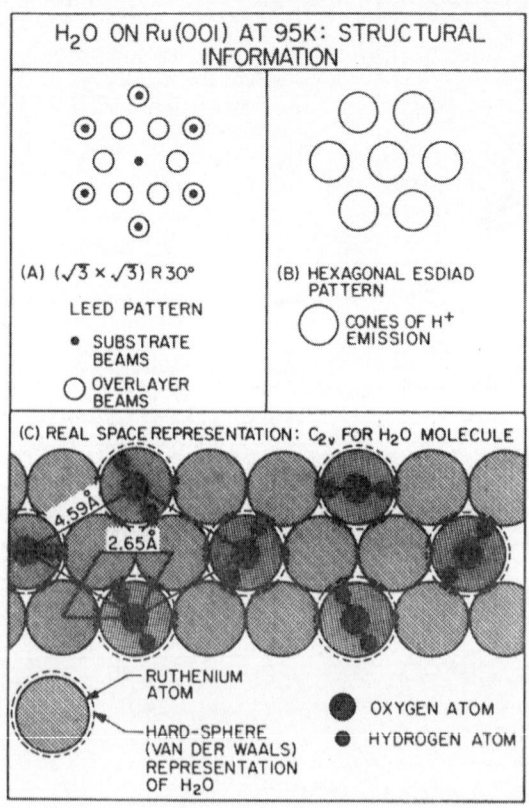

Fig. 5 - Structural information for H_2O on Ru(001) at 90 to 95 K.

1.37, whereas the mode at 370 to 400 cm^{-1} is the frustrated translation of the molecule against the metal surface, with an experimental isotopic frequency ratio of 1.04. This again supports the
assignments in this region discussed previously.

An investigation of the scattering mechanism for a surface
prepared in this manner confirms that in this case, also, the
librational modes are dipole-enhanced whereas the O-H stretching
mode is not. Figure 8 illustrates the intensities of these modes
as functions of scattering angle, which are discussed in detail
elsewhere[11]. One possible explanation for the differences in the
vibrational spectra taken at the two temperatures of adsorption
is that at 165 K, the adsorbed H_2O is chemically different, perhaps dissociated. In fact there is evidence that a small fraction
of water dissociates as the surface is heated beyond 200 K[1],[11].
However, the results of photoemission experiments, shown in Fig. 9,

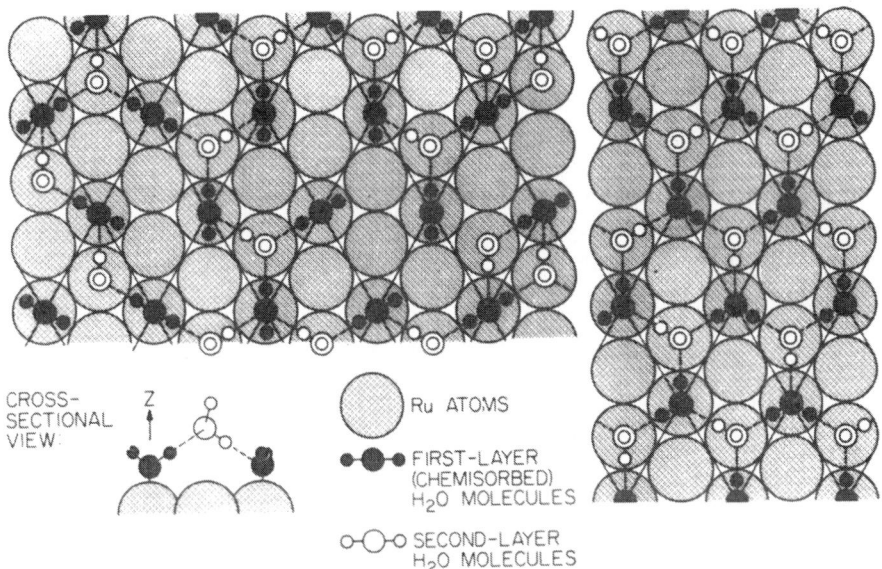

Fig. 6 – Adsorption of a second-layer H_2O molecule, hydrogen-
bonded to the chemisorbed H_2O, provides a model which is
consistent with both the LEED and ESDIAD data of Fig. 5.
Possible structures are shown here.

do not support the hypothesis that significant dissociation occurs
at 165 K. In these experiments, photoelectron intensities were
analyzed at two different exit angles for comparison. In the HeI
photoemission spectra of gaseous H_2O[14] and also of ice[15,16], three
features are observed which are ascribed to the antibonding O-H
orbital ($1b_2$), the bonding O-H orbital ($3a_1$), and the nonbonding
oxygen orbital ($1b_1$). Possible overlap of the latter feature with
emission from the ruthenium substrate at 5 to 6.5 eV in these
spectra makes identification of the $1b_1$ orbital ambiguous for
exposures of 2 L or less. At higher exposures (cf. Fig. 9(D)),
the UP spectrum resembles that of ice. The similarity of spectra B
and C, taken following exposure to 2 L of H_2O at 95 K and following
saturation with H_2O at 165 K, respectively, indicates that major
differences in the chemical composition of the adsorbate layers
prepared under these two sets of conditions are not present.

A more probable hypothesis is that, following adsorption at
165 K, the vibrational spectra reflect a surface with a higher
proportion of first-layer (chemisorbed) H_2O molecules, and that
the structure of this adsorbed layer is different at saturation
than at lower temperature.

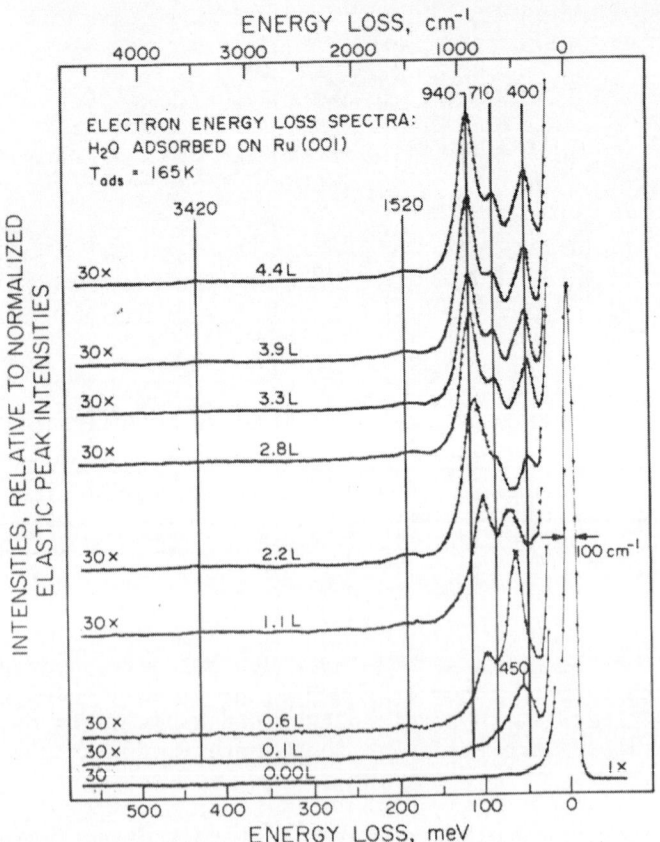

Fig. 7 - Vibrational spectra of H_2O adsorbed on Ru(001) at 160 to
170 K.

The results of this investigation can be summarized as follow :

1. The direct interaction of water with the metal surface is
   characterized by molecular adsorption through the oxygen atom.
2. At 95 K, clusters of H_2O molecules form even at submonolayer
   coverages via hydrogen bond interactions. The evidence which
   supports this conclusion is :

   a. The O-H stretching frequency at 3320 to 3380 cm^{-1};
   b. Analysis of the scattering mechanism and selection rules for
      the librational modes; and
   c. Analysis of the LEED and ESDIAD data, which provides a phy-
      sical model for the structures of the clusters.
3. Following adsorption at 165 K, the vibrational spectra reflect
   a greater proportion of chemisorbed (first-layer) water molecules

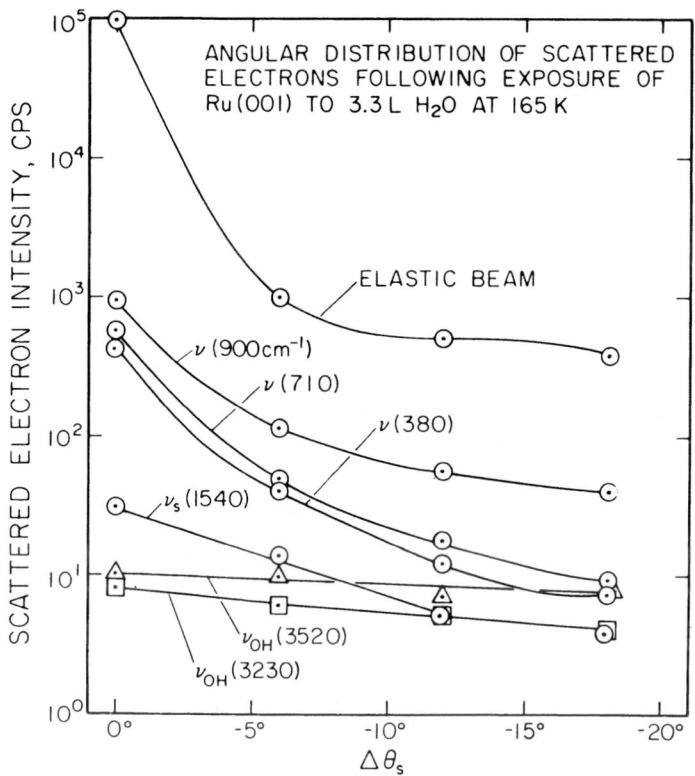

Fig. 8 — Intensities of electron energy loss features as functions
of polar scattering angle, θ_s, following exposure of the
Ru(001) surface to 3.3 L H_2O at 160 to 170 K. In the
specular direction, $\Delta\theta_s \equiv 0°$.

and possible structural changes in the overlayer, relative to
adsorption at 95 K.

REFERENCES

1. T.E. Madey, J.T. Yates,Jr., Chem. Phys. Letters 51 (1977) 77.
2. S. Andersson, J.W. Davenport, Solid State Commun. 28 (1978) 677.
3. G.B. Fisher, B.A. Sexton, Phys. Rev. Letters 44 (1980) 683.
4. H. Ibach, S. Lehwald, Surface Sci. 91 (1980) 187.
5. F. Franks, Ed., Water : A Comprehensive Treatise, Plenum Press,
   New York (1972).
6. P. Schuster, G. Zundel, C. Sandorfy, Eds., The Hydrogen Bond,
   North-Holland Publishing Co., Amsterdam (1976).
7. K. Nakamoto, Infrared and Raman Spectra of Inorganic and Coordination Compounds, Wiley-Interscience, New York (1978) 226.

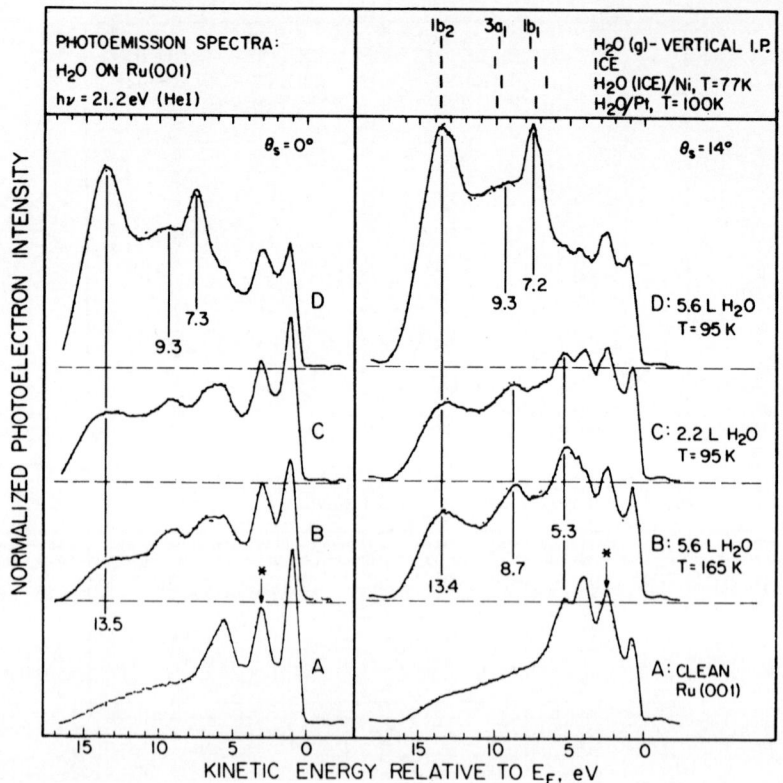

Fig. 9 – Photoemission spectra taken following exposure of the
Ru(001) surface to H_2O. The angle of incidence of the HeI
radiation is 80° from the surface normal. Experimental
details are discussed in 17. The three molecular orbitals
shown in the upper right panel are referenced to the $1b_2$
orbital in energy. From top to bottom, the energy values
are taken from 14, 15, 16, and 3. The spectra have been
normalized to the substrate feature of the clean surface
indicated by (★).

8. J.R. Ferraro, Low-Frequency Vibrations of Inorganic and Coordination Compounds, Plenum Press, New York (1971) 65-76.
9. M. Van Thiel, E.D. Becker, G.C. Pimentel, J. Chem. Phys. 27
   (1957) 486.
10. S.Y. Tong, C.H. Li, D.L. Mills, Phys. Rev. Letters 44 (1980)
    407.
11. P.A. Thiel, F.M. Hoffmann, W.H. Weinberg, in preparation.
12. J.J. Zinck, W.H. Weinberg, J. Vacuum Sci. Technol. 17 (1980) 188.
13. L.E. Firment, G.A. Somorjai, Surface Sci. 84 (1979) 275; J.
    Chem. Phys. 63 (1975) 1037.
14. D.W. Turner, C. Baker, A.D. Baker, C.R. Brundle, Molecular

Photoelectron Spectroscopy, Wiley-Interscience, London (1970) 77.
15. I. Abbati, L. Braicovich, B. De Michelis, Solid State Commun. 29 (1979) 511.
16. P.J. Page, D.L. Trimm, P.M. Williams, J. Chem. Soc. Far. Trans. I 70 (1975) 1769.
17. G.E. Thomas, W.H. Weinberg, Rev. Sci. Instrum. 50 (1979) 497.

ON THE EXCITATION MECHANISM OF VIBRATIONAL MODES OF C_2H_4 CHEMISORBED

ON Pt(111) AND Ni(111) BY EELS

A.M. Baró[*] , S. Lehwald and H. Ibach

Institut für Grenzflächenforschung und Vakuumphysik
Kernforschungsanlage Jülich, Postfach 1913
D-5170 Jülich, West Germany

ABSTRACT

 The angular dependence of the vibrational loss intensities of
C_2H_4 chemisorbed on Pt(111) and Ni(111) have been studied by non-
specular EELS measurements. The analysis of the angular profiles
permits the separation of the contributions from dipole and impact
scattering. For Pt(111) all the angular range is dominated by
impact scattering except for the C-Pt stretch mode which is dipole.
For Ni(111) dipole scattering is important and dominates the
intensity contribution in the specular direction. In both cases
impact scattering is found to increase at low energies (2 eV).
The values of the differential cross section for impact scattering
are calculated and compared with those observed by electron impact
on free ethylene. The implications of these results for the
interpretation of the spectra are outlined. We present also the
vibrational spectrum of the room temperature phase of ethylene chemisorbed on Pt(111) recorded in and out of specular, to illustrate
the importance of the excitation mechanism for chemical analysis.
The analysis of both spectra indicates that two different species
are present on the surface each one being excited by a different
scattering mechanism.

---

[*]Permanent address : Departamento de Fisica Fundamental, Universidad
 Autónoma de Madrid, Cantoblanco (Madrid), Spain.

INTRODUCTION

        The study of the vibrational modes of several hydrocarbons
adsorbed on transition metal surfaces has been done in the last
few years using EELS[1] .   By analysis of the number and frequency
values of some characteristic modes a substantial knowledge of the
bonding and adsorption geometry can be obtained[2] .   In doing that
it has been generally assumed that the "dipole selection rule" is
valid and therefore only the totally symmetric modes are excited.
It has been shown recently, also, that for some systems a short
range mechanism, termed impact scattering, has an important cross
section for vibrational excitation.   A basic theory of impact
scattering from adsorbate vibrations has been formulated and
applied to a c(2x2) overlayer of CO on Ni(100) recently[3] .   By
applying time reversal symmetry one is able to show that for the
situation of specular reflection and $\hbar\omega \ll$ E_o the cross section
for vibrational excitation vanishes for vibrations parallel to the
surface, which are odd under C_2 rotation[3] .   The vibrational
excitation by electron impact of gas phase benzene has also been
found to show a strong selectivity and selection rules have been
formulated[4].   In this case, however, the inelastic cross section
is dominated by negative ion resonances resulting by temporary
capture of the incident electron in an orbital of the molecule.
The excitation selectivity is then in addition related to the
symmetries of the orbitals involved in the resonance process and to
the nuclear motion of the molecules during the vibrations.   It
is therefore interesting to investigate the vibrational excitation
mechanism for an adsorbed molecule.   This can be done by measuring
the non-specular loss intensities.   The dipole contribution can
be determined in many cases by subtracting the off-specular from
the in-specular intensity.

        In this paper we report intensity measurements of the vibratio-
nal losses of ethylene chemisorbed on Pt(111) and Ni(111) under
non-specular conditions.   Ethylene is one of the few molecules which
have been studied by electron impact in the gas phase[5] and a
direct comparison between the free and adsorbed phase can be made.
The analysis of the results allows us to draw conclusions concerning
the excitation mechanism and the adsorption geometry.   The implications of the excitation mechanism on chemical analysis are also
outlined.

EXPERIMENTAL

        The data reported here were performed in two different UHV
systems.   Two electron spectrometers of the cylindrical type were
used : a single pass for Pt(111) and a double pass for Ni(111).
To analyse the data for impact scattering it is necessary to know

the primary beam current and the angle of acceptance of the analyzer
These are measured in the direct beam position (primary beam
entering directly into the analyzer)[6] .   For both the single and
double pass spectrometer we measured the same angle of acceptance of
0.7°.   The primary current was about four times larger for the single
than for the double pass.   For dipole scattering the cut off angle
of the elastic beam has to be measured.   We did not find any
appreciable difference between the cut off angles of both systems.
These measurements are needed in order to compare the data taken with
different spectrometers.

The standard cleaning procedure for Pt(111) was oxygen treatment
at 1250 K in 5x10^{-8} Torr of oxygen for five minutes followed by a
flash in UHV at 1500 K about 30 sec long[7] .   The cleaning procedure for Ni(111) was argon bombardment at 700° C and flashing to
1000° C.   For Ni(111) the cleanliness of the sample could be tested
by AES.

RESULTS

a) Ethylene on Pt(111), low temperature : The low temperature
vibration spectrum of ethylene adsorbed on Pt(111) is shown in Fig. 1.
By working at low energy a new vibrational mode at 650 cm^{-1} which
was not observed in previous investigations can be resolved. A complete assignment of the vibrational modes of adsorbed ethylene is
given in [8].

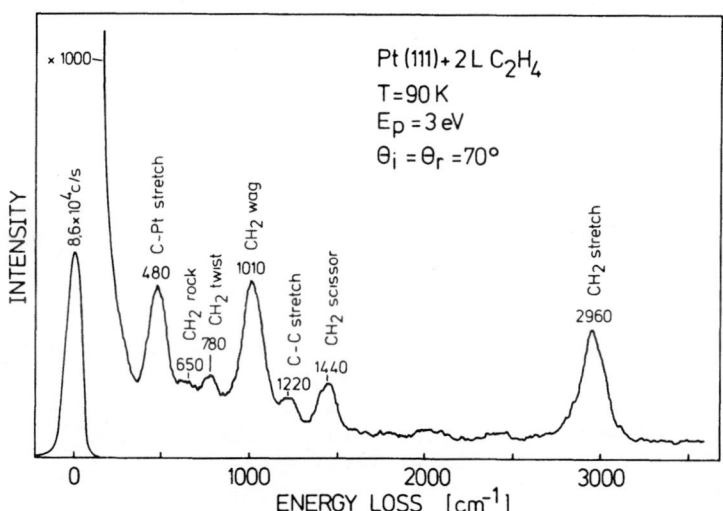

Fig. 1. : Vibrational spectrum of a 2L dose of
            ethylene on Pt(111) at T = 90 K.

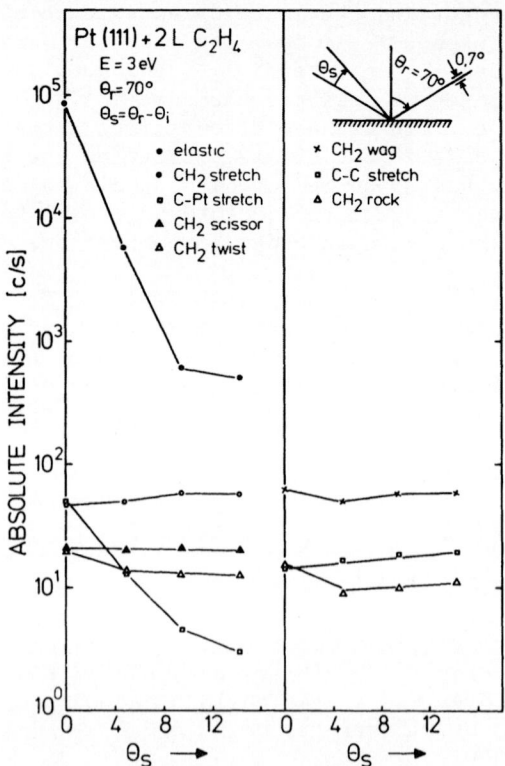

Fig. 2. : Angular dependence of the elastic and inelastic
intensities for the different ethylene modes
when adsorbed on Pt(111).

We attribute the new mode at 650 cm^{-1} for Pt(111) to the excitation
of the CH$_2$ rock vibration. This mode is in fact observed for C$_2$H$_4$
adsorption on Ni(111)[8]. The angular dependence of the elastic and
inelastic intensities has been plotted in Fig. 2. for a primary
energy of 3 eV. Only one of the modes, the C-metal stretch (470 cm^{-1})
is peaked in the specular direction and by going out of specular
the inelastic intensity goes down steeply. The CH$_2$ scissor (1420
cm^{-1}), CH$_2$ stretch (2940 cm^{-1}) and C-C stretch (1230 cm^{-1}) show a
constant intensity when going out of specular. The three other
modes CH$_2$ rock (650 cm^{-1}), CH$_2$ wag (990 cm^{-1}) and CH$_2$ twist (790 cm^{-1})
show a small increase in the specular direction. Except for the
C-metal mode, we infer that the ethylene modes are excited mainly
by a non-dipole scattering process. This result has consequences
with respect to the analysis of the molecule adsorption geometry
which will be discussed in the next section. Concerning the excitation process we conclude that impact scattering is dominant.

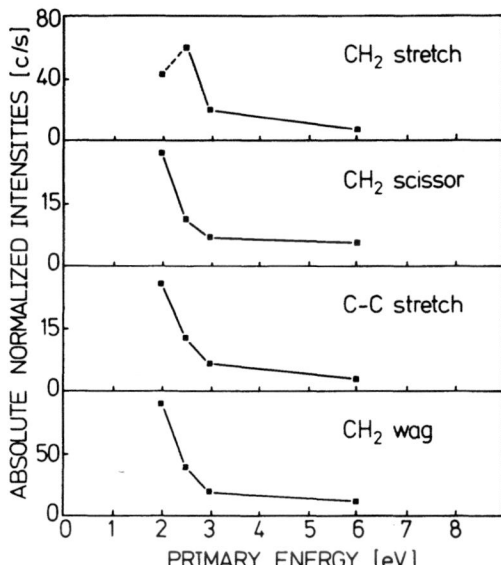

Fig. 3. : Absolute intensity of several ethylene modes as a function
         of primary energy. The intensities are normalized to a
         primary intensity of 1×10^{-11} Amp.

     Further evidence for impact scattering can be obtained from
the dependence of the inelastic intensities on primary energy.
In Fig. 3. we have plotted the off-specular intensities of some
vibrational modes as a function of primary energy.
     There is an increase of the intensity for low energies. The ine-
lastic intensities plotted in Fig. 3. are normalized with respect
to the "primary intensity", i.e. with respect to the intensity of the
direct beam measured at the detector for the certain primary energy.
This quantity therefore contains the primary beam current and the
transmission of the analyzer. We could not go to primary energies
below 2 eV because the primary intensity decreased too much. The
normalized elastic intensity shows, however, a completely different
energy dependence. This means that the inelastic intensities are
not correlated with the elastic intensity. This indicates impact
scattering[3].

     In order to decide whether a dipole contribution exists for
some of the ethylene modes we measured the inelastic angular
dependence at 6 eV. At this energy the impact scattering cross
section is low (see Fig. 3.) and at the same time the elastic
intensity is high favouring dipole scattering. By doing these
measurements we have found that the CH_2 wag and CH_2 rock modes
have a non negligible dipole contribution and therefore the
"dipole selection rule" can be applied to both modes.

   b) Ethylene on Ni(111), low temperature : The vibrational
spectrum of the low temperature phase of ethylene on Ni(111) is
reported in Fig. 4. for specular reflection and two angles off specular. We refer to previous work[8] for the assignment of the different modes. The angular dependence of the elastic and inelastic
intensities is given in Fig. 5. All the modes show a decrease in
intensity when going off specular. Only the intensity of the CH_2
stretch vibration is above 8° of specular higher than in specular.
The dipole contribution is predominant for the 1100 cm^{-1} (CH_2 wag)
and 450 cm^{-1} (C-metal stretch) modes. It has a non-negligible
value for the 740 cm^{-1} (CH_2 rock), 1430 cm^{-1} (CH_2 scissor), 880 cm^{-1}
(CH_2 twist), 2970 cm^{-1} (CH_2 stretch), and 1200 cm^{-1} (C-C stretch).
From this we conclude that all the observed modes have a dipole
contribution.

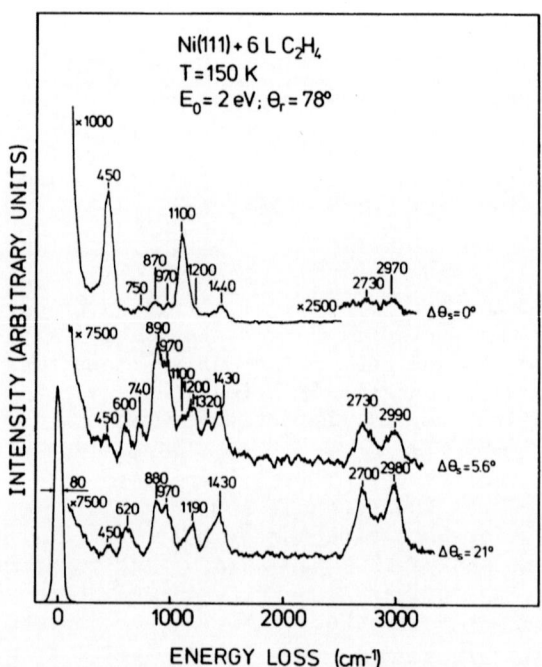

Fig. 4. : Vibrational spectrum of low temperature adsorption of
        ethylene on Ni(111) for several angles of incidence θ_i.
        $\Delta\theta_s = \theta_i - \theta_r$, θ_r = angle of reflection. The assignment
        of the losses is as follows : 2970 cm^{-1}, CH_2 stretch(s) ;
        2730 cm^{-1}, CH_2 stretch with H bonding ; 1430 cm^{-1}, CH_2
        scissor ; 1200 cm^{-1}, C-C stretch ; 1100 cm^{-1}, CH_2 wag ;
        870 cm^{-1}, CH_2 twist ; 740 cm^{-1}, CH_2 rock ; 610 and 450
        cm^{-1}, C-metal stretch. The losses at 970 and 1320 cm^{-1}
        resolved off specular are probably due to CH_2 scisssor
        and CH_2 twist with H bonding, respectively.

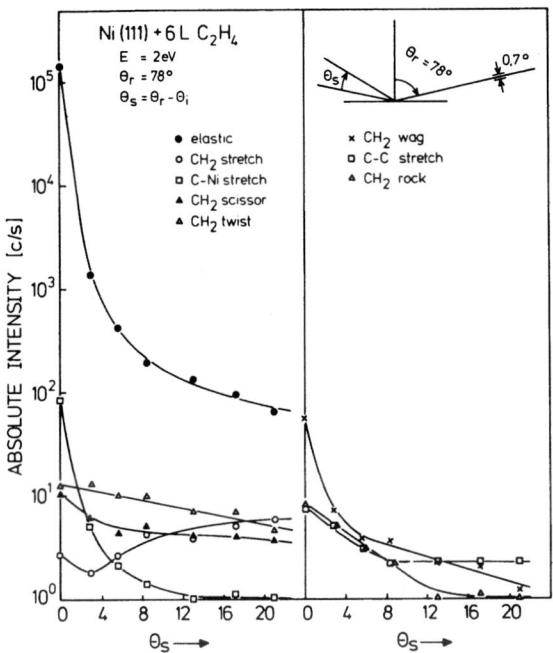

Fig. 5. : Angular dependence of the elastic and inelastic intensities for ethylene chemisorbed on Ni(111).

The slow decrease in intensity when going out of the specular direction, and the increase in the case of CH_2 stretch, however, indicate that there is also excitation by impact scattering. The impact scattering cross section is also energy dependent and is larger for 2 eV than for 4 eV (we did not measure at other primary energies).

c) Ethylene on Pt(111), room temperature phase : By thermal processing adsorbed ethylene to room temperature, it converts to a new phase which nature is still the subject of controversy. At least four different species have been proposed and we refer to[9] for a more detailed description of the subject. Here we want to show the vibrational spectra of the room temperature phase taken in and out of the specular reflection (Fig. 6.).

The interesting result is that apparently different vibrational modes are excited in and off-specular. The in-specular spectrum has been interpreted as CH_3-CH[10] . The off-specular spectrum, however, shows frequency values closely related to those reported for the low temperature phase and characteristic of CH_2 groups.

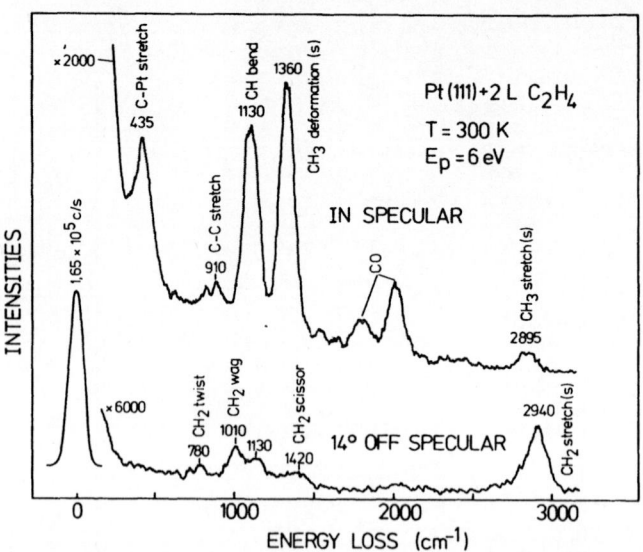

Fig. 6 – Vibrational spectrum of the room temperature phase of
         ethylene on Pt(111) taken in and out of specular. The off-
         specular spectrum makes apparent new modes which are hid-
         den in the in-specular spectrum.

This may be an indication that two different species are simultaneously present on the surface. An alternative explanation would
attribute the off-specular bands to modes of the room temperature
phase which have no dipole moment perpendicular to the surface.
For CH_3-CH the asymmetric stretch and deformation modes of the CH_3
group could be responsible for the 2940 cm^{-1} and 1420 cm^{-1} bands.
The 1010 cm^{-1} band could be attributed to the CH_3 rock mode. We
cannot find a suitable mode with a frequency value of 780 cm^{-1}.
The observation of the two different vibrational spectra for one
single species would however imply that all modes excited by dipole
scattering are not excited by impact scattering. This would be at
variance with the general selection rule. We also note that the
data reported in this paper for the low temperature phase show just
the opposite behaviour.

     Therefore we interpret the non-specular observed modes of the
room temperature spectrum as due to another species present simultaneously on the surface with CH_3-CH. The in-specular species is
excited by dipole scattering and therefore not seen out of specular.
The off-specular species is excited by impact scattering and the
observation in-specular is obscured by the vibrational bands of
the in-specular species. This mode fits also quite well with thermal desorption data as has been described elsewhere[9].

The analogy with the low temperature ethylene phase and the similarity with acetylene transformation on the same surface[10] leads us to interpret the off-specular species as $C = CH_2$.

DISCUSSION

a) The excitation mechanism : The angular dependence of the vibrational losses of ethylene chemisorbed on Pt(111) and Ni(111) show that impact scattering gives an important contribution to the vibrational cross section. For Pt(111) a dependence of the inelastic intensity on primary energy characteristic of impact scattering could be found. In the following we make an analysis of the intensities observed for both surfaces and compare with the free molecule. As the spectra have been taken in different spectrometers the intensities have to be normalized. For impact scattering we normalize to the "primary intensity" (as described in 1§) and for dipole scattering to the elastic intensity. We have also taken the values reported by Backx et al.[11] for ethylene adsorbed on Ag(110). In this case, however, some uncertainty exists due to the fact that only the primary current measured through the crystal is reported.

From the value of the crystal current we estimate that the primary intensity should be comparable with the value used in the platinum experiments and therefore we have normalized to this value. We assume also identical coverages for all the surfaces, which is reasonable as all the data are taken at nearly saturation. The data of Backx refer to ethylene spectra with preadsorbed oxygen, which has been found only to promote the adsorption of ethylene on Ag(110) compared to the clean surface[11].

The data concerning impact scattering are reported in Table I. As can be seen the impact scattering cross section is higher for Pt than for Ni and Ag with a sequence Pt > Ni > Ag. The strong difference between the three metals indicates that the impact scattering process is very sensitive to chemical bonding. This also indicates that impact scattering analysis may be very interesting to obtain information on the bonding process. As discussed in 4b §) ethylene chemisorption on Pt(111) is rather similar. The impact scattering data, however, are significantly different. Tle same is true for ethylene adsorption on Ag(110) as compared with free ethylene. On Ag(110) ethylene is π-bonded and therefore with a configuration similar to the free molecule. The impact scattering data indicate, however, a significant decrease in the cross section for the adsorbed phase.

Table I.

Impact scattering intensities for ethylene adsorption on Pt(111), Ni(111) and Ag(110). The intensities (counts/sec) are normalized to a primary intensity of 1 x 10^{-11} Amp. For Ag(110) we assume a primary intensity equal to that on Pt(111). For Ni(111), as the inelastic intensity for some modes drops down slowly with increasing angle, there is some uncertainty in the impact data, because it is difficult to distinguish between the drop down of the dipole and the angular dependence of the impact intensity. The intensities in the Table correspond to the largest angle out of specular.

Surface	Energy (eV)	CH$_2$ rock	CH$_2$ twist	CH$_2$ wag	C-C stretch	CH$_2$ scissor	CH$_2$ stretch
Pt(111)	2	24	24	90	26	27	43
Ni(111)	2	4	15	8	10	15	15
Pt(111)	2.5	13	13	39	13	11	60
Ag(110)	2.6	not observed	not observed	0	1	1	4

We can also compare the data of free[5] and adsorbed ethylene
on Pt(111) as they are rather complete. The inelastic intensities
can be converted to differential cross sections[6] and compared to
the free molecule data[5] (Table II). For doing that we have taken
an ethylene coverage of 0.25. Even if ethylene does not give a LEED
structure at low temperatures it makes a (2x2) structure upon heating
to room temperature[10] . This result together with a consideration
of the van der Waals diameter of the molecule makes the coverage
value of 0.25 very reasonable. We see in Table II that the cross
section for two vibrations is larger for the adsorbed than for
the free molecule. Also the selectivity is changed and particularly the CH$_2$ wag is observed in the adsorbed case. This shows that
the selection rules for the free molecule and the adsorbed species
are different. Concerning the energy dependence, in both cases
the inelastic intensity is larger for $\sim$ 2 eV although in the
adsorbed case we could not observe a definite peak due to the lack
of data at lower energies. The data of energy dependence might
have to be regarded with some caution, however, because due to
focussing conditions changes in the transmission of the analyzer
may be possible for the case of reflection from the crystal compared
to the direct beam. Nevertheless, a short-lived shape resonance
observed in the gas phase molecule at 7.5 eV[5] is absent in the
adsorbed case.

Table II

Differential cross section of impact scattering for C$_2$H$_4$ free and
adsorbed on Pt(111). Values are given in Å^2 Sr^{-1}.

System	CH$_2$ stretch	CH$_2$ scissor	C-C stretch	CH$_2$ wag
Adsorbed C$_2$H$_4$	0.14	0.08	0.06	0.36
Free C$_2$H$_4$	0.04	0.09	0.12	not observed

Therefore concerning the nature of the non-dipole scattering
process for adsorbed ethylene the experimental results seem to
favour impact scattering rather than a negative ion resonance,
since we observe the CH$_2$ wag mode for the adsorbed species in impact
scattering.

Table III.

Dipole scattering intensities (Iloss/Ielastic x 10^5) for ethylene adsorption on Pt(111), Ni(111) and Ag(110). The impact scattering contribution has been subtracted. For Ag(110) the cut off angle is about two times larger than for Pt and Ni.

System	Energy(eV)	C-metal	CH$_2$rock	CH$_2$twist	CH$_2$wag	C-Cstretch	CH$_2$scissor	CH$_2$stretch
Ni(111)	2	59	4.7	4	37	2.7	4	-
Pt(111)	6	71	3.2	-	23	-	-	-
Ag(110)	2.6	-	-	-	300	-	-	-
Ni(111)	4	71	-	5	64	-	11	9

The analysis of the dipole intensities of the different
modes gives information on the orientation of the molecule on the
surface[12] .   The dipole intensities $I_{loss}/I_{elastic}$ are reported
in Table III.
They have been obtained by taking the difference of the in and off-
specular intensities.   For Pt(111) they are difficult to measure
as the in and off-specular intensities are nearly equal.
The Ag(110) intensity is not directly comparable to Pt and Ni as
the cut off angle is about two times larger.
But nevertheless, on silver only the CH_2 wag mode has dipole scattering intensity[11].   This may be indicative of parallel planar
adsorption of C_2H_4 on Ag(110) which is expected for π-bonding
and no or nearly no breaking of the D_{2h} symmetry.   The latter is
supported by the observation that the vibrational frequencies of
the adsorbed ethylene are nearly unshifted compared to the gasphase[11]
On Pt(111) and Ni(111) ethylene is rehybridized to nearly sp^3 and
we expect non planar adsorption of the molecule.   The lower value
of the CH_2 wag dipole intensity for Pt(111) would indicate a larger
C-C-H angle.

b) The bonding and adsorption geometry : From the analysis
of the frequency values of some specific modes like the CH_2 stretch
and the C-C stretch it was concluded that ethylene is di-σ bonded
to both Pt and Ni with a rehybridization state near to sp^3 [8,10].
More insight into the adsorption process has been traditionally
obtained by using the surface selection rule.   For systems where
impact scattering is dominant, the applicability of the surface
selection rule may be restricted to high impact energies.   In
these cases, the surface selection rule can be applied when a
dipole contribution has been found by making the appropriate angular
analysis.

With this in mind, let's now apply the dipole selection rule
to ethylene adsorption on the Pt and Ni surfaces.   The highest
possible symmetry is C_{2v} and then the dipole active modes are
CH_2 stretch, C-metal stretch, CH_2 scissor, CH_2 wag and C-C stretch.
These modes are dipole active in any geometry but may have a very
small intensity.   Hence, only the CH_2 rock and twist modes are
to be taken into account.   An interesting observation reported
previously[10] is that the vibration spectrum is coverage dependent
for ethylene adsorbed on Pt(111).   Some of the modes appear only
at high exposures.   We have quantified this observation by
plotting the in-specular inelastic intensity for some modes as
a function of the exposure for both Pt(111) and Ni(111)(Fig. 7.).

For Pt(111) we see that the C-C stretch, CH_2 rock and CH_2
twist only appear at high coverage, whereas for Ni(111) they show
a monotonic increase with exposure.   Therefore, for Ni(111) all
the observed modes are dipole active (see Fig. 5.) even at the
lowest coverage.   For Pt(111) the situation is not so clear.

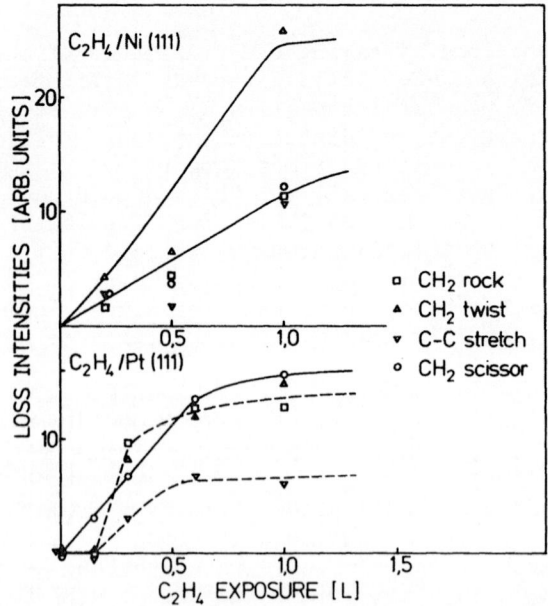

Fig. 7. : Loss intensities ($I_{loss}/I_{elastic}$) for several modes of
ethylene adsorbed on Pt(111) and Ni(111) as a function
of exposure. The broken lines for Pt(111) correspond to
the modes (CH$_2$ rock, CH$_2$ twist and C-C stretch) which are
not excited at low coverage. The exposure values may not
be comparable, and are probably higher for Pt than for Ni
because they are measured in different UHV-systems.

Nothing can be said concerning the CH$_2$ twist mode as we could not
find any dipole contribution. The CH$_2$ rock mode is dipole active
but appears only at high exposures.

The question is : what can be deduced from these data about
the adsorption geometry. For Ni(111) we conclude that the
symmetry point group is C$_2$ or C$_1$. A twisting of the molecule
about the C-C axis would lead to C$_2$. Such a geometry has been
in fact suggested previously[13,14]. This geometry would place
two hydrogen atoms near to the surface which is expected from the
observation of hydrogen bonding.[1] For Pt(111) the adsorption
geometry is different for low and high exposures. For low

exposures as we do not see any CH2 rock intensity we conclude that the point group symmetry is C$_{2v}$. At high exposures it depends whether of not we assume a dipole character for the CH$_2$ twist mode. If not, the highest symmetry group would be C$_s$ with a plane of symmetry perpendicular to the C-C axis. The results of Fig. 7. indicate that the adsorption geometry for Ni (C$_2$ or C$_1$) is imposed by the bonding itself. Fot Pt, however, the adsorption geometry is C$_{2v}$ and it becomes distorted to lower symmetry at high coverage.

The characteristics of ethylene adsorption on both surfaces can be summarized as follows : Ethylene is di-σ bonded on Pt and Ni(111) surfaces with rehybridization near to sp^3. On Ni it is H bonded in addition. The adsorption geometry gives a symmetry point group C$_2$ or C$_1$ on Ni and C$_{2v}$ on Pt. This comparison is important to understand the different behaviour upon thermal processing of ethylene. Whereas on Ni(111) ethylene dehydrogenates to acetylene, on Pt(111) it goes through C = CH$_2$. This is likely to play a key role in the chemistry of both surfaces.

c) Chemical analysis : The results reported on the thermal processing of ethylene, show the interest of angular studies for chemical analysis. The situation observed is not likely to occur for simple adsorption studies which are usually done up to now in surface laboratories, but certainly is going to occur in more complex adsorption processes and chemical reactions. The observation of both dipole and impact scattering can be then a useful tool to distinguish between the products, radicals and intermediates which stay on the surface. This adds a new aspect to the capability of EELS to do the chemical analysis of surface reactions.

ACKNOWLEDGEMENTS

We are grateful to Dr. C. Backx for giving us results of his Ag work prior to its publication. One of us (A.M. Baró) gratefully acknowledges the support of the Alexander von Humboldt Stiftung.

REFERENCES

1. J.E. Demuth, H. Ibach, S. Lehwald, Phys. Rev. Lett. 40 (1978) 1044.
2. H. Ibach, H. Hopster, B. Sexton, Appl. Surface Sci. 1 (1977) 1.
3. C.H. Li, S.Y. Tong, D.L. Mills, Phys. Rev. B 21 (1980) 3057.
   S.Y. Tong, C.H. Li, D.L. Mills, Phys. Rev. Lett. 44 (1980) 407.

4. S.F. Wong and G.J. Schulz, Phys. Rev. Lett. 35 (1975) 1429.

5. I.C. Walker, A. Stamatovic, S.F. Wong, J. Chem. Phys. 69 (1968) 5532.

6. A. M. Baró and H. Ibach, to be published.
   H. Ibach and S. Lehwald, to be published.

7. A.M. Baró, H. Ibach and H.D. Bruchmann, Surface Sci. 88 (1979) 384.

8. S. Lehwald and H. Ibach, Surface Sci. 89 (1979) 425.

9. A.M. Baró and H. Ibach, to be published.

10. H. Ibach and S. Lehwald, J. Vac. Sci. Technol. 15 (1978) 407.

11. C.Backx, C.P.M. de Groot and P. Biloen, to be published.

12. H. Ibach, H. Hopster and B. Sexton, Appl. Phys. 14 (1977) 21.

13. K. Horn, A.M. Bradshaw and K. Jacobi, J. Vac. Sci. Technol. 15 (1978) 575.

14. J.E. Demuth, Surface Sci. 76 (1978) L603.

VIBRATIONS MEASURED AT METAL SURFACES BY EELS : A REVIEW TABLE [†]

P. Thiry

Laboratoire de Spectroscopie Electronique
Facultés Universitaires N.D. de la Paix
61, rue de Bruxelles, B-5000 Namur, Belgium

Although there has been, in high resolution electron energy loss spectroscopy (EELS) a great interest in semiconductors and ionic solids during the early seventies[1], and again recently[2,3,4], most of the experiments reported on this technique have been performed on metals. In this paper, we have reviewed the experimental observations of extrinsic phonons (adsorbate vibrations) as well as intrinsic phonons (surface vibrations) that have been made on metallic surfaces by EELS. Table I gives a systematic overview and proceeds from a "geographical" point of view. The figures therein refer to the different research laboratories where the experiment has been performed and published. These figures are listed in Table II. For each metal studied, we have tabulated in Table III the adsorbed gases, and the corresponding publications are given thereafter in chronological order. This review is intended to be updated up to october 1980. A similar work, by H. Weinberg, has already been done up to april 1980[5].

This tabulation has become possible thanks to the collaboration of the research groups working in the field of EELS. We would like to thank them for sending exhaustive information about their publications and present field of research.

[†] Work performed under the auspices of the I.R.I.S. project (Institute for Research in Interface Sciences), sponsored by the Belgian Ministry for Science Policy, and supported by a F.R.F.C. contract from the National Science Foundation (FNRS).

Table 1

	Al	Fe	Stainl Steel	Ni				Cu	
	(100)	(110)	Steel	(100)	(110)	(111)	Step[d]	(100)	(110)
Surface Phonons				2		2		5	
Hydrogen				5		6,12			
Nitrogen									
Oxygen	1	2	4	2,5,6		2		10	9
Sulfur				5					
Carbon Monoxide		2		5,6	6	2,6	2	5,10	9
Nitric Oxide				2		2			9
Carbon Dioxide									
Water			4	5				10	
Ammonia			4						
Acetylene				6		2,6	2		
Ethylene						2,6	2		
Cyclopropane									
Benzene				6		2,6			
Cyclohexene						6			
Cyclohexane						2	2		
Methanol			4			2		10	
Ethanol			4					10	
Formic Acid								10	
Acetic Acid								10	
Acetone			4						
Pyridine									

Table 1, con't

Ru	Rh	Pd	Ag		W			Pt			Pt$_{10}$
(100)	(111)	(100)	(110)	(111)	(100)	(110)	(111)	(100)	(111)	Step[d]	Ni$_{90}$
										2	
8					1,2,3 4,7	7	7		2	2	
					1,7						
8			13		2,4				10		
8	11	8,15			1,2,3 4	7	7	2	2	2	6
8					3			2	2,10		
	11										
8					1			2	10		
	11					7	7		2		
	11		13						2		
8											
				14					2		
8									2		
				14							

TABLE II. REFERENCES FOR TABLE I

1. Coordinated Science Laboratory and Department of Physics, University of Illinois at Urbana-Champaign, Urbana, Illinois 61801, U.S.A.

2. Institut für Grenzflächenforschung und Vakuumphysik der Kern-forschungsanlage, Postfach 1913, D-5170, Jülich, West Germany.

3. Service de Physique Atomique, Section d'Etudes des Interactions Gaz-Solides, Centre d'Etudes Nucléaires de Saclay, B.P. 2, F-91190, Gif-sur-Yvette, France.

4. Centre de Recherches sur les Atomes et les Molécules, et Département de Physique, Faculté des Sciences et du Génie, Université Laval, Québec, Canada G1K 7P4.

5. Department of Physics, Chalmers University of Technology, S-41 296, Göteborg, Sweden.

6. Institut de Recherches sur la Catalyse, CNRS, 2, av. Albert Einstein, F-69626, Villeurbanne Cédex, France.

7. Surface Physics Group, Astronomy Division, European Space Research and Technology Center, Noordwijk, Holland.

8. Division of Chemistry and Chemical Engineering, California Institute of Technology, Pasadena, California 91125, U.S.A.

9. Solid State Division, Oak Ridge National Laboratory, Oak Ridge, Tennessee 37830, U.S.A.

10. Physical Chemistry Department, General Motors Research Labora-tories, Warren, Michigan 48090, U.S.A.

11. Materials and Molecular Research Division, Lawrence Berkeley Laboratory and Department of Chemistry, University of California, Berkeley, California 94720, U.S.A.

12. Department of Physics, University of Pennsylvania, Philadelphia, Pennsylvania 19104, U.S.A.

13. Koninklijke / Shell - Laboratorium, Postbus 3003, NL-1003 AA, Amsterdam, The Netherlands.

14. I.B.M. Thomas J. Watson Research Center, P.O. Box 218, Yorktown Heights, New York 10598, U.S.A.

15. Fritz-Haber Institut der Max Planck Gesellschaft, Faradayweg 4-6, D-1000, Berlin 33, West Germany.

TABLE III

Metal Surface	Adsorbed gas	References
Al(100)	oxygen	4,7
Fe(110)	oxygen	136
	carbon monoxide	132
Stainless Steel	clean	3
	oxygen	33,35,53
	ammonia	33,53
	water	33,53
	methanol	33,53
	ethanol	33,53
	acetone	33,53
Ni(100)	surface phonons	121
	hydrogen	25,38
	oxygen	8,14,24,27,52,66,121
	sulphur	52,66
	carbon monoxide	12,24,25,27,62,134
	nitric oxide	138
	water	62
	acetylene	20
	benzene	10,28,92
Ni(110)	carbon monoxide	113,134
Ni(111)	surface phonons	101
	hydrogen	41,69,96
	oxygen	101
	carbon monoxide	30,31,32,41,44,46,57,69,71
	nitric oxide	114
	acetylene	34,46,48,50,55,63,79,90,
	acetylene	110,133
	ethylene	34,39,76,90,125
	benzene	28,49,64,92
	cyclohexene	26
	cyclohexane	39,46,90
	methanol	67
Ni[5(111) x ($\bar{1}$10)]	carbon monoxide	77
	acetylene	74,90
	ethylene	90
	cyclohexane	90
Cu(100)	surface phonons	120,126
	oxygen	80
	carbon monoxide	78,80,93,107,120,126
	water	80
	methanol	88
	ethanol	88

(TABLE III continued on next page)

Cu(100)	formic acid	89,97
	acetic acid	84,97
Cu(110)	oxygen	137
	carbon monoxide	73,124
	nitric oxide	124
Ru(100)	hydrogen	130
	oxygen	51,60,68
	carbon monoxide	42,51,68,72,75,82
	nitric oxide	61,82,94,130
	water	117,127
	cyclopropane	135
	cyclohexane	131
Rh(111)	carbon monoxide	83,100,116
	carbon dioxide	83,100,116
	acetylene	106,116
	ethylene	106,116
Pd(100)	carbon monoxide	87,118
Ag(110)	oxygen	109,115,129
	ethylene	109,115
Ag(111)	benzene	128
	pyridine	128
W(100)	hydrogen	1,2,5,16,19,33,40,45,47,65,
	hydrogen	70,91,96,111,122
	nitrogen	1,108
	oxygen	6,22,33,122
	carbon monoxide	1,2,9,13,33,54,122
	nitric oxide	2
	water	1
W(110)	hydrogen	11,16
	carbon monoxide	29
	acetylene	11,23,29,36
W(111)	hydrogen	15,16
	carbon monoxide	29
	acetylene	15
Pt(100)	carbon monoxide	43
	nitric oxide	81
	water	99
Pt(111)	hydrogen	37,86
	oxygen	98,112
	carbon monoxide	9,17,22,43,58,59,95
	nitric oxide	56,104
	water	85,103,105
	acetylene	18,21,22,37,50
	ethylene	21,22,37,39,119,125

(TABLE III continued on next page)

Pt(111)	benzene	49,64
	cyclohexane	39
Pt[6(111) x ($\bar{1}$11)]	surface phonons	59
	hydrogen	102
	carbon monoxide	43,58
$Pt_{10}Ni_{90}$(111)	carbon monoxide	123

REFERENCES FOR TABLE III

1. F.M. Propst, T.C. Piper $W(100) + H_2, N_2, CO, H_2O$
   J. Vac. Sci. Technol. 4, 53 (1967).

2. Y. Ballu, R.A. Armstrong, J. Lecante $W(100) + H_2, NO, CO$
   C.R. Acad. Sci. (Paris) B 274,
   718 (1972).

3. A. Adnot, Y. Ballu, J.D. Carette,   Stainless Steel
   J. Appl. Phys. 43, 2796 (1972).

4. J.F. Wendelken, $Al(100) + O_2$
   Ph.D. Thesis, Univ. Illinois, Urbana
   (1974).

5. H. Froitzheim, H. Ibach, S. Lehwald $W(100) + H_2$
   Phys. Rev. Lett. 36, 1549 (1976).

6. H. Froitzheim, H. Ibach, S. Lehwald, $W(100) + O_2$
   Phys. Rev. B 14, 1362 (1976).

7. J.F. Wendelken, F.M. Propst, $Al(100) + O_2$
   Rev. Sci. Instrum. 47, 1069 (1976).

8. S. Andersson, $Ni(100) + O_2$
   Solid State Commun. 20, 229 (1976).

9. H. Froitzheim, H. Ibach, S. Lehwald $W(100) + CO$
   Proc. Int. Symp. Photoemission, $Pt(111) + CO$
   Noordwijk, 1976, p. 277.

10. J.C. Bertolini, G. Dalmai-Imelik, $Ni(100) + C_6H_6$
    J. Rousseau,
    Proc. Int. Symp. Photoemission,
    Noordwijk, 1976, p. 285.

11. C. Backx, B. Feuerbacher, B. Fitton $W(110) + H_2, C_2H_2$
    R.F. Willis, Proc. Int. Symp. Photo-
    emission, Noordwijk, 1976, p. 291

12. S. Andersson $Ni(100) + CO$
    Solid State Commun. 21, 75 (1977).

13. H. Froitzheim, H. Ibach, S. Lehwald $W(100) + CO$
    Surf. Sci. 63, 56 (1977).

14. G. Dalmai-Imelik, J.C. Bertolini, $Ni(100) + O_2$ [NiO]
    Surf. Sci. 63, 67 (1977).

15. C. Backx, B. Feuerbacher, B. Fitton $W(111) + H_2$, C_2H_2
    R.F. Willis, Surf. Sci. $\underline{63}$, 193
    (1977).

16. C. Backx, B. Feuerbacher, B. Fitton, $W(100) + H_2$; $W(110) + H_2$
    R.F. Willis, Phys. Lett. $\underline{60A}$, 145 $W(111) + H_2$
    (1977).

17. H. Froitzheim, H. Hopster, H. Ibach, $Pt(111) + CO$
    S. Lehwald, Appl. Phys. $\underline{13}$, 147
    (1977).

18. H. Ibach, H. Hopster, B. Sexton $Pt(111) + C_2H_2$
    Appl. Phys. $\underline{14}$, 21 (1977).

19. A. Adnot, J.D. Carette, $W(100) + H_2$
    Phys. Rev. Lett. $\underline{39}$, 209 (1977).

20. J.C. Bertolini, G. Dalmai- Imelik, $Ni(100) + C_2H_2$
    J. Rousseau, Le Vide, les Couches
    Minces, Suppl. n° $\underline{185}$ (1977).

21. H. Ibach, H. Hopster, B. Sexton, $Pt(111) + C_2H_2$, C_2H_4
    Applications Surf. Sci. $\underline{1}$, 1 (1977).

22. H. Ibach $W(100) + O_2$
    Proc. 7th I.V.C. and 3rd I.C.S.S., $Pt(111) + CO, C_2H_2, C_2H_4$
    Vienna, 1977, p. 743.

23. C. Backx, R.F. Willis, $W(110) + C_2H_2$
    Proc. 7th I.V.C. and 3rd I.C.S.S.,
    Vienna, 1977, p. 751.

24. S. Andersson, B.I. Lundqvist, J.K. $Ni(100) + O_2$, CO
    Nørskov, Proc. 7th I.V.C. and 3rd
    I.C.S.S., Vienna, 1977, p. 815.

25. S. Andersson $Ni(100) + H_2$, CO
    Proc. 7th I.V.C. and 3rd I.C.S.S.,
    Vienna, 1977, p. 1019.

26. G. Dalmai-Imelik, J.C. Bertolini, $Ni(111) + C_6H_{10}$
    J. Massardier, J. Rousseau, B. Imelik,
    Proc. 7th I.V.C. and 3rd I.C.S.S.,
    Vienna, 1977, p. 1179.

27. S. Andersson, $Ni(100) + O_2$, CO
    Solid State Commun. $\underline{24}$, 183 (1977)

28. J.C. Bertolini, G. Dalmai-Imelik,   Ni(100) + C_6H_6
    J.Rousseau, Surf. Sci. <u>67</u>, 478 (1977). Ni(111) + C_6H_6

29. C. Backx, R.F. Willis, B. Feuerbacher,   W(110) + CO, C_2H_2
    B. Fitton, Surf. Sci. <u>68</u>, 516 (1977). W(111) + CO

30. J.C. Bertolini, G. Dalmai-Imelik,   Ni(111) + CO
    J. Rousseau, Surf. Sci. <u>68</u>, 539
    (1977)

31. J.C. Bertolini, G. Dalmai-Imelik, J.   Ni(111) + CO
    Rousseau, J. Microsc. Spectrosc.   Ni(100) + NiO
    Electron. <u>2</u>, 575 (1977).

32. J.C. Bertolini, G. Dalmai-Imelik,   Ni(111) + CO
    J. Rousseau, C.R. Acad. Sci. (Paris)
    C <u>285</u>, 409 (1977).

33. A. Adnot   W(100) + H_2, O_2, CO
    Ph.D Thesis, Univ. Laval, Québec   St. Steel + O_2, H_2O, NH_3,
    (1977) CH_3OH, C_2H_5OH, CH_3COCH_3

34. J.C. Bertolini, J. Massardier,   Ni(111) + C_2H_2, C_2H_4
    G. Dalmai-Imelik, C.R. Acad. Sci.
    (Paris) C <u>285</u>, 515 (1977).

35. A. Adnot, J.D. Carette   Stainless Steel + O_2
    J. Microsc. Spectrosc. Electron. <u>3</u>,
    35 (1978).

36. C. Backx, R.F. Willis,   W(110) + C_2H_2
    Chem. Phys. Lett. <u>53</u>, 471 (1978).

37. H. Ibach, S. Lehwald,   Pt(111) + H_2, C_2H_2, C_2H_4
    J. Vac. Sci. Technol. <u>15</u>, 407
    (1978).

38. S. Andersson   Ni(100) + H_2
    Chem. Phys. Lett. <u>55</u>, 185 (1978).

39. J.E. Demuth, H. Ibach, S. Lehwald,   Ni(111) + C_2H_4, C_6H_{12}
    Phys. Rev. Lett. 40, 1044 (1978).   Pt(111) + C_2H_4, C_6H_{12}

40. W. Ho, R.F. Willis, E.W. Plummer,   W(100) + H_2
    Phys. Rev. Lett. <u>40</u>, 1463 (1978).

41. J.C. Bertolini, B. Imelik,   Ni(111) + H_2, CO
    Ned. Tijdschr. Vac. Techn. <u>16</u>,
    24 (1978).

42. G.E. Thomas, W.H. Weinberg,   Ru(100) + CO
    Ned. Tijdschr. Vac. Techn. 16,
    57 (1978).

43. G. Pirug, H. Hopster, H. Ibach,   Pt(111) + CO; Pt(100)+CO
    Ned. Tijdschr. Vac. Techn. 16,   Pt [6(111) x ($\bar{1}$11)] + CO
    152 (1978).

44. W. Erley, H. Ibach, H. Wagner,   Ni(111) + CO
    Ned. Tijdschr. Vac. Techn. 16,
    154 (1978).

45. R.F. Willis, W. Ho, E.W. Plummer,   W(100) + H_2
    Ned. Tijdschr. Vac. Techn. 16,
    357 (1978).

46. H. Ibach,   Ni(111) + CO, C_2H_2, C_6H_{12}
    Proc. Int. Conf. Vibrations in   Pt(111) + CO, C_2H_2, C_6H_{12}
    Adsorbed Layers, Jülich, 1978,
    p. 64.

47. R.F. Willis,   W(100) + H_2
    Proc. Int. Conf. Vibrations in
    Adsorbed Layers, Jülich, 1978,
    p. 76.

48. J.C. Bertolini, G. Dalmai-Imelik,   Ni(111) + C_2H_2
    J. Rousseau,
    Proc. Int. Conf. Vibrations in
    Adsorbed Layers, Jülich, 1978,
    p. 82.

49. S. Lehwald, H. Ibach, J.E. Demuth   Ni(111) + C_6H_6
    Proc. Int. Conf. Vibrations in   Pt(111) + C_6H_6
    Adsorbed layers, Jülich, 1978,
    p. 86.

50. J.E. Demuth,   Ni(111) + C_2H_2
    Proc. Int. Conf. Vibrations in   Pt(111) + C_2H_2
    Adsorbed Layers, Jülich, 1978,
    p. 96.

51. G.E. Thomas, W.H. Weinberg,   Ru(100) + O_2, CO
    Proc. Int. Conf. Vibrations
    in Adsorbed Layers, Jülich, 1978,
    p. 97.

52. S. Andersson,   Ni(100) + O_2, S
    Proc. Int. Conf. Vibrations in
    Adsorbed Layers, Jülich, 1978,
    p. 103.

53. A. Adnot,
    Proc. Int. Conf. Vibrations in
    Adsorbed Layers, Jülich, 1978,
    p. 109.

    St. Steel + O_2, NH_3, H_2O
    CH_3OH, C_2H_5OH, CH_3COCH_3

54. A. Adnot, J.D. Carette
    Surf. Sci. $\underline{75}$, 109 (1978).

    W(100) + CO

55. J.C. Bertolini, J. Massardier,
    G. Dalmai-Imelik, J. Chem. Soc.
    Far. Trans. I, $\underline{74}$, 1720 (1978).

    Ni(111) + C_2H_2

56. H. Ibach, S. Lehwald,
    Surf. Sci. $\underline{76}$, 1 (1978).

    Pt(111) + NO

57. J.C. Bertolini, J. Billy,
    J. Massardier, Le Vide, les
    Couches Minces, n° spécial,
    1978, p. 55.

    Ni(111) + C, CO

58. H. Hopster, H. Ibach,
    Surf. Sci. $\underline{77}$, 109 (1978).

    Pt(111) + CO
    Pt[6(111) x $(\overline{1}11)$] + CO

59. H. Ibach, D. Bruchmann,
    Phys. Rev. Lett. $\underline{41}$, 958 (1978).

    Pt[6(111) x $(\overline{1}11)$]
    Pt(111) + CO

60. G.E. Thomas, W.H. Weinberg,
    J. Chem. Phys. $\underline{69}$, 3611 (1978).

    Ru(100) + O_2

61. G.E. Thomas, W.H. Weinberg,
    Phys. Rev. Lett. $\underline{41}$, 1181 (1978).

    Ru(100) + NO

62. S. Andersson, J.W. Davenport,
    Solid State Commun. $\underline{28}$, 677 (1978).

    Ni(100) + CO, H_2O
    NiO(111) + OH

63. J.E. Demuth, H. Ibach,
    Surf. Sci. $\underline{78}$, L238 (1978).

    Ni(111) + C_2H_2

64. S. Lehwald, H. Ibach, J.E. Demuth,
    Surf. Sci. $\underline{78}$, 577 (1978).

    Pt(111) + C_6H_6
    Ni(111) + C_6H_6

65. M.R. Barnes, R.F. Willis,
    Phys. Rev. Lett. $\underline{41}$, 1729 (1978).

    W(100) + H_2

66. S. Andersson,
    Surf. Sci. $\underline{79}$, 385 (1979).

    Ni(100) + O_2, S

67. J.E. Demuth, H. Ibach,
    Chem. Phys. Lett. $\underline{60}$, 395 (1979).

    Ni(111) + CH_3OH

68. G.E. Thomas, W.H. Weinberg,   Ru(100) + CO, O_2
    J. Chem. Phys. 70, 954 (1979).

69. J.C. Bertolini, B. Imelik,   Ni(111) + H_2, CO
    Surf. Sci. 80, 586 (1979).

70. R.F. Willis, W. Ho, E.W. Plummer,   W(100) + H_2
    Surf. Sci. 80, 593 (1979).

71. W. Erley, H. Wagner, H. Ibach,   Ni(111) + CO
    Surf. Sci. 80, 612 (1979).

72. G.E. Thomas, W.H. Weinberg,   Ru(100) + CO
    J. Chem. Phys. 70, 1437 (1979).

73. J.F. Wendelken, M.V.K. Ulehla,   Cu(110) + CO
    J. Vac. Sci. Technol. 16, 441 (1979).

74. S. Lehwald, W. Erley, H. Ibach,   Ni[5(111) x ($\bar{1}$10)] + C_2H_2
    H. Wagner
    Chem. Phys. Lett. 62, 360 (1979).

75. G.E. Thomas, W.H. Weinberg,   Ru(100) + CO
    Rev. Sci. Instrum. 50, 497 (1979).

76. J.C. Bertolini, J. Rousseau,   Ni(111) + C_2H_4
    Surf. Sci. 83, 531 (1979).

77. W. Erley, H. Ibach, S. Lehwald,   Ni[5(111) x ($\bar{1}$10)] + CO
    H. Wagner,
    Surf. Sci. 83, 585 (1979).

78. B.A. Sexton,   Cu(100) + CO
    Chem. Phys. Lett. 63, 451 (1979).

79. J.E. Demuth, H. Ibach,   Ni(111) + C_2H_2
    Surf. Sci. 85, 365 (1979).

80. B.A. Sexton,   Cu(100) + O_2, CO, H_2O
    J. Vac. Sci. Technol. 16, 1033
    (1979).

81. G. Pirug, H.P. Bonzel, H. Hopster, H. Pt(100) + NO
    Ibach, J. Chem. Phys. 71, 593 (1979).

82. P.A. Thiel, W.H. Weinberg, J.T.   Ru(100) + CO, NO
    Yates Jr.,
    J. Chem. Phys. 71, 1643 (1979).

83. L.H. Dubois, G.A. Somorjai, $Rh(111) + CO, CO_2$
    Surf. Sci. <u>88</u>, L13 (1979).

84. B.A. Sexton, $Cu(100) + CH_3COOH$
    Chem. Phys. Lett. <u>65</u>, 469 (1979).

85. B.A. Sexton, $Pt(111) + H_2O$
    G.M. Res. Publ. GMR - 3078 (1979).

86. A.M. Baró, H. Ibach, H.D. Bruchmann, $Pt(111) + H_2$
    Surf. Sci. <u>88</u>, 384 (1979).

87. R.J. Behm, K. Christmann, G. Ertl, $Pd(100) + CO$
    M.A. Van Hove, P.A. Thiel, W.H.
    Weinberg,
    Surf. Sci. <u>88</u>, L59 (1979).

88. B.A. Sexton, $Cu(100) + CH_3OH, C_2H_5OH$
    Surf. Sci. <u>88</u>, 299 (1979).

89. B.A. Sexton, $Cu(100) + HCOOH$
    Surf. Sci. <u>88</u>, 319 (1979).

90. S. Lehwald, H. Ibach, $Ni(111) + C_2H_2, C_2H_4 . C_6H_{12}$
    Surf. Sci. <u>89</u>, 425 (1979) $Ni[5(111) \times (\bar{1}10)] + C_2H_2$
                                          C_2H_4, C_6H_{12}

91. R.F. Willis, $W(100) + H_2$
    Surf. Sci. <u>89</u>, 457 (1979).

92. J.C. Bertolini, J. Rousseau, $Ni(100) + C_6H_6$
    Surf. Sci. <u>89</u>, 467 (1979). $Ni(111) + C_6H_6$

93. S. Andersson, $Cu(100) + CO$
    Surf. Sci. <u>89</u>, 477 (1979).

94. P.A. Thiel, W.H. Weinberg, $Ru(100) + NO$
    J.T. Yates Jr.,
    Chem. Phys. Lett. <u>67</u>, 403 (1979).

95. A.M. Baró, H. Ibach, $Pt(111) + CO$
    J. Chem. Phys. <u>71</u>, 4812 (1979).

96. W. Ho, N.J. DiNardo, E.W. Plummer $W(100) + H_2$
    J. Vac. Sci. Technol. <u>17</u>, 134 $Ni(111) + H_2$
    (1980).

97. B.A. Sexton $Cu(100) + HCOOH, CH_3COOH$
    J. Vac. Sci. Technol. <u>17</u>, 141 (1980).

98. G.B. Fischer, B.A. Sexton, J.L. Gland,
    J. Vac. Sci. Technol. <u>17</u>, 144 (1980). $Pt(111) + O_2$

99. H. Ibach, S. Lehwald,
    Surf. Sci. <u>91</u>, 187 (1980). $Pt(100) + H_2O$

100. L.H. Dubois, G.A. Somorjai,
     Surf. Sci. <u>91</u>, 514 (1980). $Rh(111) + CO, CO$

101. H. Ibach, D. Bruchmann,
     Phys. Rev. Lett. <u>44</u>, 36 (1980). $Ni(111) + O_2$, surface phonons.

102. A.M. Barò, H. Ibach,
     Surf. Sci. <u>92</u>, 237 (1980). $Pt[(6(111) \times (\bar{1}11)] + H_2$

103. G.B. Fischer, B.A. Sexton,
     Phys. Rev. Lett. <u>44</u>, 683 (1980). $Pt(111) + H_2O$

104. J.L. Gland, B.A. Sexton,
     Surf. Sci. <u>94</u>, 355 (1980). $Pt(111) + NO$

105. B.A. Sexton,
     Surf. Sci. <u>94</u>, 435 (1980). $Pt(111) + H_2O$

106. L.H. Dubois, D.G. Castner, G.A. Somorjai,
     J. Chem. Phys. <u>72</u>, 5234 (1980). $Rh(111) + C_2H_2, C_2H_4$

107. S. Andersson, B.N.J. Persson, T. Gustafsson, E.W. Plummer,
     Solid State Commun. <u>34</u>, 473 (1980). $Cu(100) + CO$

108. W. Ho, R.F. Willis, E.W. Plummer,
     Surf. Sci. <u>95</u>, 171 (1980). $W(100) + N_2$

109. C. Backx, C.P.M. de Groot, P. Biloen
     Ned. Tijdschr. Vac. Techn. <u>18</u>, 35 (1980) $Ag(110) + O_2, C_2H_4$

110. S. Lehwald, H. Ibach,
     Ned. Tijdschr. Vac. Techn. <u>18</u>, 71 (1980). $Ni(111) + C_2H_2$

111. W. Ho, R.F. Willis, E.W. Plummer,
     Phys. Rev. B <u>21</u>, 4202 (1980). $W(100) + H_2$

112. J.L. Gland, B.A. Sexton, G.B. Fisher
     Surf. Sci. <u>95</u>, 587 (1980). $Pt(111) + O_2$

113. J.C. Bertolini, B. Tardy, $Ni(110) + C, CO$
     Proc. E.C.O.S.S. 3 (Cannes, Le
     Vide, les Couches Minces, suppl.
     201, 209 (1980).

114. S. Lehwald, J.T. Yates Jr., $Ni(111) + NO$
     H. Ibach,
     Proc. E.C.O.S.S. 3 (Cannes), Le
     Vide, les Couches Minces, suppl.
     201, 221 (1980).

115. C. Backx, C.P.M. de Groot, P. Biloen, $Ag(110) + O_2, C_2H_4$
     Proc. E.C.O.S.S. 3 (Cannes), Le
     Vide, les Couches Minces, suppl.
     201, 248 (1980).

116. M.A. Van Hove, L.H. Dubois, R.J. $Rh(111) + CO, CO_2, C_2H_2,$
     Koestner, G.A. Somorjai, C_2H_4
     Proc. E.C.O.S.S. 3 (Cannes), Le
     Vide, les Couches Minces, suppl.
     201, 287 (1980).

117. P.A. Thiel, F.M. Hoffmann, W.H. $Ru(100) + H_2O$
     Weinberg,
     Proc. E.C.O.S.S. 3 (Cannes), Le
     Vide, les Couches Minces, suppl.
     201, 307 (1980).

118. A. Ortega, A. Garbout, F.M. $Pd(100) + CO$
     Hoffmann, W. Stenzel, R. Unwin,
     K. Horn, A.M. Bradshaw,
     Proc. E.C.O.S.S. 3 (Cannes), Le
     Vide, les Couches Minces, suppl.
     201, 335 (1980).

119. A.M. Baró, H. Ibach, $Pt(111) + C_2H_4$
     Proc. E.C.O.S.S. 3 (Cannes), Le
     Vide, les Couches Minces, suppl.
     201, 458 (1980).

120. S. Andersson, B.N.J. Persson, $Cu(100) + CO$, surface
     Phys. Rev. Lett. 45, 1421 (1980).   phonons.

121. S. Lehwald, H. Ibach, $Ni(100) + O_2$, surface
     Proc. Int. Conf. Vibrations at   phonons
     Surfaces, Namur, 1980. p. xxx.

122. A. Adnot $W(100) + H_2, O_2, CO$
     Proc. Int. Conf. Vibrations at
     Surfaces, Namur, 1980, p. xxx.

123. J.C. Bertolini, B. Tardy, $Pt_{10}Ni_{90}(111)$ + CO
     Proc. Int. Conf. Vibrations at
     Surfaces, Namur, 1980, p. xxx.

124. J.F. Wendelken,   Cu(110) + CO, NO
     Proc. Int. Conf. Vibrations at
     Surfaces, Namur, 1980, p. xxx.

125. A.M. Barò, S. Lehwald, H. Ibach,   Ni(111) + C_2H_4
     Proc. Int. Conf. Vibrations at   Pt(111) + C_2H_4
     Surfaces, Namur, 1980, p. xxx.

126. S. Andersson,   Cu(100) + CO, surface
     Proc. Int. Conf. Vibrations at   phonons
     Surfaces, Namur, 1980, p. xxx.

127. P.A. Thiel, F.M. Hoffmann, W.H.   Ru(100) + H_2O
     Weinberg,
     Proc. Int. Conf. Vibrations at
     Surfaces, Namur, 1980, p. xxx.

128. J.E. Demuth, P.N. Sanda, J.C. Tsang   Ag(111) + C_6H_6, C_5H_5N
     Proc. Int. Conf. Vibrations at
     Surfaces, Namur, 1980, p. xxx.

129. C. Backx, C.P.M. de Groot, P. Biloen   Ag(110) + O_2
     Surf. Sci. (to be published).

130. P.A. Thiel, W.H. Weinberg,   Ru(100) + H_2, NO
     J. Chem. Phys. (1980), in press.

13⊦. F.M. Hoffmann, T.E. Felter, P.A.   Ru(100) + C_6H_{12}
     Thiel, W.H. Weinberg,
     J. Vac. Sci. Technol. (1981), to be
     published.

132. W. Erley,   Fe(110) + CO
     J. Vac. Sci. Technol. (1981), to
     appear.

133. H. Ibach, S. Lehwald,   Ni(111) + C_2H_2
     J. Vac. Sci. Technol. (1981) to
     appear.

134. J.C. Bertolini, B. Tardy   Ni(100),(110),(111)+C,CO
     Surf. Sci., in press.

135. T.E. Felter, F.M. Hoffmann,   Ru(100) + C_3H_6
     P.A. Thiel, W.H. Weinberg,
     in preparation.

136. H. Ibach, al., $Fe(110) + O_2$
     Solid State Commun., submitted.

137. J.F. Wendelken, $Cu(110) + O_2$
     to be published.

138. J.F. Wendelken, S. Lehwald, al., $Ni(100) + NO$
     work in progress.

REFERENCES

1. See, e.g. H. Froitzheim, in "Electron Spectroscopy for Surface
   Analysis",ed. H. Ibach, Springer-Verlag, 177, p. 205.
2. G. Dalmai-Imelik, J.C. Bertolini, J. Rousseau, Surf. Sci. 63, 67
   (1977).
3. R. Matz, H. Lüth, Proc. E.C.O.S.S. 3 (Cannes), Le Vide, les
   Couches Minces, suppl. 201, 762 (1980).
4. H. Onuki, H. Iwamoto, R. Onaka, Solid State Commun. 34, 941 (1980).
5. H. Weinberg, in "Experimental Methods of Surface Physics", ed. R.
   L. Park, (Academic Press, New York, 1980), to appear.

HIGH RESOLUTION ELECTRON SPECTROSCOPY : INSTRUMENTATION AND SOME RESULTS ON A TUNGSTEN SURFACE

A. Adnot

ISA RIBER, Paris, France

ABSTRACT

General principles for the design of electron spectrometers will be briefly given with emphasis on the 127° cylindrical and 180° hemispherical types. A high resolution electron spectrometer can be used for studying several aspects of surfaces and applications for LEED work (diffraction patterns and I-V curves) and Auger work will be demonstrated.
Finally, some adsorbate induced vibration spectra will be presented for a tungsten W(001) surface covered with H_2, CO and O_2.

THEORY OF ELECTRON ENERGY LOSS SPECTROSCOPY

Marijan Sunjic

Institute of Physics, University of Zagreb

41001 Zagreb, Croatia-Yugoslavia

ABSTRACT

Basic concepts of the Electron Energy Loss Spectroscopy are reviewed and applied to the investigation of the vibrational modes at crystal surfaces, in particular the adsorbed atom or molecule vibrations. The nature of the very low energy electronic states near transition metal surfaces and their coupling to the vibrations (resonant scattering, dipolar interaction) are discussed and compared with the gas phase situation.

Theoretical results for the differential cross sections are presented and analyzed, especially their characteristic energy dependence, angular distributions and selection rules which may serve to identify the rôle of the different excitation mechanisms when compared to experimental observations.

The origin of the possible selection rules ("dipole", "specular scattering") are discussed and their quantitative consequences for the observability of vibrational modes are discussed. Some recent advances and prospects for future theoretical work are also reviewed.

ADSORPTION SITES OF CO ON THE (111) FACE OF $Pt_{10}Ni_{90}$ SINGLE-CRYSTAL ALLOY

J.C. Bertolini, B. Tardy

Institut de Recherche sur la Catalyse
2, Avenue Albert Einstein
69626 Villeurbanne, France

ABSTRACT

The surface vibrations of CO chemisorbed at room temperature on the close-packed (111) face of Pt-Ni single crystal alloy (10 atom % Pt) have been investigated by vibrational electron energy loss spectroscopy. During adsorption two well-separated CO stretching modes develop at respectively 1820 ± 10 and 2070 ± 10 cm^{-1} while, in the low frequency region, a well defined loss peak at 440 ± 10 cm^{-1} and a shoulder near 365 cm^{-1} appear simultaneously. The 2070 cm^{-1} loss peak, lower than the 1820 cm^{-1} peak for small exposures (< 0.35 L), dominates the spectrum at saturation in our experimental conditions (exposure $\geqslant$ 1.5 L, 300 K, P_{CO} < 10^{-8} Torr) : $I_{2070}/(I_{2070} + I_{1820})$ = 64 %. These vibrations are attributed to CO molecules adsorbed linearly (2070 and 440 cm^{-1}) and multi-bonded (1820 and ~ 365 cm^{-1}) respectively.

Besides CO adsorption state closely similar to that obtained on Ni(111) (ν_{CO} = 1820 cm^{-1}) - i.e. multibonded CO species -, alloying with small amount of platinum makes the surface able to adsorbe large amounts of CO in on-top position.

A COMPARISON OF THE REFLECTION-ABSORPTION INFRARED METHOD WITH THE ELECTRON ENERGY LOSS METHOD FOR STUDYING VIBRATIONS AT SURFACES

Robert G. Greenler

Department of Physics and Laboratory for Surface Studies
University of Wisconsin-Milwaukee
Milwaukee, Wisconsin 53201

INTRODUCTION

Our original insight into the geometric structure of molecules resulted from an understanding of their infrared absorption spectra, which was interpreted in terms of the vibrations and rotations of the molecules. Although the vibrational spectra of molecules can be used for a wide variety of investigations, the primary information yielded by a spectrum concerns the geometric structure of the molecule and the nature of the bonds between the atoms. When we obtain the vibrational spectrum of molecules adsorbed on a solid surface, we obtain the same kind of information; but now the geometric structure includes the surface atoms to which the molecule is attached and we get information not only about the intramolecular bonds but also about the bonds between the adsorbed molecule and the surface atoms.

I will restrict my remarks to two methods for obtaining the vibrational spectra of adsorbed molecules on extended metal surfaces, reflection-absorption infrared spectroscopy (RAIRS) and electron energy loss spectroscopy (EELS). Traditionally Raman spectroscopic data has been used with the infrared data to provide very useful information about complex molecules and it may be that such a combination will continue to provide insight into surface problems. Surface-enhanced Raman spectroscopy is in such a developing state that is is not yet clear what its sensitivities and limitations will be and so I will not include it in this comparison.

REFLECTION-ABSORPTION INFRARED SPECTROSCOPY

The RAIRS experiment can be described very simply. Infrared
radiation is reflected at a high angle of incidence from a metal
surface on which there are adsorbed molecules. The reflected beam
enters a spectrometer and the resulting spectrum shows absorption-
type bands resulting from vibrations of the surface molecules[1].

An understanding of the mechanism by which the infrared energy
is transferred to the molecule is important to a comparison of the
two techniques. To help foster that understanding I will belabor
the mechanism by describing it in three different ways, different
ways to consider the same process.

1. Infrared excitation of a vibrational mode of a molecule requires
   that the vibration has associated with it an oscillating dipole
moment and that the electric field (from the infrared electromagnetic
wave) has a component along the axis of the oscillating dipole.

Fig. 1A shows the results of representative calculation of how
the absorption intensity (represented by ΔR) depends on angle of
incidence. ΔR is defined by fig. 1B. The geometry of the E vectors
for the effective polarization is shown in fig. 1C. If the experiment
is done using the high angles of incidence prescribed by fig. 1A, it
is clear from 1C that the vibrating E field (which is the combination
of the fields of the incident and reflected waves with the appropriate
reflection phase shift) is essentially normal to the surface. It
follows that only those vibrations that have a component (of the
oscillating dipole) normal to the surface can be excited. This additional restriction on vibrational excitation has been named by
Pearce and Sheppard[2], the "surface selection rule".

2. A more general way to arrive at the same result uses the boundary
   condition that the tangential component of the E field at an
interface is continuous across the interface; i.e. it has the same
value immediately on either side of the boundary. Insofar as the
metal can be considered to be a perfect metal at these infrared
frequencies, the E field inside the metal is zero. These two considerations combine to indicate that the tangential E field outside
a perfect conductor is zero; or that the E field is always normal
to the surface of the conductor. The conclusion again, is that
only vibrations with a component perpendicular to the surface can
be excited by infrared radiation. Viewed in this way, the restriction
seems to be of a fundamental nature rather than the result of a
particular experimental arrangement.

The boundary-condition argument does not distinguish between
E-field produced by electromagnetic radiation or by electric charges.
The conventional image-charge approach to determining the field
resulting from a point charge near a conducting plane is illustrated

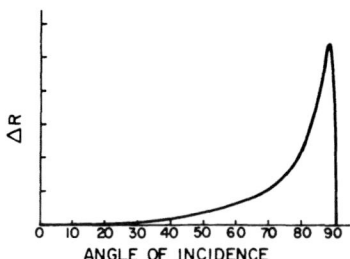

Fig. 1A - Calculation of the dependence of the absorbance (ΔR) of
          a thin layer on a metal surface as a function of angle
          of incidence. The optical constants used in the calcu-
          lation are for a metal similar to the copper at 2000 cm^{-1}.
          The result is for the p-polarization (polarization parallel
          to the plane of incidence).

in Fig. 2. The field outside the metal is the same as would result
from the charge plus that of its image, a charge of opposite sign
located on the other side of the interface. From this construction
it is clear that the field is normal to the surface.

3. Another way to treat this same problem is to use a trick familiar
   to workers in optics. Sometimes it is useful, rather than to
reflect light rays from a surface, to let the rays continue directly through the surface, and, instead, reflect the object space
through the mirror surface. The two approaches are illustrated in
Fig. 3A and B. Note that the appropriate optical image is a charge
of the same sign as the object, unlike the electrical image charge
of Fig. 2. An important physical feature in the reflection problem
is the phase shift in the electromagnetic wave that accompanies
reflection. The phase shifts for the two polarizations are shown in
Fig. 4. This can be incorporated into the model of Fig. 3B by

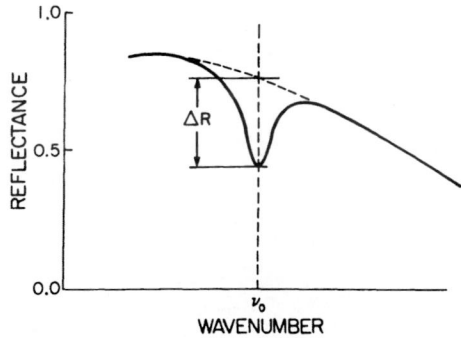

Fig. 1B - Definition of ΔR as the depth of an RA band superimposed
          on the background reflectance curve.

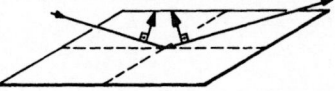

Fig. 1C – The E vectors for p-polarized radiation reflected at a
         high angle of incidence.

introducing an abrupt phase shift in the waves as they cross the
dotted line representing the mirror position. Note that the infrared
wavelengths are very long compared with the molecular dimension so
we can consider the instantaneous phase of the wave to be the same
over all parts of the molecule above the dotted line and shifted by
the amount shown in Fig. 4 for all parts of the molecule below the
dotted line. The maximum sensitivity of the RAIR experiment occurs
for angles of incidence that give a phase shift near 90°. So the
physical model (represented in 3B) is of two dipole oscillators,
constrained to move together (i.e., contract and expand in phase)
but driven by oscillating E fields that are 90° out of phase from
one oscillator to the other. To neglect this phase difference or to
approximate it by 0 or 180° is not reasonable under the conditions
of this experiment.

     If we think of driving an oscillator parallel to the surface,
we must consider an s-polarized wave (with E vector perpendicular
to the plane of incidence). It is reasonable to approximate the
phase shifts of the s-polarized wave as 180°. Fig. 5 attempts to
show the counterpart of Fig. 3B for an s-polarized wave interacting
with a dipole oscillator parallel to the surface. When the field is
trying to compress the dipole above the plane, it is trying to
expand the one below the plane because of the 180° phase shift.
Because the oscillators must move together, the forces cancel and
the oscillation parallel to the surface is not excited.

     It should be possible to treat the effect of the surface using
electrical image charges (a positive charge imaged as negative and
vice versa) rather than the optical image, but then the two oscillators will not move in phase, but with a phase given by the complex

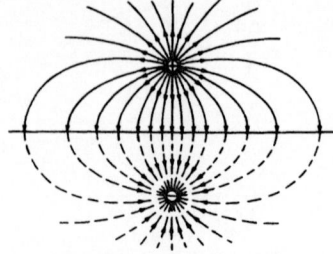

Fig. 2 – The image-charge model that gives the electric field lines
        between a point charge and a flat conducting plane.

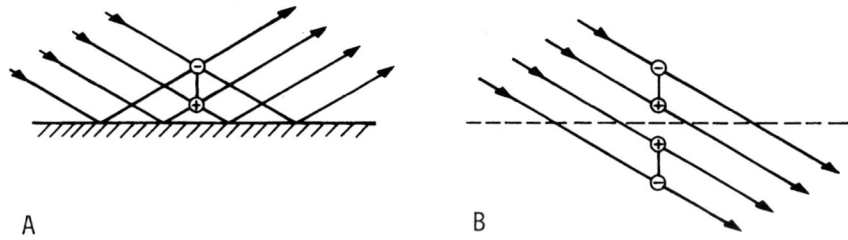

A   B

Fig. 3A - A dipole near a surface is affected by the fields of
         both the incident and reflected radiation.

Fig. 3B - An alternate way to view the interaction of 3A : the
         radiation interacts with the dipole and its optical image.
         There is an abrupt phase shift as the waves cross the
         dashed boundary.

polarizability of the metal. This model seems more complicated and
I will not pursue it here. The upshot of all these different points
of view is that infrared radiation, coupling to the oscillating
dipole moment of a molecule can only excite molecular vibrations that
have a component of the oscillating dipole perpendicular to the
surface - the surface selection rule.

ELECTRON ENERGY LOSS SPECTROSCOPY

     The EELS experiment can also be described simply. A beam of
monoenergetic, low-energy electrons is reflected (scattered) from
the metal surface on which molecules are adsorbed. The scattered
electrons are energy-analyzed and discrete energy losses are associated with the energies required to excite vibrations in the
adsorbed molecules. The energy-loss peaks give the vibrational
spectrum of the adsorbed molecule[3].

     At first glance the process looks like a Raman process : an
incident particle loses some energy to excite vibrational motion,
and it is the energy-loss - not the initial energy - that is
important. However, this analogy is misleading; we have to look at
the excitation mechanism to make a comparison with excitation by
electromagnetic waves. One possibility is that the changing electric
field is produced by the moving electron coupled with the dipole
moment of the molecule. For a few-volt electron, the time required
to move a molecular dimension is about 10^{-16} seconds; thus the field
observed by a molecule from such an itinerant electron would be a
pulse of half width a few times greater than 10^{-16} seconds. If we
Fourier-analyze such a pulse to determine its frequency spectrum we

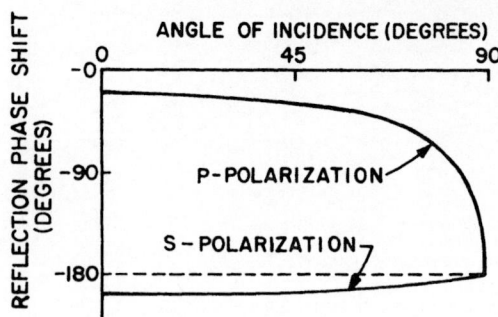

Fig. 4 - A typical dependence on angle of incidence of the reflection
phase shift occurring for a metal in the infrared spectral
region. δ_p is the phase shift for p-polarized radiation
(parallel to the plane of incidence) and δ_s is the phase
shift for s-polarized radiation (perpendicular to the
plane of incidence).

will find components of all lower frequencies (i.e., of all waves
with longer periods than the pulse width). Molecular vibrations
have periods typically in the range 10^{-13} to 10^{-14} seconds, corresponding to frequencies occurring in the frequency spectrum of
the E-field pulse. So the process of dipole excitation would really
correspond to the infrared absorption case. Furthermore, if we use
the same classical treatment as before, the electric field produced
by the electron, (as long as it is several molecular dimensions
away from the surface) will be normal to the surface in the region
occupied by the molecule and would therefore excite only vibrations
that followed the same selection rule as in the infrared case. This
seems to explain data obtained by EELS for electrons with a reflection
angle near the incident angle; i.e. for electrons diffracted near
the (0,0) direction. For such vibrational processes, we would expect
the same restriction as the normal infrared vibrational selection
rules plus the surface selection rule[4].

There is experimental evidence[5] and theory proposed to explain[6]
some other types of interaction. If the electron moves inside the
molecule, various short-range interactions can occur - different
from the relatively long-range dipole interaction already discussed.
These short-range forces give rise to excitation processes in which
the electron may be scattered through large angles. The transitions
are not restricted by the infrared selection rule and may involve
vibrations with no oscillating dipole moment or with a dipole moment
that oscillates parallel to the surface. Such vibrations may be allow-
able Raman excitations or they may even be vibrations that are
inactive for both Raman and infrared. Unlike the dipole excitation
process, these short-range interactions are not restricted to
electrons scattered near the specular direction and it seems

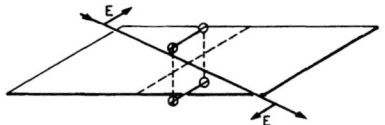

Fig. 5 -   s- polarized radiation interacts with a dipole oriented
           parallel to the surface and its optical image. The
           (approximate) 180° phase shift results in no net driving
           force on the dipole-image dipole system.

promising that we may be able to study such vibrations using
electron loss peaks measured in off-specular directions.

COMPARISONS BETWEEN THE TWO TECHNIQUES

     At first look the RAIRS and EELS show that they both yield
the vibrational spectra of adsorbed molecules and so might be
expected to be nearly equivalent techniques. A second look reveals
a number of significant differences. Each technique has its strengths
and its limitations. Some of the limitations are of a fundamental
nature whereas others are state-of-the-art limitations that may
be expected to change. I will compare these features where the
techniques differ.

A. Resolution

     Until recently, the best spectral resolution obtainable by
the EELS technique was about 55 cm^{-1} (7 mev)[7]. At this conference
examples have been shown[8] of loss peaks with 28 cm^{-1} (3.5 mev)
resolution. The limit in the present instruments results from space
charge effects on electron imaging and can be reached only with low
currents through the monochromator and analyzer. Thus the ultimate
in resolution is achieved by sacrificing the ultimate in sensitivity.

     A similar trade-off is made in the infrared case. Commercial
spectrophotometers can easily obtain resolutions better than 1 cm^{-1}
by using narrow spectrometer slits. To get a reasonable sensitivity
in RAIRS, however, resolution is usually traded for an improved
signal-to-noise ratio by opening the slits - typically to give a
resolution in the range 4 to 8 cm^{-1} (0.5 to 1.0 mev)[9]. The better
resolution of the RAIR technique may be significant in identifying
closely-spaced bands (for example stretching vibration of CH_2 and
CH_3 groups) or for accurately measuring small band shifts.

B. Sensitivity

The sensitivity of either technique results from a combination of operating compromises and is difficult to specify accurately. I will use the detection of adsorbed CO as a measure for comparison. On work to date, the detection limit for CO with RAIRS has been typically a coverage of a few percent of a monolayer[9], although a recent paper[10] shows a RA band for a coverage of CO on Ru(001) of 0.3 %. The EELS data show a detection capability of an order of magnitude lower coverage[3,7]. This difference is very significant as it allows the technique to be applied routinely to the study of hydrocarbon bands that are typically an order of magnitude weaker than the CO stretching vibrations.

There is the expectation that the RAIR sensitivity will increase in the next few years. Current state-of-the-art commercial spectrophotometers with modern data-handling methods have not been applied to this problem and their application will bring a significant improvement. The use of specially-constructed infrared spectrometers with supercooled bolometer detectors promise an additional improvement over commercially produced instruments.

C. Spectral (Energy) Range

The EELS technique covers the range of molecular vibrations out to a low frequency limit imposed by the rising background of the elastically scattered peak. The limit depends on the intensity of the loss peak but appears to be in the range 20-25 meV $(160 - 200 \text{ cm}^{-1})$. Commercially available, dispersive infrared spectrophotometers cover a similar range; e.g., Beckman model 4260 : 4000 to 200 cm^{-1}, and Perkin-Elmer model 580 : 4000 to 180 cm^{-1}. In the past, the spectral range of RAIRS has been limited in some investigations by the restricted range of available spectrometers, by the use of photoconductive detectors that limit the low frequency response, or by the problems of making an appropriate ultrahigh vacuum seal with a window of suitable infrared transparency. Satisfactory solutions to the window problem have been found[11].

D. Selection Rules

The molecular vibrations that can be studied by RAIRS are those that are infrared active and satisfy the surface selection rule. The technique is blind to vibrations where the dipole oscillation is parallel to the surface. One of the significant potential advantages of the EELS technique is the possibility for looking near the specular direction at the dipole excitations that are subject to the surface selection rule, and also looking in an off-specular direction at other vibrations that are restricted by

neither the surface selection rule nor the requirement that they
be infrared active.

E. Pressure Region

    As with all of the electron spectroscopies, EELS must operate
at a gas pressure no higher than about 10^{-5} torr. The infrared
technique can operate over any pressure range. Interference of
gasphase absorption bands can be minimized by using a polarization
modulation technique[12] that should eliminate gas phase spectra.
Often there is concern over whether reactions studied on surfaces
under UHV conditions are the significant ones for catalytic
processes that are normally operated at much higher pressures.
Potentially one of the important features of the RAIR technique is
that it may be able to bridge the experimental gap between the UHV,
controlled, clean-surface studies and the important, industrial
processes that take place at (or above) atmospheric pressure.

F. Nature of Samples

    Till now all the work with EELS has been done with single-
crystal metals. Such samples provide a concentration of scattered
electrons near the specular direction. With sufficient sensitivity,
it may be possible to detect loss peaks in the widely scattered
electrons from non-ordered surfaces. Any specularly reflecting
metal sample can be used as a RAIR sample. This includes a polished
polycrystalline metal or alloy, an evaporated metal or alloy film,
as well as a single-crystal metal sample.

G. Availability of Apparatus

    Through the considerable efforts of Ibach and his co-workers,
EELS systems have benefitted from careful design and engineering
development and are commercially available. The RAIRS system is in
a do-it-yourself state, requiring the marriage of three techniques
to produce a working system : ultrahigh vacuum, surface analysis
techniques, and infrared spectroscopy. The resulting investment of
experimental effort to set up a RAIRS system is considerable.

    I expect the advantages and limitation of each technique to
result in the continuing fruitful application of each to problems
involving adsorption and reactions at metal surfaces.

REFERENCES

1. The following reviews give a general description of the RAIR
   method : R.G. Greenler in CRC Critical Rev. Solid State Sci. 4,
   415 (1974); J. Pritchard in Dechema-Monographien, Moderne Ver-
   fahren der Oberflächenanalyse 78, 231 (1975); H.G. Tompkins in
   Methods of Surface Analysis, Vol. 1, Methods and Phenomena
   p. 472 (Elsevier, 1975); J. Pritchard and T. Catterick in
   Experimental Methods in Catalytic Research 3, 281 (1976); and
   M.J. Dignam and J. Fedyk, Appl. Spectrosc. Rev. 14, 249 (1979).
2. H.A. Pearce, N. Sheppard, Surface Sci. 59, 205 (1976).
3. H. Froitzheim, "Electron Energy Loss Spectroscopy in Electron
   Spectroscopy for Surface Analysis" ed. by H. Ibach (Springer
   Verlag, 1977).
4. H. Ibach, H. Hopster, B.A. Sexton, Appl. Surf. Sci. 1, 1 (1977).
5. W. Ho, R.F. Willis, E.W. Plummer, Phys. Rev. Lett. 40, 1463
   (1978) ; S. Andersson, J.W. Davenport, Solid State Commun. 28,
   677 (1978).
6. J.W. Davenport, W. Ho, I.R. Schrieffer, Phys. Rev. B17, 3115
   (1978) ; C.H. Li, S.Y. Tong, D.L. Mills, Phys. Rev. B21, 3057
   (1980).
7. H. Ibach in Conference Proceedings of Vibrations in Adsorbed
   Layers, held in Jülich, June 12-14, 1978, p. 64.
8. See papers by S. Lehwald and H. Ibach, and by S. Andersson in
   the Proceedings of this Conference.
9. See, for example : J.C. Campuzano, R.G. Greenler, Surface Sci.
   83, 301 (1979) ; A.M. Bradshaw, F.M. Hoffmann, Surface Sci. 72,
   513 (1978) ; A. Crossley, D. King, Surface Sci. 95, 131 (1980);
   P. Hollins, J. Pritchard, Surface Sci. 89, 486 (1979).
10. H. Pfnür, D. Menzel, F.M. Hoffmann, A. Ortega, A.M. Bradshaw,
    Surface Sci. 93, 431 (1980).
11. P. Hollins, J. Pritchard, J. Vac. Sci. Technol. 17, 665 (1980) ;
    J.T. Yates, T.M. Duncan, S.D. Worley, R.W. Vaughan, J. Chem.
    Phys. 70, 1219 (1979).
12. A.M. Bradshaw, F.M. Hoffmann, Surface Sci. 52, 449 (1975).

APPLICATIONS OF IR ELLIPSOMETRIC SPECTROSCOPY TO SURFACE STUDIES

M.J. Dignam

Department of Chemistry
University of Toronto - Toronto, Canada M5S 1A1

ABSTRACT

Several methods whereby the polarizing properties of a planar interface may be used to determine the vibrational spectrum of adsorbed species are presented, including two recently proposed near null techniques based on the principle of orthogonal reflections. A particularly promising approach to surface IR spectroscopy, called Fourier transform reflection ellipsometry (FTRE), is described which involves the use of a Fourier transform spectrometer equipped with a special polarizing interferometer.

Data obtained on a rotating polarizer ellipsometric spectrometer are presented to illustrate the potential of such techniques, but particularly of FTRE, for studying surface reactions. The data chosen are for CO adsorption on Ni(110) and Ni(100), CO and H_2 reaction on Ni(110), and the decomposition of CH_3OH on Ni(100).

INTRODUCTION

The importance of vibrational spectroscopy in surface science is well established. The optimum technique for obtaining such spectra, if indeed one exists, is still an open question. At present, however, the honours for such distinction appear to belong to high resolution electron energy loss spectroscopy (EELS), followed by infrared reflection spectroscopy (IRRS), including polarization and wavelength modulation variations. Other techniques, such as surface Raman vibrational spectroscopy, while extremely important in certain applications, have not yet proven themselves of general applicability.

The primary position of EELS derives mainly from its ability
to cover a wide energy range with high sensitivity, sensitivity
roughly ten times higher than has currently been achieved using
IRRS. The strength of IRRS, on the other hand, is derived mainly
from the high resolution achievable (10 to 100 times higher than
EELS) and its ability to handle high ambient pressures. Its weaknesses relative to EELS are lower sensitivity, already mentioned,
and the difficulty of covering the entire vibrational region,
particularly the far IR.

A more complete discussion of the strengths and weaknesses of
various surface vibrational spectroscopic techniques can be found
elsewhere[1-3]. Here, we first examine some strategies for over-
coming the weaknesses of IRRS, then present some data that empha-
sizes its strength.

COMPARISON OF DIFFERENT IRRS STRATEGIES

Good IRRS sensitivity has been obtained to date by each of the
following three techniques : wavelength modulation (WM)[4,5], polarization modulation (PM)[6-8], and Fourier transform (FT)[9] spectroscopies. Of the two modulation techniques, PM has the major potential
advantage over WM that, with proper optical design, the resulting
spectra are very insensitive to absorption anywhere in the optical
train except close to the test surface. Certain forms of PM spectroscopy called ellipsometric spectroscopy (ES)[1] , or spectroscopic
ellipsometry, yield dispersion as well as absorption information,
which could be of some advantage, but it is the former advantage
of PM that sets it appart and makes it in principle the preferred
technique for in situ studies of heterogeneous catalysis[3].

The advantages of FTIR spectroscopy over IR spectroscopy
performed using a dispersive instrument are well known[10]. In the
present context, they may be summarized as a range advantage (large
wavelength range covered in a single scan) and speed advantage. One
difficulty of applying FTIR to surface studies is that the speed
advantage is not always transformable into a sensitivity advantage
due to limitations on the accuracy of the analogue-to-digital
signal conversion process. However, Baker and Chesters[9] have
obtained excellent results via specular reflection FTIR.

To reap the full potential advantages of FTIR in reflection
spectroscopy, one would like to null out much of the reflected
light in such a way as to leave unchanged the absolute intensity
drop arising from the absorption properties of the adsorbed
species. There is no known way of accomplishing this ideal,
but two related approaches have been proposed recently by Dignam

and Baker[11-12] that match the above ideal in different aspects. One accomplishes the optical null, but at the expense of the absolute intensity change[12], while the other preserves the absolute intensity change, but the optical null is in the modulated components only[11]. Both make use of a reference surface and the principle of orthogonal reflections.

NEAR NULL FTIR REFLECTION SPECTROSCOPY

    As any null technique involving FTIR must be achromatic in nature, the only way of accomplishing a near null condition appears to be through the use of a reference surface, matched as closely as possible to the test surface, but subjected to constant environmental conditions. A direct approach along these lines is to use a dual beam FTIR spectrometer, such as that employed by Kuehl, Gomez-Taylor, and Griffiths[13]. Such an instrument generates two beams, each modulated by the special Michelson interferometer, but 180° out of   phase with each other. For identical samples in the two arms of the spectrometer, a time independent signal (zero FT signal) is produced when the two beams are superimposed at the IR detector. The above authors have achieved a 10 -fold increase in sensitivity over single beam operation for weakly absorbing solutions, by placing pure solvent in the reference arm. A similar increase in sensitivity can be anticipated for dual beam reflection FTIR. A superior approach exists, however, as it at once combines the potential advantages of dual beam reflectance FTIR and IR ellipsometric spectroscopy (IRES). The key to this approach is the polarizing Michelson interferometer (PMI), proposed originally by Martin and Puplett[14] for conventional far IR FT spectroscopy, and recently demonstrated in this application by Burton and Akimoto[15].

    A conventional scanning Michelson interferometer, for use in FTIR spectroscopy, is converted to a PMI by replacing the beam splitter by a metal grid polarizing beam splitter, and the planar retromirrors by rooftop mirrors oriented to rotate the beam (hence the polarization direction) by 90° (see Fig. 1). When polarized light is incident on such an interferometer, the recombining light beams, being orthogonally, linearly polarized, interfere to produce a beam of constant intensity (as the interferometer is scanned) but of varying polarization state. If this beam is reflected from a planar surface, with the plane of incidence set close to $\pm$ 45° with respect to the interferometer axis (grid direction of the polarizing beam splitter), the modulated portion of the reflected intensity for any given wavelength component, I^{ac}, is given by[11]

$$I^{ac} = \frac{1}{2} \mid K_{PMI} \mid^2 I_o \left[\mid R_s \mid^2 - \mid R_p \mid^2\right] \cos\omega t \qquad (1)$$

where K_{PMI} is a constant ~ 1, I_0 the light intensity incident on the PMI, ω the modulation angular frequency, determined by the interferometer scanning rate and the wavelength, and R_s and R_p the complex valued reflection coefficients for the reflecting surface. Insertion of a grid polarizer before the reflecting sur- face, oriented to produce s-polarized light at the surface, leads to the following expression for the modulated component of the intensity[11],

$$I_o^{ac} = \frac{1}{2} |K_{PMI}|^2 I_o T |R_s|^2 \cos\omega t, \tag{2}$$

so that one may write,

$$I^{ac} = I_o^{ac} T^{-1} (1 - |R_p|^2 / |R_s|^2), \tag{3}$$

where T is the transmittance of the polarizer for the strongly transmitted component. Thus upon letting $I_o^{ac} T^{-1}$ represent the effective intensity incident on the surface (T can be determined independently), the resulting FT spectrum yields $1 - |R_p|^2 / |R_s|^2$ versus wavenumber. Other polarizer arrangements will allow the simultaneous determination of the spectra of $|R_p|^2 / |R_s|^2$ and Δ the relative phase change for the interface[11]. The ellipso- metric absorption spectrum is derived entirely from $|R_p| / |R_s|$, however, with the ellipsometric absorbance A_e, defined according to [1-3],

$$A_e = \log_{10} [(|\bar{R}_p|^2 / |\bar{R}_s|^2)/(|R_p|^2 / |R_s|^2)] \tag{4}$$

$$= \log_{10}(|\bar{R}_p|^2 / |R_p|^2) - \log_{10}(|\bar{R}_s|^2 / |R_s|^2)$$

$$= A_p - A_s$$

where A_p and A_s are the conventional reflectance absorbances for p- and s-polarized light respectively, and the bar denotes the "bare" surface or reference surface state. For highly reflecting substrates, such as metals in the IR, A_s is essentially zero up to coverages of several monolayers so that A_e and A_p become in- distinguishable.

By using a PMI, a FTIR instrument can operate in an ellipso- metric mode and thereby gain all the advantages of both FTIR and ellipsometric surface vibrational spectroscopy. This technique will be referred to as Fourier transform reflectance ellipsometry (FTRE).

To gain the advantage that goes with dual beam FTIR, the
single reflection, described above, must be replaced by a pair
of "orthogonal" reflections, one from the test surface, the other
from a reference surface. We define orthogonal reflections as
having the same angle of incidence, but orthogonal planes of incidence, so the component of the amplitude vector that is p-polarized
for one surface is s-polarized for the other surface. To take account
of the reflection from the reference surface, R_p is replaced by
$R_p R_s{}^o$, and R_s by $R_s R_p{}^o$, giving in place of eq. (3), the following,

$$I^{ac} = I_o{}^{ac} \, T^{-1}[\; 1 \; - \; (|R_p|^2/|R_s|^2)/(|R_p{}^o|^2/|R_s{}^o|^2)] \; , \tag{5}$$

where the superscript ^{o} refers to the reference surface. For the two
surfaces exactly matched, $I^{ac} = 0$, as for a dual beam instrument.
Otherwise, A_e is determined as for the single reflection case. In
either case, A_e is given by

$$A_e = \log_{10}(1 - T\bar{I}^{ac}/\bar{I}_o{}^{ac}) - \log_{10} (1 - TI^{ac}/I_o{}^{ac}) \tag{6}$$

but when a closely matched reference surface is used, this reduces
approximately to

$$A_e \simeq (T/2.303)(I^{ac}/I_o{}^{ac} - \bar{I}^{ac}/\bar{I}_o{}^{ac}). \tag{7}$$

FTRE with orthogonal reflections achieves a near null condition,
but it is only the modulated component of the intensity on the IR
detector that is near null, the time-independent component is very
large. Frequently, this will prevent use of the detector with the
highest detectivity for the spectral range of interest, due to
detector saturation. A technique operating close to a full optical
null condition is clearly desirable, and readily achievable, but at
a high price[12]. Thus a combination of a conventional FTIR instrument,
with the orthogonal reflections occurring between two linear polarizers, leads to the following expression for the total intensity
(both constant and modulated components) incident on the detector[12],

$$I_\pm = \tfrac{1}{8}T^2 I_o{}' |R_p R_s{}^o \pm R_s R_p{}^o|^2 \; , \tag{8}$$

where $I_o{}'$ is the total intensity incident on the first polarizer.
The + sign is for parallel polarizers set $\pm 45°$ to the planes of
incidence, while the − sign is for crossed polarizers likewise set
$\pm 45°$ to the planes of incidence. For closely matched surfaces

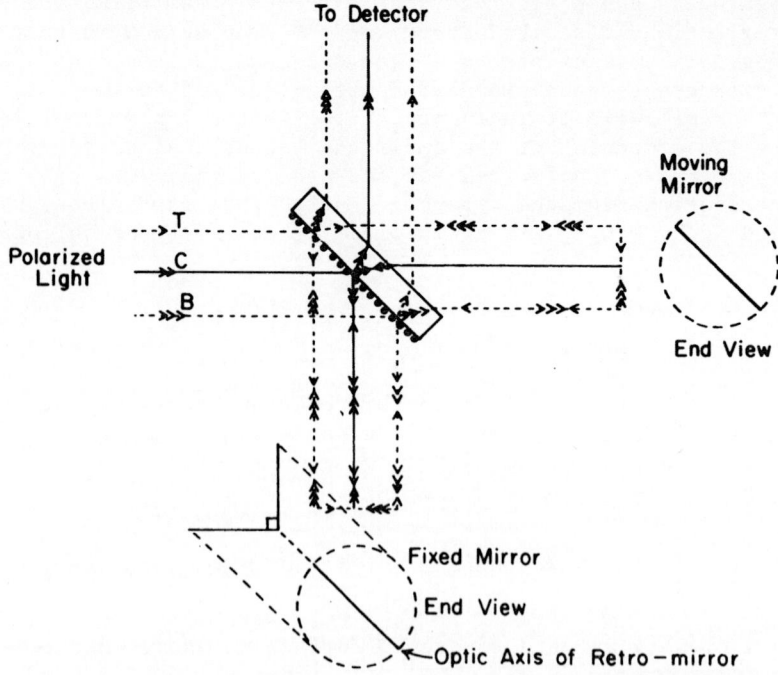

Fig. 1 - Layout of polarizing Michelson interferometer reproduced
from Dignam and Baker[11] by permission of J. Appl. Spec.
The metal grid lines on the front face of the beam splitter
are exaggerated for illustrative purposes. The rays designated T and B lie in the top right and bottom left of
the incident beam, and are interchanged by the roof-top
retro-mirrors. The ray designated C is centrally positioned
and sent back along itself by the retro-mirrors.

eq. (8) leads to

$$A_e \simeq (4/2.303) \left| (I_-/I_+)^{1/2} \pm (\bar{I}_-/\bar{I}_+)^{1/2} \right|, \tag{9}$$

while for $\left| I_- - \bar{I}_- \right|/\bar{I}_- = \left| \Delta I_- \right|/\bar{I}_- \ll 1$

this in turn reduces to,

$$A_e \simeq (2/2.303)\rho_{ex}^{1/2}(\left| \Delta I_- \right|/\bar{I}_-), \tag{10}$$

where the extinction ratio $\rho_{ex} \equiv \bar{I}_-/\bar{I}_+$. In conventional reflectance spectroscopy, the reflectance absorbance is related to the intensity change according to,

$$A_p = -(1/2.303)(\Delta I_p / \bar{I}_p) \tag{11}$$

from which it follows that introducing the reference surface and crossed polarizers leads to an increase in the relative intensity change, $|\Delta I|/I$, by a factor of $1/(2\rho_{ex}^{1/2})$. However, this is accompanied by a decrease in the base intensity from $\bar{I}_p$ to $\bar{I}_-$, i.e. of about a factor of $\frac{1}{4}\rho_{ex}$, and hence by a decrease in the absolute intensity change (from $|\Delta I_p|$ to $|\Delta I_-|$) of about $\frac{1}{4}\rho_{ex}/(2\rho_{ex}^{1/2}) = \frac{1}{8}\rho_{ex}^{1/2}$. A sensitivity gain will be achieved, therefore, only if this loss in intensity change due to adsorbed species can be made up either by increasing the optical aperture, or the detectivity of the detector, or a combination of the two. While this could frequently be done, there are other problems with this approach arising from the non-linear relationship between reflected intensity and absorption. It appears, therefore, that FTRE offers the most promising approach to IR surface vibrational spectroscopy, an approach made even more appealing by the achromatic nature of polarizing, as opposed to conventional beam splitters. Thus using modern lithographic techniques, it should be possible to deposit metal gratings of spacing ~ 0.1 μm on substrates such as KBr (perhaps even CsI), thus making possible polarization FTIR instruments covering 25,000 cm^{-1} to $\gtrsim 3$ cm^{-1} in two ranges, 25,000 cm^{-1} to 450 cm^{-1} using KBr supported polarizers, and 600 to $\gtrsim 3$ cm^{-1} using unsupported grid polarizers. UHV windows made of CsI would cover the range down to 250 cm^{-1}. For frequencies below this, a combination of organic polymer windows and differential pumping would be required.

In summary, FTRE should provide sensitivities higher than conventional reflection FTIR, giving sensitivities and speeds rivaling those of EELS with resolution ~ 1 cm^{-1}. Since EELS covers down to ~ 100 cm^{-1} in a single range, while even with CsI optics, FTRE would cover down to only 250 cm^{-1} in a single range, the range advantage of EELS would to some extent be retained. However, FTRE would cover down to well below 100 cm^{-1} (i.e. ~ 10 cm^{-1} for polyethylene optics) in only 2 ranges. It would also possess the main advantage of polarization modulation spectroscopy, that of insensitivity to the absorption properties of the ambient gas phase.

As we are only just beginning to develop FTRE, the data presented in the following section, were obtained by IR ellipsometric spectroscopy (IRES), and chosen to illustrate the potential IRES, and hence by inference FTRE, for surface studies in general, but particularly for studying surface reactions.

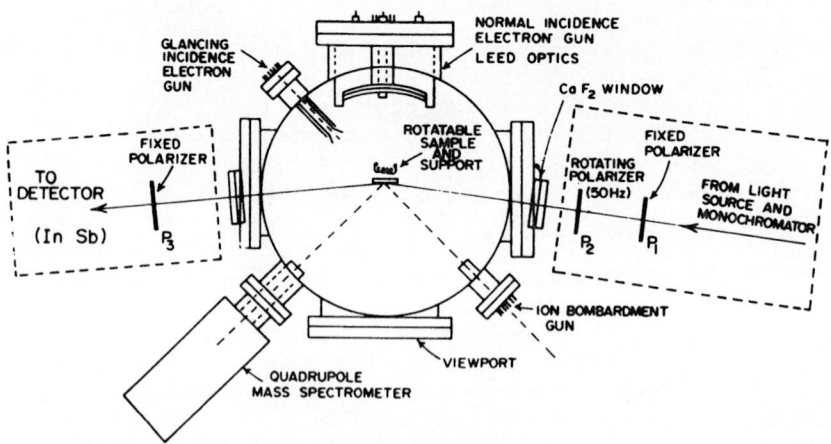

Fig. 2 - Diagram of UHV chamber showing also in schematic form the
         IR ellipsometer, reproduced from Baudais et al[17] by per-
         mission of Surface Sci.

SOME APPLICATIONS OF IR ELLIPSOMETRIC SPECTROSCOPY

IR ellipsometer

The Toronto instrument, described in detail elsewhere[16],
consists of the following optical train (see Fig. 2) : a glow bar
source; $\frac{1}{2}$ meter, grating monochromator with a 7.5 cm grating; a
stationary, grid polarizer; a grid polarizer rotating at 50 Hz; the
reflection cell; a third polarizer, stationary; and a detector
(LN$_2$- cooled InSb detector for all of the data reported here). The
grating and slit spacing give a resolution of about 6 cm^{-1} at
2000 cm^{-1} . This configuration leads to a detector signal that is
modulated at both 100 and 200 Hz. The electronics determine the
phases of these signals relative to the rotating polarizer (using
level crossing detectors and digital timers) to a precision of
about 10^{-3} degrees for one second averaging. From these phase values,
both the modulus and argument of R_p/R_s ($\tan\Psi \equiv |R_p|/|R_s|$ and Δ res-
pectively) can be calculated, and hence changes in these due to
adsorbed species. The change in the former leads to the reflectance
absorbance spectrum of the adsorbed species through eq. (4). The
dispersion spectra, $\bar{\Delta}$ - Δ vs. wavenumber, are not reported here.

Experimental Procedure

The data presented involve two different UHV chambers and associated equipment, both of which are described in detail elsewhere[17,18]. That used for the Ni(100) data is shown schematically in Fig. 2, while the chamber used for the Ni(110) data[18] was designed to minimize the optical path length in the chamber so that substantial pressures ($\sim$ 1 atm) of absorbing gases would not lead to serious losses in signal level.

For the Ni(110) data[18], dosing was achieved through control of the ambient pressure, while for the Ni(100) data[3], effusive molecular beam dosing was used which established pressure condition at the crystal surface about 100 times higher than that at the ionization gauge.

Details of the crystal preparation and cleaning procedures can be found elsewhere[3,17,18]. For both Ni(100) and Ni(110), the most sensitive test of surface cleanliness was found to be thermal desorption measurements for low coverage of CO. The Ni(100) surface was taken to be clean when the desorption spectra agreed within a few K with those of Benzinger and Madix[19], while the data of Falconer[20], Taylor and Estrup[21], and Madden et al[22] were similarly used in assessing the cleanliness of the Ni(110) surface.

Typically, spectra are calculated from two scans, one for the bare surface, or reference state, the other either a repeat for the bare surface, or following the addition of the adsorbate, etc. The data are all for 80° angle of incidence, and a scan speed of about 1 $cm^{-1}s^{-1}$. The absorption data are recorded as % absorption ($\simeq$ 2.303 A_e x 100 for A_e << 1) being essentially the % reduction in p-polarized light intensity due to the adsorbate. They have been subjected to a sliding polynomial smoothing procedure chosen so as not to degrade the 6 cm^{-1} resolution.

CO on Ni(110)

Fig. 3 presents spectra typical of those reported by Mahaffy and Dignam[18], chosen here as they are the only set of spectra that were subjected to band resolution analysis. They show two main sets of bands, a high frequency set appearing in the range 2030 to 2160 cm^{-1} and assigned to linearly bonded CO, and a low frequency set appearing in the range 1840 to 1980 cm^{-1} and assigned to CO on two-fold bridge sites. These assignments are consistent with the findings for CO on other Ni surfaces, and additionally correlate extremely well with independent LEED studies, as indicated in the following.

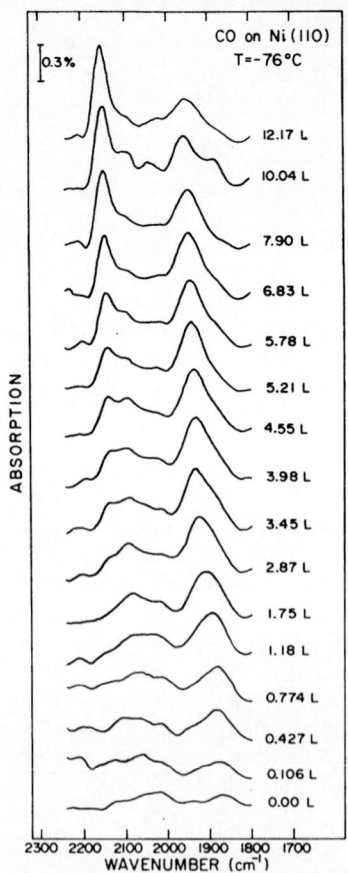

Fig. 3 - IRES absorption spectra for CO in Ni(110) held at - 76°C,
due to Mahaffy and Dignam[18] and reproduced by permission
of Surface Sci. The CO pressure was maintained constant at
1.1×10^{-9} torr throughout this series, with the dose
designations referring to the start of the spectral scan
in question, each scan requiring about 300 S.

   The spectra of Fig. 3 were fitted to a sum of Cauchy-Gauss
product functions and it was found that a minimum of 5 such functions
was required to represent the data satisfactorily, i.e. provide a
good fit to all of the spectra, and a smooth shift of the component
band areas and frequencies with CO dose. The results of the fit are
shown in Fig. 4 to 6. Data for higher temperatures were not subjected to a band resolution analysis due to poorer band resolution,
but appear to show the same general trends with dosing[18].

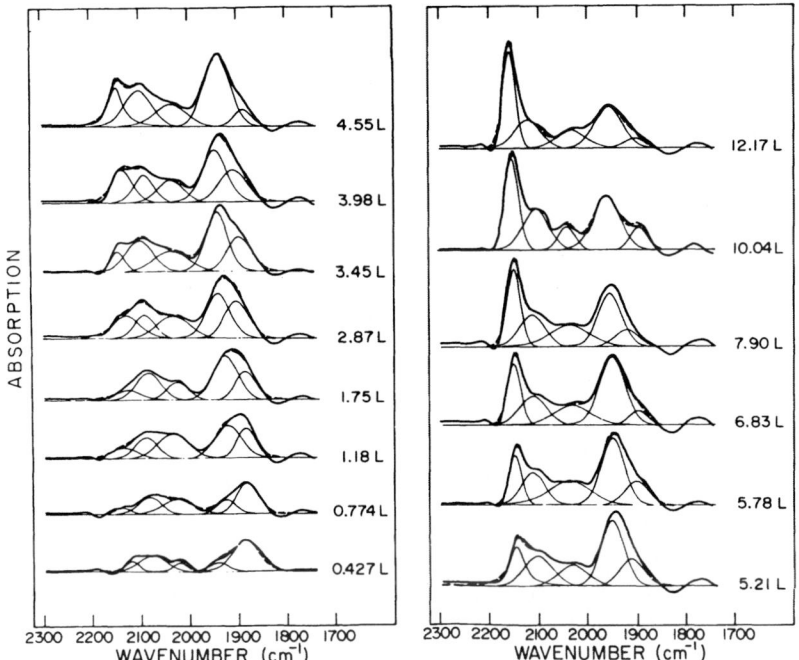

Fig. 4 - Band resolution of the spectra of Fig. 3, reproduced from
Mahaffy and Dignam[18] by permission of Surface Sci. The
measured spectra (corrected for any linear baseline slope)
are plotted as solid lines, while the five band sums of
Cauchy-Gauss product functions are plotted as dashed lines.
The component bands are also shown as solid lines.

The spectral results were interpreted in the light of the LEED
results reported by Madden, Kuppers and Ertl[22] and Taylor and
Estrup[21] , the interpretation being summarized in Fig. 7 and the
accompanying caption. The interpretation leaves no conflict between
the LEED and IR data. Furthermore, on this interpretation. band
areas appear to be proportional to population within experimental
scatter. Thus, there is no indication in the data of the large
band area differences for equal populations of bridged and linear
CO reported by Erley et al for CO on Ni(111)[23] and by Andersson
for CO on Ni(100)[24] , both in connection with EELS results.

Accepting the interpretation of the data, it follows that one
can distinguish between linearly bonded CO within different surface
structures. Thus the band B1, attributed to linearly bonded CO in
a c(2x1) overlayer structure (Fig. 7d), has a frequency $\sim$ 35 cm^{-1}
higher than B2, attributed to linearly bonded CO in the "line
structure" in which all of the CO molecules are restricted to linear

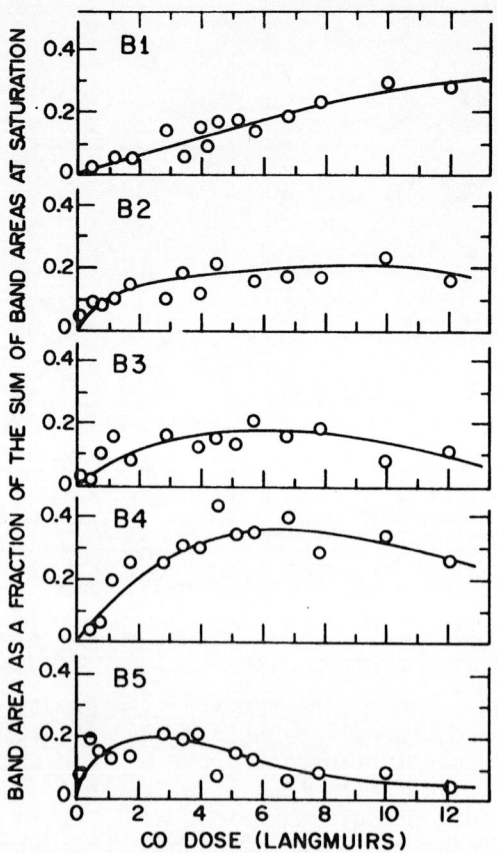

Fig. 5 – Individual band areas, expressed as fractions of the total
saturation area, versus CO dose for the resolved bands of
Fig. 4 are reproduced from Mahaffy and Dignam[18] by permission of Surface Sci.

and short-bridge positions on the ridges of the Ni(110) plane
(Fig. 7 b). A similar, substantial frequency difference exists
between B5 and B4, attributed to short bridge bonded CO in a c(2x2)
overlayer structure (Fig. 7a) and line structure (Fig. 7b), respectively.

Note that the bands that represent a single type of surface
species (e.g. linearly bonded CO) in a well defined overlayer
(structures a, b and d, Fig. 7) namely B1, B2 and B5, nevertheless
show a shift in frequency with total coverage (Fig. 6). Such a
shift clearly cannot be attributed to changes in short range interaction between CO molecules. Dynamic dipole-dipole interaction is

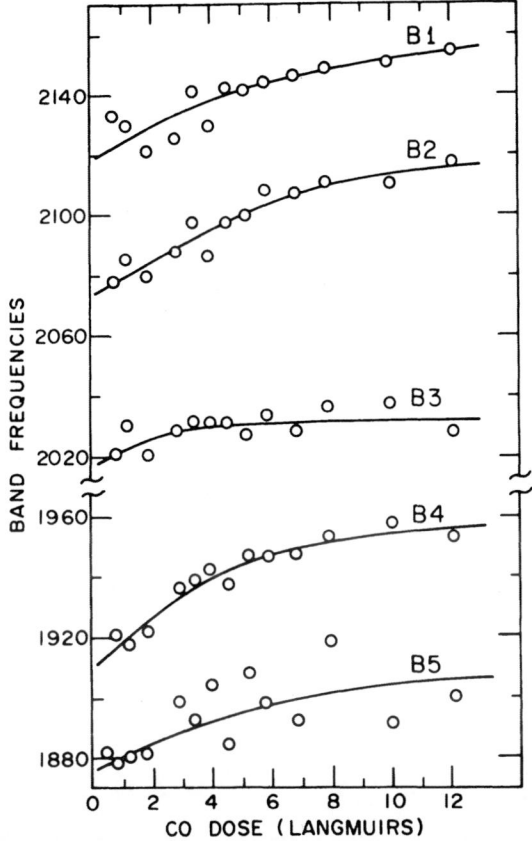

Fig. 6 - Individual band frequencies versus CO dose for the resolved bands of Fig. 4, reproduced from Mahaffy and Dignam[18] by permission of Surface Sci.

also not likely responsible for these shifts, since the frequencies continue to increase after the band areas have saturated or are decreasing (B2 and B5). The fact that all three bands shift in frequency at about the same rate with total coverage suggests that all of the adsorbed CO molecules are in some way equally responsible for all of the shifts, for example, through their contribution to the static surface field or work function. In fact, such can be shown to lead to a frequency shift of about the observed magnitude, arising from the effect of the static surface field on the equilibrium C-O separation, and thence on the vibrational frequency due to anharmonicity in the molecular potential[25]. That the frequencies for B3 and B4 vary in a different way with total coverage presumably reflects the composite nature of these bands.

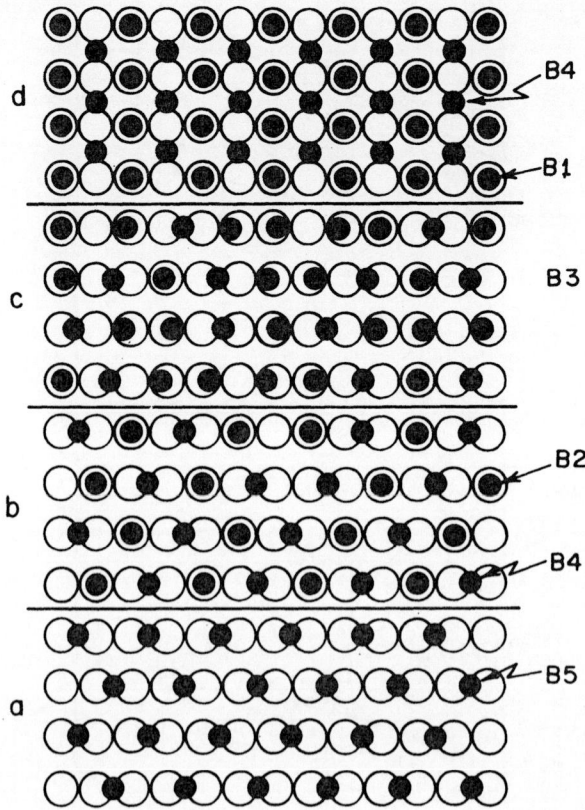

Fig. 7 - Sequence of surface structures according to Mahaffy and
Dignam[18] , with Ni represented by the open circles, CO by
the solid circles, and increasing coverage being in the
direction a to d. Structures a and d are completely ordered and correspond to c(2x2) and c(2x1) structures
respectively, while b and c are ordered in one dimension
only giving lines in the LEED patterns. Structure b is
representative of a range of structures above a in coverage,
saturation presumably at a coverage of 2/3. Structure c is
representative of the linear compressional structure lying
above 2/3 coverage structure, before formation of the
c(2x1) structure. The band assignments are indicated in
the figure.

Stretching frequencies for adsorbed CO as high as 2150 cm^{-1}
(B1, Fig. 6) have generally been associated with oxygen on the
surface in question. It is possible that the main effect of oxygen
on the C-O stretching frequency is again through the static surface

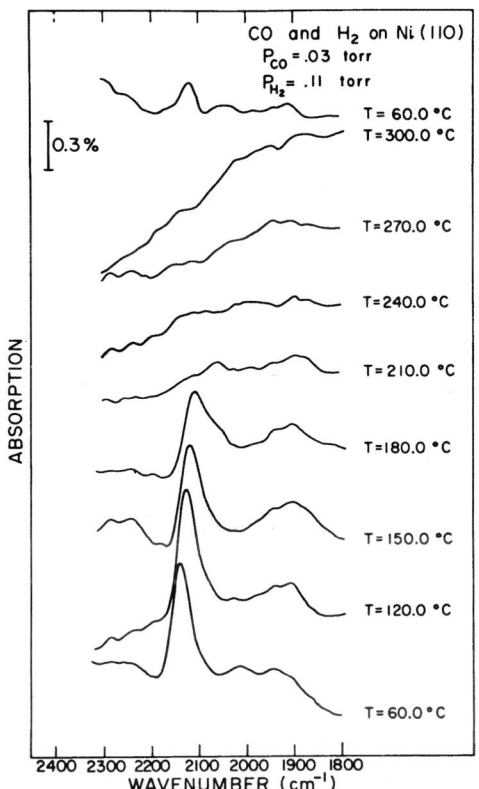

Fig. 8 - IRES absorption spectra showing the CO stretching bands
under methanation conditions for CO and H_2 on Ni(110),
according to Mahaffy and Dignam[26]. The line sequence is
from bottom to top.

field, or work function, as above.

Preliminary Results for H_2 and CO on Ni(110)

A series of spectra for a clean Ni(110)surface, exposed continuously, but initially at 60°C, to a gaseous mixture, 0.11 Torr
H_2, 0.03 Torr CO, are reproduced in Fig. 8 from data due to Mahaffy
and Dignam[26]. As the temperature is raised, the CO vibrational
spectrum undergoes a substantial change from 150°C to 180°C, with
little change in total band area, followed by a virtual elimination
of the bands at 210°C. The band reappearing on cooling the crystal
to 60°C is much weaker than the original, due presumably to the
formation of a carbide layer on the Ni surface above 210°C. The

spectra are in accord with the results of Madey et al[27] that at
elevated temperatures, CO and H_2 react on Ni through a surface
carbide, with no surface oxygen present, while surface oxygen is
involved at lower temperatures and is responsible for the different
product distribution observed at high and low temperature.

Such experiments can easily be done at pressures up to 1 atm
or more. The prospect of performing such experiments with FTRE,
covering not only the C-O stretch, but as well other modes such as
the C-H stretch, Ni-C stretch, etc. is an exciting one.

CO on Ni(100)

Figs. 9 to 11 present IR spectra for CO on Ni(100) due to
Fedyk and Dignam[3] . As for the Ni(110) data, they show two main
sets of bands : a high frequency set appearing in the range 2068
to 2154 cm^{-1}, due presumably to linearly bonded CO, and a low frequency set covering a broad region centered on 1900 cm^{-1}, due presumably to bridge bonded CO. As dosing was done using an effusive
molecular beam source, the absolute exposures were not known. The
effective exposures were obtained from the relative exposures by
choosing a multiplying factor that leads to near saturation coverage
at 298 K for 1L effective exposure (see Fig. 9).

Unlike the Ni(110) spectra that were readily reconciled with
earlier LEED and coverage data, these data appear to be in conflict
even with the very basic and generally accepted conclusion that CO
saturates at a coverage $\sim$ 0.7 on (100) metal surfaces in general,
and on Ni(100) in particular.

The band at 2068-2080 cm^{-1}, Fig. 9, correlates with the
development of the c(2x2) LEED pattern, and hence at saturation
corresponds presumably to a coverage of 0.5. Assuming that band
areas are proportional to coverage, with the same proportionality
constant applying to all of the bands, leads then to a saturation
coverage at 298 K of $\sim$ 0.7, in agreement with the results of
Tracey[28] and others. However, at lower temperatures (e.g., 180 K,
Fig. 10), the bands centered near 1900 cm^{-1} continue to develop
in area well beyond that for saturation coverage at 298 K, though
the bands in the range 2080-2150 cm^{-1} do not change appreciably in
area. Thus on the basis of the same assumption concerning band
areas, a saturation coverage at 180 K well in excess of one monolayer is deduced. In order to avoid this conclusion, and deduce
instead a saturation coverage $\sim$ 0.7, it must be assumed that the
oscillator strengths for the carbonyl species giving rise to the
bands centered around 1900 cm^{-1} are about 7 times higher than for
those centered around 2100 cm^{-1}. Andersson[24] accounted for his
observation of high EELS band areas for bridge-bonded species on
Ni(100) in this way. However, no such assumption is required to

account for the IRES data for CO on Ni(110). Furthermore, thermal
desorption data appear to confirm the apparent high coverages
deduced from the IR data[3]. Other published data also suggest the
attainment of coverages greater than 1 monolayer for effusive
molecular beam dosing of Ni(100) with CO below room temperature,
details of which will be presented elsewhere, along with a much
more complete discussion of this coverage problem than there is
time for here. Note, however, that Ni is one of only two metals
that react directly in the metallic state with CO to form carbonyls,
and it does this only at temperatures below ~ 250°C and above
~ 1 Torr CO pressures[29]. Perhaps the relatively high CO pressures
established at the Ni(100) surface when dosing with an effusive
molecular beam leads to surface structures closely related to the
precursor of $Ni(CO)_4$, commonly considered to be $Ni(CO)_2$ on the
surface[29].

We turn now to a consideration of the structures within both
the high and low frequency groups of bands at 298 K. As the band
attributed to linearly bonded CO grows in, it shifts from a low
coverage value of 2068 cm^{-1} to around 2080 cm^{-1}, most likely due
mainly to dipole-dipole interactions. As this band saturates,
however, a broad band with apparently considerable structure,
develops in the approximate range 1800 to 2000 cm^{-1}. At about 1 L
effective exposure, the most prominent feature in this 1800 to 2000
cm^{-1} region is a peak at ∿ 1830 cm^{-1}, a situation which remains
constant with further dosing at 298 K (Fig. 9), but at 180 K
(Fig. 10), by 2 L effective exposure, two well-developed bands are
present at 1830 and 1870 cm^{-1}. Further dosing at 180 K leads to
dominance of the 1870 cm^{-1} band. During growth of the 1830 and
1870 cm^{-1} bands, the band at 2080 cm^{-1} decreases in area, being
replaced in turn by bands at about 2115 and 2154 cm^{-1}, with the
combined area of the three bands remaining more or less constant.
The 2154 cm^{-1} band is only clearly visible in this series of
spectra for the case in which the Ni(100) was exposed continuously
to 1 x 10^{-7} Torr of CO. In other special series not shown here,
however, at high coverages it becomes the dominant peak in the
high frequency region[3].

A comparison of the band frequencies and band areas for CO
saturation on the Ni(110) surface at 223 K with those on the
Ni(100) surface at 180 K is interesting, in that the frequencies
for the linear and bridged species are the same for the two surfaces within about 10 cm^{-1}. However, for the Ni(110) surface, the
ratio of the band areas for the bridged to that of the linear
species is close to unity, in accord with the proposed structure
(Fig. 7d), while for the 180 K, Ni(100) surface, the ratio is
close to 2. This suggests that for the 180 K, Ni(100) surface, the
saturation condition corresponds to occupation of half the linear,
and half the two-fold bridge sites. However, only the (1x1) LEED
spots were obtained at these coverages.

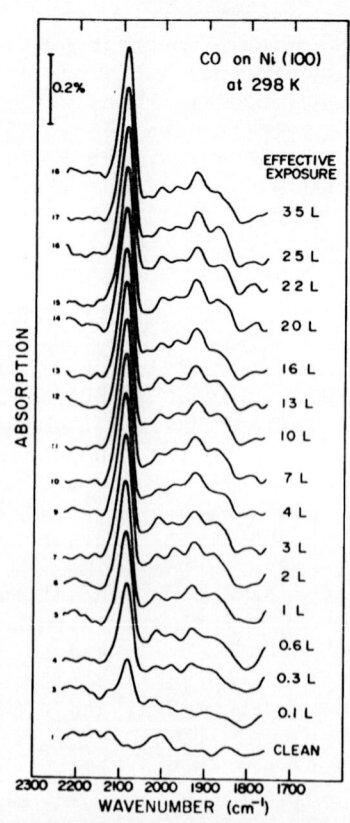

Fig. 9 - IRES absorption spectra for CO dosed via effusive molecular
beam onto Ni(100) at 298 K. The effective dose was
obtained by scaling to give saturation coverage for 1 L
effective dose at 298 K. The data are reproduced from
Fedyk and Dignam[3] by permission of The American Chemical
Society.

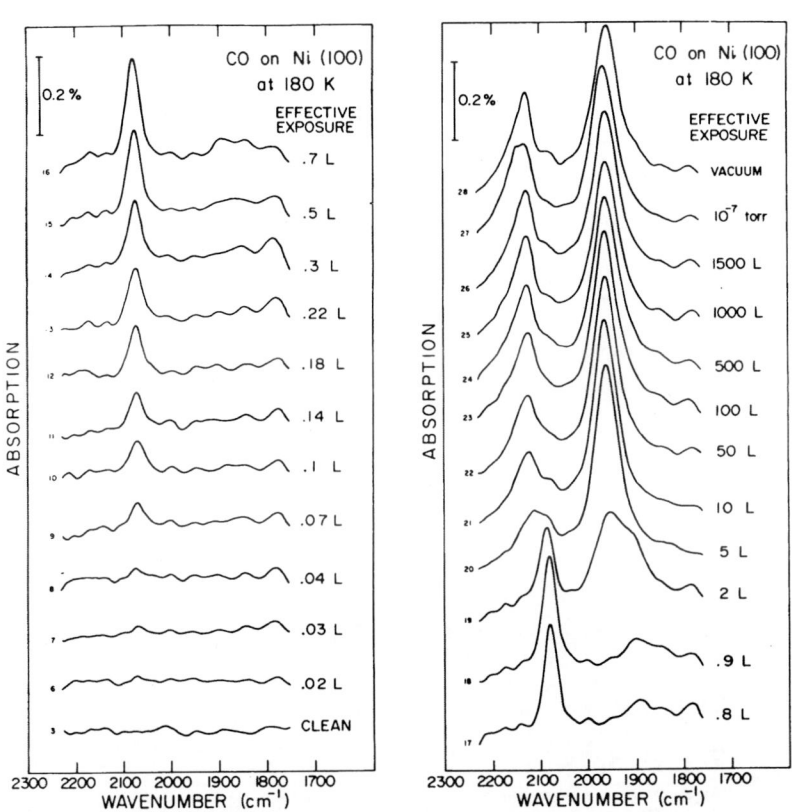

Fig. 10 - IRES absorption spectra as in Fig. 9, but for a sub-strate temperature of 180 K, reproduced from Fedyk and Dignam[3] by permission of The American Chemical Society.

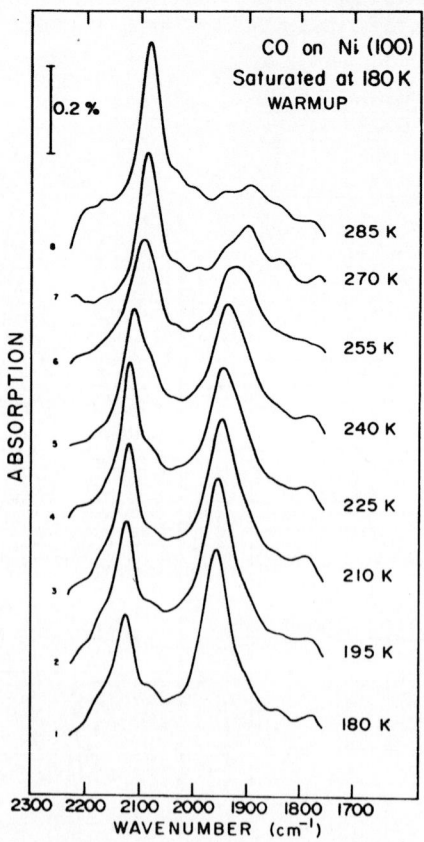

Fig. 11 - IRES absorption data obtained by the stepwise heating in
vacuo of the CO-covered Ni(100) surface formed in the
process of generating the spectra of Fig. 10, the line
sequence being from bottom to top. Reproduced from Fedyk
and Dignam[3] by permission of The American Chemical Society.

Extending this line of speculation to the Ni(111) surface, with
CO occupying 1/3 of the linear, and 1/3 of the three-fold bridge
sites, for a total coverage of 1/3, leads to a surface structure
which appears to be the ideal precursor state for $Ni(CO)_4$ formation,
since certain Ni atoms with linearly bonded CO would be associated
with precisely the proper number of CO molecules, and in a configuration not far removed from that of $Ni(CO)_4$. De Groot, Coulon,
and Dransfeld[29] have shown that carbonyl formation on Ni(111) most
likely proceeds according to a Langmuir-Hinshelwood mechanism
(reaction between adspecies) and on Ni(100) and Ni(110) leads to
(111) faceting of the surface, suggesting that this plane reacts
most rapidly, in contrast to the usual situation in which the close-

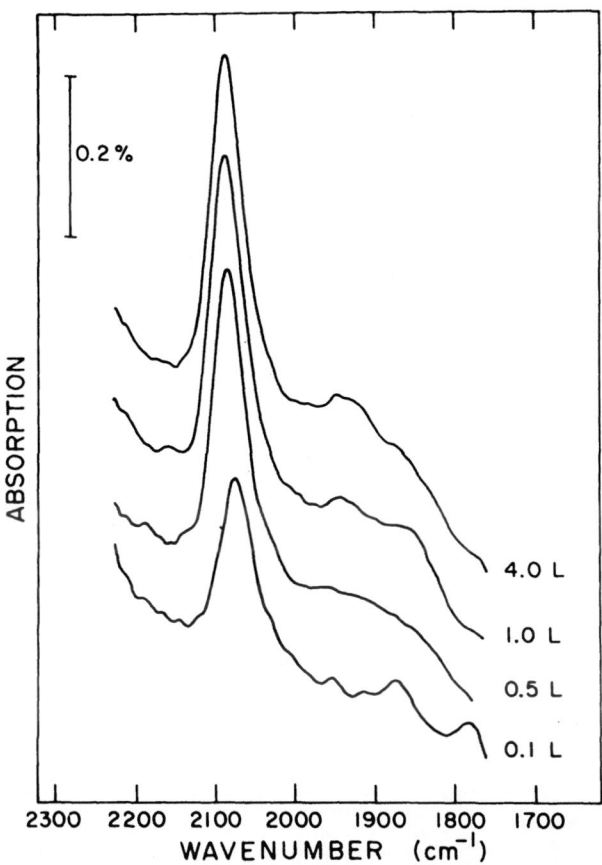

Fig. 12 – IRES absorption spectra generated following successive doses of CH$_3$OH on Ni(100) at 293 K via molecular beam dosing. The total doses indicated are relative only, the actual dosing level being perhaps as much as 100 times higher. Reproduced from Baudais et al[17] by permission of Surface Sci.

packed (111) surface of fcc metal crystals is the least reactive one.

It seems possible that these data may provide a bridge between the low pressure surface physics of CO on Ni(100) and the high pressure surface chemistry of Ni(CO)$_4$ formation.

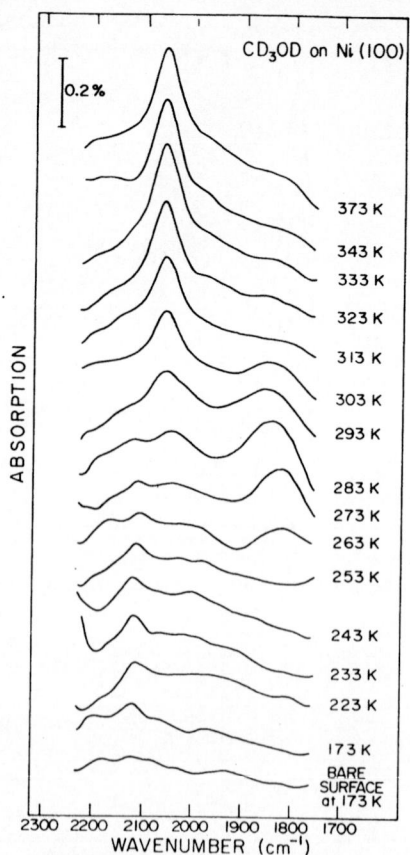

Fig. 13 - IRES absorption spectra generated on adsorbing CD_3OD into Ni(100) at 173 K to saturation, followed by warming up in successive 10 K steps. Reproduced from Baudais et al[17] by permission of Surface Sci.

Decomposition of Methanol on Ni(110)

Figs. 12 and 13 present IRES data, due to Baudais et al[17], for the decomposition of methanol on Ni(100). Fig. 12 shows the development of a band at around 2080 cm^{-1} with increasing relative dose of CH_3OH at 298 K, the band presumably being that for linearly bonded CO. The new result, however, is seen in Fig. 13, showing spectra for Ni(100) dosed to saturation with CD_3OD at 173 K. then warmed in stages.

Following dosing at 173 K, a weak band attributed to C-D stretching modes appears at about 2120 cm^{-1}. By 263 K, a broad band is present, centred at about 1830 cm^{-1} which continues to grow in area (and shift in frequency) up to 283 K, at which point the C-D stretching band is appreciably attenuated, and a band at about 2080 cm^{-1}, attributed to linearly bonded CO, makes its appearance. With further temperature increase, the lower frequency band (1800 - 1880 cm^{-1}) decreases in area, vanishing by 313 K, along with the 2120 cm^{-1} band attributed to C-D stretch. Thermal desorption results indicate that this final stage is accompanied by D$_2$ desorption[17]. The results for CH$_3$OH decomposition are similar,. except that the 2120 cm^{-1} band is absent[17].

Though several possibilities were considered, the explanation for the low frequency band offered as the most likely one was that it was generated by surface formyl moieties. According to this interpretation, on saturating the surface with CH$_3$OH at 173 K and warming, the methoxy species, present initially, first decomposes to a formal species without loss of surface species. Desorption only begins with the decomposition of the formyl species to linearly bonded CO, this process being accompanied by H$_2$ desorption. On this basis, it would appear that the formyl species is stabilized by a surface saturation condition.

Unlike the other examples given, data at least as good as these could readily be obtained by EELS. That Demuth and Ibach[30] did not observe the 1800-1880 cm^{-1} band for CH$_3$OH decomposition on Ni(111) is presumably due either to their failure to obtain spectra in the right temperature range, or to differences in the behaviour of the (100) and (111) surfaces. The reason this example is presented here is to give some notion of the potential of FTRE for such studies. The speed advantage of FTRE would presumably allow one to obtain spectra such as those in Fig. 13, except more complete both in spectral range and temperature resolution, during application of a temperature ramp $\sim$ 0.1 to 1 K s^{-1} to the crystal. Thus one could obtain, simultaneously, surface vibrational spectra and thermal desorption spectra, a combination that would greatly enhance one's likelihood of correctly determining reaction mechanisms.

ACKNOWLEDGEMENTS

    The author wishes to express his appreciation to the Natural Sciences and Engineering Research Council of Canada for supporting his research on IRES and applications, and to his collaborators in this same effort, in chronological order : the late B. Rao, R.W. Stobie, M. Moskovits, J.D. Fedyk, P. Mahaffy, A.J. Borschke, F.L. Baudais, and M.D. Baker.

REFERENCES

1. M.J. Dignam, J.D. Fedyk, Appl. Spectrosc. Rev., 14 (1979) 249.
2. J.D. Fedyk, P. Mahaffy, M.J. Dignam, Surface Sci., 89 (1979) 404.
3. J.D. Fedyk, M.J. Dignam, in Vibrational Spectroscopy Applied to Adsorbed Species, A.C.S. Symposium Series, in press.
4. K. Horn, J. Pritchard, Surface Sci. 52 (1975) 437.
5. P. Hollins, J. Pritchard, Surface Sci. 89 (1979) 486.
6. M.J. Dignam, B. Rao, R.W. Stobie, Surface Sci., 46 (1974) 308.
7. F. Hoffman, A.M. Bradshaw, Surface Sci., 72 (1977) 513.
8. W. Golden, D.S. Dunn, J. Overend, J. Phys. Chem., 82 (1978) 843.
9. M.D. Baker, Ph.D. Dissertation, University of East Anglia, Nov. 1979. Also M.D. Baker, M.A. Chesters, this Conference proceedings.
10. P.R. Griffiths, "Chemical Infrared Fourier Transform Spectroscopy" (Wiley-Interscience, New York, N.Y., 1975).
11. M.J. Dignam, M.D. Baker, Submitted to Appl. Spectrosc.
12. M.J. Dignam, M.D. Baker, unpublished results.
13. D. Kuehl, P.R. Griffiths, Anal. Chem. 50 (1978) 418. Also M.M. Gomez-Taylor, P.R. Griffiths, Anal. Chem., 50 (1978) 422.
14. D.H. Martin, E. Puplett, Infrared Physics, 10 (1969) 105.
15. C.H. Burton, Yoshiaki Akimoto, Infrared Phys. 20 (1980) 115.
16. R.W. Stobie, B. Rao, M.J. Dignam, Appl. Opt., 14 (1975) 999.
17. F.L. Baudais, A.J. Borschke, J.D. Fedyk, M.J. Dignam, Surface Sci., in press.
18. P.R. Mahaffy, M.J. Dignam, Surface Sci., in press.
19. J.B. Benzinger, R.J. Madix, Surface Sci., 79 (1979) 394.
20. J. Falconer, Ph.D. Dissertation, Stanford University, 1974.
21. T.N. Taylor; P.J. Estrup, J. Vac. Sci. Techn., 10 (1973) 26.
22. H.H. Madden, J. Kuppers, G. Ertl, J. Chem. Phys., 58 (1973) 3401.
23. W. Erley, H. Wagner, H. Ibach, Surface Sci., 80 (1979) 612.
24. S. Andersson, Solid State Commun., 20 (1976) 229; also 21 (1977) 75.
25. M.J. Dignam, paper in preparation.
26. P.R. Mahaffy, M.J. Dignam, paper in preparation.
27. T.E. Madey, D.W. Goodman, R.D. Kelley, J. Vac. Sci. Techn., 16 (1979) 433.
28. J.C. Tracy, J. Chem. Phys., 56 (1972) 2736.
29. P. de Groot, M. Coulon, K. Dransfeld, Surface Sci. 94 (1980) 204.
30. J.E. Demuth, H. Ibach, Chem. Phys. Letters 60 (1979) 395.

REFLECTION-ABSORPTION INFRARED SPECTROSCOPY USING A FOURIER TRANSFORM SPECTROMETER : A STUDY OF CARBON MONOXIDE ADSORBED ON Pt(111)

M.D. Baker * and M.A. Chesters

School of Chemical Sciences, University of East Anglia

Norwich   NR4 7TJ, U.K.

ABSTRACT

The major advantage of infrared spectroscopy over EELS for the study of adsorbate vibrations lies in superior resolution. This factor is particularly important when bandwidths are inherently low (< 5 cm^{-1}) and adsorbate-adsorbate interactions lead to splitting of vibrational bands on a scale of 10 cm^{-1} or less.

We have been investigating the application of infrared interferometry to the reflection-absorption experiment while studying the adsorption of carbon monoxide on a Pt(111) recrystallised foil using a Digilab FTS-14 spectrometer. In this paper the advantages and limitations associated with the interferometric technique will be outlined and the ultimate sensitivity of currently available FT-IR instruments discussed. The results for the platinum/carbon monoxide system will be used to illustrate the desirability of operation at high (2 cm^{-1}) resolution and the advantage of scanning a wide spectral range.

INTRODUCTION

Virtually all reflection-absorption infrared (RAIR) studies reported to date have been carried out with dispersive spectrometers. The approach adopted is to scan a narrow spectral range (100-300 cm^{-1}) at high sensitivity and best results have been achieved using

---

* Present address : Lash Miller Chemical Laboratories, University of Toronto, 80 St. George Street, Toronto, Ontario, Canada M5S 1A1.

wavelength or polarization modulation techniques[1-4]. However, if bands are very broad, or if they turn up in unexpected places, they may be missed. Ideally one should scan as much of the mid-infrared range as possible and here the well-known 'multiplex' advantage of Fourier transform infrared spectroscopy (FT-IR) becomes very significant[5-7].

The advent of electron energy loss spectroscopy (EELS) has emphasized the desirability of scanning a wide spectral range, particularly for adsorbed hydrocarbons, but the poor resolution (~ 60 cm^{-1}) is a serious limitation. Most RAIR work has been carried out at ~ 5 cm^{-1} resolution but the results described below illustrate that even this may sometimes be insufficient for analysis of line shapes and fine structure. In general FT-IR spectrometers give optimum performance at higher resolution, e.g. the FTS-14 is optimized (throughput matched[5]) at 2 cm^{-1} resolution.

We shall describe the results obtained for CO/Pt(111), which illustrate some of the points mentioned above, and then go on to estimate the ultimate sensitivity available from current FT-IR spectrometers.

EXPERIMENTAL

The spectrometer used in this work was a Digilab FTS-14 rapid scanning interferometer interfaced to a Nova 1200 minicomputer. A magnetic tape unit was used for long-term storage of interferograms and spectra. The spectrometer was operated with a germanium on potassium bromide beam splitter which allowed the range 400-4000 cm^{-1} to be covered but the mercury cadmium telluride (MCT) detector ($D\star \sim 10^9$) had a cut-off at ~ 700 cm^{-1}. For the spectra presented in this paper the spectrometer was scanned at 2 cm^{-1} resolution (RES 2) giving transformed spectra with one data point per wavenumber.

An optical schematic of the spectrometer plus reflection-absorption system is shown in Fig. 1. The whole of the optical path was purged with dry air to reduce the effect of atmospheric water vapour absorption bands. The glass vacuum system was mercury diffusion pumped and, following bakeout at 250°C, a background pressure $< 2 \times 10^{-10}$ torr was routinely achieved. The cell windows were sodium chloride, sealed with silver chloride via short silver tubes[8]. The image of a 'V' shaped tungsten strip source was focussed at the leading edge of the sample, which was a 30 x 10 x 0.0125 mm platinum foil, and the range of angles of incidence was 84 ± 6°. The reflected light was focussed into the spectrometer through a 20 mm circular aperture and plane mirror which blocked light from the internal source.

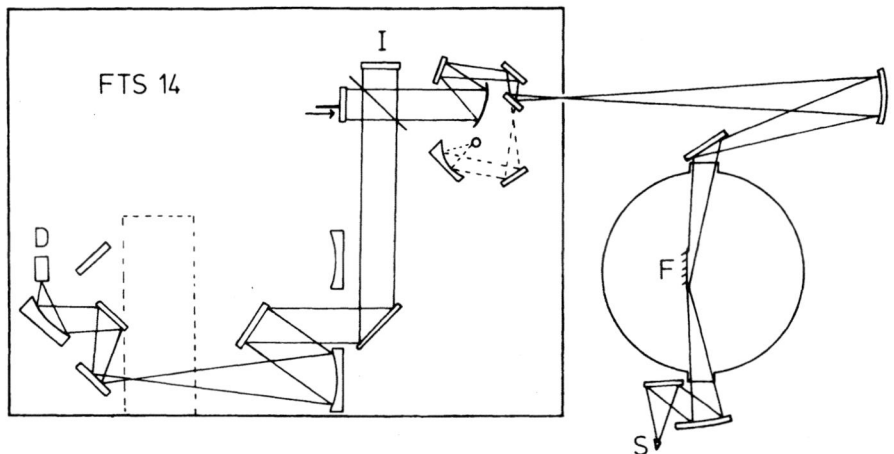

Figure 1 – Schematic of the optical system. S : 'V' shaped
tungsten strip source, F : platinum foil, I : Michelson
interferometer, D : MCT detector.

The platinum foil was cleaned by cycles of oxidation and
annealing at 1500 K. This procedure was checked by Auger
spectroscopy on a foil cut from the same sheet mounted in a separate chamber.

The spectrometer, Fig. 1, consists essentially of a Michelson
interferometer, I, in which a collimated light beam is split
into two components by a beam splitter. Half the beam is reflected
by a fixed plane mirror, the other half by a plane mirror moving
at constant velocity. A second interaction then produces two exit
beams, one of which returns towards the source and one is transmitted to the detector via a sample well which is unused in this
experiment. Ideally, 50 % of the light falling on the detector
has been phase shifted (retarded) by an amount dependent on the
position of the moving mirror. The resulting detected signal as
a function of retardation (i.e. the interferogram) is shown in
Figure 2. For a wide band source this will consist of an intense
zero retardation peak (at the point where light of all wavelengths
is in phase) which rapidly diminishes to a weak oscillatory signal
with increasing retardation. Sharp features in the spectrum produce
oscillations in the interferogram which persist at high retardation
and the spectral resolution achieved is proportional to the maximum
retardation in a scan. Data is accumulated by coaddition of interferograms and the resultant is Fourier transformed to give the
single beam spectrum, Fig. 3, which is basically the emission
spectrum of the source as modified by the transmission character-

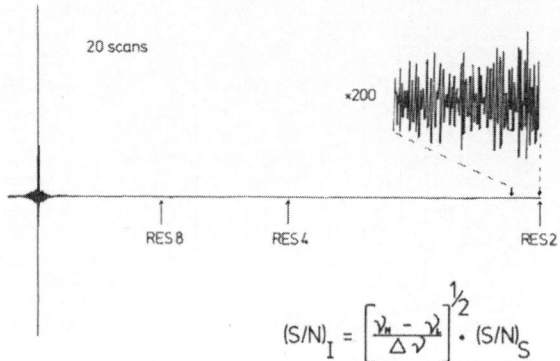

$$(S/N)_I = \left[\frac{\nu_H - \nu_L}{\Delta \nu}\right]^{1/2} \cdot (S/N)_S$$

Figure 2 - Interferogram resulting from coaddition of twenty scans.
Maximum retardation points for RES 8, 4 and 2 are arrowed.

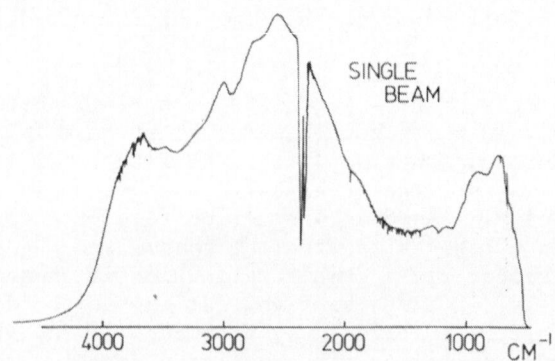

Figure 3 - Single beam spectrum.

istics of the spectrometer. There is an intense absorption band due
to atmospheric CO_2 but water vapour bands have been reduced to a
low level after several hours purging with dry air.

Results will be presented as ratios of single beam spectra
taken before and after adsorption on the sample surface which
ideally would produce adsorbate absorption bands on a flat background. The ratio of two, clean surface, single beam spectra is
shown in Fig. 4 in which the baseline drifts by $\pm$ 0.25 % over
3000 cm^{-1}, probably due to fluctuations in the source temperature.
There is also some miscancellation of the atmospheric CO_2 bands
but this does not interfere with our measurements.

The peak-to-peak (ptp) noise level in the RES 2 spectrum
(228 scans = 15 minutes for each single beam) varies from $\sim$ 0.1 %
ptp in mid-range to $\sim$ 0.2 ptp at the extremes. A gaussian smoothing
programme was available which smoothed spectra to effective resolution increasing in units of 2 cm^{-1}. The figure shows the same
spectrum after smoothing to RES★ 8.

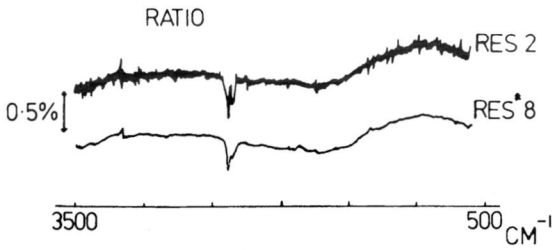

Figure 4 - Ratio of two clean surface, single beam spectra at
           RES 2 and after smoothing to RES★ 8.

RESULTS

The spectrum of the platinum surface saturated with carbon
monoxide at 300 K is shown in Fig. 5. The 4 % absorption band at
2100 cm^{-1} has a half-width of 6 cm^{-1} and the second band at 1860 cm^{-1}
is readily detected despite its low intensity ($\sim$ 0.15 %) and large
half-width ($\sim$ 40 cm^{-1}). The broad band is defined more clearly
without distortion after digital smoothing to RES★ 8 but recording
of the sharp peak at poorer resolution, or smoothing, results in
broadening and loss in intensity. The series of spectra at increasing coverage, Fig. 6, shows evidence for 2-D island formation in

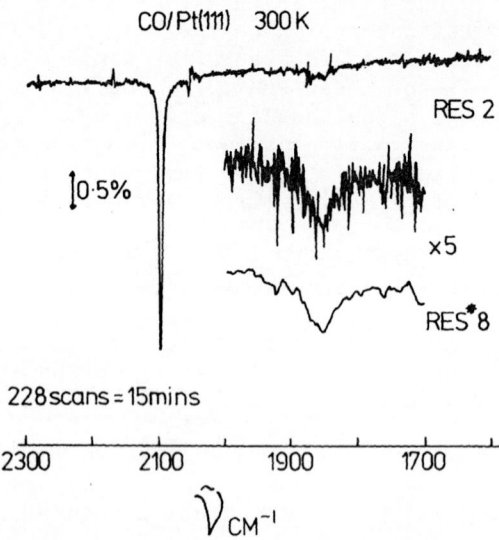

Figure 5 – Ratio spectrum Pt(111)/10^{-7} Torr carbon monoxide/300 K.

the fine structure of the high frequency band which is split into two at intermediate coverages. This will be discussed in detail elsewhere but here it serves to emphasize the need for high spectral resolution, particularly as the two bands merge together. The broad band at 1860 cm^{-1} appears before the high frequency band has fully developed into a single peak at 2090 cm^{-1} and continues to grow while the linear band shifts from 2090 to 2100 cm^{-1}.

These results illustrate the combined advantages of high spectral resolution and a wide spectral range but they do not represent the optimum performance achievable in the RAIR experiment by current FT-IR spectrometers. The instrument we have used is ten years old and in that time there have been improvements in both spectrometers and detectors. In the next section we discuss the limitations of the FT-IR – RAIR experiment and estimate the sensitivity available using current commercial spectrometers.

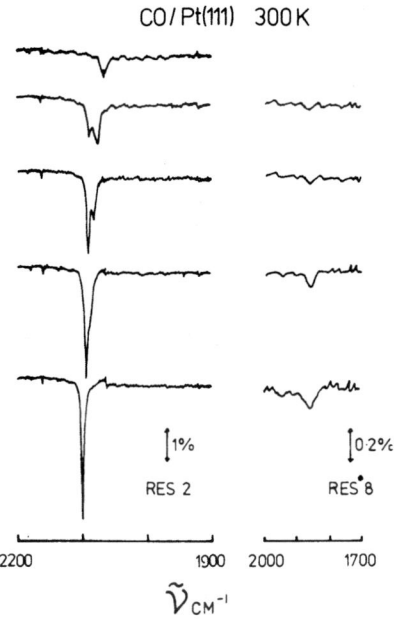

Figure 6 - Ratio spectra showing increasing coverage of CO/Pt(111)/
         300 K.

DISCUSSION

The major sensitivity advantages of FT-IR spectroscopy may be
summarized as : 1) Multiplex or Fellgett advantage
                2) Throughput or Jacquinot advantage.

The multiplex advantage follows from the fact that light of
all wavelengths is sampled simultaneously which, in the ideal
limit, gives a $N^{1/2}$ times better S/N than for a dispersive spectrometer operating over the same wavenumber range and for the same
scanning time, where N is the number of spectral elements in the
range[10],[11]. The throughput advantage derives from the fact that

narrow input and exit slits need not be used in FT-IR spectrometers to define a resolution element. However, these two factors
suffer some limitations when recording RAIR spectra.

Dynamic Range Limitation

For a single beam spectrum of high S/N or dynamic range the
resolution of the analog-to-digital converter (ADC), which
digitizes the interferogram, may become limiting. The ADC in the
FTS-14 has a dynamic range of 15 bits (i.e. 32000) which is about
the present state-of-the-art. If an interferogram is to be
represented accurately (including noise) then the peak-to-peak
noise level should at least be equivalent to the least significant
bit of the ADC[5,12]. This puts an upper limit on the dynamic range
of the interferogram (expressed here as signal to ptp noise) of
$\sim$ 16000. We can estimate the $(S/N)_S$ of the corresponding single
beam spectrum from the following expression :

$$(S/N)_I = N^{1/2} (S/N)_S$$

Given a spectral range of 400-4000 cm^{-1} and 2 cm^{-1} resolution,
N = 1800 and the best $(S/N)_S$ for a single scan is $\sim$ 300, i.e.
ptp noise $\sim$ 0.3 %. Clearly our results described above (1.5 %
ptp noise for a single scan) are well short of this limit. There
is no serious limit to the S/N of the coadded interferograms
since the computer dynamic range is 16 bits and this is extended
to 32 bits by the use of double precision arithmetic[10].

Throughput Limitation

The throughput of the FTS-14 at $\Delta\bar{\nu}$ = 2 cm^{-1} is 7 x 10^{-2} cm^2 sr
but that of the reflection-absorption system is considerably
smaller. This restriction arises through the need to reflect
light off the sample surface over a narrow range of angles around
the optimum grazing angle. In our case using a foil sample (3 cm x
1 cm) the throughput of the reflection-absorption system is $\sim$ 7 x
10^{-3} cm^2 sr but for a typical single crystal sample this would be
reduced to $\sim$ 2 x 10^{-3} cm^2 sr.

Of the above two restrictions the ADC dynamic range limit is
by far the most important. The reduced throughput can be overcome
by using a high temperature source and a high detectivity, wide
range MCT detector, and it appears that current commercial FT-IR
spectrometers are capable of operating at the ADC limit for the
RAIR experiment, even on small single crystals. For example,
Mattson[13] has reported a noise level of $\sim$ 0.4 % ptp for a single
scan with a Nicolet spectrometer using an MCT detector with D*
$\sim$ 10^{11} and cut-off $\sim$ 400 cm^{-1} while operating at throughput

comparable to a RAIR system (2×10^{-3} cm^2 sr). Another important
advance has been in rapid scanning such that accumulation rates
of the order of 10 scans s^{-1} are now available at 2 cm^{-1} resolution.
Thus for 15 minutes scanning time we can estimate a noise level of
4×10^{-3} % ptp or a sensitivity of $\sim 10^{-5}$ (rms) absorbance units
covering most of the mid-infrared range, which is comparable to the
best reported results using wavelength modulation or ellipsometric
techniques and dispersive spectrometers scanning only $\sim$ 300 cm^{-1}
at $\Delta\tilde{\nu}$ = 4 cm^{-1} [1-4]. It is important to note that, while for a FT-IR
spectrometer at constant throughput and for a constant scanning
time S/N $\propto \Delta\tilde{\nu}$, for a dispersive spectrometer S/N $\propto \Delta\tilde{\nu}^{5/2}$. This
means that the S/N using a dispersive spectrometer would be reduced
by 1/6 on increasing the resolution from 4 cm^{-1} to 2 cm^{-1} .

A sensitivity of 10^{-5} absorbance units is possibly sufficient
to tackle monolayer quantities of weak absorbers, such as hydrocarbons, where a wide spectral range will be necessary. However
further improvements in performance are possible and will be needed
to compete with the projected sensitivity of dispersive spectrometers using ultra-sensitive detectors such as the liquid helium
cooled bolometer[17].

How is FT-IR performance to be improved further ? Apart from
mechanical difficulties, faster scan rates are likely to run into
limits set by the maximum sampling frequency of the ADC[14]. Therefore it is necessary to overcome the ADC dynamic range problem and
increase the S/N for one scan. Techniques do exist to achieve this
such as dual beam interferometry in which both exit beams of the
interferometer (which ideally are out of phase) are directed to the
detector. Griffiths has recently demonstrated this technique successfully[14,15] but it suffers the disadvantage that the high light
levels falling on the detector can lead to saturation problems.

Perhaps the ideal optical system for the FT-IR - RAIR combination would allow no light to reach the detector except when the
sample is perturbed by adsorption of gas. Such a concept is now
under development by Dignam and co-workers in Toronto using ellipsometric techniques[16].

In conclusion, we feel that Fourier transform spectroscopy is
likely to become increasingly important to reflection-absorption
infrared studies both of monolayers of weak absorbers and of
strongly absorbing transient species.

ACKNOWLEDGEMENTS

It is a pleasure to record our thanks to Professor Norman
Sheppard for the use of the FTS-14 spectrometer (provided by an
SRC equipment grant) and for many stimulating discussions. We are

also indebted to Mr. D.H. Chenery who provided considerable assistance in the operation of the spectrometer in addition to many
useful discussions.

   M. D. Baker is grateful to the SRC for a research studentship.

REFERENCES

1. J. Pritchard, T. Catterick, R. Gupta, Surface Sci. 53 (1975) 1.
2. H. Pfnür, D. Menzel, A.M. Bradshaw, A. Ortega, F.M. Hoffmann,
   Surface Sci. 93 (1980) 431.
3. J.D. Fedyk, P. Mahaffey, M.J. Dignam. Surface Sci. 89 (1979) 404.
4. M.J. Dignam, J. Fedyk, Appl. Spectrosc. Rev. 14 (1978) 249.
5. 'Chemical Fourier Transform Infrared Spectroscopy', P.R. Griffiths
   Wiley-Interscience Publications 1975.
6. 'Introductory Fourier Transform Spectroscopy', R.J. Bell,
   Academic Press, New York, 1972.
7. 'Transform Techniques in Chemistry', P.R. Griffiths (Ed.),
   Plenum Press 1978.
8. R.C. Lord, R.S. McDonald, Rev. Sci. Instr. 23 (1952) 442.
9. M.D. Baker, M.A. Chesters, C.M. Deeley,to be published.
10. N. Sheppard, R.G. Greenler, P.R. Griffiths, Appl. Spectrosc. 31
    (1977) 448.
11. P.R. Griffiths, H.J. Sloane, R.W. Hannah, Appl. Spectrosc. 31
    (1977) 485.
12. Tomas Hirschfeld, Appl. Spectrosc. 33 (1979) 525.
13. David R. Mattson, Appl. Spectrosc. 32 (1978) 335.
14. Donald Kuehl, Peter R. Griffiths, Anal. Chem. 50 (1978) 418.
15. G.J. Kemeny, P.R. Griffiths, Appl. Spectrosc. 34 (1980) 95.
16. M.J. Dignam, M.D. Baker, to be published.
17. J.F. Blanke, S.E. Vincent, J. Overend, Spectrochimica Acta 32A,
    (1976) 163.

COLLECTIVE VIBRATIONAL MODES IN ISOTOPIC MIXTURES OF CO ADSORBED ON Cu(100)

B.N.J. Persson and R. Ryberg

Institute of Theoretical Physics and Department of
Physics, Chalmers University
S-412 96 Göteborg, Sweden

Isotopic mixtures of CO adsorbed on Cu(100) have been studied
by infrared spectroscopy. For mixtures of $^{12}C^{16}O/^{12}C^{18}O$ at a
constant coverage of 0.5 it was found that the molecules constitute
a collective vibrational system. The C-O stretching vibrations were
studied as a function of composition of the overlayer and the data
are successfully interpreted by a dipole-dipole coupling theory.

INTRODUCTION

The C-O stretching vibration frequency of a free CO molecule
is 2143 cm^{-1}. When a single CO molecule is adsorbed on a metal
surface this frequency decreases typically 50-200 cm^{-1}. Part
of this decrease arises from a weakening of the C-O bond strength
due to a partial filling of the antibonding 2 π^* orbitals, but
there are also several other effects, e.g. "the self-image shift",
which affect the frequency. Furthermore, there is in general an
additional, coverage dependent, shift usually towards higher
frequencies and of varying magnitude (0-130 cm^{-1}). This shift
must be due to interaction between the molecules, either directly
via the dipole fields of the vibrating molecules or via an interaction mediated by the metal. The large spread in the reported
frequency shifts indicates that there are probably several contributing effects. One type of experiment that keeps at least one
parameter fixed is the study of isotope mixtures whereby the influence on the vibrational modes due to changes in the chemical envi-
ronment of the molecules is excluded. In this paper, we present
experimental and theoretical results for the absorption spectra
of different mixtures (at constant coverage Θ = 0.5) of $^{12}C^{16}O$
and $^{12}C^{18}O$ adsorbed on a Cu(100) metal surface.

299

EXPERIMENTAL RESULTS AND THEORY

In an IR reflection absoption experiment, one measures the change in reflectance, $\Delta(\omega)$, defined by

$$\Delta(\omega) = \frac{I' - I}{I'}$$

where I' and I are the intensities of the electromagnetic wave reflected from the metal surface without and with adsorbed molecules, respectively. Using this definition it is straightforward to show that

$$\Delta(\omega) \sim n\omega \ \text{Im} \ \alpha_0(\omega) \tag{1}$$

where n is the number of adsorbed molecules per unit surface area and $\alpha_0 = \sum_i p_i/NE_0$ is an average polarizability. p_i is the induced dipole moment of the molecule at site $\vec{x}_i$, N is the total number of sites within the overlayer and E_0 is the external electric field.

Figure 1. shows our experimental data for $\Delta(\omega)$ at different mixtures of $^{12}C^{16}O/^{12}C^{18}O$ adsorbed on a Cu(100) metal surface in the ordered structure c(2x2). Actually, what is recorded experimentally is not $\Delta(\omega)$ but its second derivative $\Delta''(\omega)$.

The main feature of Fig. 1. is that the molecules obviously constitute a coupled vibrational system in which the high frequency mode is dominant at nearly all compositions. The low frequency mode decreases very rapidly with increasing $^{12}C^{16}O$ concentration. The properties of this system are summarized in Fig. 2.

In the following, we discuss a theoretical interpretation of the experimental data presented above. The crucial problem lies in the appropriate description of the disorder within the monolayer due to the presence of two kinds of CO molecule isotopes. We consider three different approximations (A, B, and C) of increasing accuracy of which only the third can reproduce all of the experimental data.

We assume that at site $\vec{x}_i$ on the metal surface there is an adsorbed molecule with polarizability α_i and induced dipole moment p_i. We allow the polarizabilities α_i to vary from site to site corresponding to having different types of adsorbed molecules. (Note that by choosing $\alpha_i=0$ for some $\vec{x}_i$, we get as a special case

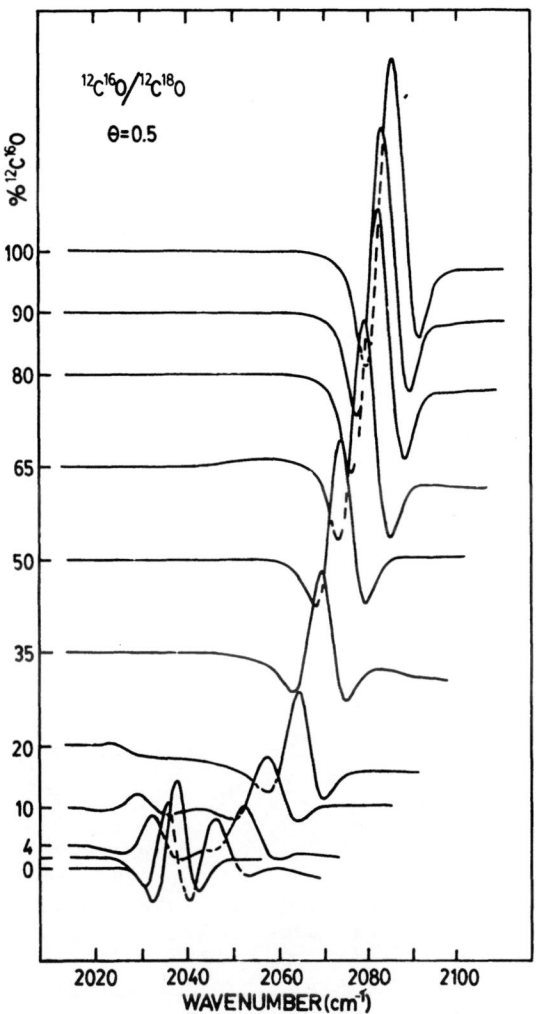

Fig. 1. : The spectrometer signal for different compositions of
$^{12}C^{16}O/^{12}C^{18}O$ on Cu(100) at 100 K in the c(2x2)
structure.

a partly filled monolayer). However, we assume that the molecules
are stochastically distributed among the adsorption sites. This
assumption should be satisfied for an ordered structure of an
isotopic mixture.

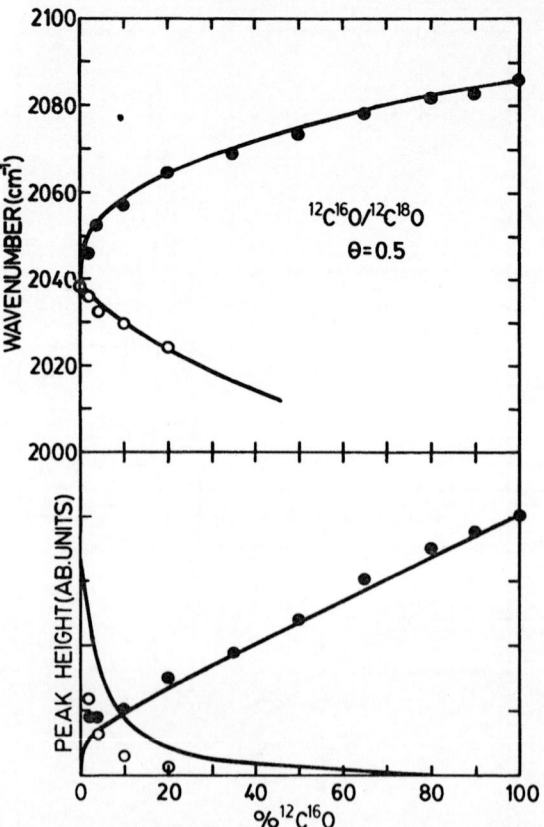

Fig. 2. : Peak height and position for the absoption peaks
of mixtures of $^{12}C^{16}O/^{12}C^{18}O$.   Full circles correspond to the high frequency mode, open circles to
the low frequency mode., The curves are calculated
from eqs. (9) and (10).

The basic equation from which p_i (and therefore also $\Delta(\omega)$
according to eq. (1)) can be calculated is

$$p_i = \alpha_i (E_0 - \sum_j U_{ij} p_j) \quad ; \quad U_{ii} = 0 \qquad (2)$$

where $-U_{ij}p_j$ is the field at $\vec{x}_i$ due to the dipole p_j and its image. Eq. (2) can be iterated

$$P_i = \alpha_i E_0 + \sum_j \alpha_i (-U_{ij}) \alpha_j F_0 + \ldots \tag{3}$$

We interprete each term in this expansion as follows : the molecules at $\vec{x}_i$ can be excited either directly by absorption of a photon (the first term) or a photon may first excite the molecules at $\vec{x}_j$, followed by a jump (due to the dipole-dipole coupling $-U_{ij}$) of the excitation to the molecule at $\vec{x}_i$ (the second term) and so on.

The infinite sum in eq. (3) can only be summed exactly if all molecules are identical. In the presence of an isotopic mixture, approximate methods for the evaluation of eq. (3) have to be used. In the following, we discuss three approximations and compare their predictions with the experimental data.

A. Noninteracting molecules.

If the interaction between the molecules on the metal surface is ignored (i.e. $U_{ij}=0$), then according to eq. (3)

$$P_i = \alpha_1 E_0$$

and consequently

$$\alpha_0(\omega) = \sum P_i/NE_0 = c_A \alpha_A + c_B \alpha_B \tag{4}$$

where c_A (c_B) is the concentration and α_A (α_B) the polarizability of a molecule of type A (B). One may write

$$\alpha_A = \alpha_e + \frac{\alpha_v}{1 - \dfrac{\omega}{\omega_A}\left(\dfrac{\omega}{\omega_A} + i\gamma\right)} \tag{5}$$

where α_e and α_v are the electronic and vibrational polarizabilities, respectively, and γ a small number determined by the finite lifetime of the vibrational mode due to its interaction with the metal.

Substitution of eqs. (4) and (5) into (1) gives (for $\gamma \to 0$)

$$\Delta(\omega) \quad \sim \quad c_A \omega_A^2 \; \delta(\omega - \omega_A) + c_B \omega_B^2 \; \delta(\omega - \omega_B) \tag{6}$$

i.e. the absorption spectrum consists of two delta functions centred around the frequencies ω_A and ω_B with weights proportional to the concentration of the respective species. Obviously, this absorption spectrum (dashed lines in Fig. 3.) is quite different from the experimentally observed spectrum, i.e. the interaction between the adsorbed molecules must be taken into consideration.

B. The average T—matrix approximation (ATA).

According to eq. (2), the ensemble average of Σp_i (denoted by a bar) is given by

$$\overline{\sum_i p_i} = \overline{\sum_i \alpha_i} \; E_0 + \sum_{ij} (-U_{ij}) \; \overline{\alpha_i \alpha_j} \; E_0 + \dots \tag{7}$$

We define

$$\overline{\alpha_i} = \overline{\alpha} \quad , \quad \overline{\alpha_i^2} = \overline{\alpha^2} \quad , \quad \overline{\alpha_i^3} = \overline{\alpha^3} , \; \dots$$

since the quantities on the LHS in these equations are independent of i. We then have

$$\overline{\alpha_i \alpha_j} = \overline{\alpha}^2 + \delta_{ij} (\overline{\alpha^2} - \overline{\alpha}^2)$$

$$\overline{\alpha_i \alpha_j \alpha_k} = \overline{\alpha}^3 + (\delta_{ij} + \delta_{ik} + \delta_{jk}) \; \overline{\alpha}(\overline{\alpha^2} - \overline{\alpha}^2)$$

$$+ \delta_{ij}\delta_{ik}(\overline{\alpha^3} - 3\overline{\alpha^2} \; \overline{\alpha} + 2 \; \overline{\alpha}^3)$$

and so on.   Now, within the so-called "average T-matrix approximation" one keeps only the first terms on the RHS of these equations.
This means neglecting correlation corresponding to processes in which
the excitation jumps between any two molecules more than once.
The series (7) is then simply a geometric series in $X = \sum_i U_{ij} \, \bar{\alpha}$
and can therefore be summed :

$$\overline{\sum_{1} P_i} \approx N\bar{\alpha}E_0 \, (1 + X + X^2 + \dots) = \frac{N\bar{\alpha}E_0}{1-X}$$

and so

$$\Delta(\omega) = n\omega \ \mathrm{Im} \ \frac{\bar{\alpha}}{1 + \sum_j U_{ij} \, \bar{\alpha}} \tag{8}$$

where

$$\bar{\alpha} = c_A \alpha_A + c_B \alpha_B$$

The solid lines in Fig. 3. are calculated using this formula.
Obviously, there is very good agreement with the experimental
data for the frequency shift but not for the peak height.   Consequently, the approximation of neglecting the correlation terms in the
expansion 7 is too crude.

C. The coherant potential approximation (CPA).

This approximation takes into account a large amount of the
neglected correlation terms in the expansion [7].   However, the
CPA approximation has also a simple physical interpretation which
makes it very appealing.   Here, we only give the final expression
for $\Delta(\omega)$ (see ref. 1. for the derivation) :

$$\Delta(\omega) = n\omega \ \mathrm{Im} \ \frac{\alpha(\omega)}{1 + \alpha(\omega) \, \tilde{U}(0)} \tag{9}$$

where

$$\alpha = \sum_{\mu=A,B} \frac{c_\mu \alpha_\mu}{1 + (\alpha_\mu - \alpha) \frac{1}{A^\star} \int d^2 q \frac{\hat{\mathcal{U}}(\vec{q})}{1 + \alpha \hat{\mathcal{U}}(\vec{q})}} \tag{10}$$

The integral is over the first Brillouin zone (area $A^\star$) of the adsorbate lattice and

$$\hat{\mathcal{U}}(\vec{q}) = \sum_j U_{ij} e^{i\vec{q}\cdot\vec{x}_j}$$

Equation (10) was solved for α by iteration and from eq. (9) we obtained the change in reflectance. The solid lines in Fig. 2. show the result which obviously is in very good agreement with experiment.

Notice that α obtained from eq. (10) has in general a finite imaginary part. This means that $\Delta(\omega)$ is nonzero for finite intervals in ω, i.e. the disorder introduces broadening of the spectra features. In the case of approximations A and B, $\Delta(\omega)$ consists of a sum of two delta functions (or Lorentz curves if the damping γ is taken finite), i.e. no broadening due to disorder is obtained. This is the main reason for why approximation B fails to reproduce the experimental peak heights shown in Fig. 3.

The CPA predicts an increase of the full width at half maximum of about 1 cm^{-1} for the 50% $^{12}C^{16}O/^{12}C^{18}O$ mixture as compared to a 100%$^{12}C^{16}O$ adsorbate layer. This is in good agreement with our experimental findings (Ref. 1).

---

[*]There exists a sum rule which says that $\int d\omega \Delta(\omega) \approx$ const., independent of mixture, where the integral is over the two absorption peaks. Nevertheless, we see from Fig. 2. that the sum of the peak heights of the two absorption peaks varies strongly with mixture. This is possible only if the shape of the absorption peaks changes with coverage so that the sum of the peak heights is not a good measure of $\int d\omega \Delta(\omega)$.

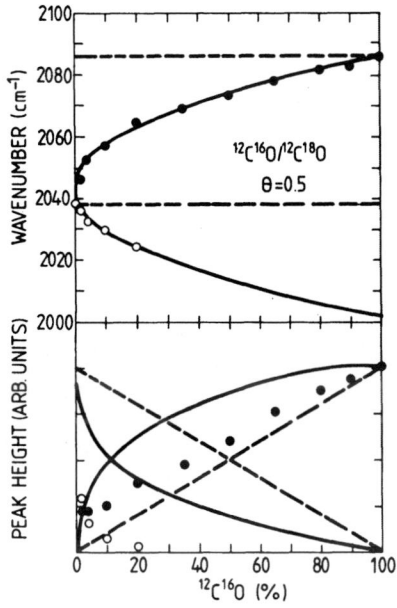

Fig. 3. : The solid curves are calculated from eq. (8)
         and the dashed curves from eq. (6).

ACKNOWLEDGEMENTS

We would like to thank A. Liebsch for several useful discussions.

REFERENCE

1. For a detailed discussion of the experiments and the theoretical
   analysis, see B.N.J. Persson and R. Ryberg, submitted to Phys.
   Rev. B.

INTRINSIC LINE WIDTHS OF CO ADSORBED ON Cu(100)

Roger Ryberg

Department of Physics, Chalmers University of Technology

S-421 96 Gothenburg, Sweden

ABSTRACT

A systematic investigation of the intrinsic line widths of the C-O stretch vibration for different configurations of CO adsorbed on Cu(100) has been made, using infrared spectroscopy. The line width of $^{12}C^{16}O$ in the ordered c(2x2) structure was found to be 4.5 cm^{-1} and for $^{12}C^{18}O$ 4.0 cm^{-1}. Disorder, structural or isotopic, increases the line width by about 1 cm^{-1}, whereas there is no additional broadening for partial coverages. The results are correlated with existing theories for damping and molecular interactions.

INTRODUCTION

Rather little attention has previously been paid to systematic investigations of the vibrational line widths of molecules adsorbed on a metal surface. This is perhaps not so surprising, as such studies have to be done with infrared spectroscopy and there exist only a few spectrometers with sufficient sensitivity. Furthermore, most systems that have been studied are CO adsorbed on a transition metal, where the intrinsic line width is about 10 cm^{-1}, which masks the rather small effects due to disorder or difference in vibrational frequencies.

However, the vibrational damping of molecules adsorbed on a noble metal surface is weaker and should allow studies of such effects. In this paper an investigation of different configurations of CO adsorbed on Cu(100) is reported. The results are correlated with theoretical calculations of vibrational damping and molecular interactions.

309

EXPERIMENTAL

   The measurements were made with an infrared spectrometer using the wavelength modulation technique. All experimental details are given elsewhere[1].

RESULTS AND DISCUSSION

   The intrinsic line widths of different configurations of CO on Cu(100) are given in table 1. A detailed presentation of the work will be given elsewhere[1]. The line widths are mainly due to life time broadening caused by interaction with the metal. Persson and Persson[2] have recently made a theoretical review of the three most probable damping mechanisms :

1. The molecular vibration is damped through creation of phonons in the metal. However, as the C-O vibration frequency is about eight times the highest phonon frequency, this mechanism is probably inefficient.

2. The oscillating dipole field of the vibrating molecule excites electron-hole pairs in the metal causing vibrational damping. The theory of Persson and Persson[3] gives

$$\frac{1}{\tau} \propto \Omega \rho^2 (e_F)$$

Table 1 - The intrinsic line width (FWHM) for different configurations of CO on Cu(100). Θ is the coverage with respect to the copper surface atoms (c(2x2)$\Rightarrow$ $\Theta = 0.5$)

SYSTEM	PEAK POSITION (cm^{-1})	PEAK WIDTH FWHM (cm^{-1})
$^{12}C^{16}O$ $\Theta=0.5$	2086	4.5±0.5
$^{12}C^{16}O$ $\Theta=0.25$	2079	4.5
$^{12}C^{18}O$ $\Theta=0.4$	2038	4.0
$^{12}C^{16}O$ $\Theta=0.57$	2088	5.5
$0.5\,^{12}C^{16}O$ $0.5\,^{12}C^{18}O$ $\Theta=0.5$	2074	6.0

where τ is the lifetime, Ω the vibration frequency and $\rho(\varepsilon_F)$
the density of states at the Fermi level. The appropriate value
for CO on copper is

$$\tau = 1.10^{-10} \text{ s.}$$

Kozhushnev et al.[4] have made another calculation for the same
mechanism but with a different description of the metal, giving

$$\tau = 2.10^{-11} \text{ s.}$$

3. As discussed elsewhere[5,6] , there is an increase in the dynamical
   dipole moment of the CO molecule upon adsorption on a Cu surface.
This is due to a charge flow into molecular orbitals from the
metal. When the molecule is vibrationally excited, this causes a
charge oscillation between these orbitals and the metal, which
produces a damping of the vibration. The theoretical expresssion
from[2] is

$$\frac{1}{\tau} \propto \Omega \; (\delta n)^2$$

where δn is the magnitude of the oscillating charge.
By estimating δn from the measured increase in the dynamical dipole
moment for CO on Cu(100) the authors find

$$\tau \approx 2.10^{-12} \text{ s.}$$

These figures will be compared with the measured peak width
of $^{12}C^{16}O$ at $\Theta = 0.5$, table 1, which corresponds to $\tau = 1.3 . 10^{-12}$s.
This is surprisingly close to the value in 3 and between one and two
orders of magnitude smaller than the estimations in 2.

In order to study the <u>vibration frequency dependence</u> of the
damping, consider the line width of the $^{12}C^{18}O$ isotope. The peak
width in both model 2 and 3 is proportional to the vibration frequency. For the $^{12}C^{18}O$ isotope, this would give a decrease of
0.1 cm^{-1} , whereas the measured one is 0.5 cm^{-1} (the accuracy in
the relative values is better than $\pm$ 0.2 cm^{-1}). However, the
vibrational amplitude is smaller for $^{12}C^{18}O$ and as the density of
states around ε_F for the molecular orbitals in model 3 is not
known, this could account for the remaining decrease.

Introducing <u>disorder</u> into the adsorbate layer gives a contribution to the line width in addition to that caused by lifetime
broadening. This is due to the interaction between the molecules.
A calculated spectrum for an isotopic mixture of 50 % $^{12}C^{16}O$ and
50 % $^{12}C^{18}O$ at $\Theta = 0.5$, assuming dipole-dipole interaction[6],
shows an additional broadening of about 1 cm^{-1}, in quite good
agreement with the experimental data in table 1.

About the same increase in line width is found experimentally for the compressed structure of $^{12}C^{16}O$ ($\Theta = 0.57$). Instead of an isotopic disorder this probably arises from site disorder. There are namely strong indications[1,7,8], that the structure can be regarded as a mixture of 75 % terminal bonded and 25 % bridge-bonded molecules, with different vibration frequencies. The analogy to the isotopic mixture is therefore obvious.

The third disordered system studied was fractional monolayers of $^{12}C^{16}O$. Calculations[6] similar to those for the isotopic mixture show an additional broadening at $\Theta = 0.25$ of about 1 cm^{-1}. Mahan and Lucas[9] previously derived the following expression for the increase in peak width due to dipole-dipole interaction :

$$\Delta \propto \frac{\Theta(\frac{1}{2} - \Theta)^{1/2} \alpha_v}{(1 + A\Theta)^2}$$

where α_v is the vibrational polarizability and A is a constant. Table 1 shows that no additional broadening is detected for $\Theta = 0.25$. One may therefore assume that α_v is coverage dependent, that is α_v decreases with decreasing coverage.

CONCLUSIONS

Measurement of the intrinsic line width of CO on Cu(100) and comparison with theoretical calculations indicates that the dominating vibrational damping mechanism is not the excitation of electron-hole pairs in the metal by the oscillating dipole field of the molecule. Furthermore, a theory considering the damping due to charge oscillations between molecular orbitals and the metal is able to account for the measured value. The observed additional broadening arising from disorder is consistent with calculations for the dipole-dipole interaction between the adsorbed molecules. The absence of such broadening for partial coverages of one isotope can be attributed to a coverage dependence of the vibrational pola-rizability.

ACKNOWLEDGEMENT

I am grateful to B.N.J. Persson for useful discussions.

REFERENCES

1. R. Ryberg, submitted to Phys. Rev. B.
2. B.N.J. Persson, M. Persson, Solid State Comm., in press.
3. B.N.J. Persson, M. Persson, Surf. Sci., in press.

4. M.A. Kozhushnev, V.G. Kustarev, B.R. Shub, Surf. Sci. $\underline{81}$, 261 (1979).
5. B.N.J. Persson, R. Ryberg, Solid State Comm., in press.
6. B.N.J. Persson, R. Ryberg, submitted to Phys. Rev. B.
7. J. Pritchard, Surf. Sci. $\underline{79}$, 231 (1979).
8. S. Andersson, Surf. Sci. $\underline{89}$, 477 (1979).
9. G.D. Mahan, A.A. Lucas, J. Chem. Phys. $\underline{68}$, 1344 (1978).

"IN SITU" CHARACTERIZATION OF SURFACE COMPLEXES OF PROPENE ON OXIDATION CATALYSTS (I.R. SPECTROSCOPY)

René Delobel, Michel Le Bras, Michel Traisnel and Jean-Marie Leroy

Laboratoire de Catalyse et Physicochimie des Solides E.N.S.C.L., U.S.T.L., B.P. 40, 59650 Villeneuve d'Ascq

INTRODUCTION

Complete or selective catalytic oxidation of olefins requires two different types of intermediate complexes which initiate two different processes. The knowledge of the surface complexes may allow the understanding of the characteristic of the catalysts surfaces which play a leading part in catalysts activities.

The absorbed form of propene on oxide catalysts has already been studied. Dent and Kokes[1] were the first to observe the formation of π-allylic complexes on ZnO using I.R. spectroscopy. It should be noted that this π-allyl species differs in spectral characteristics from the allylic complexes discovered later[2]. Thus, for the latter, ν_{as} CCC lies in the region from 1400 to 1460 cm^{-1}, whereas on ZnO this frequency is about 1545 cm^{-1}. Haber[3] suggested that the allylic species transform into allyl species σ-bonded to the lattice oxygen ions of the active centers ($\nu_{c=c}$ = 1580 – 1600 cm^{-1}), and that σ-bonded species desorb in the form of aldehyde leaving an oxygen vacancy on the surface of the catalysts.

In a previous paper[4], we reported on a study of the fixation of C_3H_6 on Co_3O_4. It was demonstrated that the allylic fixations lead to a σ-allyl, intermediate species for the C_3H_4O and C_2H_4O formation, and that dissociative fixations lead to uni- and bidentate carbonates, intermediate forms for complete oxidation.

The aim of this work is to elucidate the nature of adsorbed propene forms which appear on the surface of $Cu_{0.24}Co_{2.76}O_4$ and USb_3O_{10}, catalysts for selective oxidation into acrolein, and of $Cu_{0.72}Co_{2.28}O_4$, catalyst for complete oxidation.

EXPERIMENTAL

Infra-red spectroscopy studies of a surface during catalytic work are only significant when they are performed in conditions closely related to the reactor ones. For this purpose, a differential cell with thermostatic control of its body was used (Figure 1).

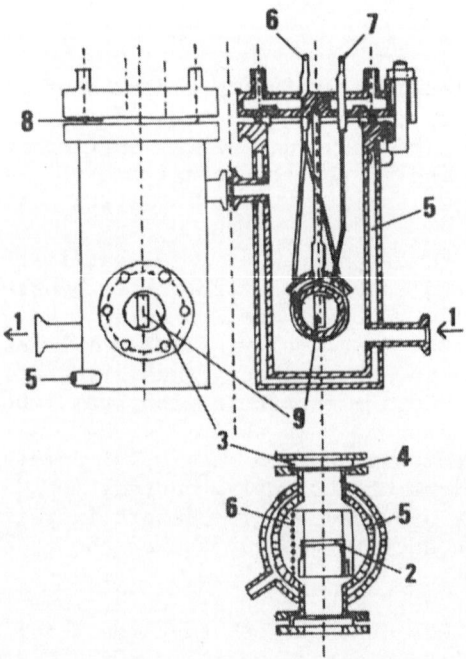

Figure 1 - IR Cell - 1 : Vacuum, He or reactant gases inlet (pneurop NW 25); 2 : pellet of sample or powdered sample laying on a AgCl window; 3 : AgCl windows; 4 : teflon vacuum seal; 5 : jacket for thermostatic control; 6 : stainless steel furnace with "thermocoax" heating resistance; 7 : "thermocoax" seal; 8 : pernunan seal; 9 : i.r. beam.

The symmetrical paths of the IR beams lead to a strong attenuation or the disappearance of the reactant gases absorptions and the thermostatic control of the body avoids the condensation of reactants or products and allows quenchings (circulation of a cryogenic fluid). The cell allows previous catalysts treatments : reduction, oxidation, outgassing (10^{-7} Torr) or regeneration of the studied samples. It may be used as a reactor in static conditions or with reactants flow, in the temperature range : 20 - 600°C ; a chromatographic apparatus being used to analyze reactants and products. All IR spectra were systematically recorded at reaction temperatures and at room temperature, after quenching, with a PERKIN ELMER 21, a BECKMAN IR 12 or a BECKMAN IR 20AX spectrophotometer, respectively in the 4000 - 700 and 4000 - 250 cm^{-1} spectral ranges. Studied catalysts were mechanically deposited in the form of a thin layer on a AgCl window and baked into the furnace of the cell. The method of tableting of these powdered catalysts did not give good results owing to marked absorption and scattering of IR radiation by the specimens. The procedure for preparation of the catalysts, their characteristics and their catalytic performances have been previously reported[5-6]. Blank tests were performed before each experiment because we often noted changes of the spectra of solids in vacuum or in gases mixtures, reactions of the oxides with their support and gases adsorption on the stainless steel furnace, in a given temperature range. Homogeneous processes never took place, in our experimental conditions, in temperature ranges lower than 320°C.

RESULTS AND DISCUSSION

After exposing the solids to C_3H_6 , the gaseous products observed are CO_2, C_3H_4O (ν C=O : 1680 - 1710 cm^{-1}), C_2H_4O (ν C=O : 1735 - 1740 cm^{-1}) and HCHO (ν C=O : 1725 - 1735 cm^{-1}). Introduction of reactant mixtures leads to the formation of the products afore mentioned without HCHO. Besides we specially define the species fixes on the solid.

$Cu_{1-x}Co_{2+x}O_4$:

Table 1 and 2 present the species observed when $Cu_{0.24}Co_{2.76}O_4$ and $Cu_{0.72}Co_{2.28}O_4$ are respectively submitted to the following treatments : C_3H_6(A), sequence C_3H_6-vacuum (B), sequence C_3H_6-vacuum-O_2 (C), (O_2 + C_3H_6) (D), sequence (O_2 + C_3H_6) - vacuum (E), sequence (O_2 + C_3H_6) - vacuum - O_2 (F), with P_{O_2} + $P_{C_3H_6}$ = 10 torr, O_2/C_3H_6= 4, at T = 200°C. Treatment of the catalysts with O_2 before all described studies does not lead to an evolution of the spectra previously described.

TABLE 1

Frequencies (cm^{-1})	Interpretation	A	B	C	D	E	F
3580 – 3680	$Co \leftarrow O \stackrel{H}{<}_{H}$	x	–	–	x	+	+
3090	as. $= CH_2$	x	+	–	x	–	–
3010	$= CH -$	x	+	–	x	–	–
2930 – 2970	as. CH_3, CH_2	x	–	–	x	–	–
2320 – 2370	CO_2 (adsorbed form)	+	–	+	+	–	+
2130 – 2310	$CO{-}{-}{-}O \equiv C \equiv O$	+	–	–	+	–	–
2030 – 2080	$Co - C \equiv O$	+	–	–	–	–	–
1640	$C = C$	x	+	–	x	+	–
1590 – 1630, 1285 – 1280, 1020 – 1050, 935	$Co\underset{O}{\overset{O}{<}}C = O$	–	–	–	–	–	–
1540 – 1550, 1340, 1030 – 1050, 810 – 820	$Co{-}{-}{-}O \equiv C \stackrel{O}{\underset{O}{<}}$	+	–	+	+	–	+
1580 – 1595, 1375 – 1380 cm^{-1}	Formate groups	–	–	–	+	–	–

The $\pi-$ allyl species, generally proposed as responsible for C_3H_6 oxidation into C_3H_4O or C_2H_4O are not found on these samples. Meanwhile, the presence of C_3H_6 fixed on their surfaces is always proved by the presence of monocarbonaceous products after C or F. The amount of adsorbed C_3H_6 on $Cu_{0.72}Co_{2.28}O$ seems to be very small because its absorption spectrum never appears, which may explain the bad selectivity for C_3H_4O on this oxide. Some treatments of $Cu_{0.24}Co_{2.76}O_4$, followed by evacuation, reveal the spectrum of fixed C_3H_6. The absorption 1640 cm^{-1} characterizes a π complex, hence fixed propene is not subjected to destructive fixation. Ozaki[7] noted that the formation of this complex is the first step for the selective oxidation of C_3H_6 via a surface allyl. It is well known that selective oxidation of olefin is favoured by fixation of the hydrocarbon on cations able to have a d^0 or a d^{10} shell[8]. We may assume that surface Cu^{2+} ions act as acceptors, the formation of the complexes occurring by electron donation from the double bond of the hydrocarbon. Its reduction leads to Cu^+ ions with the d^{10} shell.

TABLE 2

Frequencies (cm^{-1})	Interpretation	A	B	C	D	E	F
	fixed C_3H_6	x	-	-	x	-	-
3580 - 3680	Co ← O<^H_H	x	-	-	x	+	+
2320 - 2370	CO_2 (adsorbed form)	+	-	+	+	-	+
2130 - 2310	Co---O ≡≡ C ≡≡ O	+	-	-	+	-	-
2030 - 2080	Co - C ≡ O	+	-	-	-	-	-
1590 - 1630, 1285 - 1280 } 1020 - 1050, 935	Co<^O_O>C = O	+	-	+	+	-	+
1540 - 1550, 1330 } 1020 - 1050, 810	Co--O ≡ C<^O_O	+	-	+	+	-	+
1580, 1370 - 1380	Formate groups	-	-	-	+	-	-

Irreversible forms of propene adsorbed are formates and carbonates of two types : uni- and bidentate. Davydov et al[9] connected the formation of formates on the surface of oxides to their activities for complete oxidation. Their presence on Co_3O_4 are not detected; so, their bearing seems to be due to the Cu surface ions. Previous i.r. studies showed that $Cu_{1-x}Co_{2+x}O_4-CO_2$ interactions did not allow, in similar experimental conditions, to obtain uni- and bidentate carbonates[9] : these species we determined before may be considered as transition compounds formed by a dissociative chemisorption of the olefin. A comparison of their vibrational frequencies and those previously reported on Co_3O_4 leads to conclude that they are mainly formed on Co surface ions. Bidentate carbonates are not found on $Cu_{0.24}Co_{2.76}O_4$, catalyst for C_3H_6 oxidation in acrolein, it may be assumed that they play the main part in CO_2 formation, maybe because its desorption is easier than the one of the other fixed monocarbonaceous species.

The 3580 - 3680 cm^{-1} absorption band is attributed, by analogy[4] to Co ← O<^H_H compounds which result from the fixation of water in anionic vacancy. It may be postulated that this species is not an hydroxyl coming from the allyl formation because it has never been observed when C_3H_6 reacts when brought alone into the presence of the oxides.

TABLE 3

Frequencies (cm^{-1})	Interpretation	A	B	C	D	E	F	G
3750	Free hydroxyl group	x	-	-	-	+	-	-
3600	ν_{OH}	x	-	-	-	+	-	+
3090 - 3100	$\nu_{as} = CH_2$ ⎫	x	+	+	-	+	-	-
3010 - 3030	$\nu = CH-$ ⎪	x	-	-	-	+	-	-
2930 / 2980 }	ν_{CH} } in π-allyl complexes	x	+	+	-	+	-	-
1540	$\nu_{as} CCC$ ⎪	x	+	+	-	+	+	-
1260 - 1275	β CH ⎭	x	-	+	-	+	+	-
1680	adsorbed C_3H_4O	x	-	-	-	+	-	+
2380	adsorbed CO_2	+	-	-	+	+	-	+
2315 - 2335	M - O ≡≡ C ≡≡ O	+	-	-	+	+	-	+
2060 - 2080	M - C ≡ O	+	-	-	-	-	-	
1790 - 1800 or 1740	C = O in aldehydes	+	-	+	-	+	-	+
1540 - 1580, 1340 - 1360	Formate groupe	-	-	+	-	+	-	-
1020 - 1040, 1360 - 1380 / 1460 - 1480 }	Unidentate carbonate	-	-	-	+	+	-	+
1080		+	+	+	+	+	+	+
625		+	+	+	+	+	+	+

USb_3O_{10}

The observed species are compiled in table 3, the sample
being submitted to the following treatments: C_3H_6 (A), sequence C_3H_6-
vacuum (B), sequence C_3H_6-He (C), sequence C_3H_6 - vacuum - O_2 (D),
sequence m.r. - He (E), sequence m.r. - vacuum (F), sequence m.r. -
vacuum O_2 (G) (C_3H_6 : C_3H_6 flow ($P_{C_3H_6}$ = 1 Atm), m.r. : reactant
mixture flow (P_{O_2} + $P_{C_3H_6}$ + P_{He} = 1 Atm, $O_2/C_3H_6/He$ = 425.6/106.4/
288), He : evacuation by He flow (P_{He} = 1 Atm), vacuum : 10^{-3} - 10^{-4}
torr range) at T = 300° C.

As with the Cu-Co-O samples, the presence of fixed C_3H_6 is
proved (D and G treatments). The bands attributed to a π-allyl
species do not disappear of the spectra after desorption. The

absorption maximum at 1540 cm^{-1} characterizing ν_{as} CCC in the fragment C_3H_5 is always found. It is rather widely believed that oxidation of propene to acrolein takes place via participation of such a π-allylic complex of the olefin. This species transforms into a σ-allyl transition species whose desorption leads to acrolein[3]. The latter is never found : Davydov[10] proposed that this intermediate form is removed from the surface by desorption at T > 200°C. Sb ions are generally given as responsible of the allyl groups fixation on selective catalysts, it should be postulated that they are active site for the olefin fixations during selective oxidation. The absorption with maximum 1680 cm^{-1} may belong to the C = O vibration in acrolein fixed on the surface as a result of the reaction[2] and 1790 – 1800 and 1740 cm^{-1} band to other fixed aldehydes (HCHO C_2H_4O) which desorb as CO_2 and C_2H_4O.

Fixed monocarbonaceous species are the same than those observed on $Cu_{0.24}Co_{2.76}O_4$ during corresponding treatments. These intermediates desorb easily in the 200 – 300°C range. We may conclude that they do not play a main part in the complete oxidation mechanism because the catalysts are only selective for allylic oxidation. Two maxima of absorption, 3750 and 3600 cm^{-1} , are assigned to ν_{OH}, the first one to OH occurring with the hydrogen separation during allyl complexes formation and the second one to H_2O fixation on the surface. Absorptions at 1080 and 625 cm^{-1} are observed for all treatments. They do not disappear after strong desorption (10^{-7} torr, 300°C) or high temperature oxidation (P_{O_2} = 1 Atm, 600°). They are characteristic of a structural change of the mixed oxide : formation of a new solid solution of Sb oxides in $USbO_5$ after the sharp reduction of Sb^{5+} ions of USb_3O_{10}, leading to its degradation[11].

CONCLUSION

On the basis of the results of the present work as well as from literature data cited above, the transformation of C_3H_6 on oxidation catalysts can be represented as follows:

with M_1 = Cu and Sb ions, M_2 = uncharacterized Cu ions on USb_3O_{1}),
M_3 = uncharacterized Co ions on USb_3O_{10}.

This model does not specify the oxidation states of the catalysts, but, it is well known that reduced catalysts are obtained
after product desorption (Cu^+, Co^{2+}, Sb^{3+} state), O_2 of the reactant
gases mixture oxidizing this reduced form.

REFERENCES

1. A.L. Dent and R.J. Kokes, J. Amer. Chem. Soc., 1970, 92, 6709.
2. A.A. Davydov, I. Tichy and A.A. Efrenov, React. Kinet. Catal.
   Lett., 1976, 5, 353.
3. J. Haber, Pure and Appl. Chem. 1978, 56, 923 - 40.
4. H. Baussart, M. Le Bras and J.M. Leroy, Z. Phys. Chem., Neue
   Folge, for editing.
5. M. Le Bras, H. Baussart, R. Delobel and J.M. Leroy, J. Chem.
   Soc., Faraday Trans. I, 1979, 75, 1337 - 45.
6. H. Baussart, R. Delobel, M. Le Bras, CR Acad. Sc., Ser. C,
   1978, 286, 605 - 8.
7. A. Ozaki, S. Tan, and Y. Moro-Oka, J. Catal., 1970, 17, 132 - 42.
8. J.E. Germain, Intra Sci. Chem. Rep., 1972, 6, 101.
9. M. Le Bras, Thèse Lille, 1977.
10. A.A. Budneva, A.A. Davydov and V.G. Michal'chenko, Kin. Kat.
    1975, 16 (2), 480-485.
11. H. Baussart, R. Delobel, J.M. Leroy, J.P. Bonnelle and L. Gen-
    gembre, Assemblée Générale de la SCF, Journées de Chimie du
    Solide, Bordeaux (Fr.), 10-11-12/09/1980.

INFRA-RED CRYOSPECTROSCOPIC STUDY ON THE INTERACTION OF CO WITH

A (111)Ni- AND NaCl-FILM AT LOW TEMPERATURE UNDER ULTRAHIGH VACUUM

J. Heidberg and I. Hussla

Institut für Physikalische und Theoretische Chemie der
Universität Erlangen-Nürnberg, D-8520 Erlangen
Bundesrepublik Deutschland

ABSTRACT

The interaction of carbon monoxide with clean nickel and sodium chloride surfaces has been investigated for the first time by infrared cryospectroscopy under ultra-high vacuum conditions at low adsorbent temperatures. The adsorbents were NaCl films, deposited onto an air cleaved (100)NaCl face used as such or covered by a (111)Ni film.

When nickel and NaCl surfaces at $T \leqslant 25$ K are exposed simultaneously to CO, no infra-red absorption due to Co-Ni adsorption is observable but extraordinarily sharp infra-red bands of a new metastable phase, μ-CO-NaCl and of a metastable solid-like phase $\bar{\alpha}$-CO are detected at 2153 and 2138 cm^{-1}, respectively.

Infra-red absorption due to Co-Ni chemisorption first appears at 2094 cm^{-1} on warming the nickel adsorbent to $\sim$ 35 K.

At very low temperature, $T \leqslant 15$ K, no infra-red adsorption of μ and α-CO-NaCl adsorption systems is detectable. Only the IR band of $\bar{\alpha}$-CO is observed. Evidence is presented that surface migration of CO on NaCl is strongly hindered up to temperatures $T \leqslant 40$ K.

An ultrahigh vacuum helium cryostat was developed allowing salt and metal evaporation in the 10^{-8} Pa (10^{-10}mbar) range as well as optical measurements in the spectral region between 220 nm and 50 µm. Cold alkali halide adsorbents turned out to be useful in trapping and analyzing desorbates from nickel surfaces acting as catalyst in CO conversion. Thus UHV cryospectroscopy may be applied to the analysis of desorbates from surfaces of any optional temperature under the clean and defined conditions of ultrahigh vacuum.

INTRODUCTION

     Low temperature experiments on adsorption may be directed
towards the study : first, of the interaction of gas and solid
surfaces, where either the solid surface or the gas or both may
have a low temperature. Or secondly, of low temperature properties
of adsorption phases, possibly prepared at higher temperatures.
Investigations of the interaction between gases and solid surfaces
at low temperature may yield valuable information on transients,
not detectable at higher temperature, as well as on sticking coefficients, surface migration,and barrier to formation of adsorption
phases[1,2]. Experiments belonging to the second group may be designed to selectively induce climbing up the manifold of excited
levels of specified modes by taking advantage of the reduction of
relaxation rates at lower temperatures. Then energy may be concentrated along the reaction coordinates relevant for the desired
consecutive processes[3]. This work is concerned mainly with experiments of the former type. As a first attempt, the interaction of
carbon monoxide with nickel and sodium chloride is studied in this
work , since especially the first system has been extensively
investigated at higher temperatures[4].

EXPERIMENTAL

     Although the theoretical description of radiation absorption
appears to be straightforward for the IR transmission technique,
the advantage of enhanced sensitivity is offered by the reflection
technique under a large angle of incidence. We chose the infra-red
transmission absorption technique (which we could implement more
easily), although previous relevant experience with this technique
is limited, since only a few IR transmission studies of the
adsorption on single crystal films have been performed under ultra
high vacuum. We considered that the disadvantage of lower sensitivity could be partially compensated by an increase at low temperatures of the metal surface enhancement of the molecular infrared absorption[4].

     Ultrahigh vacuum cryostats for infra-red spectroscopy have
been described for temperatures T $\geqslant$ 77 K, low temperature cryostats
are available for pressures $\geqslant 10^{-5}$ Pa. In the course of this work, a
liquid helium cryostat has been developed which allows salt and
metal film evaporation as well as optical studies in the spectral
range from 220 nm to 50 μm, with base pressures in the low 10^{-8} Pa
range. The (111) Nickel film was obtained by evaporation onto an
ultrathin NaCl film (< 100 Å thick) deposited on NaCl (100) air
cleaved substrates at 77 K, at a rate of 1 Å s^{-1} under a total
pressure of 6.10^{-8} Pa (6 x 10^{-10} mbar), the main component of the

residual gas being H_2 and He. The film structure was determined by X-ray diffraction and transmission electron micrography ex situ at room temperature. The pressure values given in the figures are the values at the measuring time. The pressure drops after gas admission. The temperatures given in Fig. 1 are sample holder temperatures, those indicated in Figs. 2 and 3 are sample temperatures. Details of the construction of this cryostat will be reported elsewhere[5]. The spectroscopic cell compatible with the cryostat and the optical system (Perkin Elmer 225 grating spectrophotometer with on line signal averager 1074 Nicolet) is described in the Appendix.

In an extension of the well-proven matrix isolation technique, ultrahigh vacuum cryospectroscopy employing highly transparent alkali halide adsorbents may be used in order to trap and identify desorbates from catalytic surfaces at any temperature under the clean and highly controllable conditions of ultrahigh vacuum. First endeavors in this direction have been made.

RESULTS AND DISCUSSION

The infra-red absorption in the spectral range between 2000 and 2200 cm^{-1} by the clean (111)Ni film at a sample temperature of 25 ± 5 K and at 4.10^{-8} Pa total pressure is shown in Fig. 1a. Admission of carbon monoxide at room temperature up to 1.10^{-6} Pa for 10^2 s onto the (111)Ni and NaCl film, both being in the infrared path at 25 K, had the surprising result presented in Fig. 1b. Only infra-red bands at 2138 cm^{-1} and 2153 cm^{-1} were appearing. They belong to metastable solid-like $\bar{\alpha}$-CO and to a new metastable adsorption phase, μ-CO-NaCl, respectively. (A detailed report on the temperature and coverage dependence of the vibrational bands of the adsorption systems will be presented in the near future) . The second CO admission up to 4.10^{-5} Pa for 100 s under the same conditions as the first one had the same result (Fig. 1c.) : no infra-red absorption due to CO-Ni-chemisorption was observed. Only the intensity of the $\bar{\alpha}$-CO infra-red band increased, in agreement with results obtained in the pure CO-NaCl system. It may be concluded that no chemisorption of CO on Ni occurs at T ⩽ 25 K.

Infra-red absorption at 2092 cm^{-1} of the adsorption system CO-Ni was first observable after warming the nickel film by 10 K to ca. 35 K for 10^2 s, thereby producing a CO pressure pulse up to 10^{-5} Pa. The intensity of the $\bar{\alpha}$-CO band also increased, and the sharp absorption of $\bar{\alpha}$-$^{13}C^{16}O$ became visible at 2091 cm^{-1} together with the very weak low frequency phonon side band. Complete removal of $\bar{\alpha}$-CO was accomplished by warming to 40 K for 5.10^2 s. Cooling to 25 K, leading to formation of $\bar{\alpha}$-CO again, yielded spectrum 1h. Depletion of phase $\bar{\alpha}$-CO took place on warming and pumping, while the IR absorption due to $\bar{\mu}$-CO-NaCl and CO-Ni remained essentially

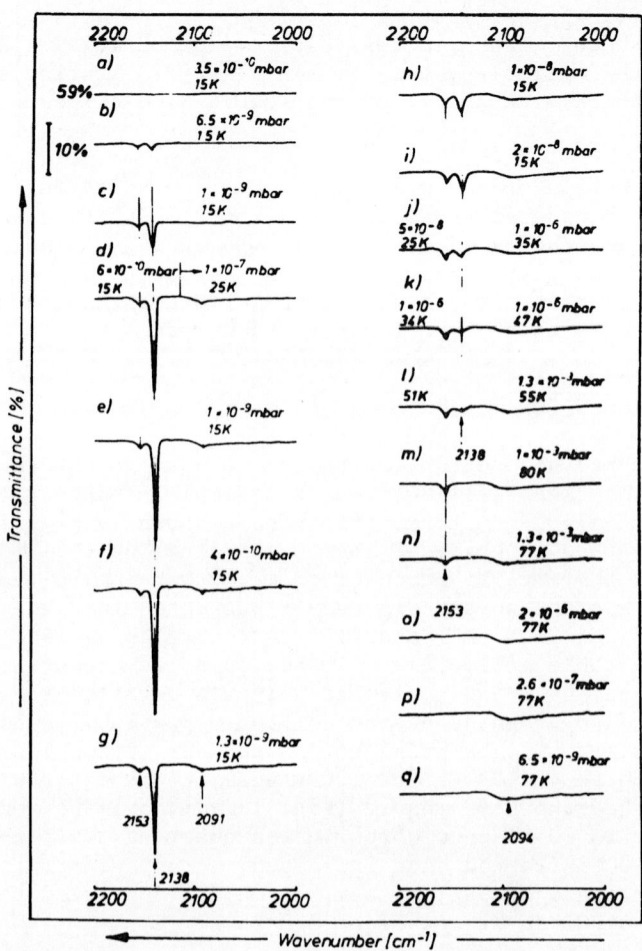

Fig. 1 - Infrared absorption by the systems CO (111) nickel film
         and CO NaCl at low temperature. Sample holder temperatures
         are given.

unchanged (Spectra li - ll). The band at 2153 cm^{-1} of CO-NaCl
disappeared at 77 K on pumping. (Spectra l m - 2o). Spectra lp and
lq show the remaining absorption of CO-(lll)Ni film at 77 K which
cannot be eliminated by pumping[7,8].

A corresponding sequence of experiments was carried out at
higher temperatures and revealed a rather different interaction
of CO with Ni and NaCl surfaces at 77 K (Fig. 2). The surface areas
of the adsorbents were larger in these experiments than before.
No infrared absorption of CO-NaCl is detectable at CO pressures of
5.10^{-7} Pa, while the infrared band of CO-Ni chemisorption system
is fully developed at 2090 cm^{-1} (Spectrum 2b). At higher pressure,
beside the CO-Ni absorption, the infrared band at 2156 cm^{-1} of
stable σ-CO-NaCl emerged (4 % absorption) growing to almost 47 %
absorption and shifting to a wavenumber of 2152 cm^{-1} at a CO
pressure of 7.5 Pa at 77 K (Spectra 2c and 2d). The spectra 2e and
2f were obtained after warming the system to 160 K and further to
250 K, respectively. The infra-red band of CO-NaCl disappeared as
in the preceding experiments.

In analogous measurements on the pure CO-NaCl system neither
the IR band of the stable adsorption phase σ-CO-NaCl nor that of
the metastable μ-CO-NaCl were detected on admitting CO onto the
NaCl film at very low temperatures T $\leqslant$ 15 K. Only the band of
solid-like $\bar{\alpha}$-CO was observed under these conditions. Its line width
was broader than that of solid $\bar{\alpha}$-CO. On warming the system to ca.
20 K, the metastable adsorption phase μ-CO-NaCl was formed (compare
spectrum lb). This phase had the same vibrational wavenumber,
2153 cm^{-1}, at low surface coverage (spectra to be published) as
the stable phase σ-CO-NaCl at high coverage (spectra 2d and 3). The
wavenumber at constant coverage of the latter is independent of
temperature. $\mu \rightarrow \sigma$ conversion occurred between 40 to 50 K on a time
scale of minutes.

These observations strongly suggest, that island formation
takes place and that a large hindrance to surface migration of CO
on the NaCl surface exists at temperatures $\leqslant$ 40 K. Experiments,
in which the inverse approach is made to the important problem of
surface migration are underway : in a uniformly covered surface a
vacant spot is generated by a short laser pulse and the time is
measured in which at constant temperature the surface is covered
again, the gas pressure being sufficiently low.

The infra-red absorption of σ-CO-NaCl at constant temperature
as a function of surface coverage is displayed in Fig. 3. The
vibrational wavenumber of this stable adsorbate decreases with
increasing coverage as opposed to the corresponding variation in
CO-Ni. The shift upon adsorption from the gas to form both μ- and
σ-CO-NaCl is to higher wavenumbers[9], which is also opposite to
the shift upon adsorption on nickel. In the limit of very low

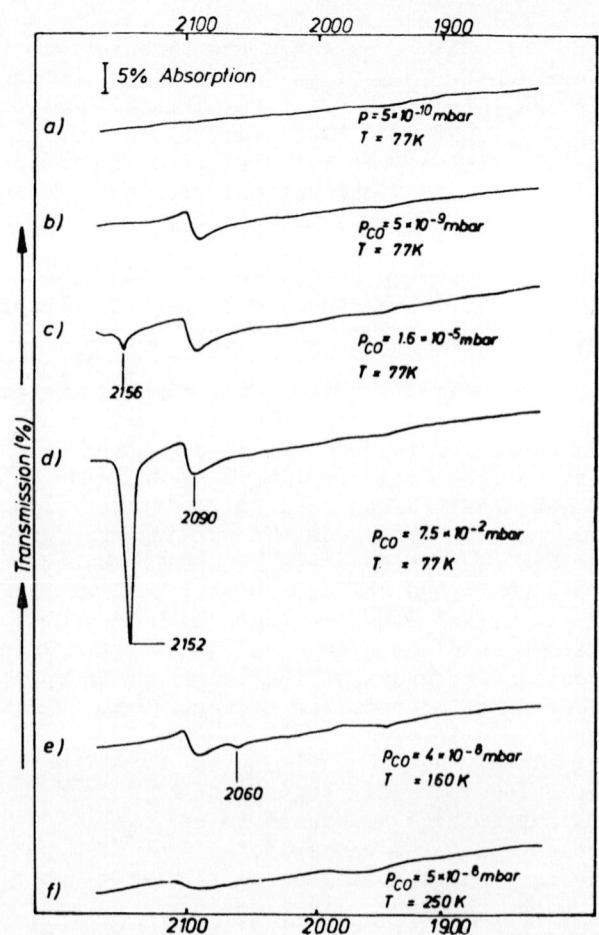

Fig. 2 - Infra-red absorption by the adsorption systems CO-(111) nickel film and σ-CO-NaCl at 77 K, 160 K, 250 K.

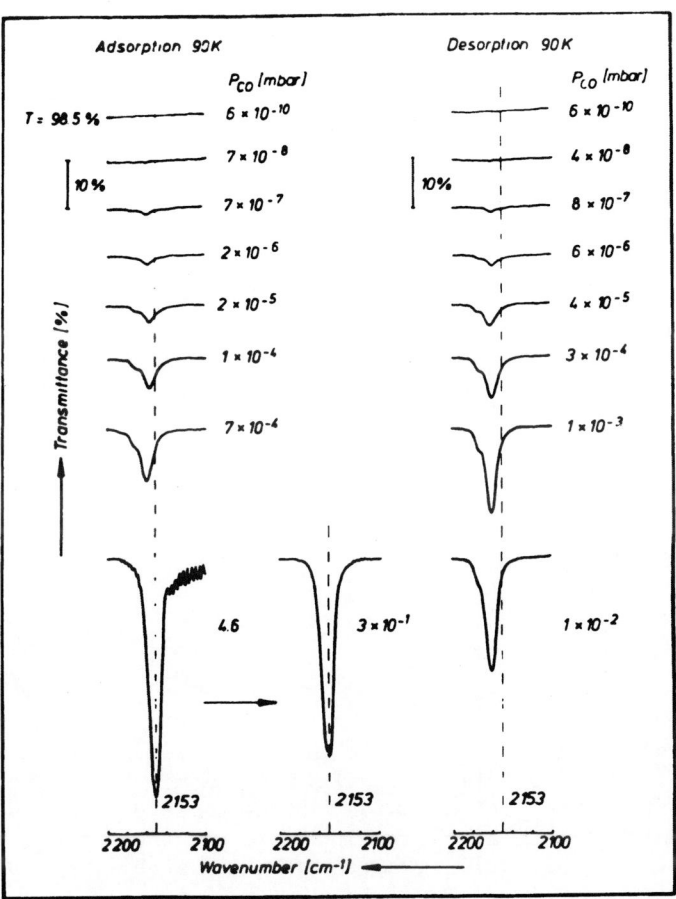

Fig. 3 – Infra-red absorption by σ-CO-NaCl at constant temperature (90 K) and varied coverage.

coverage the wavenumber of σ-CO-NaCl is 2162 cm^{-1}, whereas at maximum coverage it is 2153 cm^{-1} (compare spectra 3,2,1).

As a conclusion, by using alkali halide adsorbents as isolation matrix, UHV cryospectroscopy provides a method for studying the desorption from surfaces at any temperatures under clean and well-defined conditions. First results of the application of this method have been obtained for the system carbon monoxide nickel, where CO_2 was detected as product of the catalytic conversion. Particularly promising is laser induced desorption which may be achieved at surface coverage, temperatures and rates approaching conditions of technical significance.

ACKNOWLEDGEMENTS

The authors should like to express their gratitude to Prof. Dr. Wissmann and Dr. H. Zitzmann for carrying out the X-ray diffraction measurements, to Mr. Puppel for taking the transmission electron micrographs, and to Mr. Wallner and Mr. Bergfeld for outstanding work in building the UHV cryostat. It is a pleasure to thank Dr. H. Stein for many helpful discussions as well as efficient and kind support in experimental work.

Grants from Deutsche Forschungsgemeinschaft, Fonds der chemischen Industrie and Universitätsbund Erlangen-Nürnberg are gratefully acknowledged.

APPENDIX

Spectroscopic Cell.

Fig. 4 shows the spectroscopic cell used in this work. Evaporators for salt and metal film preparation are movable up to 8 cm, from the substrate through a Varian UHV manipulator (DDF, CF 16), which is connected with a copper block holding the evaporators over a stainless steel driving screw (GA) and stabilizer rod (GS). A tungsten wire coil (wire $\emptyset$ 0.1 mm) heats (0.4 A, 16 V) the alkali halide salt in a Knudsen cell made of quartz glass (QR). Nickel was evaporated from a helix ($\emptyset$ 0.5 cm) (MW) of highly pure nickel wire ($\emptyset$ 0.3 mm, Materials Research, MARZ), which is melted onto glass isolated tungsten holders fixed in the copper block. In back position the Ni helix was cleaned by pre-evaporation during several days. Shields (B) made of stainless steel protect the KBr windows ($\emptyset$ 52 mm, 6 mm thick, Dr. Korth, Kiel) from Ni condensation. The recently developed UHV sealing system for salt single crystals in metal flanges employing O-rings made of Kalrez (Trade mark of Du-Pont)[12] and indium to be described in detail elsewhere[13], was successfully applied. The sealing proved UHV suitable, withstanding out-

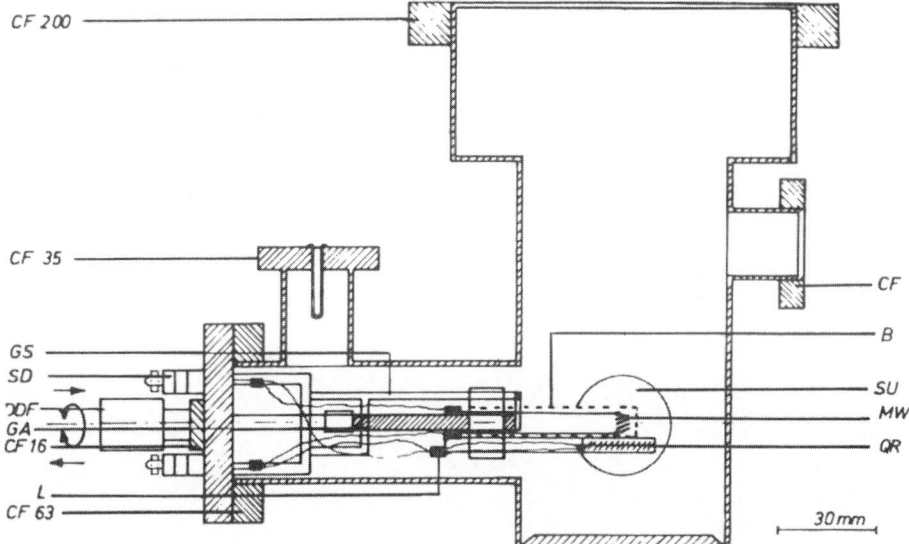

Fig. 4 - Spectroscopic Cell.

baking under careful temperature control. Helium tightness is controlled and residual gas analysis carried out by means of a quadrupole mass analyzer (Q 200 Leybold-Heraeus). Using various cryogenics, e.g. liquid helium, solid nitrogen, liquid nitrogen and liquid oxygen, stable temperatures were achieved at the sample holder and at the substrate. They were measured in situ by Au (Fe)-Chromel (Heraeus) thermocouples.

REFERENCES

1. J. Heidberg, H. Stein and I. Hussla
   Ber. Bunsenges. Phys. Chem. 83, 1238 (1978).
2. P.R. Norton, R.L. Tapping and J.W. Goodale,
   Surface Sci. 72, 33 (1978).
3. J. Heidberg, H. Stein and E. Riehl, these Proceedings.
4. G. Wedler, Chemisorption, An Experimental Approach,
   Butterworth London 1976.
   J. Pritchard, in Chem. Phys. Solids & Surfaces, Vol. 7, 1978.
   N. Sheppard and T.N. Nguyen in Advances in Infrared and Raman
   Spectroscopy, Vol. 5, Eds. R.J.H. Clark and R.E. Hester, Heyden,
   London 1978.
   W. Erley, H. Wagner and H. Ibach, to be published.
5. J. Heidberg and H. Stein, to be published in Rev. Sci. Instrum.
   J. Heidberg, I. Hussla, E. Hoefs, H. Stein and A. Nestmann,
   Rev. Sci. Instrum. 49, 1571 (1978).
6. J. Heidberg, H. Stein and I. Hussla, to be published.
   J. Heidberg, and I Hussla, to be published.
7. A.M. Bradshaw and O Vierle,
   Ber. Bun. Gesell. 74, (7) 630 (1970).
8. J.C. Campuzano and R.G. Greenler,
   Surface Sci. 83, 301 (1979).
9. R. Gevirzman, Y. Kozirovsky and M. Folman,
   Trans. Faraday Soc. 65, 2206 (1969).
10. R. Gevirzman and Y. Kozirovski,
    Trans. Faraday Soc. 67, 2686 (1971).
11. B. Rao and M.J. Dignam,
    J.C.S. Faraday Trans. II 70, 492 (1974).
12. L. de Chernatony, Proc. 7th Intern. Vac. Congr. & 3rd Intern.
    Conf. Solid Surfaces (Vienna 1977), Vol. I, p. 255.
13. J. Heidberg and I. Hussla, to be published.

INFRARED STUDY OF COMPARATIVE ADSORPTION OF SOME ALCOHOLS AND THIOLS

ON γ- ALUMINA

J. Travert, O. Saur, M. Benaissa, J. Lamotte and J.C.
Lavalley

Laboratoire de Spectrochimie, Groupe Structure et Réac-
tivité d'Espèces Adsorbées, ERA 824, I.S.M.R.A.
Université de Caen
14032 CAEN CEDEX FRANCE

Among the spectroscopic methods used for the characterization
of the chemical nature of chemisorbed species, infrared transmission
spectroscopy has certainly found the most frequent application.
Adsorption of alcohols on alumina has been studied for many years[1,2].
Despite this fact, even most fundamental questions concerning the
system have not been yet answered. A most striking example is the
process of adsorption (dissociation or coordination) and the nature
of the adsorption sites on γ-Al_2O_3 surface. To solve this problem,
we have undertaken a series of infrared experiments using non-deu-
terated and deuterated alcohols, the vibrational spectra of which
had been previously reported in the gaseous and liquid state[3] .
Some experiments have been also carried out with methanethiol for
comparison. A very detailed analysis of the spectra allows us to
determine the nature of irreversible and reversible species.
Poisoning experiments precise the mechanism of the adsorption
process.

IRREVERSIBLE SPECIES

Alcohols

Infrared studies of alcohols on alumina at room temperature
have shown that at least two types of chemisorbed species occurred:
alkoxide species, resulting from a dissociative chemisorption[1] and
coordinated species, formed by chemisorption onto Lewis acid sites[2].

333

Carboxylate species are generally formed at higher temperatures[4] .
To distinguish between dissociative and coordinative chemisorption,
the spectra have been carefully examined in two ranges :

a) in the 1500 - 1000 cm^{-1} range, comparison between vibrational
spectra of adsorbed ROH and ROD allows us to solve the problem :

  - in $R'CH_2OH$ or $\overset{R'}{\underset{R''}{\diagdown}}CHOH$ spectra, the $\delta(OH)$ mode is coupled with
    the CH_2 or CH bending modes [3].   If these alcohols are disso-
    ciatively adsorbed these couplings are suppressed and the spec-
    tra of the adsorbed species might be similar to those given
    by the corresponding ROD alcohols.   This has been effectively
    observed for propargyl alcohol, and trichloro-2,2,2 or trifluo-
    ro-2,2,2 ethanols [5];

  - in the case of tertiary alcohols, comparison between adsorbed
    ROH and ROD spectra in this range is also very helpful as
    shown for $(CF_3)_3COH$:

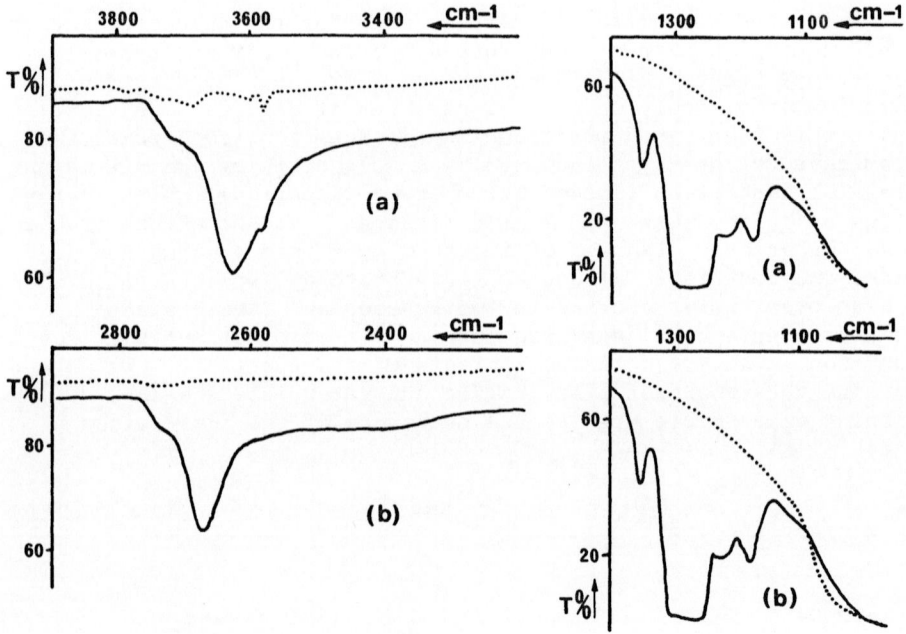

Fig. 1. : Comparative IR spectra of
        a) $(CF_3)_3COH$
        b) $(CF_3)_3COD$
        irreversibly adsorbed on alumina (Degussa-C) dehydro-
        xylated at 1150 K. Dotted line : backround.

the spectra given by $(CF_3)_3COH$ and $(CF_3)_3COD$ irreversibly adsorbed at room temperature are similar (Fig. 1.).   In particular, no $\delta(OH)$ band is observed when $(CF_3)_3COH$ is adsorbed.   Considering that the $\delta(OH)$ band is at 1380 cm^{-1} in solution [6] and that coordination does not strongly affect it [7], we again conclude that $(CF_3)_3COH(D)$ is dissociatively adsorbed.   Note that the $\delta(OH)$ vibration of hydroxyl groups newly created by alcohols adsorption is expected below 1000 cm^{-1} cm^{-1} [8] i.e. in the region of the lattice vibrations of the adsorbent, opaque in the infrared transmission technique.

b) in the 3000 - 2800 cm^{-1} range, for primary or secondary alcohols we have shown that dissociative adsorption generally shifts the $\nu(CH)$ wavenumbers to lower values whereas coordinative adsorption increases the $\nu(CH)$ frequencies.   Due to Fermi resonance interactions between $\nu(CH_2)$ fundamentals and $2\delta(CH_2)$ or $2\omega(CH_2)$ overtones in the case of $R'CH_2OH$ alcohols, deuterated $R'CHDOH$ compounds have been used [9].

     Studies of these two frequency ranges show that fluoroalcohols $[CF_3)_3COH$, $(CF_3)_2CHOH$, $CF_3CH_2OH]$, CCl_3CH_2OH and $HC\equiv CCH_2OH$ are dissociatively chemisorbed on alumina.   Experiments with a highly dehydroxylated sample of alumina (pretreated at 1150 K) allow to point clearly the frequencies of the newly created hydroxyl groups :

alcohol	$\nu(OH)$ in solution	$\nu(OH)$ created
$(CF_3)_3COH$	3589^{xx} cm^{-1}	3630 cm^{-1}
$(CF_3)_2CHOH$	3628^{xx} cm^{-1} 3589^{xx} "	3620 cm^{-1}
CF_3CH_2OH	3640^x ep " 3622^x "	3620 cm^{-1}
CCl_3CH_2OH	3600^x "	3550 cm^{-1}
$HC\equiv CCH_2OH$	3624^x "	3530 cm^{-1}

$^x CCl_4$ solution; ^{xx}heptane solution.

     The low wavenumber and the broadness of the band of some new OH groups could be due to an hydrogen bond.   According to this interpretation, we propose two kinds of species :

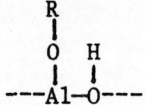

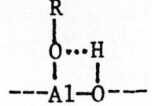

      fluoroalcohols CCl_3CH_2OH, $HC{\equiv}CCH_2OH$

Thiols.

The nature of the species could be directly deduced from the study of the 3700 – 3500 cm^{-1} and 2600 – 2400 cm^{-1} ranges.   In the case of CH_3SH for instance[10], it had been shown that OH groups were created (band at 3690 cm^{-1}) whereas the S–H bond were broken [the $\nu(SH)$ band near 2500 cm^{-1} has disappeared].   From the 3690 cm^{-1} wavenumber, we deduce the following structure :

Note that in any case (alcohols and thiols) a high increase of alumina transparency from 1100 to 1000 cm^{-1} is observed suggesting, in agreement with MORTERRA [11], that the surface Al–O modes are responsible for the alumina absorption in this range.

REVERSIBLE SPECIES

Their nature may be deduced by substracting from the spectra of total adsorbed amount the contribution of irreversible species[12]. Although the reversible adsorbed amounts are very small in the case of $(CF_3)_3COH$ and $(CF_3)_2CHOH$, their spectra show the same characteristics as those presented when CF_3CH_2OH, CCl_3CH_2OH or $HC{\equiv}CCH_2OH$ were adsorbed :

- negative absorption differences ΔA near 3650 – 3600 cm^{-1}
  provide evidence that a part of the OH groups resulting from
  the dissociative irreversible adsorption acts as proton donors;

- negative absorption differences in the 1300 – 1100 cm^{-1} range
  show that the $\nu(CO)$ stretching vibrations of some irreversible
  species are perturbed when larger quantities of alcohol are
  added;

- broad bands near 3200 cm^{-1} characterize species held by
hydrogen bonding.

Comparative experiments with highly dehydroxylated samples of
alumina (pretreated at 1153 K) and less activated samples (pretreated at 873 K) show that the residual OH groups on Al_2O_3 are not
involved in the reversible adsorption. This leads us to consider
three possible reversibly adsorbed species :

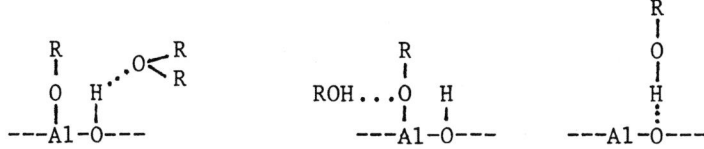

In the case of CH_3SH, the same type of hydrogen bonds are
considered since the spectrum (Fig. 2.) shows negative differences
ΔA near 3680 cm^{-1} and broad bands near 3200 cm^{-1} $[\nu$ (OH) and
2600 - 2300 cm^{-1} $\nu(SH)]$. Moreover, coordinated species, characterized by higher $\nu(CH_3)$ wavenumbers, do occur[10] :

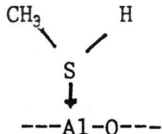

From this study, it appears that the new OH groups resulting
from the dissociative chemisorption play a very important role in
the formation of hydrogen-bonded species. They certainly change
the acidic or basic properties of the surface according to the
nature of the adsorbate. Such results agree with the study of
cyclopropane isomerization by KNOZINGER et all[12] : the rate of
reaction was enhanced by preadsorption of isobutyl alcohol on alu
mina.

ADSORPTION SITES

Poisoning experiments performed with pyridine had shown that
adsorption of H_2S and CH_3SH could occur through Lewis acid sites
(I) but also through strong basic oxygen sites (II)[10]. Some
experiments have been carried out with alcohols adsorbed on alumi-

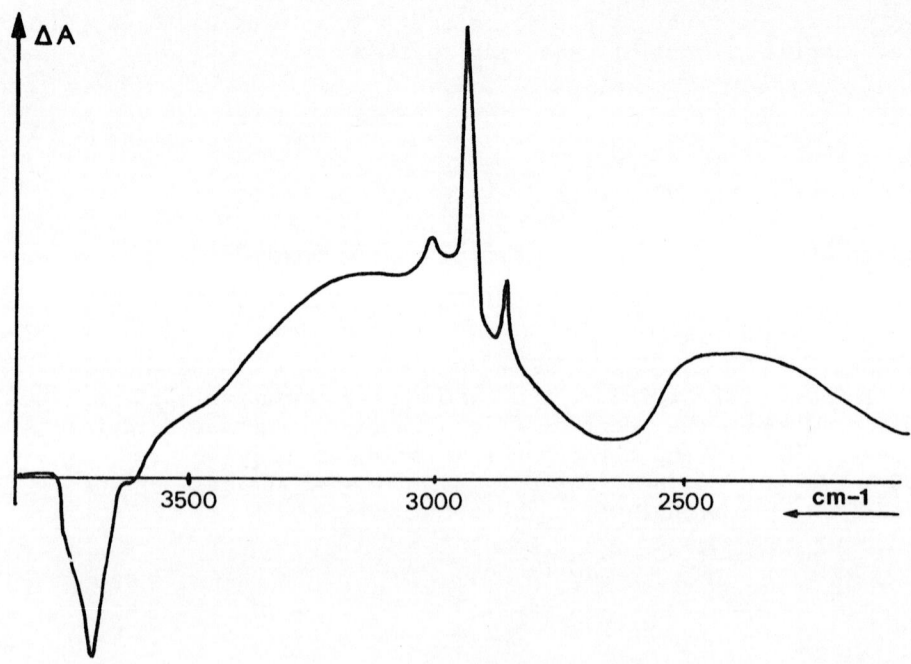

Fig. 2. : IR spectrum of reversible species given by CH_3SH
        on alumina.

na samples poisoned by pyridine or dimethylether which has been
recently used as a sensitive probe[13].   They show that two types
of sites should be involved in the process of adsorption.   The
alcoholate or thiolate species could be formed on these two
different sites via two different mechanisms involving coordinatively
bound molecules or hydrogen bonded·species :

with Z = 0 or S.   Contribution of the sites (I) and (II) depends
on alcohol nature and acid-base strength of the pair sites.

REFERENCES

1. R.G. Greenler, J. Chem. Phys., 37, 2094 (1962).
2. H. Knozinger and B. Stubner, J. Phys. Chem., 82, 1526 (1978).
3. J. Travert and J.C. Lavalley, Spectrochim. Acta, 32A, 637
   (1976).
   J. Travert, J.C. Lavalley and D. Chenery, Spectrochim. Acta,
   35A, 291 (1979).
4. R.O. Kagel, J. Phys Chem. 71. 844 (1967).
5. J. Travert, J.C. Lavalley and O. Saur, J. Chim. Phys. (in press).
6. J. Murto, A. Kivinen, J. Korppi- Tommola, R. Viitala and J. Hyo-
   maki, Acta Chem. Scand., 27, 107 (1973).
7. J.P. Gallas and C. Binet, private communication.
8. H.E. Evans and W.H. Weinberg, J. Chem. Phys., 71, 4789 (1979).
9. J.C. Lavalley, J. Caillod and J. Travert, J. Phys. Chem., 84,
   2084 (1980).
10. O. Saur, T. Chevreay, J. Lamotte, J. Travert and J.C. Lavalley
    J. Chem. Soc. Faraday Trans I (in press)..
11. C. Morterra, Proc. Int. Congr. Catal. 6th, 1976, 194 (1977).
12. J.C. Lavalley, J. Travert and J. Lamotte, J. Chim. Phys. (in
    press).
13. J. Caillod and J.C. Lavalley, J. Chim. Phys., 77, 379 (1980).

INFRARED SPECTRA OF SPECIES ADSORBED ON FILMS OR WELL-DEFINED

SURFACES : A TABULATION OF THE INVESTIGATED SYSTEMS [*]

J. Darville

Institute for Research in Interface Sciences
L.A.S.M.O.S. - Département de Physique
Facultés Universitaires Notre-Dame de la Paix
61, rue de Bruxelles, B-5000 Namur - BELGIUM

Valuable information on the structure and bonding types of adsorbates on a solid substrate can be obtained from infrared spectra of the chemical system so generated. The application of infrared methods to supported metal catalysts and to pressed powdered oxides and halides will not be considered here. Studies on corrosion inhibitors and organic thin films on metals are also out of the scope of this paper, as well as reports on laser-stimulated surface reactions.

Infrared spectroscopy has been used for a long time, both in ultra-high vacuum and at higher pressures, as a powerful tool for the study of adsorption on polycrystalline films, oriented films and single crystals. Conventional transmission experiments call for no special comments. On the contrary, the application of the emission[1-9], and, mainly, reflection[5,10-16] methods received special consideration as far as the basic processes are concerned. During the last decade, several review papers appeared on the subject of adsorbate vibrations investigated by infrared spectroscopy, essentially on metals, with the transmission and reflection techniques[17-30,196-198]. More recently a similar effort was done for infrared ellipsometric studies[31,32].

Theoretical investigations on the origin of the frequency shifts observed as a consequence of the existence of the surface and as a function of the adsorbate coverage have led to considerable progress in our understanding of these phenomena. As far as the second is

[*] Work performed under the auspices of the I.R.I.S. program supported by the Belgian Ministry for Science Policy.

concerned, two models are competing : the back-bonding model[33-36] and the oscillating dipole model[29,37-45]. Furthermore, a dynamical theory of the infrared line shape has been proposed[46].

This tabulation is only concerned with experimental results. Table I gives the meaning of the various symbols used in the third column of table II for mentioning the particular infrared techniques. Investigations on films or single crystals of metals are listed at the beginning of table II whereas the very few results on semi-conductors and oxides come at the end. When an oxide layer on a metallic film results from the oxidation of this film, the references on this work are introduced in the corresponding metal section. Finally table III gives a separate overview on the experiments performed on halide ionic compounds. This last set of references is entirely concerned with transmission experiments through high area evaporated films, deposited on an infrared transparent material.

The literature has been searched up to November 1980.

ACKNOWLEDGEMENTS

The research groups in the reviewed field, especially those using the RAIS method, are fully acknowledged for sending lists of their published and submitted work.

Table I

RAIS   = reflection-absorption infrared spectroscopy (amplitude
          modulation - single or multiple reflections - single beam).
RAIS-DB = reflection-absorption infrared spectroscopy (amplitude
          modulation - single or multiple reflections - double beam).
RAIS-WM = reflection-absorption infrared spectroscopy (wavelength
          modulation).
RAIS-PM = reflection-absorption infrared spectroscopy (polarization
          modulation).
RAIS-FT = reflection-absorption infrared spectroscopy (interfero-
          metric method, Fourier transform).
IRES   = infrared ellipsometric spectroscopy.
TR   = transmission infrared spectroscopy.
EM   = emission infrared spectroscopy.
INT.R. = internal reflection infrared spectroscopy.

TABLE II

ev. = evaporated   ox. = oxidized
ev.(CO)= evaporated in CO   expl. = exploded

Adsorbent	Adsorbate	Method	References
ALUMINIUM			
ev. film/glass	HCOOH	RAIS-DB	47,48
ev. film/glass	DCOOD	RAIS-DB	47,48
foil	oxide layer	RAIS	49,50
sheet	oxide layer	RAIS-DB	51,52
film,foil	oxide layer	RAIS	53
Al-Mg alloys	oxide layer	RAIS	54
foil	oxide layer	EM	55
(100)	oxide layer	RAIS-DB	56
(110)	oxide layer	RAIS-DB	56
(111)	oxide layer	RAIS-DB	56
CHROMIUM			
ev. film (CO)/CaF_2	CO	TR	57
ev. film (CO)/CaF_2	CO + O_2	TR	57
COBALT			
ev. film/CaF_2	CO	TR	57
ev. film/CaF_2	CO + O_2	TR	57
ev. film/NaCl	CO	TR	58
ev. film (CO)/NaCl	CO	TR	59
ev. film (CO)/NaCl	CO + $(CH_3)_3N$	TR	59
ev. film (CO)/NaCl	CO + C_2H_5NC	TR	59
ev. film(CO or H_2)/NaCl	CO + C_2H_2	TR	60
ev. film(CO or H_2)/NaCl	CO+CH_3CHCH_2	TR	60
ev. film/glass	CO	RAIS-DB	61

COPPER

ev. film/glass	CO	RAIS	61-66
ev. film/Al_2O_3, MgO	CO	RAIS	66
ev. film/glass	CO	RAIS-WM	67
ev. film/Ta foil	CO	RAIS-DB	68
ev. film/NaCl	CO	TR	58
ev. film/glass	CO	IRES	69,70
ev. film/glass	CO_2	IRES	70
expl. wire/NaCl	C_2H_2	TR	120
ev. film/glass	HCOOH	RAIS-DB	49
ev. film/KBr	HCOOH	TR	50
ev. film/glass	HCOOH + O_2	IRES	70
ev. film/glass	DCOOD	RAIS-DB	49,71
ev. film/glass	isoamyl nitrite	RAIS-DB	50
plate	CO	RAIS-DB	72
(100)	CO	RAIS	64,73,74
(100)	CO	RAIS-WM	75-77
(100)	$C^{16}O + C^{18}O$	RAIS-WM	77-80
(100)	$^{12}C^{16}O + ^{13}C^{18}O$	RAIS-WM	78
(110)	CO	RAIS	66,81
(110)	CO	RAIS-WM	82
(110)	$^{12}CO + ^{13}CO$	RAIS-WM	82
(110)	C_2D_6	RAIS-WM	83
(111)	CO	RAIS	64,66,74,81
(111)	$^{12}CO + ^{13}CO$	RAIS-WM	84,85
(211),(311),(755)	CO	RAIS	66
ox. ev. film/glass	CO	IRES	69,70
ox. ev. film/Ta foil	CO	RAIS-DB	68
ox. ev. film/glass	CO_2	IRES	70
ox. ev. film/Au	CH_3COOH	RAIS	165,166
ox. ev. film/glass	HCOOH + O_2	IRES	70
mirror	epoxy resin	RAIS-DB	86
film	oxide layer	RAIS	13,87,88

foil	oxide layer	RAIS	88
film	oxide layer	RAIS-DB	89,90
plate	oxide layer	EM	91
film/KRS-5	oxide layer	INT.R.	168

GOLD

ev. film/NaCl	CO	TR	58
ev. film/glass	CO	RAIS	64
ev. film/Ta	CO	RAIS-DB	92

IRIDIUM

ev. film/CaF$_2$	CO	TR	57
ev. film/NaCl	CO	TR	93
ev. film/glass	CO	RAIS-DB	94
ev. film/glass	$^{12}CO + ^{13}CO$	RAIS-DB	94
ev. film/CaF$_2$	CO + O$_2$	TR	57
ev. film/NaCl	CO + O$_2$	TR	93
ev. film/glass	CO + O$_2$	RAIS-DB	95
ev. film/NaCl	CO + H$_2$	TR	93
ev. film/NaCl	CO + C$_2$H$_2$	TR	93
ev. film/NaCl	HCOOH	TR	93
ev. film/NaCl	H$_2$CO	TR	93

IRON

ev. film(CO)/NaCl	CO	TR	59
ev. film/CaF$_2$	CO	TR	57
ev. film/NaCl	CO	TR	58
ev. film/CaF$_2$	CO + O$_2$	TR	57
ev. film/NaCl	CO+(CH$_3$)$_3$N	TR	59
ev. film/glass	NO	RAIS-DB	50
ev. film/glass	isoamyl ni-trite	RAIS-DB	50
mirror	epoxy resin	RAIS-DB	86

MANGANESE

ev. film (CO)/CaF$_2$	CO	TR	57
ev. film (CO)/CaF$_2$	CO + O$_2$	TR	57

MOLYBDENUM

expl. wire/NaCl	CO	TR	120

NICKEL

ev. film (CO)/NaCl	CO	TR	59,96
ev. film/CaF$_2$	CO	TR	57,97
expl. wire/KBr	CO	TR	98
ev. film/NaCl	CO	TR	58,96,99
ev. film/glass	CO	RAIS	65,100
ev. film/glass	CO	RAIS-DB	61,71,101
ev. film/CaF$_2$	CO + O$_2$	TR	57
ev. film (CO)/NaCl	CO+(CH$_3$)$_3$N	TR	59
ev. film(CO or H$_2$)/NaCl	CO + C$_2$H$_2$	TR	60
ev. film(CO or H$_2$)/NaCl	CO+CH$_3$CHCH$_2$	TR	60
ev. film(N$_2$)/NaCl	N$_2$	TR	102
ev. film/NaCl	N$_2$	TR	100,102
ev. film(N$_2$)/NaCl	N$_2$ + CO	TR	102
ev. film/NaCl	N$_2$ + CO	TR	102
ev. film/glass	NO, ^{15}NO	RAIS-DB	103,104
expl. wire/NaCl	C$_2$H$_2$	TR	120
ev. film/glass	C$_2$H$_2$	RAIS-DB	105,106
ev. film/glass	C$_2$D$_2$	RAIS-DB	105,106
ev. film/glass	C$_2$H$_4$	RAIS-DB	106
ev. film/glass	C$_2$H$_6$	RAIS-DB	106
ev. film/glass	DCOOD	RAIS-DB	71
ev. film/glass	isoamyl ni-trite	RAIS-DB	50
(110) ev. film/Al	CO, ^{13}CO	RAIS-DB	107-110
(111) ev. film/NaCl	CO	TR	111

(100)	CO	IRES	28,112-114
(100)	$^{12}CO + ^{13}CO$	IRES	32
(100)	CH_3OH	IRES	112,113
(100)	CD_3OD	IRES	112
(110)	CO	IRES	32,113,115-117
(110)	$CO + H_2$	IRES	113,117
(110)	N_2	RAIS-WM	118
(111)	CO	RAIS-DB	119
(111)	$C^{16}O + C^{18}O$	RAIS-DB	119
PALLADIUM			
ev. film(CO)/NaCl	CO	TR	59
expl. wire/NaCl	CO	TR	120
expl. wire/KBr	CO	TR	98
ev. film(CO)/KBr	CO	TR	121
porous film/CaF_2	CO	TR	122
ev. film/NaCl	CO	TR	93
ev. film/KBr	CO	INT.R.	123
ev. film/glass	CO	RAIS-DB	61
ev. film/glass	CO	RAIS-PM	124,125
foil	CO	RAIS-FT	126
ev. film(CO)/NaCl	$CO+(CH_3)_3N$	TR	59
ev. film/NaCl	$CO + H_2$	TR	93
ev. film/NaCl	$CO + O_2$	TR	93
foil	$CO + O_2$	RAIS-FT	126
ev. film/NaCl	$CO + C_2H_2$	TR	93
ev. film/glass	H_2	RAIS-DB	127
ev. film/NaCl	H_2CO	TR	93
ev. film/NaCl	HCOOH	TR	93
ev. film/NaCl	NO	INT.R.	167
ev. film/glass	$NO, ^{15}NO$	RAIS-DB	104
expl. wire/NaCl	C_2H_2	TR	120
ev. film/glass	C_2H_2	RAIS-DB	105,106

ev. film/glass	C_2D_2	RAIS-DB	105,106
ev. film/glass	C_2H_4	RAIS-DB	106
ev. film/glass	C_2H_6	RAIS-DB	106
ev. film/NaCl	SO_2	INT.R.	167
(100)	CO	RAIS-PM	128-130
(100)	$^{12}CO + {}^{13}CO$	RAIS-PM	130
(111)	CO	RAIS-PM	128,129,131
(210)	CO	RAIS-PM	129

PLATINUM

ev. film(CO)/NaCl	CO	TR	59
ev. film/CaF_2	CO	TR	122,132
ev. film/NaCl	CO	TR	93
ev. film/KBr	CO	TR	121
expl. wire/KBr	CO	TR	98
ev. film/glass	CO	RAIS-PM	125
ev. film/glass	CO	RAIS-DB	61
ev. film(CO)/NaCl	$CO+(CH_3)_3N$	TR	59
ev. film(CO)/NaCl	$CO+C_2H_5NC$	TR	59
ev. film(CO or H_2)/NaCl	$CO + C_2H_2$	TR	60
ev. film(CO or H_2)/NaCl	$CO+CH_3CHCH_2$	TR	60
ev. film/NaCl	$CO + H_2$	TR	93
ev. film/NaCl	$CO + O_2$	TR	93
expl. wire/KBr	$CO + O_2$	TR	98
ev. film/NaCl	$CO + C_2H_2$	TR	93
ev. film/glass	$CO + C_6H_6$	RAIS-DB	61
ev. film/NaCl	HCOOH	TR	93
ev. film/NaCl	H_2CO	TR	93
ev. film/glass	C_2H_2	RAIS-DB	106,133
ev. film/glass	C_2D_2	RAIS-DB	106
ev. film/glass	C_2H_4	RAIS-DB	106,133
ev. film/glass	C_2H_6	RAIS-DB	106,133
foil	CO	RAIS-FT	134

(111) oriented ribbon	CO	RAIS	135,136
foil, mostly (111)	CO	RAIS-PM	137,138
(111) oriented ribbon	$^{12}CO + ^{13}CO$	RAIS	136,139
(111) oriented ribbon	$CO + O_2$	RAIS	140
foil, mostly (111)	$CO + O_2$	RAIS-PM	138
(111) oriented ribbon	N_2	RAIS	141
foil, mostly (111)	NO, ^{15}NO	RAIS-PM	142,143
foil, mostly (111)	N_2O	RAIS-PM	143
foil, mostly (111)	NH_3, $^{15}NH_3$	RAIS-PM	144
foil, mostly (111)	ND_3	RAIS-PM	144
(001)	CO	RAIS	136
(001)	$^{12}CO + ^{13}CO$	RAIS	136
(111)	CO	RAIS-DB	145,146
(111)	CO	RAIS-WM	147
(111)	CO	RAIS-FT	148
(111)	$CO + C_6H_6$	RAIS-DB	146
(111)	C_2D_6	RAIS-WM	149
(111)	C_6H_6	RAIS-DB	146
(111)	CH_3OH	RAIS-DB	150
(111)	C_2H_5OH	RAIS-DB	150
POTASSIUM			
ev. film/glass	CO	RAIS-DB	61
RHODIUM			
ev. film(CO)/NaCl	CO	TR	59
porous film/CaF_2	CO	TR	122
ev. film/NaCl	CO	TR	93
ev. film/glass	CO	RAIS-DB	101
ev. film/Ta	CO	RAIS-DB	151
ev. film/NaCl	$CO + H_2$	TR	93
ev. film/NaCl	$CO + O_2$	TR	93
ev. film/NaCl	$CO + C_2H_2$	TR	93

ev. film(CO or H_2)/NaCl	CO + C_2H_2	TR	60
ev. film(CO or H_2)/NaCl	CO+CH_3CHCH$_2$	TR	60
ev. film(CO)/NaCl	CO+$(CH_3)_3$N	TR	59
ev. film(CO)/NaCl	CO+C_2H_5NC	TR	59
ev. film/NaCl	H_2CO	TR	93
ev. film/NaCl	HCOOH	TR	93
ev. film/glass	H_2	RAIS–DB	101
ev. film/glass	C_2H_4	RAIS–DB	101
(110) oriented film	CO	RAIS	108
(110) oriented film	O_2	RAIS	108

RUTHENIUM

(001)	CO	RAIS–PM	152
(001)	H_2O	RAIS–PM	153

SILVER

ev. film/NaCl	CO	TR	58
ev. film/glass	CO	RAIS	64
ev. film/glass	C_2H_4	RAIS–DB	133
ev. film	CH_3OH	IRES	154
ev. film	HCOOH	IRES	155,156
ev. film	butanol	IRES	157
(100)	CO	RAIS	158

TUNGSTEN

ev. film(CO)/CaF_2	CO	TR	57
ribbon	CO	RAIS	159–161
ev. film(CO)/CaF_2	CO + O_2	TR	57

SILICON

single crystal	H_2	INT.R.	162
single crystal	CO_2	INT.R.	163

MoO_3			
ev. film/Ag plate	CH_3OH	EM	164
ev. film/Ag plate	CH_3OH	RAIS-PM	164

MgO			
(100)	n-pentanol	INT.R.	167
(100)	H_2O	INT.R.	170

Al_2O_3			
(0001)	n-amyl alcohol	INT.R.	169
(0001)	H_2O	INT.R.	170
(11$\bar{2}$3)	n-amyl alcohol	INT.R.	169
(11$\bar{2}$3)	H_2O	INT.R.	170
(41$\bar{5}$0)	n-amyl alcohol	INT.R.	169
(41$\bar{5}$0)	H_2O	INT.R.	170

ZnO			
(0001)	H_2O	INT.R.	170
(2$\bar{1}\bar{1}$0)	H_2O	INT.R.	170

TiO_2			
(2$\bar{1}\bar{1}$0)	H_2O	INT.R.	170

TABLE III

System	References	System	References
LiF		**NaCl**	
HCl	172	CH_4	183
p-benzoquinone	175	CO	184–187
		CO_2, $^{13}CO_2$	176,178, 188–190
NaF		HCl, DCl	171,172
HCl, DCl	171,172	$HCl + H_2O$	172
HBr	172	HBr	172
p-benzoquinone	173,174,175	$HBr + H_2O$	172
		HI	172
KF		p-benzoquinone	173,175
HCl	172	aromatic compounds	191
CsF		**KCl**	
CO_2	176	NO	180
HCl	172	CH_4	183
CaF$_2$		**CsCl**	
p-benzoquinone	175	HCN	178,179
		NO	180
BaF$_2$		CO	192
p-benzoquinone	173–175	CO_2	176,178
		SO_2	193,194
LiCl		HCl, DCl	171,172
HCl	171,172	HBr	172
NaCl		**AgCl**	
H_2, D_2, HD	177	aromatic compounds	191
HCN	178,179		
NO	180		
NO_2	181		
N_2O	182		

TABLE III (Con't)

TlCl		NaI	
aromatic compounds	191	NO	180
		N_2O	182
		CH_4	183
NaBr		CO	184,185
NO	180	CO_2	176,178
N_2O	182	HCl	171,172
CH_4	183	HCl + H_2O	172
CO	184,185	HBr	172
HCl	171,172	HBr + H_2O	172
HCl + H_2O	172		
HBr	172	KI	
HBr + H_2O	172	CH_4	183
KBr		CsI	
CH_4	183	H_2, D_2, HD	177
p-benzoquinone	173	CO	185, 192
aromatic compounds	195	CO_2	176
		HCl	172
		HBr	172
CsBr			
CO	185,192	AgI	
		aromatic compounds	191
NaI			
HCN	178,179		

REFERENCES

1. M.J.D. Low and H. Inoue, Anal. Chem. 36 (1964) 2397.
2. M.J.D. Low, J. Catal. 4 (1965) 719.
3. M.J.D. Low, L. Abrams and I. Coleman, Chem. Commun. (1965) 389.
4. M.J.D. Low and I. Coleman, Spectroch. Acta 22 (1966) 369.
5. J.F. Blanke, S.E. Vincent and J. Overend, Spectroch. Acta 32A
   (1976) 163.
6. J.F. Blanke and J. Overend, Spectroch. Acta 32A (1976) 1383.
7. R.G. Greenler, Surf. Sci. 69 (1977) 647.
8. K. Makinouchi, K. Wagatsuma and W. Suëtaka, J. Spectrosc. Soc.
   Japan 29 (1980) 23.
9. K. Wagatsuma, K. Monma and W. Suëtaka, submitted to Applic.
   Surf. Sci.
10. S.A. Francis and A.H. Ellison, J. Opt. Soc. Amer. 49 (1959) 131.
11. R.G. Greenler, J. Chem. Phys. 44 (1966) 310.
12. R.G. Greenler, J. Chem. Phys. 50 (1969) 1963.
13. G.W. Poling, J. Coll. Interf. Sci. 34 (1970) 365.
14. J.D.E. Mc Intyre and D.E. Aspnes, Surf. Sci. 24 (1971) 417.
15. R.G. Greenler, Jap. J. Appl. Phys. Suppl. 2, Pt. 2 (1974) 265.
16. R.G. Greenler, J. Vac. Sci. Techn. 12 (1975) 1410.
17. J. Pritchard, in "Surface and Defect Properties of Solids"
    Vol. 1, ed. M.W. Roberts and J.M. Thomas (Specialist Periodical
    Reports, The Chemical Society, London, 1972) p. 222.
18. R. Greenler, CRC Crit. Rev. Sol. State Sci. 4 (1974) 415.
19. J. Pritchard, Dechema Monographien 78 (1975) 231.
20. J. Pritchard and T. Catterick, in "Experimental Methods in Ca-
    talytic Research", Vol. 3, ed. R.B. Anderson and P.T. Dawson
    (Academic Press, New York, 1976) p. 281.
21. D.A. King, Photoemission - Proc. Int. Symp. Noordwijk (ESA)
    ed. R.F. Willis, B. Feuerbacher, B. Fitton and C. Backx (1976)
    p. 281.
22. H. Ibach, Surf. Sci. 66 (1977) 56.
23. J. Pritchard, in "Chemical Physics of Solids and their Surfaces"
    Vol. 7, ed. M.W. Roberts and J.M. Thomas (Specialist Periodical
    Reports, The Chemical Society, London, 1978) p. 157.
24. J. Pritchard, Proc. Intern. Conf. on Vibrations in Adsorbed
    Layers, Jülich 1978, KFA Reports - Jülich, Conf. 26 (1978)
    p. 135.
25. N. Sheppard and T.T. Nguyen, in "Adv. in Infrared and Raman
    Spectroscopy" vol. 5, ed. Clark and Hester (Heyden, London,
    1978) p. 67.
26. A.M. Bradshaw, Surf. Sci. 80 (1979) 215.
27. N. Sheppard, XXI Colloquium Spectroscopicum Internationale -
    Keynote Lectures, ed. G.F. Kirkbright (Heyden, London, 1979)
    p. 23.
28. J. Pritchard, Annual Reports of the Chemical Society, A, 66
    (1970) 65.

29. P. Hollins and J. Pritchard, in "Vibrational Spectroscopy of Adsorbates", ed. R.F. Willis (Springer Series in Chemical Physics, Springer-Verlag, Heidelberg, 1980) ch. 6.
30. R.G. Greenler, Proc. Intern. Conf. Vibrations at Surfaces, ed. R. Caudano, J.M. Gilles and A. Lucas, Namur 1980, (Plenum, London, 1981).
31. M.J. Dignam and J. Fedyk, Appl. Spectr. Rev. 14 (1978) 249.
32. J.D. Fedyk, P. Mahaffy and M.J. Dignam, Surf. Sci. 89 (1979) 404.
33. G. Blyholder, J. Phys. Chem. 68 (1964) 2772.
34. G. Blyholder and M. Allen, J. Am. Chem. Soc. 91 (1969) 3158.
35. G. Blyholder, J. Phys. Chem. 79 (1975) 756.
36. E.J. Berendo and P. Kos, Mol. Phys. 30 (1975) 1735.
37. G.D. Mahan and A.A. Lucas, J. Chem. Phys. 68 (1978) 1344.
38. F. Delanye, A.A. Lucas, G.D. Mahan, Proc. 3rd Intern. Conf. on Solid Surfaces, Vienna (1977) p. 477.
39. M. Moskovits and J.E. Hulse, Surf. Sci. 78 (1978) 397.
40. M. Scheffler, Surf. Sci. 81 (1979) 562.
41. A.M. Bradshaw and M. Scheffler, J. Vac. Sci. Techn. 16 (1979) 447.
42. S. Efrima and H. Metiu, Surf. Sci. 92 (1980) 433.
43. E.E. Mola, Surf. Sci. 91 (1980) L45.
44. P. Hollins and J. Pritchard, in "Vibrational Spectroscopies for Adsorbed Species" ACS Symposium Series 137 (American Chemical Society, Washington, 1980).
45. D.A. King, in "Vibrational Spectroscopy of Adsorbates", ed. R.F. Willis (Springer Series in Chemical Physics, vol. 15) (Springer, Berlin, 1980).
46. H. Metiu and W.E. Palke, J. Chem. Phys. 69 (1978) 2574.
47. M. Ito and W. Suëtaka, Chem. Letters (1973) 757.
48. M. Ito and W. Suëtaka, J. Phys. Chem. 79 (1975) 1190.
49. R.W. Hannah, Appl. Spectr. 17 (1963) 23.
50. A.J. Maeland, R. Rittenhouse, W. Lahar and P.V. Romano, Thin Solid Films 21 (1974) 67.
51. G.A. Dorsey, J. Electrochem. Soc. 113 (1966) 169.
52. T. Takamura, H. Kihara-Morishita, U. Moriyama, Thin Solid Films 6 (1970) R17.
53. W. Vedder and D.A. Vermilyea, Trans. Faraday Soc. 65 (1969) 561.
54. J.D. Guthrie and B.J. Sparr, Thin Solid Films 43 (1977) 303.
55. D. Kember, D.H. Chenery, N. Sheppard and J. Fell, Spectrochim. Acta 35A (1979) 455.
56. F.P. Mertens, Surf. Sci. 71 (1978) 161.
57. F.S. Baker, A.M. Bradshaw, J. Pritchard and K.W. Sykes, Surf. Sci. 12 (1968) 426.
58. A.M. Bradshaw and J. Pritchard, Proc. Roy. Soc. London A 316 (1970) 169.
59. R. Queau and R. Poilblanc, J. Catal. 27 (1972) 200.
60. J. Wojtczak, R. Queau and R. Poilblanc, J. Catal. 37 (1975) 391.

61. D. Reinalda, Ph.D. thesis, Leiden (1979).
62. A.M. Bradshaw, J. Pritchard and M.L. Sims, Chem. Commun. (1968) 1519.
63. J. Pritchard and M.L. Sims, Trans. Faraday Soc. 66 (1970) 427.
64. M.A. Chesters, J. Pritchard and M.L. Sims, in "Adsorption Desorption Phenomena", ed. F. Ricca (Academic Press, London, 1972) p. 277.
65. E.F. Mc Coy and R. St. C. Smart, Surf. Sci. 39 (1973) 109.
66. J. Pritchard, T. Catterick and R.K. Gupta, Surf. Sci. 53 (1975) 1.
67. M. Moskovits, C.J. Hope and B. Jantzi, Can. J. Chem. 53 (1975) 3313.
68. H.G. Tompkins and R.G. Greenler, Surf. Sci. 28 (1971) 194.
69. R.W. Stobie, B. Rao and M.J. Dignam, Surf. Sci. 56 (1976) 334.
70. R.W. Stobie and M.J. Dignam, Can. J. Chem. 56 (1978) 1088.
71. M. Ito and W. Suëtaka, J. Catal. 54 (1978) 13.
72. D.T. Drmaj and K.E. Hayes, J. Catal. 19 (1970) 154.
73. M.A. Chesters, J. Pritchard and M.L. Sims, Chem. Commun. (1970) 1454.
74. J. Pritchard, Surf. Sci. 79 (1979) 231.
75. K. Horn and J. Pritchard, Surf. Sci. 55 (1976) 701.
76. R. Ryberg, Proc. Intern. Conf. Vibrations at Surfaces, ed. R. Caudano, J.M. Gilles and A. Lucas, Namur, 1980 (Plenum, London, 1981).
77. R. Ryberg, submitted to Phys. Rev. B.
78. B.N.J. Persson and R. Ryberg, Proc. ECOSS 3 (Cannes, 1980) Suppl. Le Vide - Les Couches Minces, n° 201 (1980) 790.
79. B.N.J. Persson and R. Ryberg, Sol. St. Commun. (in press).
80. B.N.J. Persson and R. Ryberg, Proc. Intern. Conf. Vibrations at Surfaces, ed. R. Caudano, J.M. Gilles and A. Lucas, Namur, 1980 (Plenum, London, 1981).
81. J. Pritchard, J. Vac. Sci. Techn. 9 (1972) 895.
82. K. Horn, M. Hussain and J. Pritchard, Surf. Sci. 63 (1977) 244.
83. K. Horn and J. Pritchard, Surf. Sci. 52 (1975) 437.
84. P. Hollins and J. Pritchard, Surf. Sci. 89 (1979) 486.
85. P. Hollins and J. Pritchard, Chem. Phys. Lett. 75 (1980) 378.
86. F.J. Boerio and S.L. Chen, Appl. Spectr. 33 (1979) 121.
87. A.A. Babushkin, Russ. J. Phys. Chem. (English Transl.) 38 (1964) 1004.
88. G.W. Poling, J. Electrochem. Soc. 116 (1969) 958.
89. R.G. Greenler, R.R. Rahn and J.P. Schwartz, J. Catal. 23 (1971) 42.
90. F.J. Boerio and L. Armogan, Appl. Spectr. 32 (1978) 509.
91. D. Kember and N. Sheppard, Appl. Spectr. 29 (1975) 496.
92. M.L. Kottke, R.G. Greenler and H.G. Tompkins, Surf. Sci. 32 (1972) 231.
93. J.F. Harrod, R.W. Roberts and E.F. Rissman, J. Phys. Chem. 71 (1967) 343.
94. D. Reinalda and V. Ponec, Surf. Sci. 91 (1979) 113.
95. D. Reinalda and V. Ponec, Applic. Surf. Sci. 5 (1980) 98.

96. A.M. Bradshaw and J. Pritchard, Surf. Sci. 17 (1969) 372.
97. C.W. Garland, R.C. Lord and P.F. Troiano, J. Phys. Chem. 69 (1965) 1195.
98. J.N. Bradley and A.S. French,Proc. R. Soc. London A313 (1969) 169.
99. A.M. Bradshaw and O. Vierle, Ber. Buns. Phys. Chem. 74 (1970) 630.
100. R.A. Gardner and R.H. Petrucci, J. Am. Chem. Soc. 82 (1960)5051.
101. H.L. Pickering and H.C. Eckstrom, J. Phys. Chem. 63 (1959) 512.
102. A.M. Bradshaw and J. Pritchard, Surf. Sci. 19 (1970) 198.
103. M. Ito, T. Kato and W. Suëtaka, Chem. Lett. (1976) 1337.
104. M. Ito, S. Abe and W. Suëtaka, J. Catal. 57 (1979) 80.·
105. M. Ito and W. Suëtaka, Proc. 7th Intern. Vac. Congr. and 3rd Intern. Conf. Solid Surfaces (Vienna 1977) p. 1043.
106. M. Ito, Y. Mori, T. Kato and W. Suëtaka, Applic. Surf. Sc. 2 (1979) 543.
107. H.C. Eckstrom and W.H. Smith, J. Opt. Soc. Amer. 57 (1967) 1132.
108. H.C. Eckstrom, G.C. Possley, S.E. Hannum and W.H. Smith, J. Chem. Phys. 52 (1970) 5435.
109. A.M. Bradshaw, J. Chem. Phys. 55 (1971) 1487.
110. W.H. Smith, J. Chem. Phys. 55 (1971) 1488.
111. J. Heidberg and I. Hussla, Proc. Intern. Conf. Vibrations at Surfaces, ed. R. Caudano, J.M. Gilles and A. Lucas, Namur, 1980 (Plenum , London, 1981).
112. F.L. Baudais, A.J. Borschke, J.D. Fedyk and M.J. Dignam, Surf. Sci. 100 (1980) 210.
113. M.J. Dignam, Proc. Intern. Conf. Vibrations at Surfaces, ed. R. Caudano, J.M. Gilles and A. Lucas, Namur, 1980 (Plenum, London, 1981).
114. J. Fedyk and M.J. Dignam, in A.C.S. Symposium Series : Vibrational Spectroscopy Applied to Adsorbed Species on Catalysts, Washington (1979) (in press).
115. P.R. Mahaffy and M.J. Dignam, Surf. Sci. 97 (1980) 377.
116. P.R. Mahaffy and M.J. Dignam, Proc. ECOSS 3 (Cannes, 1980), Suppl. Le Vide - Les Couches Minces n° 201 (1980) 335.
117. P.R. Mahaffy and M.J. Dignam, J. Phys. Chem. 84 (1980) 2683.
118. M. Grunze, R.K. Driscoll, G.N. Burland, J.C.L. Cornish and J. Pritchard, Surf. Sci. 89 (1979) 381.
119. J.C. Campuzano and R.G. Greenler, Surf. Sci. 83 (1979) 301.
120. C.P. Nash and R.P. de Sieno, J. Phys. Chem. 69 (1965) 2139.
121. J.N. Bradley and A.S. French, Proc. R. Soc. A 313 (1969) 169.
122. C.W. Garland, R.C. Lord and P.F. Troiano, J. Phys. Chem. 69 (1965) 1188.
123. R.W. Rice and G.L. Haller, J. Catal. 40 (1975) 249.
124. A.M. Bradshaw and F. Hoffmann, Surf. Sci. 52 (1975) 449.
125. F.M. Hoffmann and A.M. Bradshaw, J. Catal. 44 (1976) 328.
126. M. Kawai, T. Onishi and K. Tamaru, submitted to Surf. Sci.
127. I. Ratajczykowa, Surf. Sci. 48 (1975) 549.
128. F.M. Hoffmann and A.M. Bradshaw, Proc. 7th Intern. Vac. Congr. and 3rd Intern. Conf. Solid Surfaces (Vienna, 1977) p. 1167.

129. A.M. Bradshaw and F.M. Hoffmann, Surf. Sci. 72 (1978) 513.
130. A. Ortega, A. Garbout, F.M. Hoffmann, W. Stenzel, R. Unwin,
     K. Horn and A.M. Bradshaw, Proc. ECOSS 3 (Cannes, 1980),
     Suppl. Le Vide – Les Couches Minces, n° 201 (1980) 335.
131. F.M. Hoffmann and A. Ortega, Proc. Intern. Conf. on Vibrations
     in Adsorbed Layers, Jülich 1978, KFA Reports – Jülich, Conf.
     26 (1978) p. 128.
132. R.P. Eischens and W.A. Pliskin, Adv. Catal. 10 (1958) 1.
133. M. Ito and W. Suëtaka, Surf. Sci. 62 (1977) 308.
134. M.J.D. Low and J.C. Mc Manus, Chem. Commun. (1967) 1166.
135. R.A. Shigeishi and D.A. King, Surf. Sci. 58 (1976) 379.
136. A. Crossley and D.A. King, Surf. Sci. 95 (1980) 131.
137. W.G. Golden, D.S. Dunn and J. Overend, J. Phys. Chem. 82
     (1978) 843.
138. W.G. Golden, D.S. Dunn, C.E. Pavlik and J. Overend, J. Chem.
     Phys. 70 (1979) 4426.
139. A. Crossley and D.A. King, Surf. Sci. 68 (1977) 528.
140. R.A. Shigeishi and D.A. King, Surf. Sci. 75 (1978) L397.
141. R.A. Shigeishi and D.A. King, Surf. Sci. 62 (1977) 379.
142. D.S. Dunn, M.W. Severson, W.G. Golden and J. Overend, J. Catal.
     65 (1980) 271.
143. D.S. Dunn, W.G. Golden, M.W. Severson and J. Overend, J. Phys.
     Chem. 84 (1980) 336.
144. D.S. Dunn, M.W. Severson, W.G. Golden and J. Overend, subm. to
     J. Chem. Phys.
145. H.J. Krebs and H. Lüth, Appl. Phys. 14 (1977) 337.
146. R.W. Ittner and H. Lüth, Proc. 7th Intern. Congress on Catalysis
     (Tokyo, 1980)(in press).
147. K. Horn and J. Pritchard, J. de Phys. Coll. C4 38 (1977) C4 – 164.
148. M.D. Baker and M.A. Chesters, Proc. Intern. Conf. Vibrations at
     Surfaces, ed. R. Caudano, J.M. Gilles and A. Lucas, Namur, 1980
     (Plenum, London, 1981).
149. K. Horn, Proc. Intern. Conf. on Vibrations in Adsorbed Layers,
     Jülich 1978, KFA Reports – Jülich, Conf. 26 (1978) p. 140.
150. H.J. Krebs and H. Lüth, Proc. Intern. Conf. on Vibrations in
     Adsorbed Layers, Jülich 1978, KFA Reports – Jülich, Conf. 26
     (1978) p. 135.
151. M.G. Wells, N.W. Cant and R.G. Greenler, Surf. Sci. 67 (1977)
     541.
152. H. Pfnür, D. Menzel, F.M. Hoffmann, A. Ortega and A.M. Bradshaw,
     Surf. Sci. 93 (1980) 431.
153. K. Kretzschmar, J.K. Sass, P. Hofmann, A. Ortega, A.M. Bradshaw
     and S. Holloway, to be published.
154. M.J. Dignam, B. Rao, M. Moskovits and R.W. Stobie, Can. J. Chem.
     49 (1971) 1115.
155. R.W. Stobie, B. Rao and M.J. Dignam, Appl. Optics 14 (1975) 999.
156. B. Rao, R.W. Stobie and M.J. Dignam, J. Chem. Soc. Faraday
     Trans. II 71 (1975) 654.
157. M.J. Dignam, B. Rao and R.W. Stobie, Surf. Sci. 46 (1974) 308.
158. K. Horn, PhD thesis (London, 1976).

159. J.T. Yates and D.A. King, Surf. Sci. 30 (1972) 601.
160. D.A. King, C.G. Goymour and J.T. Yates,Jr., Proc. R. Soc. London A 331 (1972) 361.
161. J.T. Yates, R.G. Greenler, I. Ratajczykowa and D.A. King, Surf. Sci. 36 (1973) 739.
162. G.E. Becker and G.W. Gobeli, J. Chem. Phys. 38 (1963) 2942.
163. D.B. Novotny, J. Vac. Sci. Technol. 9 (1972) 1447.
164. M. Ito, Proc. Intern. Conf. Vibrations at Surfaces, ed. R. Caudano, J.M. Gilles and A. Lucas, Namur, 1980 (Plenum, London, 1981).
165. H.·G. Tompkins and D.L. Allara, Rev. Sci. Instr. 45 (1974) 1221.
166. H.G. Tompkins and D.L. Allara, J. Colloïd Interf. Sci. 49 (1974) 410.
167. G.L. Haller, R.W. Rice and Z.C. Wan, Catal. Rev. - Sci. Eng. 13 (1976) 259.
168. J.S. Mattson, Anal. Chem. 45 (1973) 1473.
169. G.L. Haller and R.W. Rice, J. Phys. Chem. 74 (1970) 4386.
170. R.W. Rice and G.L. Haller, Proc. 5th Intern. Congr. Catal.,ed. J.W. Hightower (North-Holland, Amsterdam, 1973) p. 317.
171. R. St C. Smart and N. Sheppard, Chem. Commun. (1969) 468.
172. R. St C. Smart and N. Sheppard, Proc. R. Soc. London A 320 (1971) 417.
173. H. Hartmann and H. Luchterhand, Zeit. Phys. Chem. NF 46 (1965) 103.
174. H. Hartmann, E. Eisenbraun and J. Heidberg, Zeits. Naturforsch. 23a (1968) 1689.
175. H. Hartmann, E. Eisenbraun and J. Heidberg, Zeits. Phys. Chem. NF 67 (1969) 51.
176. Y. Kozirovski and M. Folman, Trans. Faraday Soc. 62 (1966) 1431.
177. M. Folman and Y. Kozirovski, J. Colloïd Interf. Sci. 38 (1972) 51.
178. Y. Kozirovski and M. Folman, J. Chem. Phys. 41 (1964) 1509.
179. Y. Kozirovski and M. Folman, Trans. Faraday Soc. 62 (1966) 808.
180. A.J. Woodward and N. Jonathan, J. Phys. Chem. 75 (1971) 2930.
181. R.J. Copeland, PhD thesis (Un. of Arkansas, 1978), Univ. Microfilms nr. 782 3246.
182. Y. Kozirovski and M. Folman, Trans Faraday Soc. 65 (1969) 244.
183. S. Zehme, J. Heidberg and H. Hartmann, Fortschr. Kolloide u. Polym. 55 (1971) 65.
184. R. Gevirzman, Y. Kozirovski and M. Folman, Trans. Faraday Soc. 65 (1969) 2206.
185. B. Rao and M.J. Dignam, J. Chem. Soc. Faraday Trans. II 70 (1974) 492.
186. J. Heidberg, H. Stein and I. Hussla, Ber. Bunsenges. Phys. Chem. 83 (1978) 1238.
187. J. Heidberg and I. Hussla, Proc. Intern. Conf. Vibrations at Surfaces, ed. R. Caudano, J.M. Gilles and A. Lucas (Namur, 1980) (Plenum, London, 1981).
188. J. Heidberg, S. Zehme, C.F. Chen and H. Hartmann, Ber. Bunsenges. Phys. Chem. 75 (1971) 1009.

189. J. Heidberg, R.D. Singh, E. Hoefs and H. Stein, Ber. Bunsenges.
     Phys. Chem. 79 (1975) 1161.
190. J. Heidberg, R.D. Singh and H. Stein, Ber. Bunsenges. Phys.
     Chem. 82 (1978) 54.
191. G. Karagounis and O. Peter, Zeits. Elektroch. 63 (1959) 1120.
192. R. Gevirzman and Y. Kozirovski, Trans. Faraday Soc. 67 (1971)
     2686.
193. J. Heidberg and E. Hoefs, Ber. Bunsenges. Phys. Chem. 80 (1976)
     1233.
194. J. Heidberg, I. Hussla and E. Hoefs, submitted to Surf. Sci.
195. G. Karagounis and R.M. Issa, Zeits. Elektroch. 66 (1962) 874.
196. H.G. Tompkins, in : "Methods and Phenomena - vol. 1, Methods
     of Surface Analysis", ed. A.W. Czanderna, Elsevier, Amsterdam
     (1975) p. 447.
197. H.G. Tompkins, Appl. Spectr. 30 (1976) 377.
198. H.G. Tompkins, Appl. Spectr. 28 (1974) 335.

SURFACE ENHANCED RAMAN SCATTERING FROM CHARACTERIZED METAL SURFACES

J.C. Tsang, J.R. Kirtley and S.S. Jha

IBM Thomas J. Watson Research Center

Yorktown Heights, New York 10598

ABSTRACT

Surface enhanced Raman scattering from molecular monolayers adsorbed in tunnel junctions and at single crystal surfaces in ultra high vacuum has been studied as a function of surface topography, the adsorbate and the energy of the vibrational mode. Surface enhanced Raman scattering is shown to involve both the enhancement of the local electric fields at the rough noble metal surface due to the optical excitation of surface plasmon polaritons and the modulation of the optically induced surface charge density by the molecular vibrations.

The experimental observation that the Raman scattering efficiency of pyridine adsorbed on Ag can be 10^5-10^6 times greater than the Raman efficiency of the same molecule in solution has created considerable interest since it suggests Raman spectroscopy can be applied to problems involving adsorbates on surfaces. The study of surface enhanced Raman (SER) scattering has involved the use of samples roughened by electrochemical anodization[1,2,3], severe sputtering[4,5], evaporation onto cold substrates[6,7], the photochemical decomposition of Ag[8] or the evaporation of discontinuous films[9,10]. The quantitative characterization of the structural electronic and vibrational properties of these surfaces and the molecular monolayers adsorbed on them is extremely difficult, and this has greatly hindered the understanding of the mechanisms responsible for SER scattering. We have studied molecular monolayers 1) in tunnel junctions which can be laid down on substrates with well characterized surface profiles[11,12] and 2) on single crystals

361

with simple surface structures fabricated by x-ray photolithography[13] to obtain a detailed understanding of SER scattering from molecules adsorbed at Ag. We find that there are two distinct contributions to the SER effect. The first is a long range contribution due to the roughness assisted excitation of the surface plasmon polaritons of the noble metal. The second is a short range effect involving the direct interaction between the adsorbate and the metal surface. In this paper, we first define these two mechanisms. We then describe our experiments, theoretically demonstrate the surface plasmon polariton contribution to SER scattering and show how the vibrational modes of an adsorbate can modulate the optically induced charge density at a Ag surface to produce significant Raman scattering. Experimental evidence for the existence of both of these contributions is given and we show how the two together, can explain the experimentally observed enhancements of the Raman scattering efficiency of adsorbed molecules on Ag.

By studying surfaces with well characterized profiles, we explicitly demonstrate the contribution to SER scattering from the resonant excitation of the transverse collective electronic modes in Ag[14],[15]. In the case of a simple sinusoidal surface profile, we show that the excitation of the so called surface plasmon polaritons (SPP's) in Ag can produce two to three orders of magnitude of enhancement in the Raman scattering efficiency due to the classical electric field enhancement at the molecular site. Moreover, by working with molecular monolayers at the oxide-metal interface of an Al-AlO$_x$-M (M = metal) inelastic electron tunneling spectroscopy (IETS) junction, we can correlate our SER spectra with the tunneling spectra of the molecules[11]. This provides an independent characterization of the vibrational structure of the adsorbed molecule. We observe interesting differences in the IETS of those molecules which produce strong SER scattering and those which produce weak SER scattering.

In addition to the enhanced field associated with the resonant excitation of SPP's, the presence of a Ag surface next to a molecule can produce substantial changes in the Raman cross section of the molecule through the metal-molecule interaction[16],[17] and the resulting distortion of the electronic charge density near the surface. However, these effects are strongly localized to atomic distances from the surface, i.e. to the first adsorbed monolayer. We find that the modulation of the induced SPP charge density by molecular vibrations through the overlap of the wavefunctions of the surface charge and the molecule can produce a total short-range enhancement of the Raman scattering efficiency of the order of $10^4 - 10^7$ [17] Beyond the first monolayer, only the classical field enhancement is significant.

We have fabricated Al-AlO$_x$-M (M = Ag, Au, Cu, Pb, Sn...) junctions on thin, evaporated films of CaF$_2$ and on sinusoidal dif-

fraction gratings. The holographic gratings were made by exposing
5000 Å thick layers of photoresist to interference patterns set up
by crossing 3250 Å beams of a He-Cd laser[18]. The junctions were
doped with a variety of different carboxaldehydes and carboxylic
acids. The tunnel junctions were made by the evaporation of a 400 Å
layer of Al, its oxidation in air, doping from the solution and the
evaporation of a 200 Å thick overlayer at a temperature of 77 K[11].
The doped junctions could be characterized by IETS. The IETS
measurements were made at either 2 K or 4.2 K using a bridge circuit
and a lock-in amplifier with modulation voltages of 2-5 meV. The
Raman measurements were made either in air or in He at 2 K. The
Raman spectra were excited using the lines of either an Ar^+ or a
Kr^+ laser and the scattered light was detected by a conventional
photon counting system.

Studies of the tunneling spectra of doped junctions have
shown that the metal overlayer does not strongly perturb the
molecular monolayer[19]. In contrast, the acid or aldehyde group of
the dopant bonds to the oxide surface resulting in the disappearance
of the 1700 cm^{-1} C=O vibration and the appearance of the symmetric
and asymmetric 1350 and 1550 cm^{-1} vibrations of a surface COO
group[11]. In Fig. 1, we show a) the observed Raman spectrum of bulk
4-pyridine carboxylic acid (COOH), b) the IETS of 4-pyridine COOH
in an Al-AlO$_x$-Ag junction and c) the IETS of benzoic acid in an
Al-AlO$_x$-Pb junction. The IETS of 4 pyridine COOH shows all the
features in the Raman spectrum of the molecule. It also shows a
number of lines below 800 cm^{-1} which are only weakly Raman active
in the molecule. The presence of the Ag overlayer results in very
weak IETS scattering from the CH vibration near 3060 cm^{-1} compared
with the spectrum obtained from the Pb junction. IETS of the Ag
junction also shows a broad structure in the second derivative
(d^2V/dI^2) background from elastically tunneling electrons which
peaks near 3000 cm^{-1}. In the Pb junctions, the d^2V/dI^2 background
rises monotonically with increasing bias voltage as shown in Fig. 1c.
Modeling the electronic structure of the doped junction using the
four layer system in Fig. 2, both the weakness of the 3060 cm^{-1}
mode and the behaviour of the elastic background have been correlated
with low molecule-metal barrier heights by Coleman[20,21] (Φ_{b_2} < 1eV).

The dependence on the angle of incidence, Θ_i, of the reflectivity at 5145 Å of our doped junction laid on an 8000 Å periodicity
diffraction grating with an amplitude of ~ 600 Å is shown in Fig. 3a.
Light can couple to the surface plasmon polaritons causing a sharp
dip in the reflectivity at resonance. Similarly, SPP's can couple
out to light through the grating periodicity. In Fig.3b, we show
the light emission excited by inelastic electron tunneling from an
undoped Al-AlO$_x$-Ag tunnel junction fabricated on a similar grating
and biased at 2.5 eV[22]. The spectrum in Fig. 3b was obtained for
light emission at an angle Θ_s, of 36°. One branch of the dispersion
curve of the SPP for a Ag tunnel junction on a grating is shown

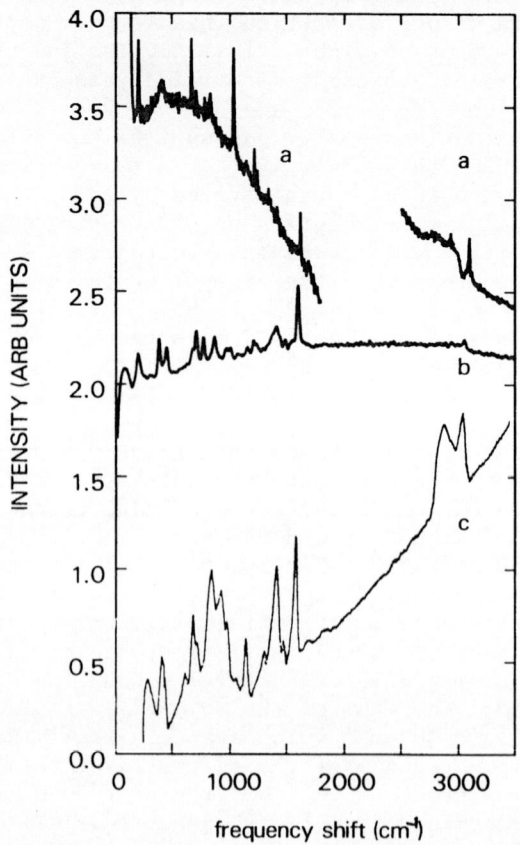

Fig. 1 - a) The Raman spectrum of 4-pyridine-carboxylic acid.
         b) The inelastic electron tunneling spectrum of a 4-
            pyridine carboxylic acid doped Al-AlO$_x$-Ag junction.
         c) The IETS spectrum of a benzoic acid doped Pb tunnel
            junction.

in Fig. 3c. The SPP dispersion shows the role of the grating periodicity in coupling light to the SPP's. SPP's are normally optically
inactive but can be excited in the presence of a periodic structure
with a wavevector g when $K_{sp} = K_{i,\parallel} \pm ng$ where $K_{i,\parallel}$ is the projection
of the wavevector of the incident light in the grating plane,
K_{sp} is the parallel wavevector of the SPP at ω_i and n = 1,2 ...

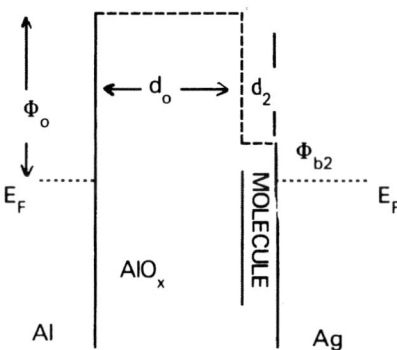

Fig. 2 - The model electronic structure of an Al-insulator-molecule-Ag
tunnel junction structure. Φ_o and d_o are the barrier height
and thickness of the oxide and Φ_{b2} and d_2 are the corresponding quantities for the molecular layer.

The linear macroscopic induced polarization P_i in an inhomogeneous system, with spatially varying fields, can be expanded in
the form

$$P_i(r,\omega) = \alpha_{ij} E_j(r,\omega) + \beta'_{ijk} \nabla_j E_k(r,\omega) + \ldots \tag{1}$$

where $E(r,\omega)$ is the electric field. Usually, the second term is
much smaller than the first. However, at a metal surface, the
normal component of the induced electric field varies rapidly on
the atomic scale so that in any microscopic calculation of P_i,
there is a significant contribution to it from this term. In
general, the intensity I_s of the Raman scattered light at
$\omega_s = \omega_i - \omega_o$, where ω_o is the vibrational frequency, can be expressed
as

$$I_s = \frac{\omega_s^4}{8\pi c^3} (\sum_{j=1}^{\infty} (\partial R_j(-\omega_o) \frac{\partial}{\partial R_j}) \, e_s \cdot \bar{\mu}(\omega_i))^2 \tag{2}$$

where $\bar{\mu}(\omega_i)$ is the induced dipole moment in the system and dR_j is
the atomic displacement of the jth atom due to a molecular vibration. Usually, this is rewritten as

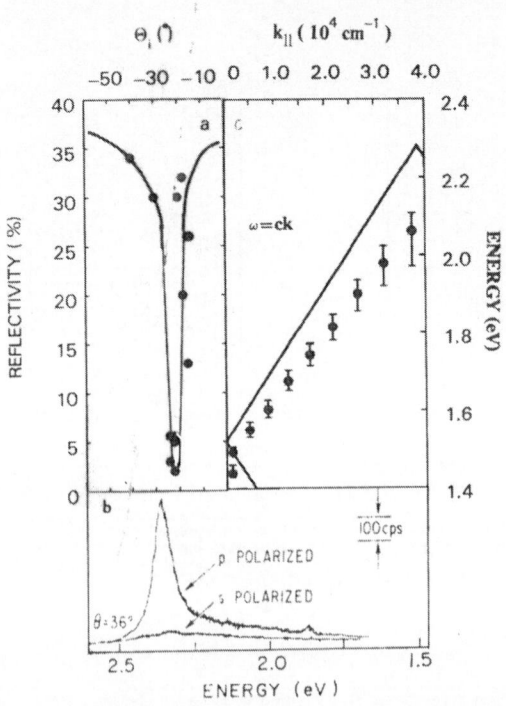

Fig. 3 - a) The reflectivity of a Ag tunnel junction on an 8000 Å
              periodicity diffraction grating.
          b) The light emission from a Ag tunnel junction on an
              8000 Å grating. Junction biased at 2.5 eV and $\Theta_s = 36°$.
          c) The dispersion of the Ag surface plasmon polariton as
              obtained from the optical emission spectrum of the
              junction[22]. Solid lines are the light line on the
              grating plane.

$$I_s = \frac{\omega_s^4}{8\pi c^3} (\sum_{j=1}^{\infty} \quad (\partial R_j(-\omega_o) \frac{\partial}{\partial R_j}) \; e_s \cdot \alpha_s \cdot E_L(\omega_i))^2 \qquad (3)$$

where α_s is the system polarizability and E_L is the average local
electric field acting on the system. E_L on a grating surface will
depend on the angle of incidence, Θ_i, the magnitude of the incident
field, E_{inc} and the dielectric function, $\varepsilon_1 + i\varepsilon_2$ of the metal. If
Θ_i corresponds to the reflectivity minimum in Fig. 3 ($\Theta(\omega_{sp})$), so
that the SPP's are optically excited, then the resonant SPP field

at a distance a from the surface is[23]

$$E_L = \xi_g K_{sp} \frac{\varepsilon_1^2}{\varepsilon_2} (\exp - (\sqrt{(K_{sp}^2 - \frac{\omega^2}{c^2})} \, a) E_{inc} \qquad (4)$$

where ξ_g is the grating amplitude. The use of values of the optical constants of bulk Ag and an 800 Å amplitude, 8000 Å periodicity grating produces a field enhancement at the Ag surface of 10^2 and therefore an enhancement of $I_s > 10^4$ [23].

In addition to the relatively long range field enhancement (4), the induced dipole moment can be further enhanced under certain conditions. If the effective barrier potential Φ_{b2} has a small value so that the product $(2m\Phi_{b2}/h^2)^{1/2}a$ is less than 1, there will be a significant contribution to I_s from a term similar to the second term in macroscopic Eq. 1. Because of tunneling, the optically induced surface charge density then has an appreciable spatial extent. When the molecule is very close to the surface, this can be strongly modulated by the molecular vibrations since

$$\Phi_{b2} = \Phi_{b20} + H'(r,t) \qquad (5)$$

where Φ_{b20} is the equilibrium electron barrier potential and $H'(r,t)$ is the change in the electronic energy due to the interaction of the electron with the atomic displacements in a vibrating molecule. This produces a large oscillating dipole moment at ω_s. We have explicitly calculated this dipole moment in Ref. 23. In Fig. 4, we show the dependence on both a and Φ_{b2} of the enhancement of the SER scattering due to this short range mechanism as compared to the Raman scattering due to the normal modulation of the molecular α_{ij}. If the barrier height is less than 1 eV, we find there can be a substantial additional contribution to the Raman scattering from the first adsorbed layer. The resonant excitation of the surface plasmons of Ag through a sinusoidal surface profile also produces the enhancement of the Raman scattering efficiency due to the enhancement of E_L (Eq. 4). We[23] have found that the total enhancement of the Raman efficiency can be as large as 10^7 when both mechanisms are operative. The Raman efficiency is largest for the first adsorbed layer where the molecular vibrations can modulate the surface charge density, so that the short range mechanism can be important. There is a smaller but still significant enhancement from more distant molecules due to the enhancement of E_L by the excitation of the SPP's.

In Fig. 5, we show two Raman spectra of an Al-AlO$_x$-Ag junction doped with nitrobenzoic acid on an 8000 Å periodicity, 600-800 Å amplitude diffraction grating. We also show the Raman

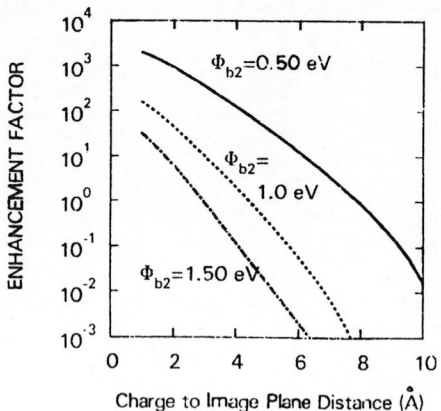

Fig. 4 - The calculated dependence on the molecular distance and
         the effective surface barrier height Φ_{b2} of the enhance-
         ment of the Raman scattering efficiency arising only from
         the modulation of the optically induced charge density
         near a metal surface and in the absence of the resonant
         field enhancement.

spectrum of a pressed pellet of nitrobenzoic acid. All three spectra
were taken using 6471 Å excitation under similar conditions. Fig. 5a
was obtained in the 180° backscattering geometry with $\Theta_s = \Theta_i = 0$
where the grating cannot couple light to the surface plasmon polaritons of the junction. Figure 5b was obtained in a scattering
geometry where $\Theta_s = -\Theta_i = \Theta(\omega_{sp})$. The coupling to SPP's of the
incident E_i when it is polarized in the plane of incidence, produces
a uniform enhancement of the Raman scattering from the vibrational
modes of the chemisorbed nitrobenzoic acid. The dependence of the
enhancement of I_s for the 1600 cm^{-1} line on both Θ_i and Θ_s is shown
in Fig. 6. The optimization of the coupling between the light and
the SPP's produces the strongest Raman-scattering. The intensity
of the SER scattering for a sample on a holographic diffraction
grating can vary by over two orders of magnitude as the scattering
geometry is changed. The position and width of the angle dependent
structures in Fig. 3 and 6 are in quantitative agreement.

     The coupling of light to the surface plasmon polaritons
of a junction can occur through the periodicity of a diffraction
grating; it can also occur through the statistical roughness of a
thin film of CaF$_2$. The amplitude (height) of the random roughness
depends on the thickness of the CaF$_2$ while its transverse correla-

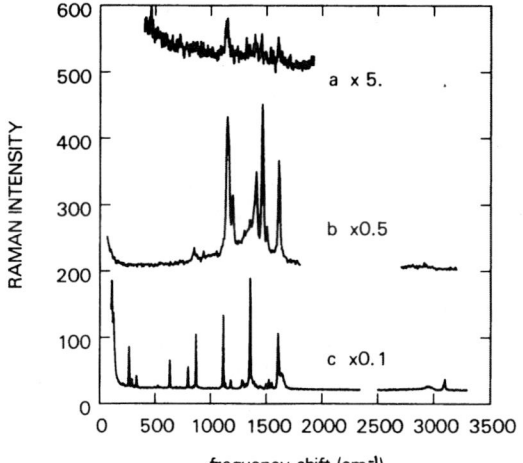

Fig.5 - a) The SER spectrum of nitrobenzoic acid in a Ag tunnel
           junction for Θ_i = 0°.
        b) The SER spectrum of nitrobenzoic acid in a Ag tunnel
           junction under the resonant excitation of the surface
           plasmon polaritons of the junction.
        c) The Raman spectrum of nitrobenzoic acid.
        Spectra obtained under 6471 Å excitation.

tion length varies from 500 to 1000 Å[24]. In Fig. 7, we show the
Raman spectrum of an Al-AlO$_x$-Ag junction doped with 4-pyridine
COOH and laid down on 1600 Å of CaF$_2$. Also shown in Fig. 7 is the
dependence of the intensity of the 1600 cm⁻mode of 4-pyridine COOH
adsorbed at the oxide in an Al-AlO$_x$-Ag tunnel junction on the thickness of CaF$_2$ under the junction. McCarthy and Lambe[24] have measured
the light emission excited by inelastic electron tunneling in
junctions laid down on similar thicknesses of CaF$_2$. The intensity
of the emission should be proportional to ξ_g^2 . The solid line in
Fig. 7b shows the dependence of the results of McCarthy and Lambe
on CaF$_2$ thickness. The intensity of the CaF$_2$ roughness induced SER
scattering follows the amplitude of the surface roughness. The
existence of residual random roughness on our grating surfaces
means that even at normal incidence, light can still couple to the
SPP's through the residual roughness. By comparing the intensity
of the SER scattering obtained from a junction on an 800 Å amplitude 8000 Å periodicity grating at the SPP resonance condition and
the scattering from a junction on a 1600 Å thick CaF$_2$ substrate,
we find that the residual roughness required to explain the scattering observed in Figs. 4 and 5 for Θ_i = 0 is about 5 Å. This is
a very conservative estimate for the residual roughness present on

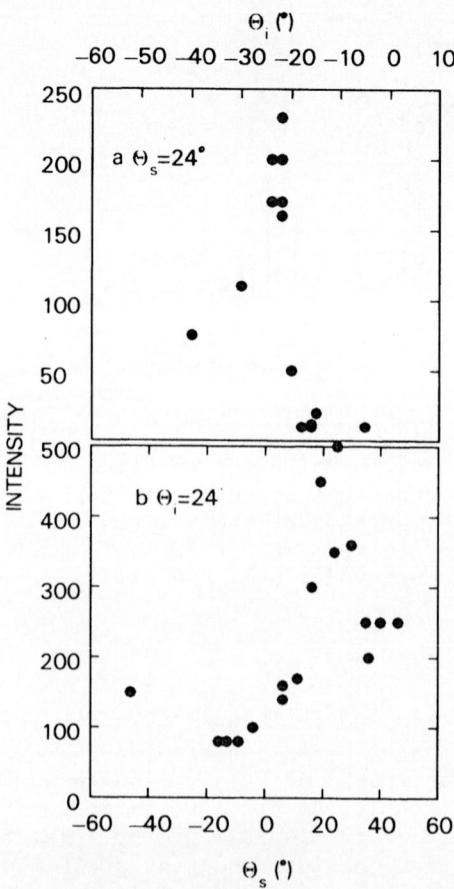

Fig. 6 – The dependence of the intensity of SER scattering from
a Ag grating on Θ_i and Θ_s.

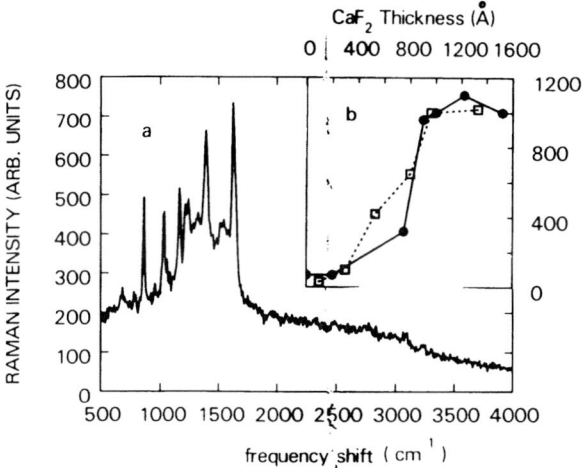

Fig. 7 - a) The SER spectrum of a 4-pyridine COOH doped Ag tunnel
             junction laid down on 1600 Å of CaF$_2$.
          b) The dependence of the intensity of the SER scattering
             (open squares and dotted line) and the tunneling
             electron pumped optical emission[24] (dots and solid
             line) on CaF$_2$ thickness.

the photoresist surface after development. We have been unable to
obtain any consistently measurable Raman scattering from doped
junctions laid down on smooth microscope slides. These results
suggest that the SER scattering seen on our grating samples for
scattering geometries where light cannot couple to the SPP's of
the Ag arises from residual random roughness on the surface.
Further support for this stems from the observations that the intensity of normal incidence scattered light from our gratings varies
greatly from sample to sample. As a result, the magnitude of the
grating induced SPP effects on the intensity of the SER scattering
must be greater than the two orders of magnitude in Fig. 5 since
the scattering observed at normal incidence is not due to the
grating. It may be as large as 10^3.

     The theoretical contribution to I_s from the enhancement of
E_L can be over 10^4 for molecules close to the surface of Ag[17,23].
We believe however, that the actual field enhancement of I_s by the
grating will be at least an order of magnitude smaller. The value of
ε_2 for a thin evaporated Ag film is between 0.6 to 1.2 in the visible and not the value of 0.25 for the bulk[25]. This is due to the
presence of surfaces and grain boundaries in 200 Å thick films. In
addition, our junctions are fabricated on Al layers which are considerably more lossy than the Ag. The use of a more realistic value
of ε_2 will produce an E_L enhancement of 30 or less, rather than 100,
and a Raman intensity enhancement of about 1000 or less. This

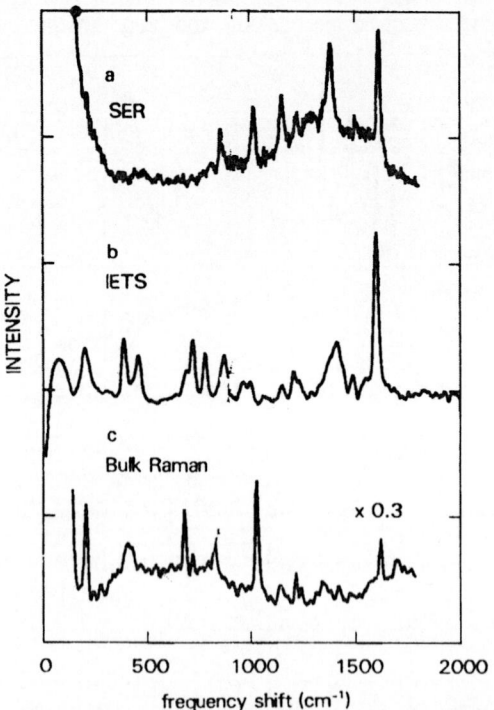

Fig. 8 - a) The SER spectrum of a 4-pyridine COOH doped Ag tunnel
            junction on an 8000 Å periodicity diffraction grating.
         b) The IETS spectrum of a 4-pyridine COOH doped Ag tunnel
            junction.
         c) The Raman spectrum of bulk 4 pyridine COOH.
         Raman spectra excited at 6471 Å.

figure is quantitatively close to what we observe if we consider
the effects of scattering from the residual surface roughness and
also the optical attenuation effects associated with the finite
thickness Ag layers.

    In Fig. 8, we show the Raman spectra of a 4-pyridine COOH
pressed pellet, the SER of 4-pyridine COOH in an Al-AlO$_x$-Ag
junction laid down on an 8000 Å periodicity, 1000-1200 Å amplitude
grating, and the IETS of a 4-pyridine COOH doped Ag junction.
Both the IETS and pellet spectra have been processed to remove
broad backgrounds to facilitate the comparison of the Raman lines.
Both Raman spectra were obtained using the same collection optics
and the tunnel junction SER spectrum was obtained for $\Theta_i = \Theta(\omega_{sp})$.

Given our scattering geometry, the number of pyridine COOH molecules excited in the pressed pellet was about 10^{17} . The number of optically excited molecules in the junction was about 5×10^{11} [26]. A comparison of particular Raman lines such as the 1600 cm^{-1} and the 991 cm^{-1} ring stretching modes show that the intensities of some, but not all of the Raman lines in the tunnel junction have been enhanced by more than 5×10^{4}. In Fig. 5, we showed the Raman spectra of a pressed pellet of nitrobenzoic acid and of nitrobenzoic acid in a Ag tunnel junction laid down on an 8000 Å periodicity, 600 - 800 Å amplitude grating. The scattering parameters in Fig. 5 were the same as those in Fig. 8. A comparison of the Raman spectra of the pellet and the junction shows that the enhancements of the Raman efficiencies of the different modes in nitrobenzoic acid on this grating appear to vary from less than 5×10^{2} to 2×10^{4} . The SER spectra of the nitrobenzoic acid (Fig. 5) and 4-pyridine COOH (fig. 7) show no sign of the C-H vibration at 3060 cm^{-1} . This means that any enhancement of this mode must be less than 5×10^{2} . Measurements of SER scattering from Ag tunnel junctions doped with benzoic acid and laid down both on 8000 Å periodicity diffraction gratings and varying thicknesses of CaF$_2$ suggest that the maximum enhancement of the Raman efficiency in benzoic acid under Ag is less than 10^{3} .

A comparison of the SER and IETS spectra is complicated by the fact that the SER and IETS can excite Raman and infrared active modes with different efficiencies. If we just consider the Raman active modes of the molecule, we find that the relative intensities seen in the SER spectrum of 4-pyridine COOH (Fig. 8) are qualitatively similar to those seen in the IETS. This pattern of weak and strong lines is qualitatively different from that seen in the Raman spectrum of the molecule in bulk. The strongest SER and IETS scattering comes from the 1600 cm^{-1} C-C ring mode. In contrast, the strongest Raman scattering from the pellets arises from the 991 cm^{-1} symmetric C-C vibration. Both the SER and IETS show very weak or unobservable scattering from the strong C-H vibration near 3060 cm^{-1}. Kuiper et al[27] studied the Raman spectrum of benzaldehyde adsorbed on alumina and found that these changes are not associated with adsorption of the molecules on the oxide surface. Therefore, they must arise from the interaction of the molecule and the metal.

One parameter describing this interaction is Φ_{b2}. The tunneling spectra of both the nitrobenzoic acid and 4-pyridine COOH doped Ag junctions feature the peaked elastic tunneling backgrounds shown in Fig. 1b. Raman studies of these and other molecules such as acetylbenzoic acid and aminobenzoic acid show that the strongest SER scattering is often observed in tunnel junctions which have the peaked d^2V/dI^2 background characteristic of a low barrier height.

   The observation of an enhancement $>5 \times 10^4$ of the Raman scattering efficiency of a molecular vibrational mode in tunnel junctions on 600-1200 Å amplitude, 8000 Å periodicity gratings means that the enhancement cannot be completely attributed to the enhancement of the incident fields under the resonant excitation of the SPP's of Ag. We have shown that the magnitude of this effect on our surfaces will be $<10^3$. The dependence of the measured enhancement on both the identity of the scattering molecule and the wavefunction of the vibration suggests the existence of a short range, local contribution to SER scattering with a magnitude between 10^2 and 10^3. This is consistent with the theoretical predictions of our model as shown in Fig. 4 [17,23].

   The existence of a specific, local, short-range interaction between the adsorbed molecule and the noble metal surface which can contribute to the SER effect for the first adsorbed monolayer. can also be seen in recent ultra high vacuum results. They[5,6,13] show that the SER efficiency is largest for the first adsorbed layer. In Fig. 9[13], we show the dependence on coverage of I_s for the 990 cm^{-1} mode of pyridine for adsorption on a Ag (111) surface at 80 K. These results were obtained from a Ag(111) surface on which a 10000 Å periodicity, 1000 Å amplitude diffraction grating had been chemically etched using a x-ray photoresist pattern. Annealing of the crystal produced the LEED pattern of the (111) surface with streaking of the spots in the direction parallel to g characteristic of a one dimensional array of steps on the surface due to the grating profile. As in the case of the tunnel junction work, I_s was a strong function of the scattering geometry with observable signals occuring only at the SPP resonance condition. Figure 9 shows that the strongest Raman scattering arises from the first layer and that the contribution to the Raman scattering from the second and subsequent layers is much weaker. The factor of 20 or more decrease in the magnitude of the enhancement from first to the second layer is consistent with a value of Φ_{b2} of between 0.5 eV and 1 eV. It is also similar to the difference in the magnitude of the enhancement seen for various lines and molecules in our tunnel junction structures. It suggests that the short range enhancement we observe here and the molecule and vibrational mode dependent effects we have observed in our junctions share a common origin.

   With the exception of the high resolution electron energy loss studies on pyridine and other molecules on Ag of Demuth[28] and of our tunneling studies of doped Al-AlO$_x$-Ag junctions, little is known about the vibrational or electronic states of the molecule-Ag interface. The electron energy loss and IETS studies suggest that the bonding between the adsorbed molecules and the Ag metal surfaces is not strong. The fact that the experimentally observed Raman enhancements on our simple grating surfaces are considerably larger than can be accounted for by the field enhancement due to

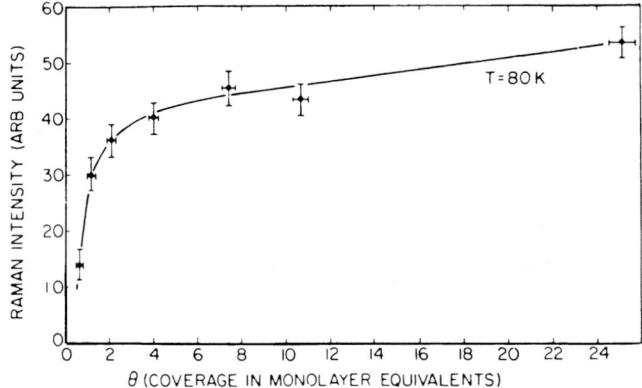

Fig. 9 - The dependence on pyridine exposure of the intensity of
SER scattering from pyridine adsorbed on a Ag(111) surface with a low amplitude grating profile[13].

the optical excitation of the surface plasmon resonances of the Ag
means that the system polarizability for the adsorbed molecules on
Ag is more than just the sum of the individual polarizabilities.
Our tunneling model[17,23] represents a simple case of a relatively
weak interaction between the conduction electrons of the metal and
the molecule. All of the features of our experimental results on
tunnel junctions and in ultra high vacuum appear to be contained
in this model. The magnitude of the surface charge density induced
by E_L is greatly enhanced by the excitation of the SPP's producing
a large enhancement of the Raman efficiency from this scattering
channel. The observation of strong SER scattering from molecules
whose IETS spectra show anomalous elastic tunneling backgrounds,
which signify low values of Φ_{b2} , arises from the fact that the
electron-phonon coupling in this model is significant only for
small values of Φ_{b2} . The observed similarity of the IETS and the
SER spectra arises from the fact that both scattering processes
involve the same electron-vibrational mode interaction. This is
different from the normal vibrational modulation of the molecular
polarizability, α_{ij}, which is responsible for Raman scattering
from the isolated molecule.

Our experimental and theoretical results show 1) that the
Raman efficiency of any molecule at a noble metal surface can be
significantly enhanced by the excitation of the surface plasmon
polariton resonances of the surface and that 2) the presence of a
metal surface within atomic distances from the molecule can introduce a new channel by which Raman scattering can proceed via the
modulation of the large induced surface charge density. This suggests that the observation of very strong SER scattering (10^5- 10^7
enhancement) is not universal for all metal molecule systems.

However, given the existing experimental capability for Raman
scattering, it suggests that Raman scattering can be done on a
variety of molecules on rough noble metal surfaces and may be performed on non-noble metal surfaces for selected molecules, even
though the enhancement in such systems may be much less than 10^4.

ACKNOWLEDGEMENT

We thank Drs. T.N. Theis and J.M. Warlaumont for the fabri-
cation of the diffraction gratings which made this work possible,
C. Aliotta for the electron micrographs of our junctions, J.A.
Bradley, M.T. Prikas and A.M. Torressen for experimental assistance
and Dr. J.E. Demuth and P.N. Sanda for their collaboration in the
ultra high vacuum experiments.

REFERENCES

1. M. Fleischmann,P.J. Hendra and A.J. McQuilla, Chem. Phys. Lett.
   26, 163 (1974).
2. D.L. Jeanmaire and R.P. van Duyne, J. Electroanal. Chem. 84, 1
   (1977).
3. M.G. Albrecht, J.A. Creighton, R.E. Hester and J.A.D. Matthew,
   Chem. Phys. Lett. 55, 55 (1978).
4. R.R. Smardzewski, R.J. Colton and J.S. Murdey, Chem. Phys.
   Lett. 68, 53 (1979).
5. G.R. Easley (private communication).
6. T.A. Wood and M.V. Klein, J. Vac. Sci. Technol. 16, 459 (1979);
   I. Pockrand and A. Otto (private communication).
7. M. Moskovits (private communications).
8. J.M. Rowe, C.V. Shank, D.A. Zwemer and C.A. Murray, Phys. Rev.
   Lett. 44, 1770 (1980).
9. E. Burstein, C.Y. Chen and S. Lundqvist, Proc. US-USSR Symposium of Light Scattering, ed. J.L. Birman, H.Z. Cummins and
   K.K. Rebane (Plenum, New York 1980). p. 479.
10. H. Seki and M. Philpott, (private communication).
11. J.C. Tsang and J.R. Kirtley, Solid State Commun. 30, 617 (1980).
12. J.C. Tsang, J.R. Kirtley and J.A. Bradley, Phys. Rev. Lett. 43,
    772 (1979).
13. P.N. Sanda, J.M. Warlaumont, J.E. Demuth, J.C. Tsang, K.
    Christmann and J.A. Bradley, (submitted for publication).
14. Y.J. Chen, W.P. Chen and E. Burstein, Phys. Rev. Lett. 36,
    1207 (1976).
15. M. Moskovits, J. Chem. Phys. 69, 4159 (1978); D.S. Wang, M.
    Kerker and H. Chew, Appl. Optics 19, 2315 (1980).
16. M. Philpott, J. Chem. Phys. 62, 1812 (1975); S. Efrima and
    H. Metiu, Chem. Phys. Lett. 60, 59 (1978); F.W. King, R.P. van
    Duyne and G.C. Schatz, J. Chem. Phys. 69, 4472 (1978).
17. J.R. Kirtley, S.S. Jha and J.C. Tsang, Solid State Commun. 35,

n° 7 (1980).

18. D. Heitmann, Optics Commun. $\underline{20}$, 292 (1977).
19. John Kirtley and P.K. Hansma, Phys. Rev. $\underline{B13}$, 2910 (1976).
20. R.V. Coleman (private communication).
21. C.S. Korman, J.C. Lau, A.M. Johnson and R.V. Coleman, Phys. Rev. $\underline{B19}$, 994 (1979).
22. J.R. Kirtley, T.N. Theis and J.C. Tsang, Appl. Phys. Lett. $\underline{37}$, 435 (1980).
23. S.S.Jha, J.R. Kirtley and J.C. Tsang, Phys. Rev. $\underline{B22}$, n° 8 (1980).
24. S.L. McCarthy and J. Lambe, Appl. Phys. Lett. $\underline{30}$, 427 (1977).
25. M. Dujardin and M. Theye, J. Phys. Chem. Solids $\underline{32}$, 2033 (1971).
26. J.D. Langan and P. Hansma, Surf. Sci. $\underline{52}$, 211 (1975).
27. A.E.T. Kuiper, J. Madema and J.J.G.M. van Bokhoven, J. Catal. $\underline{29}$, 40 (1973).
28. J.E. Demuth (this conference).

ENHANCED RAMAN SPECTROSCOPY FROM ADSORBATES ON SILVER : EVIDENCE FOR THE "ADATOM MODEL"

A. Otto, I. Pockrand, J. Billmann

Physikalisches Institut III, Universität Düsseldorf
D-4000 Düsseldorf 1
Federal Republic of Germany

INTRODUCTION

The present state of the field of surface enhanced Raman scattering (SERS) and different tentative explanations have been overviewed in Ref. [1] (representing the state of information at Oct. 1, 1980). Therefore, in this abstract, we will only shortly review the experimental work of the research group at the University of Düsseldorf and our tentative explanation.

THE "ADATOM MODEL"

We have proposed a hypothesis[2] where the roughness, important for SERS, is of atomic scale (adatoms, clusters of adatoms). This atomic scale roughness relaxes the momentum conservation rule in photon-metal electron, electron-electron and metal electron-adsorbate interaction. This leads to additional surface absorption, to an inelastic unstructured background radiation due to electron-electron interaction and to enhanced Raman scattering from molecules chemisorbed to adatoms.

The processes of absorption (a), background radiation (b), and SERS (c) are characterized by their Feynman diagrams in Fig. 1. The relaxation of momentum conservation leads to strong matrix elements for the interaction of the photon with any eh excitation. The processes (b) and (c) are resonant, concerning the photon-eh interaction, but in general not resonant concerning the eh-adsorbate vibration interaction. The outgoing radiation will be depolarized, in agreement with experiment.

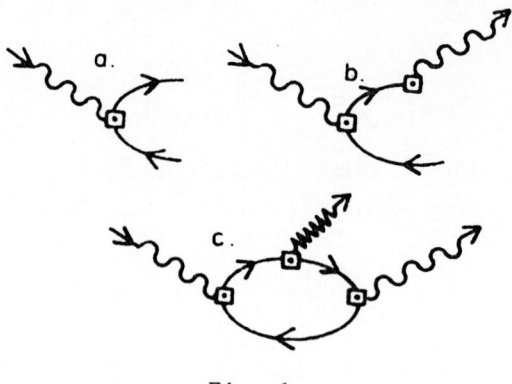

Fig. 1

The Raman selection rules for the adsorbate vibrations are relaxed, because the coupling to the adsorbate is not via photon- but via electron-adsorbate interaction.   This is in agreement with experiment.

EXPERIMENTAL EVIDENCE FOR THE INELASTIC BACKGROUND

     The inelastic background was first mentioned in [3] , and first ascribed to roughness induced electronic Raman scattering in [4].   In agreement with the "adatom model" it was found, that the background does exist without the presence of adsorbates [5].

     In an electrochemical reduction - oxidation cycle the back-ground grows with the amount of redeposited silver on silver, eventually saturating in intensity[6] , reflecting the increase of the surface concentration of atomic scale roughness.

     Besides redeposition of Ag^+ ions from the solution onto the surface, there is a second way to increase the surface concentra-tion of adatoms, namely by loosening of kinksite atoms which then diffuse onto the terraces.   As the adatom binding energy is about one third of the binding energy of a kinksite atom, this "kinksite-adatom transformation" has to be expected [7] at a potential about 2/3 of the difference between the potential of zero charge (ca. -1.0 V SCE) and the potential of anodic dissolution (at ca. 0.4 to 0.5 V SCE) which is indicated by the onset of the anodic current in Fig. 2 [7].

     The background intensity at 4200 cm^{-1} varies with potential (linear scan from - 1.0 V to 0.5 and back with 10 mV/sec), whereas the Rayleigh scattered light does not[7] - the latter effect indica-ting, that the surface roughness on supraatomic, microcrystalline

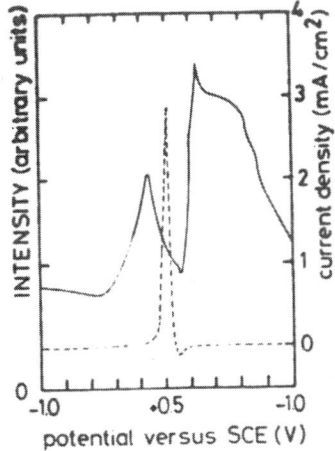

Fig. 2

scale does not change.   We ascribe the increase of the background
intensity I at about − 0.1 V to an increase in adatom concentration
c_A by the "kinksite-adatom transformation", the decrease in I
scattering at +.0.4 V to a decrease in c_A caused by the transition
of adatoms to Ag^+ ions going into solution and the increase of I
on the backsweep to redepostion of Ag^+, as discussed above.
The decrease of I, when sweeping back to −1.0 V is caused by
incorporation of adatoms at steps and kinksites.
Typical times to reach equilibrium in c_A at constant potential
are of the order of 10 minutes, as indicated in Fig. 3.
Observations of the background under UHV condition are discussed
further.

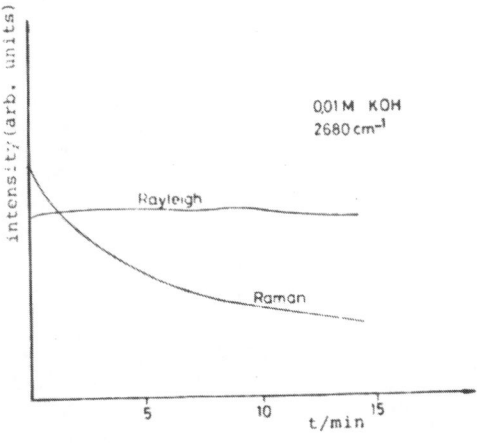

Fig. 3

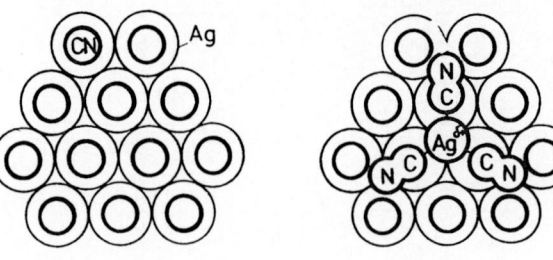

Fig. 4a   Fig. 4b

EXPERIMENTAL EVIDENCE FOR SERS FROM MOLECULES ADSORBED TO ADATOMS

The experimental observations, supporting this hypothesis are the following :

The surface of a polycristalline silver electrode in an 0.1 M Na_2SO_4/0.01 M KCN electrolyte is prepared in different ways by electrochemical means, described in[2].

One preparation, in which silver is slowly dissolved as $A_g(CN)_2$ in solution, without any redeposition of the metallic silver will lead to a surface, which has a relatively low concentration of silver adatoms and microclusters, because these are the ones most likely to go into solution. Though the surface is covered by one monolayer of cyanide, no SERS is observed [2]. A probable adsorption structure of the monolayer of CN^- is the top position (analogous to the crystal structure of bulk AgCN), as depicted in Fig. 4a.

If Ag^+ ions are redeposited onto the silver surface, one observes a SERS signal from cyanide at 2111 cm^{-1}. The signal coincides with the CN stretch frequency of the $Ag(CN)_3^=$ complex in solution[2] Also a SERS signal at 2140 cm^{-1} is found in some cases, coinciding with the CN stretch frequency of the $Ag(CN)_2^-$ complex in solution. From this we conclude, that the SERS signal is from surface complexes $Ag(CN)_3$ (depicted in Fig. 4b) and $Ag(CN)_2$. We expect only a negligible shift of the vibrational frequency between the complexes in solution and at the surface, as the surface of the bulk silver is only a second nearest neighbour to the cyanide groups of the complex[3] . The surface complexes $Ag(CN)_3$ and $Ag(CN)_2$ may be envisioned as cyanide groups, adsorbed to an adatom. The CN stretch frequency from cyanide groups simultaneously adsorbed to

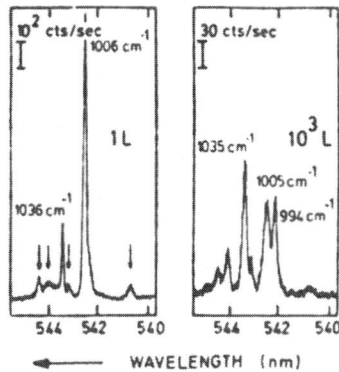

Fig. 5

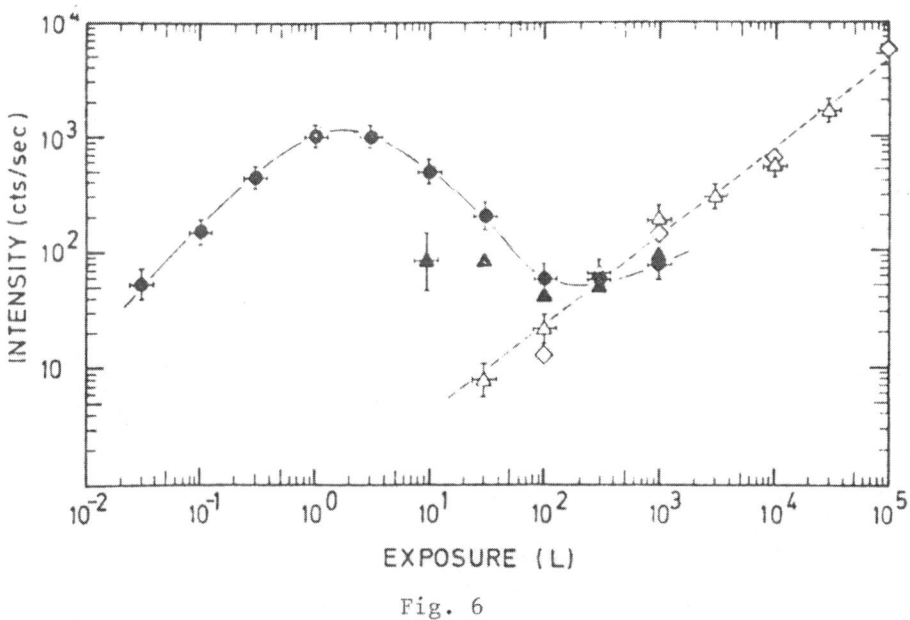

Fig. 6

the smooth surface terraces is not seen in agreement with the discussion above. Our interpretation is supported by ref.[8].

SILVER SURFACES IN UHV

SERS from pyridine on silver films, evaporated onto a cooled
copper substrate (∿130 K) has been measured as function of pyridine
exposure[9] (Fig. 5.) The strong pyridine breathing vibration (at

1006 cm^{-1}) is already seen for exposures as low as 0,03 L. This signal (full dots) reaches a maximum peak intensity at about mono-layer coverage (Fig. 6.).

At higher coverage, it looks as if this intensity is shared with a second peak at 994 cm^{-1} (full triangles). For exposures higher than 10^3 L the 994 cm^{-1} peak increases with exposure like it does in the case of ordinary Raman scattering from thick pyridine layers on annealed silver (110) surfaces (open rhombs). No indication for a silver pyridine vibration below 390 cm^{-1} was found[10].

Wood and Klein[11] observed SERS from CO adsorbed to silver films deposited on copper substrates at 120 K. After warming up to room temperature and several hours "annealing" at this temperature the films were again cooled down to 120 K and exposed to CO. No Raman signal has been observed after this procedure. These observations have been confirmed by own experiments.

More detailed results[12] on the background intensity at !006 cm^{-1} (see Fig. 7b), on the SERS intensity of adsorbed pyridine at 1006 cm^{-1} , on the intensity of a shoulder near 160 cm^{-1} (see Fig. 7a) and on the Rayleigh scattered light as a function of tempe-rature during warming up of the silver films from 120 K to room temperature during 2 to 3 hours are given in Fig. 8. The shoulder near 160 cm^{-1} was assigned to disorder induced Raman scattering from acoustical phonons in bulk silver[10]. Fig. 8a is for an unexposed silver film (open circles : background, full circles : structure at 160 cm^{-1} , triangles : Rayleigh scattered light).

The temperature range of the decay of the background signal and the 160 cm^{-1} shoulder coincides with the range, where diffusion of divacancies to voids is observed in quenched silver by DC conduc-tivity measurements[13].

Hence, we assign the background to strong electron-photon coupling by local disorder. We tentatively assign the increase of background intensity at about 180 K to a separation of divacan-cies from grain boundaries or from unstable vacancy-clusters and the irreversible decrease around 250 K to diffusion of divacancies to the film surface. Fig. 8b is for a silver film, exposed to 1 L of pyridine (open circles : background, full circles : pyridine SERS 1006 cm^{-1} , triangles : Rayleigh scattered light).
The decrease of the pyridine SERS signal sets in at lower tempera-ture than the background signal.
This is not caused by a desorption of pyridine, as is shown by the result in Fig. 8c (symbols like in Fig. 8b). Here the silver film was exposed to 1 L of pyridine, warmed up to 230 K, recooled to 120 K, exposed additionally 1 L of pyridine, and warmed up to room temperature. In the second warming up period the irreversi-ble annealing proceeds like in Fig. 8b after reaching again 230 K.

If the SERS signal of pyridine were caused by electromagnetic
resonances on a bumpy surface, the results of Fig. 8 would indicate,
that the bumps are annealed at lower temperature than the temperature, where divacancies diffuse effectively. We think this to be
unlikely. On the other hand, the annealing curve of the pyridine
SERS signal agrees well with the observed annealing of adatom-
roughness on gold films[14].

No correlation exists between Rayleigh and Raman scattered
light.

For details, the reader is referred to ref.[12].

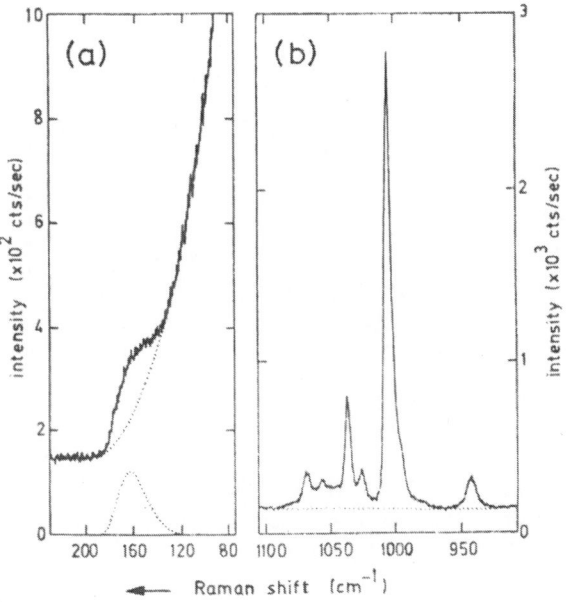

Fig. 7

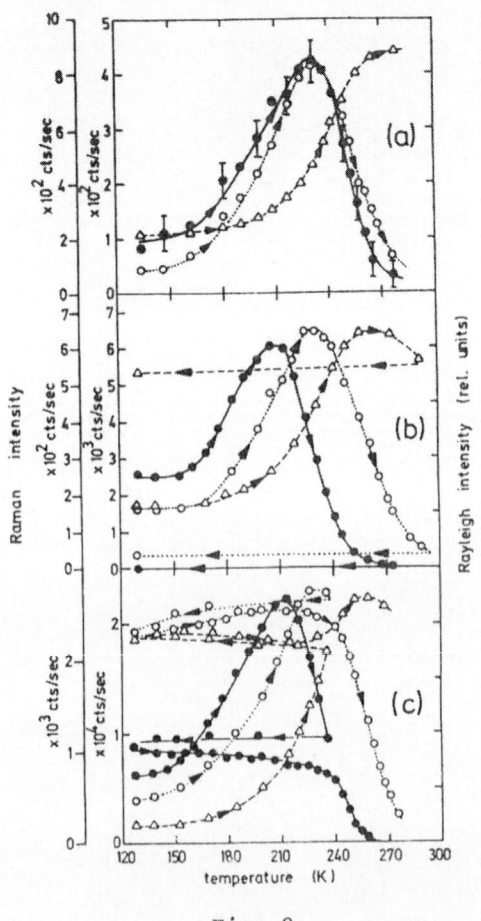

Fig. 8

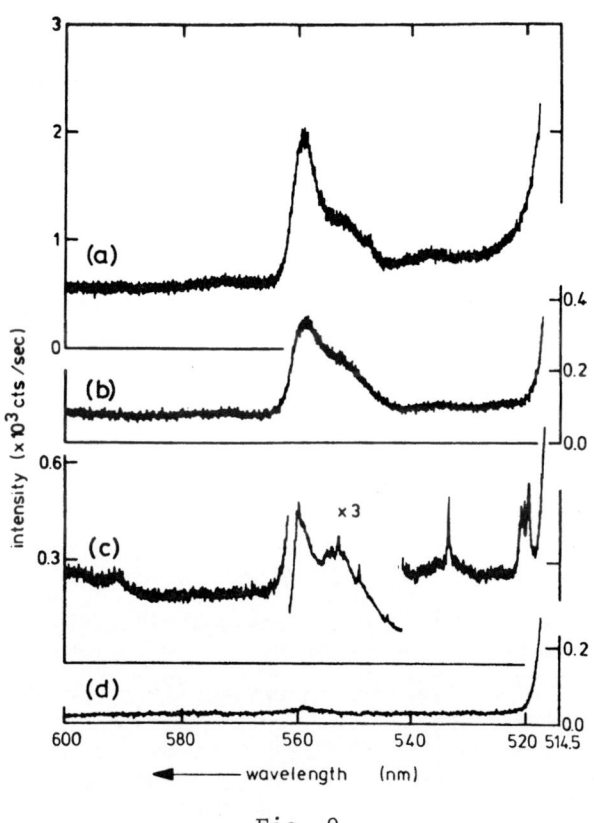

Fig. 9

THE POSSIBILITY OF A BULK RAMAN ENHANCEMENT MECHANISM

The hypothetical "carbonate structure"[3] has been investigated for silver samples in UHV[15].
It was observed :

a) Polycrystalline silver slugs, mechanically polished and cleaned
   by Ar-ion sputtering, showed pronounced Raman peaks between
   1000 cm^{-1} and 1600 cm^{-1} although AES indicated a clean surface.
   A strong, continuous background scattering was also observed
   (Fig. 9., curve (a)).

b) Annealing the sample up to 800 K for several hours and subsequent
   slight sputtering did not change the shape of the Raman spectrum,
   but results in a significant decrease of the intensity of the
   carbonate structure as well as the background intensity (Fig. 9.,
   curve (b)).

c) Extended exposure of the clean sample to CO, O_2 (T = 290 K) and
   HCOOH (T = 150 K) did not influence the carbonate structure and
   the background. Weak (CO, O_2) and prominent (HCOOH) additional
   peaks, however, have been detected due to a certain amount of
   adsorbed molecules (Fig. 9, curve (c): 7×10^4L HCOOH, T =150K),

d) A clean silver (110) surface gave a background intensity just
   above the noise level and a hardly detectable remainder of the
   carbonate structure (Fig. 9, curve (d)).

We conclude from these observations :

i) Our failure to detect carbon by AES, the insensitivity of the
   Raman peaks to additionally adsorbed molecules, and the dependence of the intensity on the crystalline structure (9b, 9d)
   suggest incorporation of the "carbonate" into the polycrystalline sample below the surface, preferably at defects (e.g. grain
   boundaries).

ii) The "carbonate" signal is enhanced.   It is always accompanied
    by a strong background scattering.

As explanation, we propose to slightly modify the adatom model for
SERS into a model for "BERS" (bulk enhanced Raman scattering).
The breaking of the translational symmetry of the metal lattice and
the strong photon-electron hole coupling induced in this way is
caused by atomic scale roughness (e.g. adatoms) for SERS and by
incorporation of impurity molecules in the bulk or by dislocations
and grain boundaries for BERS.   Hence enhanced Raman scattering
from bulk imperfections, located within the penetration depth of
light in the metal, may be observed.

     Like in the adatom model for SERS also BERS should always be
connected with a continuous inelastic background scattering, which
obviously we observe.   Annealing should cause a decrease of the
density of defects and therefore should result in a decrease of
the Raman intensity (see Fig. 9, curves (a) and (b)).

     We have already reported on similar phenomena in[5] where we
interpreted the extremely strong background from a silver slug after

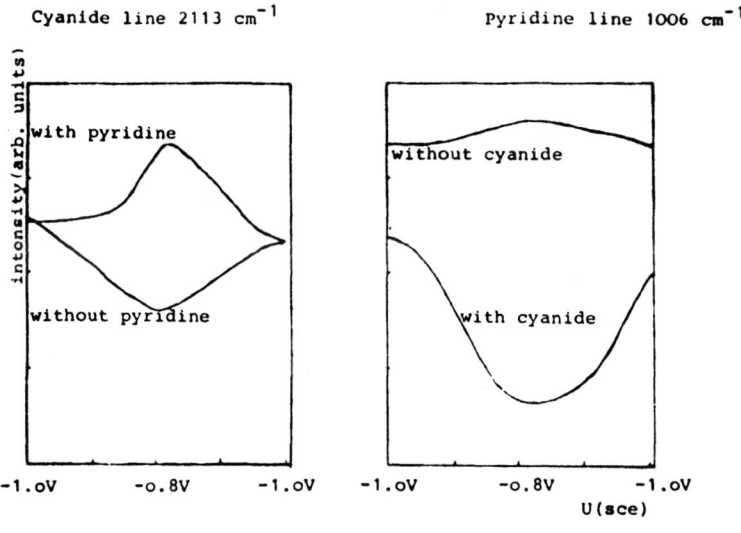

Fig. 10

heavy mechanical polishing as caused by a high density of lattice imperfections (see also discussion of the background in section V).

It should be mentioned that our experiments do not allow to unambiguously attribute the observed Raman feature to a carbonate. It may also be that the structure is due to graphitic carbon impurities as suggested recently[16].

SIMULTANEOUS SERS FROM PYRIDINE AND CYANIDE ON SILVER ELECTRODES

Potential dependent SERS spectra from polycrystalline silver electrodes in $0,1$ M Na_2SO_4 electrolyte were recorded in the presence of pyridine, of cyanide, and of pyridine and cyanide together[17]. The intensity versus potential plots of the cyanide line at 2113 cm^{-1} and the pyridine line at 1005 cm^{-1} depend very much on the presence or absence of pyridine or cyanide, respectively (Fig. 10).

This proves that the SERS signal is from the adsorbed species. The variations are interpreted with the different potential dependence of the heat of adsorption of cyanide and pyridine[17].

This observation and the absence of SERS for water in the presence of SERS for cyanide and pyridine excludes the possibility, that SERS from electrodes in aqueous solutions is mainly caused by a nonlocal mechanism.

HYPOTHETICAL NONLOCAL ENHANCEMENT MECHANISMS

A non local enhancement mechanism due to a modulation of the surface susceptibility of a metal by a vibrating molecule has been proposed in[4] . As yet, we did not find an experimental indication for such an effect[18] .

REFERENCES

1. A. Otto, to appear in Proc. 6th Solid-Vacuum Interface Conf., Delft, May 1980.
2. J. Billmann, G. Kovacs, A. Otto, Surface Sci. 92, 153 (1980).
3. A. Otto, Surface Sci. 75 , L392 (1978).
4. A. Otto, Proc. Int. Conf. Vibrations in Adsorbed Layers, Jülich, June 1978, and Surface Sci., 92, 145 (1980).
5. A. Otto, J. Timper, J. Billmann, G. Kovacs, I. Pockrand, Surface Sci. , 92, L55 (1980).
6. J. Timper, J. Billmann, A. Otto, I. Pockrand, Proc. Int. Conf. on Non Traditional Approaches to the Study of the Solid Liquid Interfaces, Snowmass Village, USA, Sept. 1979.
7. A. Otto, J. Billmann, J. Timper, I. Pockrand, Phys. Rev. Lett., 45, 46 (1980).
8. R.E. Benner, R. Dornhaus, R.K. Chang, B.L. Laube, preprint.
9. I. Pockrand, A. Otto, Solid State Commun., 35, 861 (1980).
10. I. Pockrand, A. Otto, Solid State Commun., in press.
11. T.H. Wood, M.V. Klein, Solid State Commun., 35, 263 (1980).
12. I. Pockrand, A. Otto, submitted to Phys. Rev. Lett.
13. L.J. Cuddy, E.S. Machlin, Phil. Mag. 7,745 (1962).
14. J.P. Chauvineau, Surface Sci., 93, 471 (1980).
15. I. Pockrand, A. Otto, Proc. 6th Solid Vacuum Interface Conf., Delft, May 1980.
16. M.R. Mahoney, M.W. Howard, R.P. Cooney, Chem. Phys. Lett. 71, 59 (1980).
    J.C. Tsang, J.E. Demuth, P.N. Sanda, J.R. Kirtley, preprint.
17. J. Billmann, A. Otto, Proc. 6th Solid Vacuum Interface Conf., Delft, May 1980.
18. R. Zeller, Diplomarbeit, July 1980, Universität Düsseldorf, (unpublished).

HIGH RESOLUTION ELECTRON ENERGY LOSS AND SURFACE ENHANCED RAMAN STUDIES OF PYRIDINE AND BENZENE ON Ag(111)

J.E. Demuth, P.N. Sanda ★, J.M. Warlaumont,
J.C. Tsang and K. Christmann ★★

IBM Thomas J. Watson Research Center

P.O. Box 218, Yorktown Heights, NY 10598

ABSTRACT

High resolution electron energy loss (EELS), UV photoemission (UPS), low energy electron diffraction (LEED), Auger electron, and thermal desorption spectroscopic techniques have been applied to characterize the bonding, molecular orientation and vibrations of pyridine and benzene on clean Ag(111). These results are used to interpret the surface enhanced Raman spectra of these molecules adsorbed on a clean, well-defined single crystal Ag(111) surface containing a smooth periodic modulation (1 micron wavelength, ~ 500 Å amplitude) which permits optical coupling to surface plasmons-polaritons. We observe enhanced Raman scattering (RS) for chemisorbed species only when we are at appropriate incidence angles so as to excite surface plasmons. Coverage-dependent RS studies indicate a long range field enhancement of $\sim 10^2$ for physisorbed layers and a mode-selective, short-range enhancement of $\sim 10^4$ for the symmetric ring breathing mode of pyridine. This short-range enhancement only occurs for pyridine coverages slightly above $\sim 3 \times 10^{14}$ molecules/cm^2 where changes in the UPS spectra and our UPS derived uptake-rates are also observed. Similar coverage-dependent effects are also observed in UPS and EELS for chemisorbed pyridine on the ideal Ag(111) surface. These as well as LEED results showing the modulated surface to be predominately (~ 90 %) single crystal Ag(111), encourage us to directly compare the results

† Work supported in part by the office of Naval Research.
★ Affiliated with Cornell University, Ithaca, NY.
★★ Present address : Inst. Phys. Chem., Univ. München, GFR

obtained for the two surfaces.

From EELS we determine that chemisorbed pyridine π-bonds and lies flat on the surface of Ag(111) to within $\sim 5°$ but at higher chemisorption coverages (> 3×10^{14} molecules/cm^2) becomes nitrogen-lone-pair-bonded, inclined $\sim 55°$ to the surface and rotated $\sim 30°$ about the C_{2v} symmetry axis. Benzene is found to π-bond and has vibrations characteristic of a $C_{3v}(\sigma d)$ point group symmetry. We thereby relate the strong short-range enhancement observed in RS to the occurrence of nitrogen lone-pair-bonding and the inclination of chemisorbed pyridine to Ag(111). Both the mode selectivity and structure sensitivity for *chemisorbed* pyridine are shown to be consistent with the recent surface plasmon-polariton model of enhanced Raman scattering by Kirtley, Jha and Tsang. Also observed in RS are additional vibrations which we associate with pyridine bound to step sites and at or near graphitic carbon impurities.

INTRODUCTION

The occurrence of large enhancements in Raman cross sections for molecules adsorbed at metal surfaces in electrochemical cells [1], discontinuous films[2], colloidal suspensions[3], tunnel junctions[4], and in ultra high vacuum[5-10] is well-documented. The nature and origin of this enhancement mechanism is unclear and currently in dispute. Two predominating viewpoints exist : (a) that surface roughness and the creation of collective surface excitations are important[3,4,5,8,11] ; (b) that specific ad-atom sites or arrangements of coadsorbed atoms form a type of "surface-complex" which is a strong or resonant Raman scatterer[9,12,13]. Such differing viewpoints may both be correct and could arise from the diversities of the systems under investigation. This possibility highlights a major problem of most all current surface enhanced Raman studies : *there is a lack of detailed information about the structure of the surface, its cleanliness, or the nature of bonding or orientation of the adsorbed species on the surface.*

The purpose of our study is twofold : (1) to characterize the nature of bonding, the molecular geometry and the vibrations of a variety of chemisorbed molecules on a clean Ag(111) single crystal surface using several surface science techniques; and (2) to per-

form Raman measurements on these same systems under as identical
conditions as possible. From the correlation of these two results
we hope to understand more concerning the nature of surface enhanced
Raman. In particular, we perform our Raman measurements on a clean
single crystal Ag(111) surface which has a weak periodic profile
(10,000 Å periodicity, ~ 500 Å amplitude) built into the crystal
surface. We find as observed previously for tunnel junction
structures[4] that enhanced Raman scattering also occurs from this
type of controlled roughness surface. Thus our Raman studies are
intended to address questions relating to the mechanism of enhanced Raman scattering on roughened surfaces[4,11] specifically
where surface plasmon-polaritons occur.

Here we choose to present our results for two adsorbates,
pyridine and benzene, the former being fairly extensively studied
in other enhanced Raman studies [1,5,7,9]. Benzene is chosen for
comparison to pyridine as it has a similar size, shape and
vibrations as pyridine but lacks the nitrogen-lone-pair orbital.
From previous studies[14,16] different adsorption geometries are
expected for these two molecules. Indeed, we determine that
while both benzene and pyridine can π-bond and lie flat on Ag(111),
pyridine can nitrogen-lone-pair bond at higher chemisorption
coverages. We observe a direct correspondence between these structural effects and our observed Raman signals. Evidence for similar
structural effects has been presented for the cyanopyridines in
previous electrochemical studies[17] . We also observe coverage
dependent and mode selective enhancements which are consistent with
a recent surface plasmon polariton theory by Kirtley, Jha and
Tsang[18].

EXPERIMENTAL PROCEDURES

The experimental measurements were performed in three separate
ultra-high vacuum (UHV) systems (base pressure < $1x10^{-10}$ Torr). The
first (turbomolecular pumped) system described elsewhere[19] allows
UPS, low-energy electron diffraction (LEED), Auger (AES) and TDS
to be performed. The second (ion and titanium sublimation - pumped)
system is mobile and permits *in situ* LEED, AES and Raman scattering
measurements to be performed. The third (ion- and titanium-subli-
mator-pumped) system permits EELS and work function change measurements.

The electron optics for EELS consists of two sets of 2.5 cm hemispherical deflection analyzers with associated focusing optical elements[20] so as to allow the monochromatization, reflection from a sample (total scattering angle of 90°) and energy analysis of a well-defined (< 0.2 mm dia.), collimated (< 1°), low-energy 2-100 eV electron beam. For the specular reflection of a 3 eV beam off of Ag(111), we have routinely obtained a total system resolution of 65-75 cm^{-1} (8-9 mV) with peak counting rates of 10^6 cps. Although both the monochromator and analyzer are in a fixed position, the sample rotation is arranged so as to enable the observation of specular ($\theta_{in} = \theta_{out}$) as well as off-specular ($\theta_{in} \neq \theta_{out}$) scattering events in the plane of incidence. Such off-specular angle dependent measurements are needed to assess the magnitude of resonant electron scattering and other non-dipole scattering processes in the specular scattering direction. Although such measurements have been performed and *must be* considered in the analysis of our specular scattering results, we do not present them here.

For the EELS experiments we use an ideal flat Ag(111) crystal. The Ag(111) surface was prepared by standard mechanical[19] and chemical polishing [21], then annealed and sputter cleaned as determined by AES. This clean, well-ordered Ag(111) sample was then inserted into the EELS system where further sputter cleaning and annealing was performed. Work-function change measurements performed in both UHV systems as well as EELS spectra served to verify surface cleanliness in the HREEL system. The final clean samples were mirror smooth and showed minimal optical defects or irregularities. All chemisorption experiments in the EELS system were done at temperatures of $\sim$ 140 K as monitored by a chromel alumel thermocouple.

In order to observe Raman signals from adsorbed species, we have intentionally modified the topography of a Ag(111) crystal. This surface contains a small amplitude, approximately sinusoidal modulation (amplitude $\sim$ 500 Å), which provides a well-defined surface periodicity (wavelength = 10,000 Å) to allow optical coupling to surface plasmon-polaritons, while maintaining minimal deviation from a flat Ag(111) surface. The Ag(111) sample was cut from the same boule as the aforementioned crystal, mechanically and chemically polished, then annealed and sputter-cleaned in UHV. A final chemical polish left the surface with a mirror-like finish, free of etch pits. The 10,000 Å periodic surface modulation was then fabricated into a 4x4 mm area of the 8x6 mm face of the crystal, with the modulation wave vector $\vec{K}_s$ oriented along to the (110) direction. This structure was fabricated by first creating a photoresist pattern on the sample using X-ray lithography techniques followed by chemical polishing to remove about 3000 Å of material

in the unmasked regions (50 %). The photoresist was then dissolved
and the sample repetitively Argon ion sputter-etched and annealed
to ~ 500 K in UHV for cleaning. This processing also reduces the
higher order Fourier components[22] of the profile and results in a
sinusoidal-like surface (valleys slightly wider than the peaks)
with a 10,000 Å wavelength and ~ 500 Å amplitude as estimated by
the LEED beam profiles. Photographs of the LEED pattern from the
flat and "modulated" region of this sample are shown in Fig. 1.
The modulated region of the sample shows a well-defined, low background LEED pattern, comparable to the flat portion of the sample.
Satellite lobes are observed in the beam profiles, which indicated
a distribution of steps and terraces parallel to $\vec{K}_s$. The peak in
this distribution corresponds to a terrace width to step height
ratio of about 10 to 1. The intensity of the main peak relative
to the side lobes indicates that roughly 90 % of the surface is of
(111) orientation. Although such LEED features serve as a guide
to the general nature and condition of the surface, they do not
provide specific information on all types of defects and their
distributions.

UPS measurements were performed on both samples. As determined
from these measurements, we observe negligible differences in the
UPS ionization levels of adsorbates between the flat and topologically modified surfaces. We also find no change in the chemisorption
properties for pyridine or benzene when the sample is between 80
and 140 K - the limiting values of sample temperatures in the Raman
and EELS systems, respectively.

Reagent grade pyridine and benzene (99,9 %) and their deuterated
counterparts (99 atom % D) were used in these experiments. Sample
dosing was done via the chamber ambient and directly monitored with
an ion gauge. All exposures cited here are in units of Langmuirs,
L (1 L= 10^{-6} Torr-sec) and have been corrected by a gauge factor
of 5.8[23]. The dosages required to produce both the compressional
phase of pyridine and first physisorbed layers were the same in all
three UHV systems.

EELS RESULTS

Our EELS results indicate that a structural phase transformation
occurs for chemisorbed pyridine on Ag(111) at an exposure of ~ 0.5 L.
In Fig. 2 we show the EELS vibrational loss spectra for chemisorbed
pyridine before (solid line) and after (dotted line) this transition.
Here, the relative intensities of the CH deformation modes between
400-850 cm^{-1} strongly change, especially when compared to the
relative IR-absorbances[24],[25] also shown in Fig. 2. Such IR absorbances directly reflect the dipole scattering intensities of a
randomly oriented pyridine molecule and can be compared to EELS
results for specular scattering ($\theta_i = \theta_o$). The transition is more

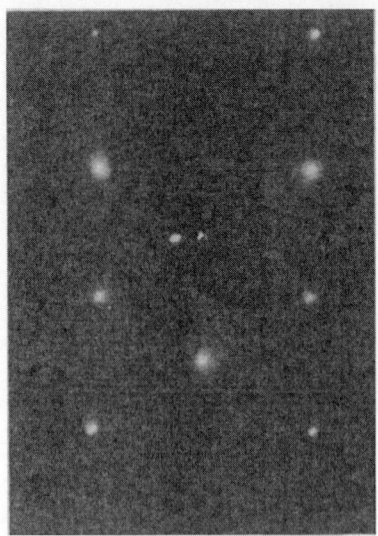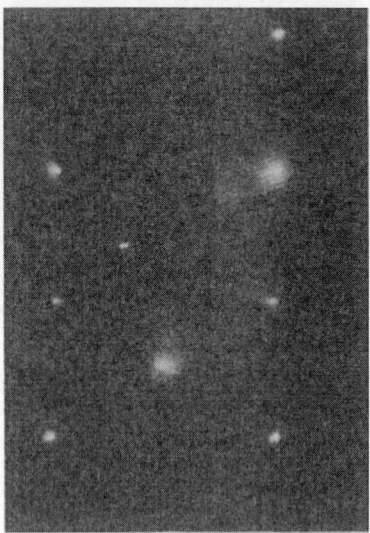

Figure 1 – LEED photograph of the Ag(111) surface (T = 80 K) at
146 eV on the grating (right) and off the grating (left).
The photo on the right of the modulated surface has been
overexposed to more clearly show the weaker satellite
spots.

striking in the coverage dependent spectra shown in Fig. 3. We
note that the strong decrease in intensity of the ~ 700 cm^{-1} feature
above 0.5 L exposure indicates that at least 93 % of the lower
coverage phase converts to a new structure. This 700 cm^{-1} feature
becomes strong again above 1 L exposures when pyridine starts to
physisorb as determined from our UPS results[8,26] .

In Fig. 4 we show the spectra of deuterated pyridine below and
above this phase change, as well as a schematic drawing of the
molecular orientations which we attribute to each phase. Note in
Fig. 4 that the symmetric ring breathing mode at 965 cm^{-1} is clearly
resolved and nearly identical to the liquid phase value of 962 cm^{-1}.
From our angle–dependent studies, we also observe strong non–dipole
scattering from the losses at 2260 (3040) cm^{-1} , 1540 (1570) cm^{-1}
and ~ 820 (~ 1000) cm^{-1} for deuterated (normal) pyridine. Finally,
we find that the vibrational frequencies for both phases of normal
or deuterated pyridine are identical within our experimental un-
certainties of ± 5 cm^{-1}.

In order to obtain structural information, the vibrational
losses must be assigned to the vibrational modes of the molecule.
Such an assignment would seem formidable based upon the relatively

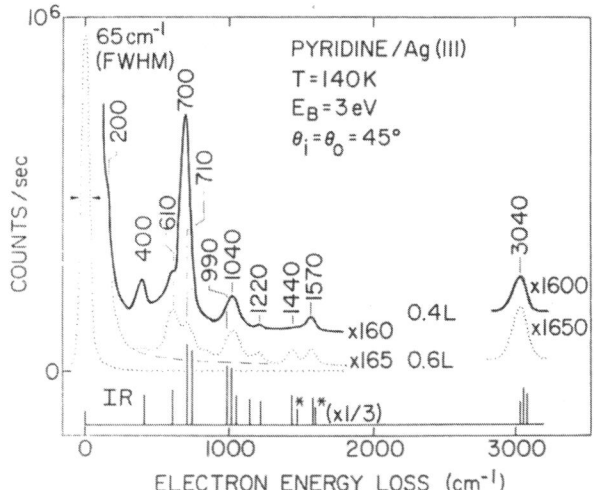

Figure 2 – Vibrational loss spectra of pyridine on Ag(111) at two
coverages above and below the compressional phase
transition. The IR absorbances[24],[25] are shown below for
comparison where the ⋆ levels are reduced by 1/3.

poor resolution of EELS and the fact that pyridine is of low symmetry and has 27 IR-active modes[27] . Fortunately, we
find that we can straightforwardly assign most all the observed
vibrational losses since (a) only a fraction of the 27 free-molecule
modes have significant dipole scattering cross-sections (i.e., IR
absorbances, see Figs. 2 and 4); and (b) pyridine is weakly chemisorbed and leads to weakly perturbed vibrations ($\Delta\nu_{avg} = \pm 6$ cm^{-1})
relative to liquid pyridine. These assignments are tabulated and
described elsewhere [26].

   Based upon the mode assignments we consider the (dipole-de-
rived)scattering intensities of in-plane and out-of-plane vibrations
relative to the IR-absorbances to obtain structural information.
Although our analysis is described elsewhere[26], we briefly comment
on the essential points here. The low coverage phase shows intense,
out-of-plane, CH deformation vibrations at 700 cm^{-1} (525 cm^{-1}) and
400 cm^{-1} (360 cm^{-1}) for normal (deuterated) pyridine which becomes

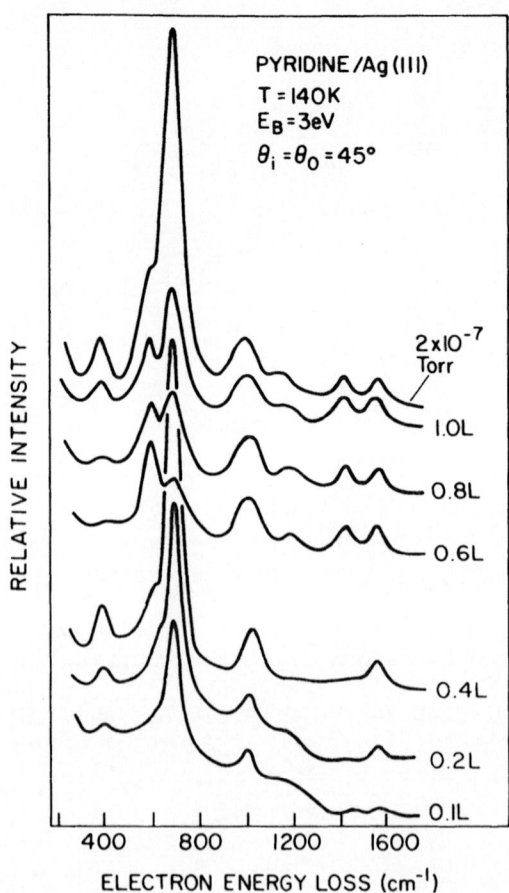

PYRIDINE /Ag (III)
T = 140K
$E_B = 3eV$
$\theta_i = \theta_0 = 45°$

Figure 3 – Coverage dependence of the vibrational spectra of
chemisorbed pyridine (< 1.0 L exposure). The spectra
taken at a 2×10^{-7} Torr ambient pressure corresponds to
the onset of condensation and the formation of the first
physisorbed layers.

suppressed after the phase change. With this phase change the in-
plane, CH-deformation vibration at 610 cm^{-1} (560 cm^{-1}) becomes more
intense. From our angle-dependent measurements we find that these
forementioned in-plane and out-of-plane modes are dominated by
dipole scattering in the specular direction which thereby permits
their use in a structural analysis. As a result of the "surface
selective rule"[28], the relative intensities of these in-plane and
out-of-plane vibrations indicate that the molecule becomes more
inclined to the surface above 0.5 L exposure. Based upon our UPS

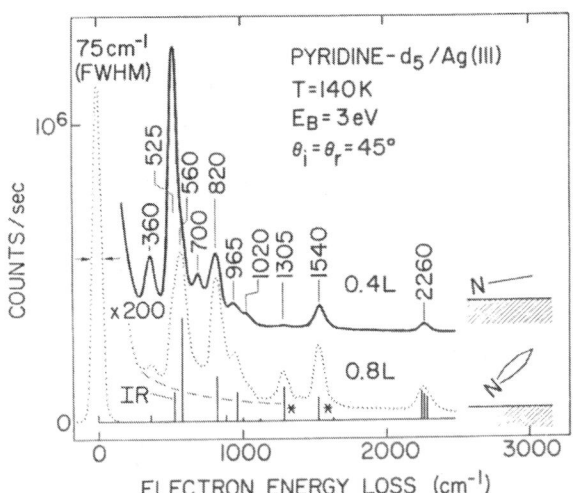

Figure 4 - Vibrational loss spectra of deuterated pyridine on
          Ag(111) at coverages above and below the compressional
          phase transformation. The IR absorbances[24,25] are shown
          below for comparison where the ★ levels are reduced by
          1/3. The insert besides each spectra schematically re-
          presents the geometry characteristic of each phase (see
          text).

results[8,26] as well as other chemical arguments and results[15,16],
the inclined phase of pyridine has the nitrogen end of the
molecule directed into the surface. Further the in-plane, CH-deformation vibration at 1440 cm^{-1} (1305 cm^{-1}) which is polarized largely
perpendicular to the C_{2v} symmetry plane of the molecule[27] increases
in intensity relative to the in-plane deformation mode at 610 cm^{-1}
which is symmetric about the C_{2v} symmetry plane. This implies that
a rotation also occurs about the inclined C_{2v} symmetry axis of the
molecule after this phase transformation occurs.

From a quantitative analysis of the such dipole scattering

intensities relative to the IR intensities of liquid pyridine[25] ,
we have determined that the low coverage phase (< 0.4 L) lies
flat on Ag(111) to within 5° and the high coverage chemisorbed
phase (> 0.6 L but < 1.0 L) lies inclined ~ 55° to the surface
and rotated ~ 30° about the molecular C_{2v} symmetry axis. Unlike
a previous analysis of EELS data to obtain quantitative geometric
information[29] , our analysis does not depend on evaluating dynamic
charges for which the detailed transmission function of the analyzers must be evaluated. Instead our analysis relies on analyzing
relative intensities of close-lying vibrational losses where we
can reasonably assume that our analyzer transmission function is
constant. Thus, our EELS results provide evidence that *chemisorbed*
pyridine initially π-bonds to Ag(111) for exposures < 0.5 L and at
higher coverages forms a compressed structure which is nitrogen-
lone-pair bonded and inclined to the surface.

Our photoemission results are also consistent with these
conclusions for chemisorbed pyridine as we observe the π_1-orbital
derived ionization feature and loose the ionization feature having
nitrogen-lone-pair character when the molecules become   inclined[26]
Our TDS results show that this inclined phase is less strongly
bound to the surface than the π-bonded phase by ~ 2 Kcal/mole and
that even the more strongly bonded phase desorbs below ~ 210 K.
Interestingly, we also find evidence that the first layers of
condensed pyridine lie "flat" on top of the compressed high
coverage phase of chemisorbed pyridine (see Fig. 3).

Our EELS results for benzene adsorption on Ag(111) indicate
only one chemisorbed phase. The EELS spectra of chemisorbed
benzene is shown in Fig. 5 where we also show the IR absorbances
for liquid benzene[25]. EELS results have previously been presented
for benzene on Ni(111) and Pt(111)[14] and also have been discussed
from a point-group-symmetry theoretic viewpoint[30]. Unlike these
previous results on transition metals the vibrations of chemisorbed
benzene on Ag(111) are remarkably similar to liquid benzene and
allow a unique opportunity to directly relate group theoretic
concepts to EELS results.

From the relative intensities of the 1480 cm^{-1} or 3030 cm^{-1}
in plane CH deformation or stretching vibration, and the out-of-
plane CH-deformation vibration at 675 cm^{-1} we conclude that benzene
lies π-bonded, flat on the surface. The losses observed at 675,
1000, 1155 and 3030 cm^{-1} can become (dipole) excited if the
molecular symmetry is lowered to the $C_{3v}(\sigma d)$ point group from the
D_{6h} point group of the free molecule (see ref. 30). These vibrations correspond to the a_1 modes which should be the *only* modes
(dipole) excited under the surface selection rule[28] . The two
weak modes at 1360 cm^{-1} and 1480 cm^{-1} must also be explained. The
1360 cm^{-1} loss may arise from an "impurity" (?) in our benzene as
it is commonly observed in the IR spectrum[25], or *more likely* to the

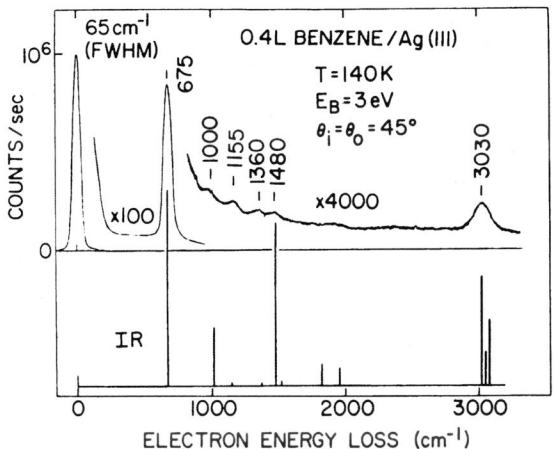

Figure 5 - Vibrational loss spectra for benzene on Ag(111) and the
corresponding IR absorbances[24]. Note that the peak at
1038 cm^{-1} is the ν_{14}, e_{iu} mode which *does not* correspond
to the 1000 cm^{-1} loss feature. The 1000 cm^{-1} IR peak is
very weak and is *not* observable on this scale.

second harmonic of the intense 675 cm^{-1}, loss feature. The very
intense 1480 cm^{-1}, peak in liquid benzene corresponds to an in-plane
vibration and may arise for chemisorbed benzene from "incomplete"
surface screening of the image charge by the substrate (i.e. a
weakening of the surface selection rule) or the occurrence of
quadruple excitations. Fortunately, the 1155 cm^{-1} loss is unique to
the $C_{3v}(\sigma d)$ point group, cannot be resonantly excited in the free
molecule[31] and from our angle-dependent studies appears to arise
from dipole scattering. This mode is also extremely weak in liquid
benzene[25] and would account for its weak intensity here. Thus,
although many of the loss features which delineate the point group
symmetry of adsorbed benzene are weak and of comparable intensity
to resonance scattering features and incompletely screened parallel
modes, we find the observed vibrations to be consistent with a
$C_{3v}(\sigma d)$ point group symmetry. Such questions regarding the excitation
mechanism or the totality of the surface selection rule underlay

applying group theoretic concepts to determining adsorption site
symmetry particularly if the new vibrations associated with the
reduced symmetry are weak. Unfortunately, one does not know *a*
priori what the intensities of such features will be.

RAMAN SCATTERING RESULTS

Enhanced Raman scattering signals were observed at surface
plasmon-polariton resonance[4,31], which was obtained by varying
the orientation of the incident radiation with respect to $\vec{K}_s$.
For the data presented here, we have used p-polarized incident
radiation and oriented the sample so that $\vec{K}_s$ is in the plane of
incidence. The angle of incidence was set to the minimum in intensity of the direct reflected beam, which corresponds to maximal
surface plasmon-polariton excitation. As expected, there was no
Raman scattered signal observed from the flat (control) portion
of the sample or when plasmon resonance conditions were not achieved. With present signal to noise ratios and sampling times we
would need an enhancement factor for chemisorbed pyridine of
~5 x 10^2 to observe any Raman signals above our noise levels.

The features found to be observable in the Raman spectrum
for chemisorbed pyridine on our "clean" modulated Ag(111) surface
occur between 850 and 1050 cm^{-1} as shown as a function of exposure
in Fig. 6a. Compared to the liquid phase spectra, for which the
symmetric (991 cm^{-1}) and the asymmetric (1030 cm^{-1}) ring breathing
modes are of about equal intensity, these spectra show selective
enhancement of the symmetric breathing mode for chemisorbed pyridine.
At higher exposures (~ 20 L) the asymmetric ring-breathing mode
becomes observable and at yet higher exposures (> 44 L), characteristic of very thick layers of pyridine, we start to observe additional Raman scattering with similar relative intensities as liquid
pyridine.

For chemisorbed pyridine on "clean" Ag(111), the carbon ring
deformation modes in the 1300 to 1600 cm^{-1} range could not be detected. Such features may be masked by the broad peaks at 1350 and
1500 cm^{-1} due to trace amounts of graphitic carbon[33,34]. Such trace
amounts of graphitic carbon always occurred in RS but were undetect-
able by AES. Also, prior to complete annealing, we observed an
additional peak at 986 cm^{-1} which we associate with pyridine bound
to step sites. Thus, we attribute the 990 cm^{-1} peak to chemisorbed
pyridine on Ag(111) which is consistent with our EELS results. We
have been unable to observe the CH stretching modes (near 3040 cm^{-1})
for chemisorbed pyridine.

In Fig. 6b we show Raman spectra obtained when greater amounts
of graphitic carbon were observed in Raman but were still undetect-
able by AES. (As described elsewhere graphitic carbon has a very
large Raman cross section and can be readily detected in small

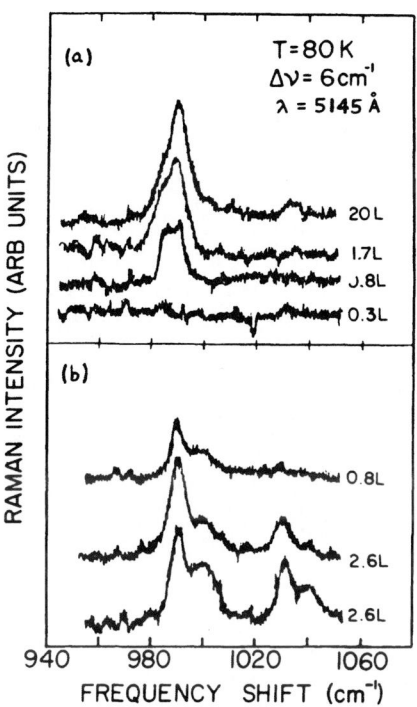

Figure 6 - Raman scattering signal as a function of increasing
          coverage for two conditions : (a) the nominally clean
          topologically modulated Ag(111) surface and (b) with
          increased carbon contamination (see text).

quantities ($\sim$ .01 monolayers) at plasmon-polariton resonance conditions[34]). For the 0.8 L and first 2.6 L spectra the graphitic carbon
Raman signal is $\sim$ 4 x larger than our nominally clean Ag(111) surface
(spectra shown in (a)) while in the second 2.6 L exposure they are
10 x more intense. As shown in Fig. 6b the presence of graphitic
carbon leads to additional Raman peaks at 1005 cm^{-1} in the chemisorption regime and at 1040 cm^{-1} for thicker condensed layers. These
extra peaks are likely associated with pyridine interacting with the
graphitic carbon but we cannot completely exclude the possibility
of an overall degradation of the grating surface topography associated with long sputtering which was attempted (unsuccessfully) to
reduce the graphitic carbon. If the former is correct, then the

Raman enhancement for pyridine-graphite/Ag is at least an order of
magnitude higher than for pyridine on Ag. Although the occurrence of
such impurities and their effect on RS are clearly important, we
restrict further discussion to our RS studies on the "clean" Ag(111)
surface.

Using our UPS results we have calibrated the relative coverages
of pyridine on clean Ag(111) from the intensity of the adsorbate-
derived ionization features and determine the coverage at which
chemisorption stops and adsorption on top of this chemisorbed layer
begins. This delineation between chemisorbed and physisorbed species
is important, and is based upon the occurrence of different relaxation effects in UPS for the chemisorbed versus physisorbed molecule
as described elsewhere[8,35]. Using coverage dependent Raman spectra
as in Fig. 6a we determine the coverage dependence of the 990 cm^{-1}
peak which is shown in Fig. 7. Here, we have calibrated our relative

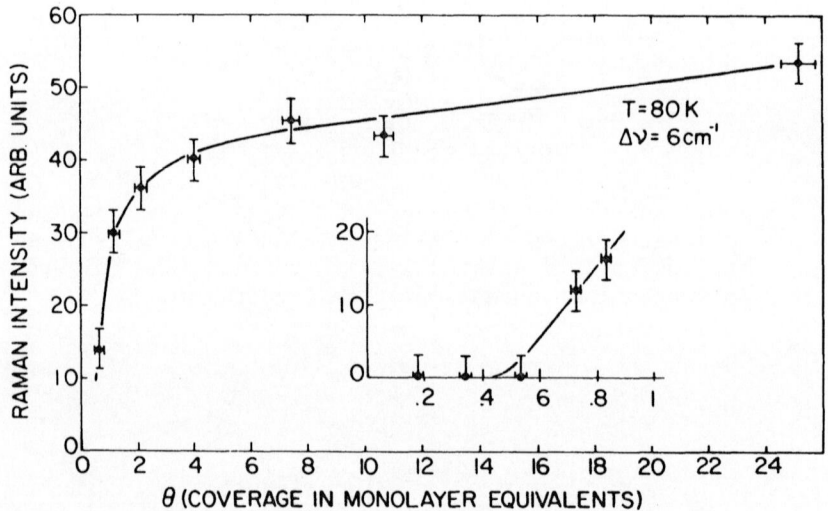

Figure 7 - Raman scattering intensity for the 990 cm^{-1} peak as a
function of coverage (in monolayer equivalents - see
text). The insert shows the detailed low coverage
behaviour (λ = 5145 Å).

coverages in terms of monolayer equivalents – the coverage at
which chemisorption is complete and condensation begins. We observe a strong signal starting about ~ 0.6 monolayer which increases less strongly above ~ 1 monolayer. Estimating the saturation coverage of pyridine to be ~ 5 x 10^{14} molecules/cm^2 , as
based on EELS results[26], and from comparisons to our Raman measurements on a known volume of liquid pyridine, we determine an
enhancement for the first monolayer of ~ 10^4 . At coverages above
~ 4 monolayers the enhancement per each additional monolayer remains
at ~ 10^2 .

These coverage dependent results are in agreement with a
theoretical model by Kirtley, Jha and Tsang[18] which predicts two
mechanisms contributing to the surface enhanced Raman scattering
process for a molecule adsorbed on a sinusoidal grating. First,
there is a long-range contribution, which extends several thousand
Angstroms away from the surface, and is due to enhancement of the
direct scattering intensity caused by the large electric field at
surface plasmon-polariton resonance. This field enhancement is
predicted by the theory to provide an increase in the Raman signal
by a factor of 10^2 to 10^4 . The second contribution is associated
with a short-range mechanism, which is very localized to the surface.
This enhancement arises from a modulation of the large oscillating
charge density in the molecular layer at surface plasmon-polariton
resonance, by the molecular vibrations, so as to produce Stokes
shifted light at this modulation frequency. Here the surface
plasmon-polariton is the intermediate state of a resonance Raman
process. At atomic distances, this theory predicts a total combined
Raman enhancement factor of between 10^4 and 10^6 .

Our Raman scattering results for benzene show a different
behaviour than observed for pyridine. Namely, we do not observe
any detectable signal until we have ~ 8 monolayer equivalents of
benzene on the surface. At higher coverages we then start to
observe a Raman spectra with similar relative intensities as in
liquid benzene. The multilayer Raman signal is enhanced by a factor
of ~ 10^2 as found for condensed pyridine. We again associated this
with a field enhancement. We do not observe any enhancements > 10^2
for chemisorbed benzene, a point we discuss later.

DISCUSSION

Our Raman results show a strong (~ 10^4) short-range enhancement
for certain modes of chemisorbed pyridine which is ~ 50 times
stronger than the expected long-range field enhancement. Our discussion here focuses on the factors which we can relate to this
strong, short-range phenomena. In particular the similarities in
chemisorption behaviour of pyridine on Ag(111) and on our topologically modified Ag(111) surface encourages us to consider a direct
comparison of the chemisorption results on these two surfaces.

Namely, each adsorbate shows ionization levels and coverage dependent features in UPS which are similar on both surfaces, while pyridine additionally shows the onset of the Raman signal where the structural phase transformation of pyridine occurs on the ideal Ag(111) surface. Such common features in the behaviour of these adsorbates on the two surfaces imply that (a) large regions of Ag(111) surface occur on the "modulated" surface and (b) that defects do not totally disrupt the nature of adsorption on these "modulated" surfaces.

We find no evidence to conclude that defect sites on our modulated surface or the terrace edge atoms are themselves dominating the strong short-range enhancement. In fact, the lack of this enhancement off surface plasmon-polariton resonance or on similar step density but nonmodulated Ag surfaces[5] strongly argues against these as the sole source of the enhancement. Further, Ag ad-atom sites which have been proposed as enhancement sites in other studies disappear above room temperature[6,7,10]. Our annealing of the sample well above room temperature would likely preclude such ad-atom sites on our surface. Indeed our observation of a sudden "turn" on of the Raman signal at higher chemisorption converges is contrary to the initially large signal observed on surfaces where ad-atoms are thought to play a rôle[9]. We thereby believe that the short range enhancement we observe (above our limits of sensitivity) is associated with the more general nature of our modulated surface.

We can consider a correlation of the adsorption properties of pyridine on the flat and modulated surface based on the "average" microscopic structure of each. Namely from our LEED results on the topologically modulated surface we expect a distribution of Ag(111) with an average terrace width of ~ 20 Å. In Fig. 8 we illustrate the atomic size of such a terrace and show how pyridine may π-bond to such a surface. Even considering the possibility that the pyridine molecules may bond differently at the step or kink sites of the terrace (i.e., the left side of Fig. 8) additional space remains on the terrace for adsorbed pyridine molecules to interact with a (111) surface and their neighbouring molecules. We expect to observe largely the same coverage-dependent interactions on these (111)-terraces as found on Ag(111). We thereby associate the onset of the Raman signal in Fig. 7 with the inclination of chemisorbed pyridine and the occurrence of nitrogen lone-pair-bonding. The molecules located at the step or kink sites of the surface may also be affected during this transition as we observe the 986 cm^{-1} peak (see Fig. 6a). We also find that adsorption on these stepped Ag surfaces is reversible - unlike what is observed on the more reactive stepped transition metal surfaces[36].

Another important feature of our enhanced Raman scattering for chemisorbed pyridine is the mode selectivity. Namely, the symmetric ring breathing mode is selectively enhanced while for liquid or thick condensed layers of pyridine the symmetric and asymmetric

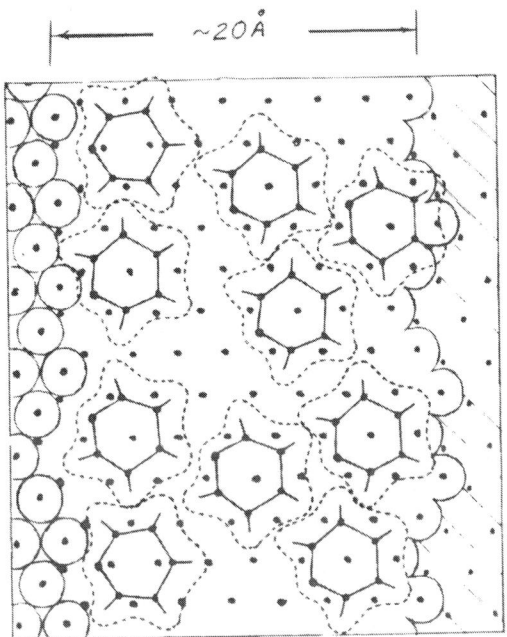

Figure 8 - Schematic model of the topologically modulated, stepped
           Ag(111) surface used in the Raman experiments. An
           average terrace width of 20 Å is determined from the
           LEED results (Fig. 1). The adsorbed pyridine molecules
           are shown to scale, bound to some arbitrary but largely
           constant local bonding sites.

modes are of comparable intensities. Any model of the enhancement
mechanism must account for such mode selectivity.

    For chemisorbed benzene, the lack of a detectable signal can
be attributed to a lower enhancement factor which, by virtue of
comparison to pyridine, we associate with the flat lying, π-bonded
molecule. Namely, we might expect to see chemisorbed benzene since
the Raman intensities of the ~ 990 cm^{-1} modes of liquid benzene
and pyridine are comparable. However, as found for pyridine, the

enhancement becomes large only when the molecule becomes inclined
to the surface.

Several origins of the enhancement mechanism have been proposed [37] and we briefly comment on their consistency with our
results. One can generally consider two related sources for the
short-range enhancement : an electronic effect (e.g., a localized
electronic excitation associated with chemisorption) and a physical effect (e.g., the orientation of the electric fields and the
molecules on the surface). In relation to the former process we
note that the symmetric-ring breathing mode is precisely the mode
excited in resonant low energy electron scattering from gaseous
pyridine and benzene[31,38]. Such a resonant state may couple to
surface plasmon-polaritons and directly or indirectly provide the
enhancement. However, this type of resonant state does not appear
to be a *sufficient condition* for the enhancement since we do not
observe a strong, short-range enhancement for chemisorbed benzene
*even though we do observe the gas phase resonances for chemisorbed
benzene* in EELS[39]. Further frequency dependent studies may help
to resolve the likelihood of this process.

The physical orientation of the molecule in the surface plasmon-polariton induced fields, with or without an image charge
component[39], also fails to account for our results. Namely, a field
enhancement only accounts for at most 3 % of the observed signal
while an image field enhancement mechanism would predict comparable
Raman intensities for both symmetric and asymmetric ring breathing
modes. As previously mentioned a localized enhancement associated
with bonding to ad-atoms on our silver surface would seem unlikely
in view of our relatively severe annealing conditions. However,
recent studies of pyridine on continuous and island Ag films
evaporated at low temperature[9,10] also show preferential enhancement of the symmetric ring breathing mode.

Our experimental results appear to be consistent with the
short-range surface plasmon-polariton enhancement mechanism proposed by Kirtley, Jha and Tsang[18]. Their enhancement contains both
electronic and structural features. Namely,(a) certain molecular
vibrations couple to the surface charge according to the nature of
the bondig and (b) the final surface charge modulation can only
occur *normal to the surface* and is thereby dependent on the
orientation of the molecule on the surface. Thus, for pyridine we
can relate the mode selectivity to the fact that the asymmetric
ring breathing mode does not strongly modulate the surface charge
density whereas the symmetric mode does. Such adsorbate-induced
surface charge modulation at the surface likely occurs for π-bonded
species as well but may be weaker or negated by the binding geometry or site symmetry. The details of such selection rules for
surface plasmon-polariton enhanced Raman scattering are discussed
in more detail elsewhere[41].

SUMMARY AND CONCLUSIONS

We have used high resolution electron energy loss spectroscopy to determine the vibrations and molecular orientation of chemisorbed pyridine and benzene on Ag(111). Chemisorbed benzene is π-bonded having a point group symmetry of $C_{3v}(\sigma d)$. Pyridine also π bonds but undergoes a compressional phase transformation to form a more weakly bound, inclined, nitrogen-lone-pair bonded phase of chemisorbed pyridine. We can relate these findings to Raman scattering results on a clean Ag(111)-oriented surface. We observe that the vibrations observed in *both* experiments on clean Ag(111) are very weakly perturbed from those of the free molecule. The symmetric ring breathing mode is within 3 cm^{-1} of its liquid phase value (RS for pyridine and EELS for d$_5$-pyridine). We also observe other frequencies for this mode in Raman which can be attributed to bonding at step sites (lower frequency) or to the presence of surface impurities (higher frequencies).

We find a strong structure dependence of our enhanced surface Raman signal for chemisorbed pyridine on clean Ag(111). Namely, we observe a ~ 10^4 enhancement of the 990 cm^{-1} mode when chemisorbed pyridine becomes inclined to the surface. For thicker layers of benzene or pyridine we observe little structural dependence and a factor of 10^2 weaker, longer-range field enhancement characteristic of randomized molecular orientations within this layer. We also observe a strong mode selective enhancement for chemisorbed pyridine. The observed enhancement occurs only at surface plasmon-polariton resonance conditions and appears to be well described within a surface plasmon-polariton model[16]. Namely, both the mode selectivity and distance dependence of the enhancement we observe are consistent with the modulated surface dipole model of surface enhanced Raman scattering. Our results imply that only certain molecular vibrations and molecular orientations of a chemisorbed species on the surface will provide a large surface plasmon-polariton derived enhancement.

ACKNOWLEDGEMENT

The authors wish to acknowledge useful discussions with J.R. Kirtley and S.S. Jha. We also thank J.A. Bradley for his technical assistance.

REFERENCES

1. M. Fleishmann, P.J. Hendra, A.J. McQuillan, Chem. Phys. Lett. 26, 163 (1974), D.L. Jeanmarie, R.P. Van Duyne J. Electroanal. Chem. 84, 1 (1977); B. Pettinger, U. Wenning, Chem. Phys. Lett. 56, 253 (1978).

2. C.Y. Chen, E. Burstein, S. Lundqvist, Solid State Commun. 32, 63 (1979).

3. J.A. Creighton, C.G. Blatchford, M.G. Albrecht, J. Chem. Soc., Faraday Transactions II 75, 790 (1979); M. Kerker, O. Siiman, L.A. Blum, D.W. Wang, Applied Optics XX, XXX (1980).

4. J.C. Tsang, J.R. Kirtley, J.A. Bradley, Phys. Rev. Lett. 43, 772 (1979).

5. J.E. Rowe, C.V. Shank, D.A. Zwemer, C.A. Murray, Phys. Rev. Lett. 4, 1770 (1980); D.A. Zwemer, C.V. Shank, J.E. Rowe, Chem. Phys. Lett. 73, 201 (1980).

6. T.A. Wood, M.V. Klein, J. Vac. Sci. Technol. 16, 459 (1979).

7. R.R. Smardzewski, R.J. Colton, J.S. Murday, Chem. Phys. Lett. 68, 53 (1979).

8. P.N. Sanda, J.M. Warlaumont, J.E. Demuth, J.C. Tsang, K. Christmann, J.A. Bradley, to be published.

9. I. Pockrand, A. Otto, to be published.

10. H. Seki, M. Philpott, J. Chem. Phys., to be published.

11. M. Moskovits, J. Chem. Phys. 69, 4159 (1978).

12. B. Pettinger, M.R. Philpott, J.G. Gordon, III, J. Chem. Phys. XX XXX (1980).

13. M. Fleishman, private communication.

14. S. Lehwald, H. Ibach, J.E. Demuth, Surface Sci. 78, 577 (1978)

15. B.J. Bandy, D.R. Lloyd, N.V. Richardson, Surface Sci. 89, 344 (1979).

16. F.P. Netzer, E. Bertel, J.A.D. Matthew, Surface Sci. 92, 43 (1980).

17. C.S. Allen, R.P. Van Duyne, Chem. Phys. Lett. 63, 455 (1979).

18. J.R. Kirtley, S.S. Jha, J.C. Tsang, Solid State Comm. 35, n° 7 (1980); S.S. Jha, J.R. Kirtley, J.C. Tsang, Phys. Rev. B., Oct. 15 (1980).

19. J.E. Demuth, Surf. Sci. 69, 365 (1977).

20. J.A. Simpson, C.E. Kuyatt, Rev. Sci. Inst. 38, 103 (1967).

21. H.J. Levinstein, W.H. Robinson, J. Appl. Phys. 33, 3149 (1962).

22. P.S. Maiya, J.M. Blakely, Appl. Phys. Lett. 7, 60 (1965).

23. This corresponds to the gauge-correction factor for benzene as supplied with our Varian ion gauge.

24. L. Corrsin, B. Fox, R.C. Lord, J. Chem. Phys. 21, 1170 (1953).

25. D.J. Pouchart, Aldrich Library of Infrared Spectra (Aldrich Chem. Co., Wisconsin, 1975).

26. J.E. Demuth, K. Christmann, P.N. Sanda, Chem. Phys. Lett., to appear.

27. D.A. Long, E.L. Thomas, Trans. Far. Soc. 59, 783 (1963).

28. D. Sokcevic, Z. Lenac, R. Brado, M. Sunjiç, Z. Phys. B 28, 273 (1977).

29. H. Ibach, H. Hopster, B. Sexton, Appl. Surface Sci, 1, 1 (1977).

30. N.V. Richardson, Surface Sci. 87, 622 (1979).

31. S.F. Wong, G.J. Schulz, Phys. Rev. Lett. 35, 1429 (1975).

32. A. Girlando, M.R. Philpott, D. Heitmann, J.D. Swalen, R. Santo, J. Chem. Phys. 72, 5187 (1980).

33. M.R. Mahoney, M.W. Howard, R.P. Cooney, Chem. Phys. Lett. 71, 59 (1980).
34. J.C. Tsang, J.E. Demuth, P.N. Sanda, J.R. Kirtley, Chem. Phys. Lett., to appear.
35. J.E. Demuth, D.E. Eastman, Phys. Rev. Lett. 32, 1123 (1974); J. Vac. Sci. and Technol. 13, 283 (1976).
36. G.A. Somorjai, J. Vac. Sci. and Technol. 1, 250 (1974).
37. See for example, E. Burstein, C.Y. Chen and S. Lundqvist in Proceedings of Joints US-USSR Symposium on the Theory of Light Scattering in Condensed Matter, eds. J.L. Birman, H.Z. Cummings and H.K. Reband (Plenum Press, NY, 1980) p. 479.
38. I. Nenner, G.J. Schulz , J. Chem. Phys. 62, 1747 (1975).
39. J.E. Demuth, to be published.
40. F. King, R.P. Van Duyne, G.C. Schulz, J. Chem. Phys. 69, 4472 (1978).
41. J.E. Demuth, P.N. Sanda, J.R. Kirtley, J.C. Tsang, S.S. Jha, to be published.

RAMAN SPECTROSCOPY OF THIN ORGANIC FILMS BY INTEGRATED OPTICAL

TECHNIQUES

J.D. Swalen and J.F. Rabolt

IBM Research Laboratory

SAN JOSE, California 95193

ABSTRACT

   Vibrational spectroscopy of molecules at the surfaces of thin
films has become increasingly important for surface studies.
Enhancement mechanisms are however needed to improve the signal to
noise ratio in order to observe these surface molecules. To this
end integrated optical techniques have been utilized to measure
the resonant and nonresonant Raman spectra of some organic molecules
in thin films ranging in thickness from 3 to 80 nm. These films
were deposited on glass optical waveguides and laser radiation was
coupled into the layered structure by prism coupling techniques. The
streak of light in the waveguide was focused onto the slit of a
JY RAMANOR HG-2S double monochromator to collect the Raman scattered
light. The increased optical intensity and scattering volume (the
optical ray makes the order of a thousand reflections) enabled us
to obtain high S/N spectra, therefore making this technique very
generally applicable to the study of organic and polymeric thin
films. Our non-resonant Raman results on thin films of poly(styrene)
and our resonant Raman results on a Langmuir-Blodgett monolayer of
squarylium and cyanine dyes will be presented as well as details
of the method.

INTRODUCTION

   The vibrational spectra of molecules on a surface can be useful for understanding molecule-surface interactions. Frequency
shifts reflecting changes in force constants as well as intensity
variations resulting from modification of the wavefunction by the
surface are expected. Raman measurements of thin films, however,

have been a challenge to the experimentalist because of the low
signal intensity and generally some means of enhancement are necessary. Consequently ATR techniques[1],low angle scattering from a
metal surface[2-4] and surface plasmon measurements have been used
to obtain infrared spectra. In Raman experiments ATR techniques[6,7],
surface plasmon methods[8,9], and optical waveguides[10-12] have been
used to enhance the signal. In this work we describe some new
results obtained with the optical waveguide technique were the
optical field strength is increased without reducing the number of
scattering molecules. In fact in the case of non-resonant Raman
experiments integrated optics have been used to increase the
scattering volume thereby allowing spectra of very thin films to
be obtained.

EXPERIMENTAL

The waveguide method for visible spectroscopy has been described in detail[13,14] and consists of coupling the laser light by
means of a prism into a thin film deposited on a substrate. The
light travels in the film and interacts with molecules in the
film or on its surface by either the evanescent wave or an extension
of the guided wave. In Figure 1 a schematic of the experimental
arrangement is shown with the streak of light in the guide being
focused onto the slit of a double monochromator. Figure 2 shows a
photograph of an experiment in progress illustrating both the
focusing and collection optics utilized. Additional details of the
method can be found in references 11 and 12.

RESULTS

In order to illustrate the sensitivity of this integrated
optical technique Raman spectra of a thin polymer film and a
Langmuir-Blodgett monolayer have been obtained. In Figure 4 is
shown the Raman spectrum of a thin film of poly(styrene) on a
sputtered glass waveguide of Corning 7059 glass, the substrate
being pyrex glass. Note that the refractive index of the optical
waveguide must be greater than either the substrate or superstrate.
The overlayer of molecules can have a higher, lower or identical
refractive index as the Corning 7059. For the case of molecules
having a higher refractive index, poly(styrene) being in this
category, the overlayer increases the guiding layer thickness. When
the molecules have a lower index the scattering is accomplished by
the evanescent wave. In Figure 3 the optical electric field intensity
is plotted for the m = 1 mode, i.e., the field configuration will
have one node. Notice that the optical field extends well into both
sub- and superstrates. It is this field which increases the Raman
scattering intensity by two orders of magnitude. Poly(styrene) is
a good Raman scatterer and the 1005 cm^{-1} band involving the C-C

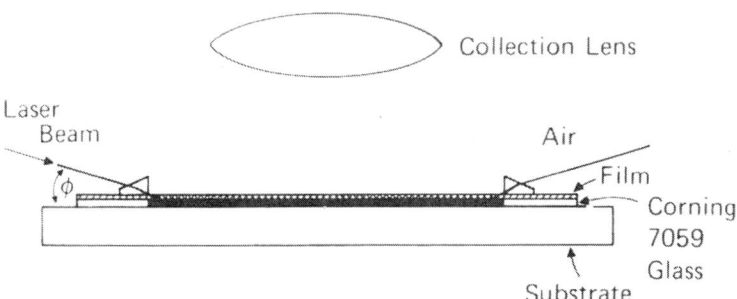

Figure 1 - Schematic diagram of the Raman experiment.

Figure 2 - Photograph of the actual experimental apparatus for recording the Raman spectrum of thin films with the integrated optical technique.

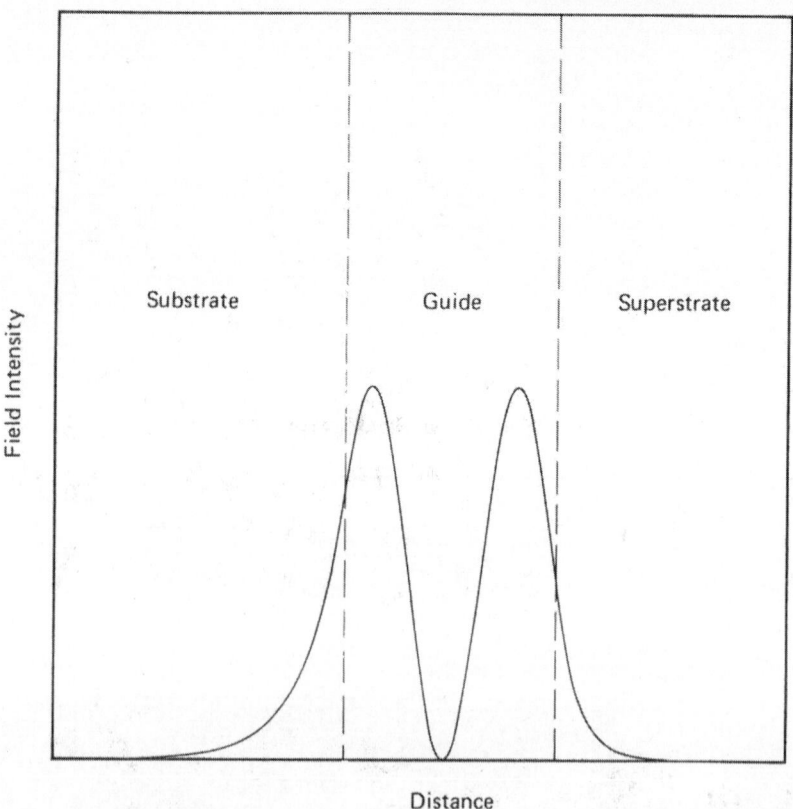

Figure 3 - Optical field intensity vs. distance in the layered
          waveguide structure.

ring breathing mode has a good signal to noise ratio after an
accumulation of 120 scans. Based on the achieved S/N it can be
estimated that a lower practical limit of film thickness would be
approximately 20nm. Since even this film would be dominated by
bulk molecules instead of surface molecules, we saw no evidence
for molecule-surface interactions.

        In order to observe influences of a surface interaction,
monolayers of organic compounds were studied and spectra obtained
for a cyanine dye, the structure of which is shown in Figure 5.
This compound was deposited on a glass optical waveguide by the
Langmuir-Blodgett technique and the Raman spectrum recorded with
the 476.5 nm line of an Ar$^+$ ion laser. This wavelength was chosen
to be at the edge of the visible absorption profile of the dye so
as to minimize absorption of the laser energy but still give rise

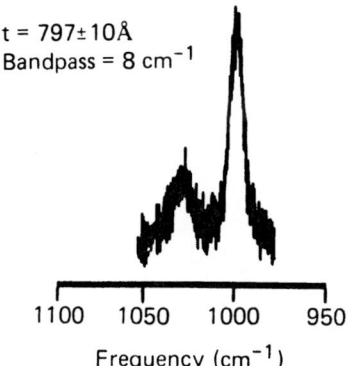

t = 797±10Å
Bandpass = 8 cm⁻¹

Figure 4 - Raman spectrum (950 - 1100 cm⁻¹)of a 70 nm film of
poly(styrene) (resolution = 8 cm⁻¹ , 120 scans).

to resonant Raman scattering from the dye chromophore. In Figure 6
the spectrum for bulk (a), bilayer (b), and monolayer (c) are
compared. The vibrations at 1366 and 1392 cm⁻¹ are attributable to
the C-C and C-N stretch modes of the quinoline moiety. The monolayer bands are shifted from those in the bulk indicating a molecule surface interaction, probably due to hydrogen bonding with the
OH groups of the glass surface (Si-OH). In the case of the bilayer
two chromophores interact with one another giving rise to a
distinctly different band profile.

For comparison a monolayer sample was produced with chromophore up on a pyrex slide. This was overcoated with a thin film of
poly(vinyl alcohol) which served as the waveguide. As with the
monolayer on the glass waveguide, this spectrum is perturbed in a

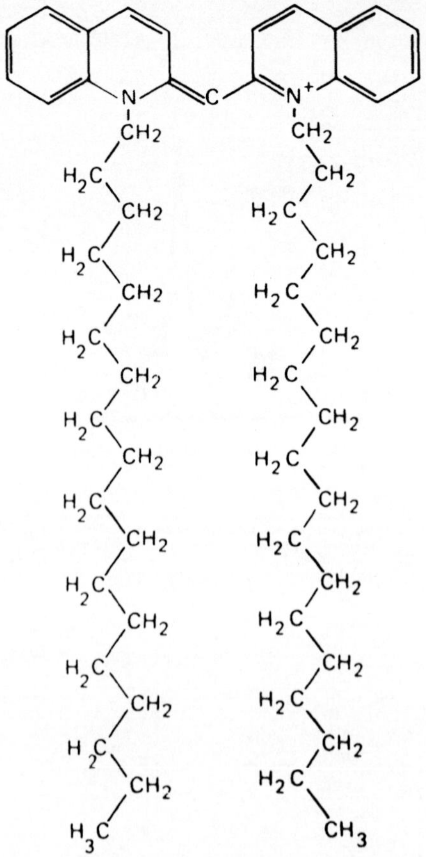

Figure 5 – Molecular formula of 1,1' dioctadecyl - 2,2' cyanine ion.

similar way but the polymer–molecule interaction is somewhat different. A third component is present and may result from a shift
in the band at 1392 cm^{-1} with a subsequent decrease in intensity
Here also the OH groups of the polymer could be hydrogen bonded to
the dye. This latter method shows that it is also possible to
overcoat the surface molecule with a thin polymeric film to observe
the interactions.

In conclusion it is our belief that the use of optical waveguides to obtain Raman spectra of thin layers is convenient and
generally applicable to a variety of surface problems. In particular
the polymeric overcoating with poly(vinyl alcohol) should provide
a useful means of investigating a large class of organic coatings.

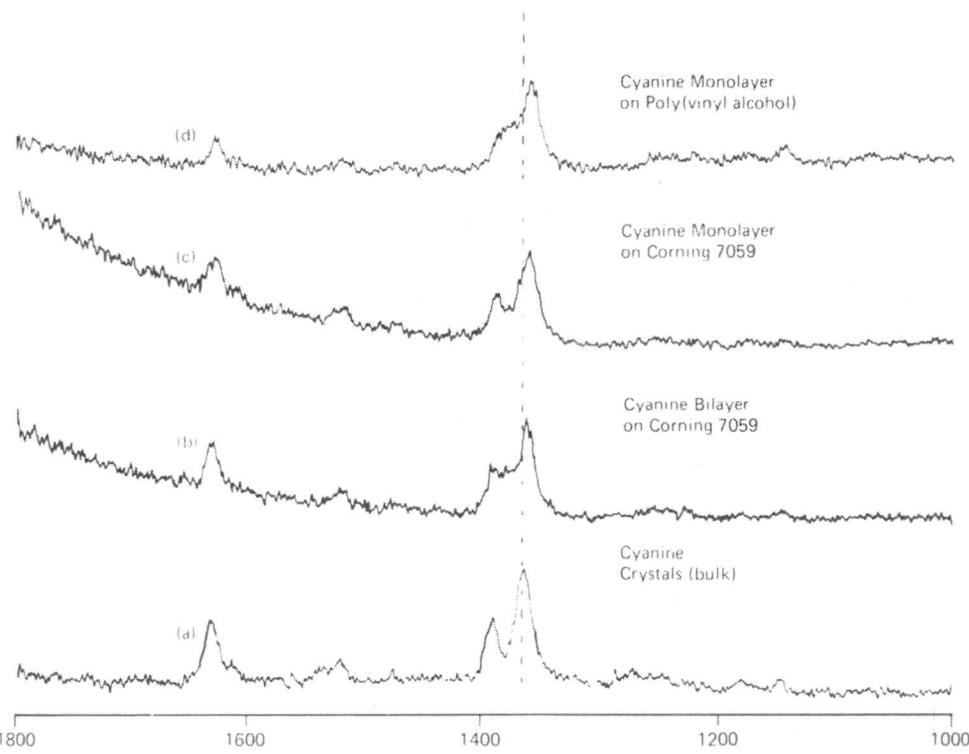

Figure 6 - Resonance Raman spectra of cyanine dye : (a) bulk
           crystals, (b) bilayer on Corning 7059 waveguide, (c)
           monolayer on Corning 7059 waveguide, and (d) monolayer
           under a poly(vinyl alcohol) waveguide (resolution =
           9 cm^{-1}, excitation at 476.5 nm and 20 to 30 scans).

REFERENCES

1. See for example, N.J. Harrick, International Reflection Spectroscopy, (Interscience Publishers, New York, 1967).
2. R.G. Greenler and J.L. Slager, Spectrochim Acta 29A, 193 (1973).
3. M.L. Howe, K.L. Watters, R.G. Greenler, J. Phys. Chem. 80, 582
   (1976).
4. D.L. Allara, A. Baca, C.A. Pyrde, Macromolecules 11, 1215 (1978).
5. A. Hjortsberg, W.P. Chen, E. Burstein, M. Pomerantz, Opt. Commun.
   11, 66 (1978).
6. J. Cipriani, S. Racine, R. Dupeyrat, H. Hasmonay, M. Dupeyrat,
   Y. Levy, C. Imbert, Opt. Commun. 11, 70 (1974).
7. M. Delhaye, M. Dupeyrat, R. Dupeyrat, and Y. Levy, J. Raman
   Spectrosc. 8, 351 (1979).
8. A. Girlando, M.R. Philpott, D. Heitmann, J.D. Swalen, R. Santo,
   J. Chem. Phys. 72, 5187 (1980).

9. A. Girlando, J.G. Gordon III, D. Heitmann, M.R. Philpott, H. Seki, J.D. Swalen, Proceeding of the conference on Non-Traditional Approaches to the Solid-Electrolyte Interface, Editors T. Furtak, K. Kliewer, D. Lynch, Snowmass, Sept. 1979, Surface Science (in press) .

10. Y. Levy, C. Imbert, J. Cipriani, S. Racine, R. Dupeyrat, Opt. Commun. 11, 66 (1974).

11. J.F. Rabolt, R. Santo, J.D Swalen, Appl. Spectrosc. 33, 549 (1979).

12. J.F. Rabolt, R. Santo, J.D. Swalen, Appl. Spectrosc. (in press).

13. J.D. Swalen, M. Tacke, R. Santo, K.E. Rieckhoff, J. Fischer, Helv. Chim. Acta 61, 960 (1978).

14. J.D. Swalen, J. Phys. Chem. 83, 1438 (1979).

ENHANCED RAMAN SCATTERING FROM MOLECULES ADSORBED AT METAL SURFACES

K. Arya, R. Zeyher

Max-Planck-Institut für Festkörperforschung

7 Stuttgart 80, Fed. Rep. Germany

ABSTRACT

A microscopic theory for surface enhanced Raman scattering from admolecules is presented. The enhancement of the cross-section due to the presence of the metal is mainly due to two sources : (a) the renormalization of the electron-photon interaction vertex mediated by the surface plasmon mode; (b) the shift and broadening of the electronic levels of the molecule due to chemisorption and charge transfer. Contributions (a) and (b) are calculated as functions of the incident photon frequency and of the model parameters appropriate for a silver surface.

In 1974 Fleischmann et al[1] reported a large increase in Raman cross-section from pyridine which has been adsorbed on a silver surface. By now this effect has been found for more than 70 molecules. Also many attempts have been made to understand the origin of this enhancement. Of particular interest is the fact that the enhancement has only been observed for Ag (enhancement $\sim 10^6$) and, for Cu,Au and Pt ($\sim 10^2 - 10^3$). Various theoretical models[2] have been proposed to explain the origin of this enhancement but so far no general agreement has been reached for the exact interpretation. For a review of the theoretical and experimental work see the recent articles by Furtak and Reyes[2] and by Otto[3].

In this paper, we outline a microscopic theory for the Raman cross-section from the admolecule and discuss also numerical results for the enhancement of the cross-section. It is shown that the presence of the metal changes the scattering from a free molecule in two ways : (a) by modifying the photon field and (b) by shifting and broadening the electronic levels of the molecule and by the possibility of the charge transfer between molecule and metal. First, we solve the Maxwell equation for the vector potential $\vec{A}(\vec{r},\omega)$ approximating the dielectric function of the semi-infinite metal (Z<0) by $\in(\omega) = 1 - (\omega^2/\omega^2)$ 0 (-Z), where ω_p is the bulk plasmon frequency of the metal. The solutions consist of surface plasmon (SP) as well as of the usual s- and p- modes[4] . The vector potential $\vec{A}(\vec{r},\omega)$ can thus be written as

$$\vec{A}(\vec{\gamma},\omega) = \sum_{\vec{Q}\lambda} \vec{B}_{\vec{Q}\lambda} (\vec{\gamma},\omega) (a_{\vec{Q}\lambda} + a^{\dagger}_{-\vec{Q}\lambda}) \tag{1}$$

where $a^{\dagger}_{\vec{Q}\lambda}$ $(a_{\vec{Q}\lambda})$ is the creation (annihilation) operator for the photon of mode λ (λ= SP, s, p). Explicit expressions for $\vec{B}_{\vec{Q}\lambda}(\vec{r},\omega)$ as well as for the eigen-frequencies $\omega(\vec{Q}\lambda)$ are given[4].

To obtain the scattering cross-section we use a polariton treatment similar to that of Mills and Burstein[5] in the case of in sulating crystals. In this approach, the cross-section is related to the imaginary part of the self-energy of a photon propagator. The infinite set of Feynman diagrams for the self-energy included in our theory is shown in Fig. 1a-1c. The wavy-line describes the photon propagator defined in terms of $\vec{A}(\vec{r},\omega)$. The dotted line denotes the vibration on the molecule. The double wavy-line, defined in Fig. 1b, describes a photon propagator renormalized by polar- ization effects in the coupled molecule-metal system. The thick and broken lines are electron propagators of the molecule-metal system. Figure 1c describes the renormalized electron-photon (e-p) vertex which appears on the two ends of Fig. 1a.

Analytically Fig. 1c corresponds to the following integral equation for the renormalized e-p vertex $V_{m1}(Q,i\omega_m)$:

$$V_{m1}(\vec{Q}\lambda,i\omega_m) = \alpha_{m1}(\vec{Q}\lambda) + \sum_{\substack{m'1' \\ m''1''}} V_{m''1''}(\vec{Q}\lambda,i\omega_m) R_{1m,1'm'}(i\omega_m)$$

$$+ \sum_{\vec{Q}'\lambda'} \alpha_{m1} (\vec{Q}'\lambda') \alpha_{1'm'}(-\vec{Q}'\lambda') D^{(o)}(\vec{Q}'\lambda',i\omega_m) \tag{2}$$

$$R_{1m,1'm'}(i\omega_m) = \beta^{-1} \sum_{i\omega_n} G_{11'}(i\omega_n) G_{mm'}(i\omega_n + i\omega_m) \tag{3}$$

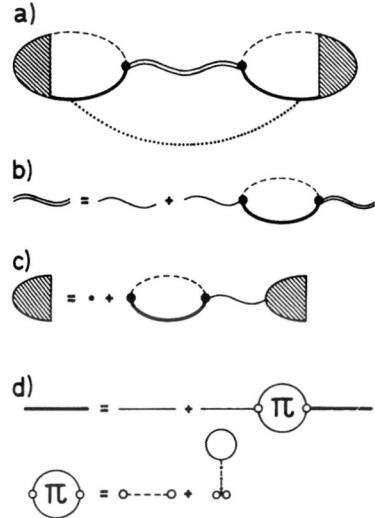

Figure 1 - (a) Self-energy of the photon propagator considered for
             Raman scattering,
          (b) Dyson equation for the photon propagator
          (c) Equation for the renormalized e-p vertex,
          (d) Anderson-Newns model of chemisorption; solid line :
             propagator for the admolecule level; dashed line :
             metal electron propagator; open circle : molecule-
             metal interaction; dash-dotted line : Hubbard
             interaction.

The indices m,l denote Wannier states which form a one-particle basis $\{\phi_m\}$ for the molecule-metal system. The bare e-p vertex α is defined by

$$\alpha_{ml}(\vec{Q}\lambda) = \langle \phi_m \mid - \frac{e}{mc} \vec{B}_{\vec{Q}\lambda} (\vec{\gamma}) \cdot \vec{P} \mid \phi_l \rangle \tag{4}$$

In case that λ = SP, the three-dimensional wave-vector $\vec{Q}$ should be replaced by its component parallel to the surface. The discrete frequencies ω_n and ω_m are given by ω_n = $(2n + 1)$ π/β and ω_m = $2m\pi/\beta$, respectively, where n and m are integers and β^{-1} is equal to $K_B T$. The photon propagator $D^{(o)}$ is given by

$$D^{(o)}(\vec{Q}\lambda, i\omega_m) = - 2\omega (\vec{Q}\lambda)/(\omega_m^2 + \omega^2(\vec{Q}\lambda)) \tag{5}$$

Equation (2) can be easily solved within the Anderson-Newns model for chemisorption[6] . The states $\{\phi_l\}$ then consist of one electronic state in the molecule and unperturbed electronic states on the metal. The corresponding propagators in Fig.1 are denoted by a thick and a broken line, respectively. The bubble in Fig. 1b or 1c describes then charge transfer transitions from the molecule to unoccupied metal states. In the Anderson-Newns model, the thick line is determined by Fig. 1d. The open circle denotes the molecule-metal interaction and the dash-dotted line denotes the Hubbard inter-action. Assuming that the molecule-metal interaction couples only the molecule with the nearest-neighbour metal atom, Eq. 3 reduces to

$$V(\vec{Q}\lambda, i\omega_m) = \alpha(\vec{Q}\lambda)/(1-X(i\omega_m)) , \tag{6}$$

$$X(i\omega_m) = R(i\omega_m) \sum_{\vec{Q}\lambda} |\alpha(\vec{Q}\lambda)|^2 D^{(o)}(\vec{Q}\lambda\ i\omega_m) , \tag{7}$$

$$R(i\omega_m) = \int_{-\infty}^{E_F} d\omega_1 \int_{E_F}^{\infty} d\omega_2 \frac{\rho_a(\omega_1) \rho_s(\omega_2)}{\omega_1 -\omega_2 + i\omega_m} , \tag{8}$$

where $\rho_a(\omega_1)$ is the spectral density of the adatom level and $\rho_s(\omega_2)$ is the local density of metal states at the nearest-neighbour metal atom. In Eqs. (6)-(8), we have omitted the redundant indices m,l.

In a similar way the equation for the renormalized photon propagator, Fig. 1b, can be solved. Using the diagrams shown in

Fig. 1a we finally obtain for the imaginary part of the self-energy
assuming Stokes scattering :

$$I_m \Sigma^{st}(\vec{Q}\lambda, \omega+i\delta) = -\pi \frac{|\alpha(\vec{Q}\lambda)|^2}{|1 - X(\omega+i\delta)|^2}$$

$$x \sum_{\vec{Q}'} \frac{|\alpha(\vec{Q}'\lambda')|^2}{|1 - X(\omega-\Omega+i\delta)|^2}$$

$$x |g T(-\omega-i\delta, \Omega)|^2 \delta(\omega-\Omega-\omega(\vec{Q}'\lambda'))), \qquad (9)$$

$$T(-\omega-i\delta, -\Omega) = -\int_{-\infty}^{E_F} d\omega_1 \int_{-\infty}^{E_F} d\omega_2 \int_{E_F}^{\infty} \frac{\rho_a(\omega_1) \rho_a(\omega_2) \rho_s(\omega_3)}{(\omega_1+\omega-\omega_3+i\delta)(\omega_2+\omega+\Omega-\omega_3+i\delta)} \qquad (10)$$

Ω is the frequency of the vibrational mode and g is the electron-
vibration interaction. In Eq. (9), we have performed the analytic
continuation from imaginary frequencies $i\omega_m$ to real frequencies
$\omega+i\delta$. Equation (9) describes the decay of the incident photon $\vec{Q}\lambda$
into a scattered photon $\vec{Q}'\lambda'$ and a vibration Ω. In our case λ and
λ' are equal to s or p.

The scattering cross-section for a free molecule is also given by
the diagrams of Fig. 1a. However, in this case, one has to consider
the free radiation field, i.e., plane waves for the $\vec{B}_{\vec{Q}\lambda}(\vec{r},\omega)$ in Eq.(1)
and no surface plasmon mode, and the true electronic levels of the
free molecule. The expression for $Im\Sigma(\vec{Q}\lambda, \omega+i\delta)$ is still given by
Eq. (9), however with the following modifications : the polariton
contribution $X(\omega+i\delta)$ is very small due to the absence of surface
plasmons and thus can be neglected; the $\alpha(\vec{Q}\lambda)$ reduces to a $\vec{p}$ matrix
element between the ground and excited states of the free molecule;
T has to be calculated with the eigenstates of the free molecule.
Taking an effective two-level model for the molecule and assuming
that the dipole matrix element is approximately equal to our matrix
element α, we obtain for the enhancement ρ of the cross-section

$$\rho(\omega) = \left[|\frac{1}{1 - X(\omega+i\delta)}|^2 |\frac{1}{1 - X(\omega-\Omega+i\delta)}|^2 \right.$$

$$\left. x |(E_2-E_1-\hbar\omega)^2 T(-\omega-i\delta, \Omega)|^2 \right. \qquad (11)$$

The energies of the two levels of the molecule are E_1 and E_2.

The enhancement $\rho(\omega)$ is caused by two different effects. The first one (first factor in Eq. 11)) is due to the renormalization of the e-p interaction vertex because of surface plasmons. The second one (last factor in Eq. (11)) is due to chemisorption : the ground state of the molecule shifts up near to the Fermi level and broadens. Furthermore, the excited states consist of a continuum of unoccupied electronic states on the metal in addition to the occupied states of the molecule.

We have calculated the two contributions to the enhancement $\rho(\omega)$ for the case of Ag. Silver has d band which is roughly 4 eV broad and which lies $\sim$ 4 eV below the Fermi level. The s-p bands are above the d-band and are about 10 eV wide (see Fig. 2). The positions of the levels of the free molecule are also shown in Fig. 2 and correspond roughly to CN^- or CO. The spectral function of the lower level of the admolecule is approximated by the semi-elliptical form[6]

$$\rho_a(\in) = \pi^{-1} \Delta_a(\in) / ((\in - \in_a - \Lambda_a(\in))^2 + \Delta_a^2(\in)) \tag{12}$$

$$\in_a = E_1 + U \int_{-\infty}^{\in_F} d\in \rho_a(\in) , \tag{13}$$

where $\Lambda_a(\in)$ and $\Delta_a(\in)$ are the real and imaginary parts of the self-energy π (see Fig. 1d) of the molecular level $\in_a$ and U is the Hubbard interaction. Equations (12) and (13) have to be solved self-consistently. For the local density of states on the metal we again use the elliptical form[6]

$$\rho_s(\in) = \frac{2}{\pi W} (1 - \frac{\in^2}{W^2})^{1/2} \tag{14}$$

where W is the half width of the s-p band. In calculating $X(\omega+i\delta)$ we find that only λ=SP gives a substantial contribution in Eq. (5). Furthermore retardation effects can safely be neglected. One thus obtains

$$X(\omega+i\delta) = R(\omega+i\delta) \frac{2\omega_s}{(\omega+\omega_s)(\omega-\omega_s+i\Delta_s)} g(Z_o) \tag{15}$$

where $\omega_s = \omega_p/\sqrt{2}$ is the surface plasmon frequency and Δ_s is the width of the surface plasmon. The function $g(Z_0)$ depends on the perpendicular distance Z_0 between the molecule and the metal surface. For large distances $g(Z) \sim 1/Z^3$ as in the simple image potential theory. For small distances $g(Z)$ approaches a constant due to the finite extent of the molecular wave-functions. We have chosen $g(Z_0)$ such that the real part of $(1-X(\omega+i\delta))$ is zero around $\omega = 2.3$ eV. The other parameters used in the calculations are $\omega_s = 3.5$ eV, $E_2 - E_1 = 15$ eV, $U = 7$ eV, $W = 5$ eV, $\Omega = 0.2$ eV, $\Delta_s = 0.15$ eV. Assuming weak chemisorption, $\Delta_a(\omega)$ is energy independent and taken to be 0.1 eV ($\Lambda_a \sim 0$). The resulting spectral density $\rho_a(\omega)$ is illustrated in Fig. 2.

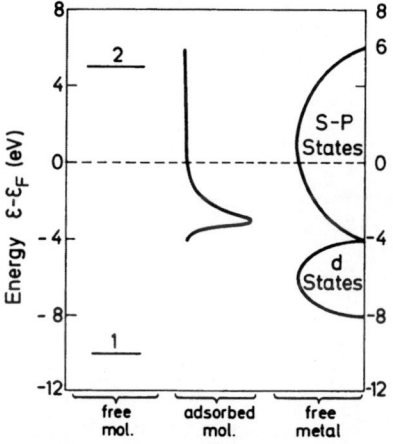

Figure 2 — Energy level scheme used for the free molecule admolecule and metal (Ag).

The numerical results for the two contributions to the enhancement are shown in Fig. 3 together with the total enhancement $\rho(\omega)$. The renormalization of the e-p vertex due to surface plasmon contributes a factor of ~ 100 to the enhancement in Ag. The shifting and broadening of electronic levels as well as the possibility of charge transfers yield an enhancement between two to three orders

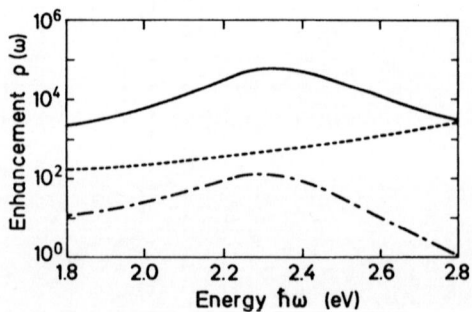

Figure 3 – Contributions to the enhancement of the Raman cross-
          section : –·–·–·–·–·– modified e-p vertex;
          –––––––– chemisorption, ————— total.

of magnitude. The total enhancement is found to be between 10^4 and 10^5. Moreover the enhancement curves depend rather weakly on the incident photon frequency. This is true for contribution (a) because $X(\omega)$ contains a large imaginary part and for contribution (b) because $\rho_s(\omega)$ is slowly varying function of ω.

Our theory explains why enormous enhancements have only been found for Ag. The experimental data[7] show that for Ag, the real part of the dielectric function $\in$ becomes zero around $\omega = 4$ eV and that the imaginary part of $\in$ is rather small at that frequency. This means that Ag has rather well defined surface plasmon with a small width $\Delta_s \sim 0.15$ eV. In contrast to Ag, surface plasmons are not well defined in Cu and Au. If we use in our calculations the corresponding large damping Δ_s we find only small enhancements due to contribution (a). We thus conclude that Ag should give enhancements which are about 2 orders of magnitude larger than for instance in Cu or Au.

REFERENCES.

1. M. Fleischmann, P.J. Hendra, A.J. McQuillan, Chem. Phys. Lett. 26, 163 (1974); D.J. Jeanmaire, R.P. Van Duyne, J. Electro anl. Chem. 84, 1 (1977).
2. See e.g., references in T.E. Furtak and J. Reyes, Surface Sci. 93. 351 (1980). Also see S.S. Jha, J.R. Kirtley and J.C. Tsang, preprint.
3. A. Otto, preprint.
4. J.M. Elson, R.H. Ritchie, Phys. Rev. B4, 4129 (1971).
5. D.L. Mills, E. Burstein, Phys. Rev. 188, 1465 (1969).
6. D.M. Newns, Phys. Rev. 178, 1123 (1969); J.P. Muscat, D.M. Newns, Progress in Surf. Sci. 9, 1 (1978).
7. P.B. Johnson, R.W. Christy, Phys. Rev. 6B, 4370 (1972).

ENHANCED RAMAN SCATTERING FROM MOLECULES ADSORBED ON RESONANT

METAL PARTICLES

C.G. Blatchford and J.A. Creighton

University of Kent
Canterbury, England

ABSTRACT

The wavelength-dependence of the intensity of Raman scattering
by pyridine molecules adsorbed on colloidal gold particles and the
Mie absorption and scattering intensities by the same particles
have been measured, along with studies of the particle size and
aggregation by electron microscopy. The Raman excitation profiles
are shown to resemble to absorption spectra of the colloids more
closely than they resemble the Mie scattering profiles. Electron
microscopy confirms that the gold colloids which show large values
of the adsorbate resonance Raman enhancement ($> 10^4$-fold) are
aggregates consisting of strings of very small particles, and it
is found that the large surface Raman enhancements result from
excitation under the longitudinal plasma resonance of the particle
strings but not from excitation under the transverse resonance.

INTRODUCTION

Surface-enhanced Raman scattering (SERS) has recently attracted
considerable interest as a potential means of obtaining information
on molecules adsorbed on some metal surfaces, including metal interfaces with liquids or high pressure gases. The Raman signal intensity is enhanced by several orders of magnitude over that expected
for normal non-resonant Raman scattering, giving vibrational spectra
with good signal/noise ratio and good resolution for monolayer
amounts of molecules on rough copper, silver and gold surfaces. The
signal enhancement for these metals is 10^4 to 10^6-fold, while
smaller enhancements have been reported for nickel, mercury and
possibly platinum surfaces. The magnitude of the SERS effect is very

dependent on the Raman excitation wavelength, and it is clear that
the enhancement is due to a resonant response of some component of
the surface to the incident light. The nature of this response is
currently a matter of considerable experimental and theoretical
attention, in part because of the need to extend the phenomenon to
a wider range of metals.

We have begun a program of measuring the dependence of the
surface Raman enhancement on the incident wavelength, in order to
ascertain whether it is determined by the metal or by the adsorbate,
and what other variables affect the resonance. Some of this work
has been done using molecules adsorbed on bulk silver electrodes[1],
but it is clear that the state of division of the surface has a
marked effect on the excitation profiles. Such surface shape
effects are most conveniently studied by investigating the Raman
scattering from molecules adsorbed on colloidal metal particles
dispersed in an aqueous medium, since (a) the Raman and elastic
(Mie) scattering excitation profiles and the colloid absorption
spectra are easily measured using standard liquid sample techniques,
and (b) the size and shape of the colloid particles or aggregates
may be controlled by chemical preparative techniques, and may be
readily determined by transmission electron microscopy.

EXPERIMENTAL AND RESULTS

Gold colloids were prepared by reduction of dilute aqueous
solutions of $H[AuCl_4]$ (2×10^{-4}M) with sodium citrate (0.02 M) at
90°C as described by Turkevich et al[2]. This procedure gives a
red colloid with spherical particles of ca. 12 nm diameter and
good monodispersicity as shown in Figure 1a. Upon addition of
pyridine (10^{-4} M), slow aggregation occurred as may be seen from
Figures 1b-c, presumably as a result of the displacement of
adsorbed ions by neutral pyridine molecules, and the colour of
the sol changed slowly from red to blue. The aggregating
particles still apparently bore some residual charge due to
adsorbed ions however, since aggregation gave rise to linear
chains of particles and ultimately to open networks, rather than
to globular clusters.

Silver colloids were prepared by reduction of silver nitrate
solution (2.5×10^{-4} M) with sodium borohydride (1.5×10^{-4} M)[3]
The sols were yellow, and on addition of pyridine (5×10^{-3} M).[3]
slow aggregation occurred to give red and finally blue colloids,
which were shown by electron microscopy to also contain stringlike
particle aggregates.

The transmission spectra of silver colloids prepared in this
way are shown in Figure 2. The unaggregated small particles show a
transmission minimum (extinction maximum) at 400 nm due to an

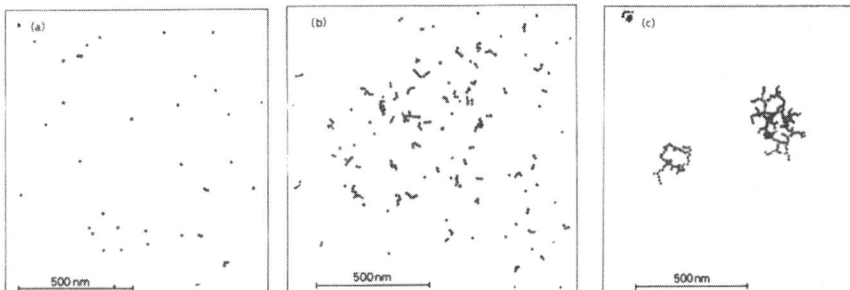

Fig. 1 – Transmission electron micrograph of colloidal gold
particles : (a) gold-citrate sol before addition of
pyridine; (b) 4 hrs. and (c) 18 hrs. after adding pyridine
(see text for concentrations).

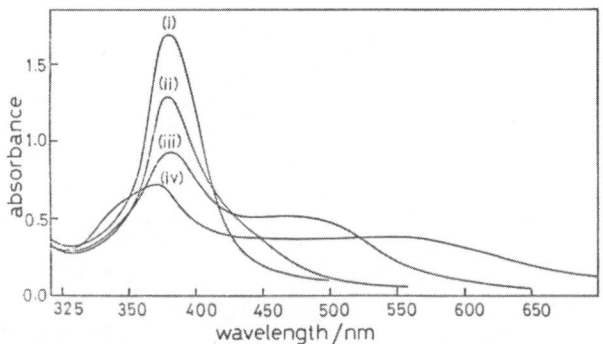

Fig. 2 – Transmission spectra of silver sols :
(i) silver-borohydride solution before addition of pyridine;
(ii) 5 mins. after adding pyridine (1.5 x 10^{-3} M);
(iii) 1 min. and
(iv) 5 mins. after adding pyridine (6 x 10^{-3} M).

absorption band at this wavelength, arising from the resonant excitation of the dipolar plasma oscillation of the spherical particles. The scattering contribution to the extinction maximum, and the contribution of higher induced multipoles to absorption and scattering, is very small for such small particles. Upon formation of the aggregate strings the 3-fold degenerate dipolar resonance splits into two components, in which the induced dipole lies respectively along and transverse to the strings. The longitudinal band lies increasingly to longer wavelengths with increase in string length (Fig. 3), while the transverse band remains near 400 nm, roughly coincident with the absorption band of residual unaggregated small particles. For the gold colloids the transmission spectra are qualitatively similar, the transmission minimum for small particles and the transverse mode of the aggregates lying near 525 nm while the longitudinal plasma mode is to longer wavelengths.

We have reported earlier the enhanced Raman scattering by pyridine adsorbed on silver and gold colloid particles [3]. It was there shown that the Raman excitation profiles over a relatively small wavelength range closely resembled the transmission spectra for a series of silver sols. In that work the particle size or aggregation was not investigated, and it was concluded from the limited data that the Raman excitation profiles followed the Mie scattering rather than the absorption spectra of the particles or aggregates, which were assumed to be essentially spherical. This paper presents more extensive measurements and reinterprets this earlier data. Thus Figure 3 shows transmission spectra of silver and gold sols aggregated into particle strings by addition of pyridine, and superimposed on these the Raman excitation profiles of the 1008 cm^{-1} bands of adsorbed pyridine, measured over a wide range of excitation wavelengths which encompass both the longitudinal and transverse plasma modes. The striking result from this figure is that there is a large SERS enhancement for excitation under the longitudinal plasma mode of the aggregate strings, but not under the transverse mode.

We have also measured the separate contribution of the Mie scattering and absorption to the extinction at various stages of aggregation of a gold colloid containing pyridine, as shown in Figure 4. The degree of aggregation of the sols (a) and (c) of Figure 4 is similar to that shown in Figure 1(b) and 1(c) respectively. The Raman excitation profiles of these same colloids were also measured, and it may be seen from Figure 4 that the Raman excitation profiles for adsorbed pyridine (1008 cm^{-1}) follows the absorption spectra of the colloids, and not the Mie scattering profiles as was concluded previously[3].

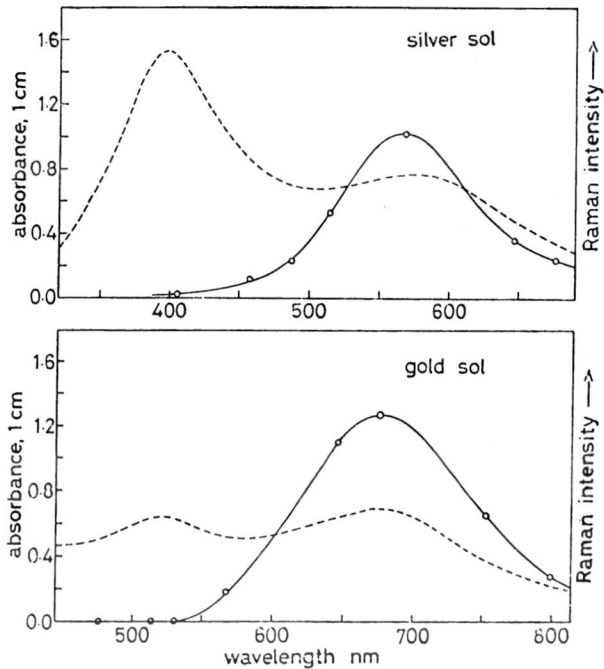

Fig. 3 - Excitation profiles (————) of 1008 cm^{-1} Raman band
of pyridine adsorbed on aggregated silver and gold sols,
and transmission spectra of the sols (------).

DISCUSSION

It is clear from the close correspondence between the aggregation dependence of the SERS excitation profile maxima and just one
component (the longitudinal plasma absorption) of the metal particle
absorption spectra, that the resonant response giving rise to
enhanced surface Raman scattering by the particles is the excitation
of metal particle dipolar plasma oscillations which are well known
to be responsible for light absorption by small metal particles.
This result is consistent with our recent measurement of the SERS
excitation profiles of molecules on roughened silver electrodes[1].
Here it was shown that although the excitation profiles for Raman
bands of pyridine and triphenylphosphine on different silver
electrodes varied considerably due to differences in surface

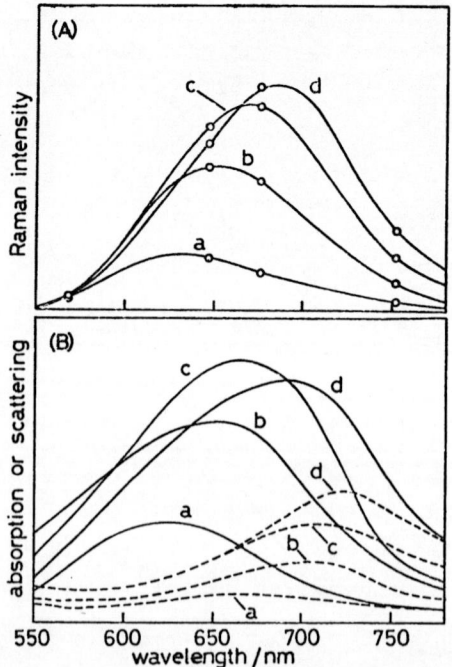

Fig. 4 - (A) Excitation profiles of 1008 cm⁻¹ Raman band of pyridine
         adsorbed on aggregated gold-citrate sols, and (B) absorption
         (———) and Mie scattering (-----) profiles. (a), (b),(c)
         and (d) are recorded respectively 3, 8, 24 and 58 hrs.
         after adding pyridine to a fresh sol (see text for concen-
         trations).

roughness, the wavelength of the profile maxima for these adsorbates
co-adsorbed on the same electrode were within experimental error
the same. Thus on the available evidence (for colourless adsorbates)
the excitation profiles are independent of the adsorbate.

    The oscillating electric field at the surface of a metal
particle illuminated on resonance is considerably enhanced. Kerker
et al.[4] , and Chang and co-workers[5], have shown that for small
spherical silver particles on resonance there is a surface field
enhancement of > 10, giving SERS enhancements[4] of > 10⁴ . We are
therefore of the view that the SERS enhancement is a surface

electromagnetic field effect, in agreement with recent UHV studies[6] which show that the enhancement extends out several molecular diameters from the surface for some relatively coarsely roughened silver surfaces. It is of particular note that in line with the results in Figure 3, Gersten[7] has very recently predicted that the SERS intensity will be particularly large for molecules adsorbed near the ends of prolate ellipsoidal metal particles illuminated in the vicinity of their longitudinal plasma resonance, due to the very large enhancement of the surface electromagnetic field which these molecules experience (lightening rod effect).

REFERENCES

1. C.G. Blatchford, J.R. Campbell, J.A. Creighton, submitted to Surface Science.
2. J. Turkevich, P.C. Stevenson, J. Hillier, Disc. Faraday Soc., 11 (1971) 58.
3. J.A. Creighton, C.G. Blatchford, M.G. Albrecht, J. Chem. Soc. Faraday II, 75 (1979) 790.
4. M. Kerker, D.S. Wang, H. Chew, Applied Optics (to be published).
5. R.K. Chang, private communication.
6. J.E. Rowe, C.V. Shank, D.A. Zwemer, C.A. Murray, Phys. Rev. Letters, 44 (1980) 1770; D.L. Allara, M. Rhinewine, C.A. Murray, Bull. Amer. Phys. Soc. 25 (1980) 425.
7. J.I. Gersten, J. Chem. Phys. 72 (1980) 5779; J.I. Gersten, A. Nitzan, J. Chem. Phys. (to be published).

ENHANCED RAMAN SCATTERING FROM MOLECULES ADSORBED ON SMALL NICKEL

PARTICLES

W. Krasser

Institut für Festkörperforschung der KFA Jülich
D-5170 Jülich 1, West Germany

ABSTRACT

We have investigated the Raman enhancement of simple molecules
like benzene, coadsorbed benzene and hydrogen, carbon monoxide, ethylene, coadsorbed carbon monoxide and hydrogen, which are bonded to
the surface of small nickel particles. The enhancement depends
strongly on the structure and diameter of the particles and on the
excitation wavelength. The maximum of the enhancement is generally
observed in the wavelength range of 5200 Å – 4500 Å and is about
10^3 – 10^4 compared with those wavelengths, where no enhancement could
be measured. If there exist several surface species or if chemical
reactions take place on the surface, the spectra of all chemisorbed
molecules are enhanced in the same way.

INTRODUCTION

Raman spectroscopy is an excellent tool for the investigation
of the structure of chemisorbed molecules and catalytic reactions,
provided the vibrational spectra are strongly enhanced. This surface enhancement is until now mainly investigated for the metals Ag,
Au and Cu[1,2,3,4,5] . Other metals like Pt , Fe and Ni, which are
important for practical catalysis have not yet been investigated.
Ag , Au and Cu are metals with a free electron behaviour in the visible region and with interband transitions in the middle ultraviolet. Nickel is an important catalyst. Therefore it seems to be interesting to measure the behaviour of the Raman enhancement of simple chemisorbed molecules. In this paper we report the conditions
for the existence of a Raman enhancement of some simple molecules,
which are thoroughly investigated with electron energy loss spectros-

copy. Generally we get Raman spectra from all catalysts, even when
no enhancement is observed, but a very long registration time is
needed, as only a few photocounts per second are registered in one
channel of the multichannel analyzer. The enhancement is never observed on plane surfaces ; a special roughness is required. Until
now, it is not yet definitely known, if particles with a diameter
smaller than the wavelength of the light are responsible for the
enhancement, or if atomic scale roughness[5] is required.

EXPERIMENTAL

We use silica supported nickel and Raney nickel as catalyst.
Silica supported nickel consists of finely divided silica particles
of about 100 Å diameter covered with 14-23% nickel. Raney nickel
consists of irregular small nickel particles with a diameter of about
100 Å. We have tested several different catalysts. We have observed, that when a catalyst shows Raman enhancement, all investigated
molecules are enhanced. A lot of catalysts never showed an enhancement. Our measurements were performed in a vacuum of 10^{-6} - 10^{-7}
torr. This vacuum is sufficient for powdered catalysts with a very
high specific surface area. The silica supported nickel sample was
reduced in a stream of hydrogen at 700°C ; the Raney nickel sample
was reduced at 200°C. This relatively low temperature is required
because of sintering at higher temperatures. Therefore, most of our
measurements were performed with silica supported nickel, but Raney
nickel shows the same enhancement as silica supported nickel. After
the reduction, the hydrogen is pumped off for about 14 h.

The reduced nickel sample is a deep black coloured powder, which
absorbs visible light strongly. In order to avoid heating of the
sample in the laser focus and additional chemical reactions on the
surface, the measurements were performed with low laser power at 100K.
Additionally, the sample was rotated in the cryostat. In order to
suppress the second order spectrum of the glass walls of the sample
cell, we used difference Raman spectroscopy. The difference Raman
equipment is described elsewhere[6] . The enhancement factor refers
to those excitation wavelengths, where no enhancement could be observed and only a few photocounts per second are registered.

EXPERIMENTAL RESULTS

Carbon monoxide adsorbed on Raney nickel

The Raman spectra of chemisorbed carbon monoxide are strongly
enhanced. We observe about 1500 photocounts per second which corresponds to an enhancement factor of about $5 \cdot 10^{3}$. In Fig. 1 the
spectrum is shown in the frequency region of the $\tilde{\nu}$ (CO)-stretching
vibrations. In fig. 2 the corresponding spectrum in the low frequency

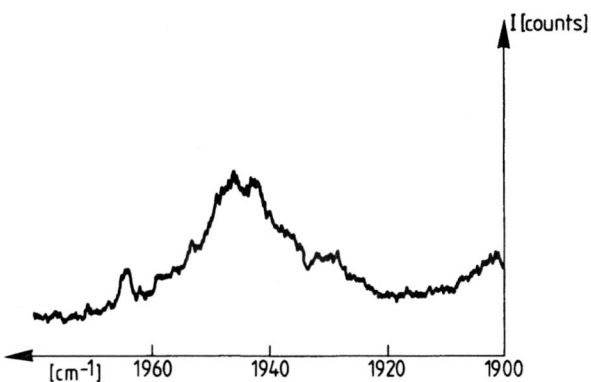

Fig. 1. Enhanced Raman spectrum of carbon monoxide adsorbed on Raney
nickel in the frequency region of the $\tilde{\nu}$ (CO) stretching vibrations

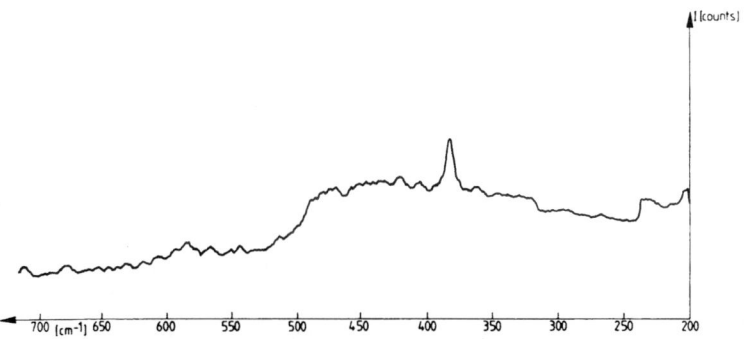

Fig. 2.   Enhanced Raman spectrum of carbon monoxide adsorbed on Raney
nickel in the low frequency region.

region is presented. The spectrum was excited with the wavelength
λ = 5145 Å from an argon ion laser.

Benzene, coadsorbed benzene and hydrogen

In order to show, that the Raman enhancement is not of pure chemical origin, we have measured the Raman spectra of adsorbed benzene
and ooadsorbed benzene and hydrogen. If benzene and hydrogen is coadsorbed, a chemical reaction takes place and cyclohexene is formed.
The vibrational spectra change drastically but the enhancement remains nearly constant. Therefore, a normal resonance Raman effect
caused only by electronic transitions in the chemisorbed molecules
can be excluded. Cyclohexene has no absorption bands in the visible
region. The interactions between cyclohexene and the surface should
not be so important to shift the electronic transitions in the visible region. The vibrational spectra of the chemisorbed molecules
are much more complicated than the corresponding spectra of the free

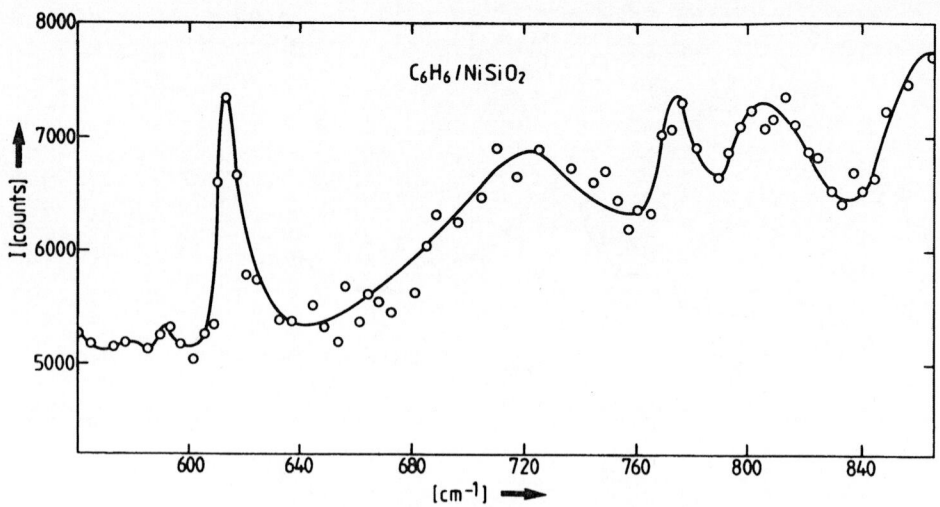

Fig. 3. Enhanced Raman spectrum of C_6H_6 / Ni-SiO$_2$

molecules. In Fig. 3 the Raman spectrum of chemisorbed benzene is presented in the frequency range of 600 - 840 cm^{-1}. In this frequency range, free benzene has only one Raman active vibration at 605 cm^{-1}, which is also observed in the spectrum of chemisorbed benzene. For comparison, also the Raman spectrum of coadsorbed C_6D_6 and D_2 is given in Fig. 4. There are several reasons for the excitation of the numerous vibrations. The normally inactive and only infrared active vibrations may become Raman active by lowering of the symmetry, there exist several sites of adsorption, a lot of bands can be interpreted as overtones which may be excited strongly by coupling to the electronic excitations in the metal. The Raman spectra are completely depolarized. If the enhancement could be explained by a normal resonance Raman effect, at least the totally symmetric vibrational bands should be polarized.

Coadsorbed hydrogen and carbon monoxide

Carbon monoxide and hydrogen are coadsorbed on Ni-SiO$_2$ at T = 180°C. At this adsorption temperature, chemical reactions take place on the surface and simple hydrocarbons should be produced. The enhancement depends strongly on the wavelength of the excitation and is observed in the wavelength range of 5200 Å - 4500 Å. The spectra are rather complex with many sharp spikes and a broad diffuse enhanced background which is also observed without adsorbate. All vibrations show the same wavelength behaviour. In Fig. 5

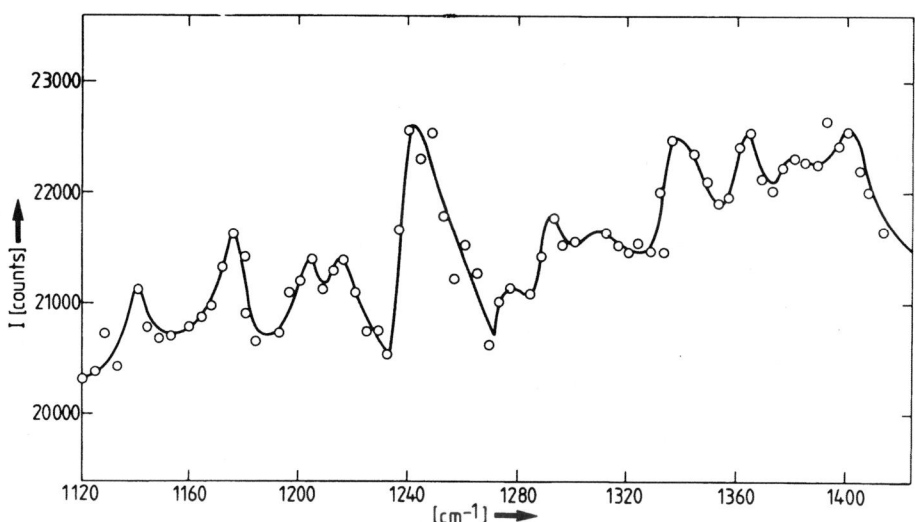

Fig. 4. Enhanced Raman spectrum of coadsorbed C_6D_6 and D_2 in the frequency region of the $\tilde{\nu}$ (C–D) vibrations.

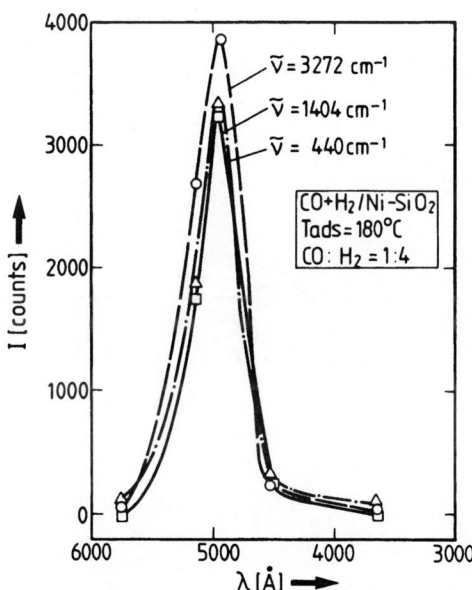

Fig. 5. Wavelength dependence of three distinct vibrations.

the wavelength behaviour for three distinct vibrations at high, medium and low frequencies is presented. A further hint, that the enhancement is not due to a normal resonance Raman scattering is the temperature dependence. We have heated the sample to 500°C and all sharp spikes resulting from vibrations disappear. Only some broad peaks resulting presumably from carbon remain. Despite the vibrational spectra change drastically, the enhancement remains nearly constant.

CONCLUSION

All investigated molecules show strong enhancement. If several surface species exist simultaneously on the surface, all vibrations of all surface species are enhanced with the same enhancement factor. The relative intensities of the different vibrational bands change drastically for different catalysts. The used metallic particles all have diameters small compared to the wavelength of the light. In this range, the scattering depends strongly on the diameter and shape of the particles and the dielectric constants of the adsorbate. Generally there exists a distribution of particles with different size. The size and shape may influence the vibrational frequencies of the adsorbate. With a distribution of particle sizes, there exists a distribution of vibrational frequencies. This may partly explain the complicated spectra. We have also investigated nickel particles with a diameter of $10^4 \text{\AA} - 5 \cdot 10^4 \text{\AA}$ and we have never observed a Raman enhancement. So we may assume, that for the Raman enhancement particles of a diameter which is smaller than the wavelength of light, are required.

REFERENCES

1. P.L. Jeanmarie and R.P. van Duyne, J. Electroanal. Chem. 84, 1 (1977).
2. R.E. Furtak, Surf. Sci. 93, 351 (1980).
3. J.A. Creighton, C.G. Blatchford and M.G. Albrecht, J.C.S. Farad. Trans. II, 75, 790 (1979).
4. J.R. Kirtley, J.C. Tsang, T.N. Theis and S.S. Jha, Proc. VII th International Conference on Raman Spectroscopy, Ottawa, 1980, North Holland.
5. A. Otto, Surf. Sci. 75, 392 (1978).
6. W. Krasser, H. Ervens, A. Fadini and A.J. Renouprez, J. Raman Spectr. 9, 80 (1980).

SURFACE ENHANCED RAMAN SCATTERING FROM DIFFERENT TYPES OF METAL SUBSTRATES

Ralf Dornhaus

I. Physikalisches Institut der RWTH Aachen

51 Aachen FRG

ABSTRACT

SERS from molecules adsorbed on different types of metal substrates including Ag- island films, Ag- and Au- sols and Ag- and Cu- electrodes has been studied. Additional absorption measurements have been performed to test several proposed explanations relating the Raman enhancement to plasmon or collective electron resonances.

INTRODUCTION

Surface enhanced Raman spectroscopy (SERS) as an interesting new tool for the study of molecules adsorbed at metal surfaces has been recognized and exploited for a few years now. SERS signals have been observed for approximately fourty different molecules and ions, ranging from very simple adsorbates like CN⁻, CO etc. to complicated molecules like adenine and others. Enhancements have been observed for different metal substrates like Ag, Au, Cu and possibly Pt, Ni, Al and Hg. It seems well established by now that roughness of the metal substrate is mandatory for maximum enhancement. There is, however, a considerable debate concerning the rôle of this surface roughness in the enhancement process.

Recently it has been argued that geometrically defined optical absorption resonances excited by the incident and scattered photons may contribute a considerable part to the total enhancement in most systems studied up to now (electrochemically reformed surfaces[1,2], metal island films[3,4], evaporated films on CaF_2[5,6] and on holographic gratings[7], metal sols[8,9]). Such resonances require roughness on a

scale of $\simeq$ 50-2000 Å. On the other hand there has been evidence
for the influence of atomic scale roughness from investigations
with electrochemically prepared substrates[10] as well as with metal
films evaporated at low temperature[11,12].

Arguments for the influence of optical absorption resonances
have been mainly based on the following observations :

- mildly electrochemically reformed evaporated Ag-, Au- and Cu-
  films exhibit marked intensity variations depending on the
  incident and scattering angles[1,2].

- SERS from monolayers separated from the metal substrate either by
  spacer layers[6] or by layers of the same type of molecule[13] was
  claimed to be observed.

- the dependence of SERS on the excitation wavelength has been
  observed to be correlated with absorption bands due to "collective
  electron resonances" of metal island films[3] or colloidal particles[8]

- the angular dependence of SERS from molecules on holographic
  gratings and covered with Ag show a strong correlation with
  surface plasmon excitation[7,14].

Despite this evidence most authors agree that the overall enhancement in many cases seems to be the result of a combination of
contributions, some of which might be quite specific of the particular metal- adsorbed molecule system[15]. This might also explain
the widely differing results that have been recently reported on
the coverage dependence of SERS- intensities in UHV-studies[11-13,16].
We do not intend to discuss these descrepancies here but rather to
present some of our own results on three different systems : metal
island films, colloidal solutions and metal-electrodes.

METAL ISLAND FILMS

As for the experiments with holographic gratings[7,14] we have
attempted to determine the contributions to SERS of surface plasmon
polariton excitaions[17]. We have compared directly the Raman intensity
from molecules adsorbed on 5- and 57-nm thick evaporated Ag films
deposited on a hemicylindrical prism which enabled direct excitation
of surface plasmon polaritons in the Kretschmann configuration (ATR)
for the thicker films. We indeed observed a strong increase in the
Raman intensity at the surface plasmon angle, making Raman observations from monolayer adsorbates possible. The exact enhancement
factor is difficult to determine since no Raman signal was detectable
away from the plasmon angle or with s- polarized light. The estimated
enhancement from similar experiments with thicker benzene layers was
at least two orders of magnitude. Comparing the Raman intensity at

the plasmon angle from the thicker film with the intensity from
the thin metal-island films (as evidenced from SEM photographs)
we noticed an additional intensity enhancement by an order of
magnitude. The thin Ag-film did not permit a direct ATR coupling
to SP polariton modes and consequently neither the polarization nor
the angle of incidence of the exciting beam had an influence on the
intensity differing from the predictions of Fresnel theory.

Using glass slides as substrates for 5 nm metal-island films
we obtained the SERS-spectra shown in Figs. 1 and 2. Fig. 1 com-
pares spectra of Pyridine-n-carboxylic-acids (n = 2,3,4) and Fig. 2
of n-Aminobenzoic-acids, a detailed spectral analysis of which will
be published elsewhere. Fig. 3 shows measured transmission curves
for a thick (48 nm) and two thin (5 nm) Ag-films, one of them
coated with a monolayer of 4-Aminobenzoic acid. The two metal-island
films display a noticeable absorption maximum around 600 nm, which
approximately corresponds to the increasing SERS intensity in the
red wavelength range (see also[3]).

Again we did not notice any appreciable angular dependence of
the SERS intensity.

We would like to mention, however, that evaporated Ag- Au- and
Cu-films of thicknesses exceeding several tens of nm always exhibit
a pronounced texture (orientation of crystallites in the plane of
the substrate), usually with (111) orientation[18]. This texture might
be an alternative explanation for the angular dependence observed
with mildly anodized evaporated films in an electrochemical envi
ronment[19].

METAL SOLS

Stimulated by recent calculations[20,21] and experiments [8,9]
on colloidal solutions, we have measured the extinction and scattering of metal sols as well as SERS intensities from adsorbed
molecules.

An Ag-sol was prepared by reducing Ag from a dilute (10^{-3} M
AgCl in NH_4OH) solution with Hydraziniumhydrate. The sol was yellow-
green in color and seemed stable in a closed bottle. Even after the
addition of 10^{-3} M Pyridine in H_2O no change in color nor any
influence on the transmission curve (Fig. 4) could be detected. TEM-
pictures revealed mostly layered, nearly ellipsoidal shaped particles
with a mean diameter of 80 nm and a nucleus of about 50 nm.

Contrary to the results reported in references[8,9] no shoulder
or second maximum on the long wavelength side in the extinciton curve
could be detected. Attempts to measure the scattering contribution
to the extinction using an Ulbricht-sphere resulted in no detectable

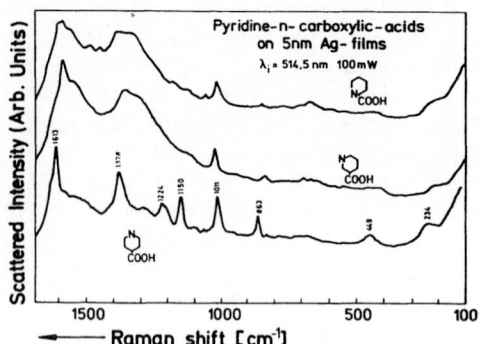

Fig. 1 - SERS spectra of Pyridine-n-carboxylic-acids (n = 2,3,4)
         adsorbed on 5 nm thick metal-island films.

scattering for wavelengths longer than 400 nm. Consequently the
extinction is almost totally due to absorption, which indicates a
considerable deviation from calculations based on Mie-theory for
spherical particles (using bulk ε-value for Ag)[21]. In spite of
the absence of the second absorption peak at longer wavelengths[8]
the measured SERS intensity still peaks in the 650-700 nm region
(Fig. 4, dashed curve) and corresponds to a maximum enhancement of
$\simeq 10^5$ (assuming monolayer coverage). A SERS-spectrum is shown in
Fig. 5. Fig. 4 also shows the transmission of an Au-sol prepared
by reducing $K(Au(CN)_2)$ with $NaBH_4$ [21]. This sol shows one strong
SERS peak due to $Au(CN)_2^-$ near 2138 cm^{-1} by using 647 nm excitation.
Also with this sol only a negligible scattering contribution to the
extinction curve could be detected. Although for both sols we observed enhancements comparable to the ones measured in other SERS
investigations, the wavelength dependence of the extinction and of
the Raman intensity differs from theoretical predictions for spherical particles[20,21].

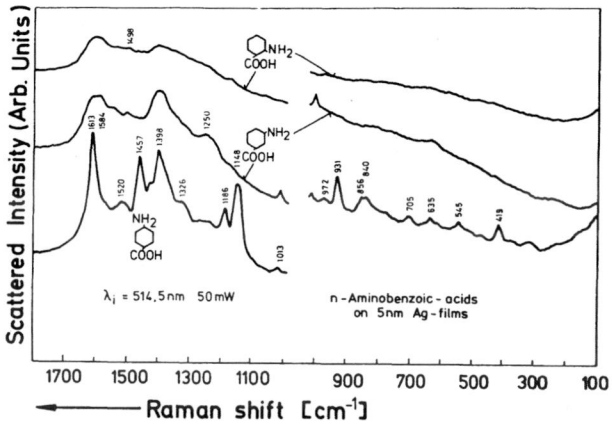

Fig. 2 - SERS spectra of n-Aminobenzoic-acids (n = 2,3.4) adsorbed on 5 nm thick metal-island films.

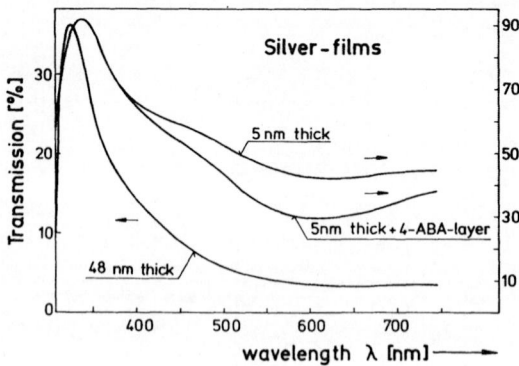

Fig. 3 – Transmission spectra of a 48 nm thick continuous and two
         5 nm thick metal-island Ag-films. One metal-island film
         has been coated with a monolayer of 4-Aminobenzoic-acid.

METAL ELECTRODES

        SERS has already proved to be a useful tool to study adsorbates
on surfaces in many different environments. We have recently demonstrated[22],[23] that it may even be used to follow reactions on
electrode surfaces. Using an optical multichennal analyser system
it was possible to monitor changes in the vibrational frequencies
of CN⁻ ions on Ag- and Cu-surfaces during the electrochemical
treatment in real time and correlate this changes with voltammogram
structures recorded simultaneously (Fig. 6).

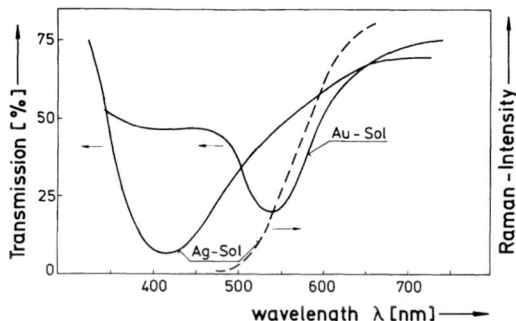

Fig. 4 – Transmission spectra of an Ag·sol and an Au-sol (solid
        lines). The dashed curve represents the dependence of
        the 1008 cm^{-1} mode of pyridine adsorbed at the Ag colloidal
        particles on the wavelength of the exciting laser beam.

CONCLUSIONS : POSSIBLE ENHANCEMENT MECHANISMS

    Any viable model or combination of models for SERS must explain
the large enhancement and the wavelength dependences for both the
discrete peaks and the background continuum. Although our own and
other recent experiments have attempted to isolate the contribution
of several possible mechanisms to the observed SERS there remain
considerable difficulties in the unambiguous identification of any
single mechanism. Especially a serious controversy remains over

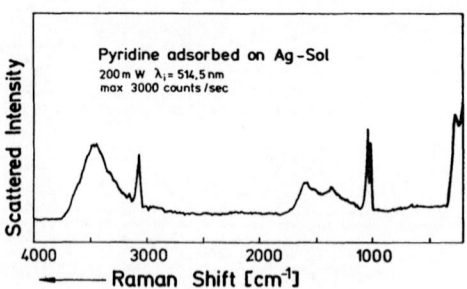

Fig. 5 – SERS spectrum of pyridine adsorbed on Ag colloidal particles.

whether the role of metallic microstructure is of a more "chemical" nature (e.g. charge transfer between the metal and adsorbate[14],[15]) or of an EM wave nature (i.e. enhancement of the local EM field intensity for the incident wave near the microstructure and enhancement of the reemission at the scattered wavelength[14],[20],[24],[25]).

The conclusions from our experimental results partly presented in this work are :

- that no single enhancement mechanism can account under realistic experimental conditions for the total enhancement of $10^5 - 10^6$ observed in favorable cases.

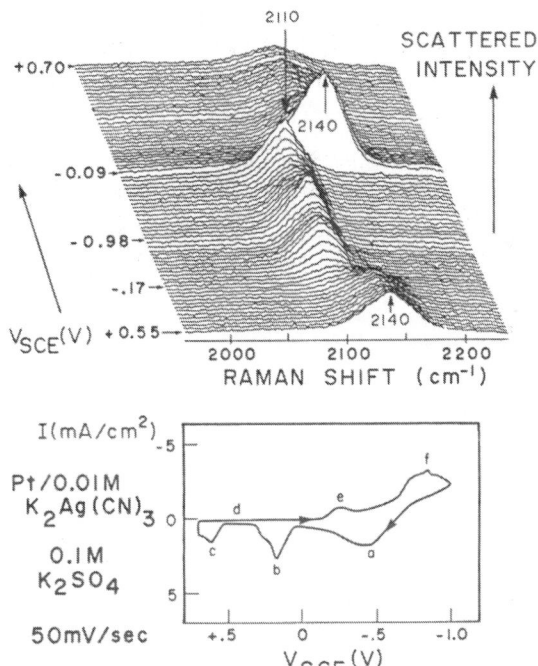

Fig. 6 – Development of SERS and voltammogram for a Pt working
electrode on 0.1 M K_2SO_4 and 0.01 M $K_2Ag(CN)_3$ aqueous
electrolyte. The voltage ramping rate is 50 mV/sec.

- that under the very different experimental conditions employed
e.g. in references[1-9] and also in this work some of the proposed
mechanisms contribute to a different extent to the total enhancement.

- that surface plasmon polariton excitation under well defined
conditions (grating-[14] or ATR-excitation) contributes a factor
of up to 10^3 to the total enhancement and that it may contribute
with a somewhat smaller factor to the enhancement in other cases
(statistical roughness).

- that realistic calculations of the excitation EM field intensity
  and the reradiation efficiency of the molecules near a metallic
  microstructure are needed taking into account the solution of the
  EM problem for eventually nonspherical, layered and aggregated
  particles and using realistic (not simply bulk) values for the
  dielectric function. Up to now the EM field enhancement mechanisms
  do not explain correctly the dependence of discrete SERS-peaks or
  of the continuum emission with respect to the energy ω_i of the
  incident radiation.

- that some of the observed experimental features (e.g. the ω_i-
  dependence of the enhancement, the influence of the nature of
  the adsorbates[21] and some selective enhancement of different
  modes[14]) may be explained by "charge transfer models"[21,14,10,11,
  15] although almost no quantitative estimates are available to
  critically test these models with the available experimental data.

ACKNOWLEDGEMENTS

    I gratefully acknowledge the extremely valuable assistance of
R. Schmitte and H. Lisken in some of the measurements. I thank
Prof. R.K. Chang and K.U. von Raben for many stimulating discussions.

REFERENCES

1. B. Pettinger, U. Wenning, H. Wetzel, Chem. Phys. Lett. <u>67</u> 192
   (1979).
2. G.R. Trott, T.E. Furtak, to be published.
3. E. Burstein, C.Y. Chen, S. Lundqvist, Light Scattering in Solids,
   ed. J.L. Birman et al. (Plenum, New York 1979), p. 479.
4. C.Y. Chen, E. Burstein, S. Lundqvist, Solid State Commun. <u>32</u>,
   63 (1979).
5. J.C. Tsang, J. Kirtley, Solid State Commun. <u>30</u>, 617 (1979).
6. C.A. Murray, D.L. Allara, M. Rhinewine, Proc. VIIth Int. Conf.
   on Raman Spectroscopy, Ottawa 1980, ed. W.F. Murphy (North
   Holland, Amsterdam 1980), p. 406.
7. J.C. Tsang, J.R. Kirtley, T.N. Theis, Solid State Commun <u>35</u>
   667 (1980).
8. J.A. Crieghton, C..G. Blatchford, M.G. Albrecht, J. Chem. Soc.
   Faraday II <u>75</u>, 790 (1979).
9. M. Kerker, D.S. Wang, H. Chew, to be published.
10. J. Billmann, G. Kovacs, A. Otto, Surf. Sci. <u>92</u>, 153 (1980).
11. I. Pockrand, A. Otto, Solid State Commun. <u>35</u>, 861 (1980).
12. T.H. Wood, M.V. Klein, Solid State Commun. <u>35</u>, 263 (1980).
13. J.E. Rowe, C.V. Shank, D.A. Zwemer, C.A. Murray, Phys. Rev.
    Lett. <u>44</u>, 1770 (1980).
14. S.S. Jha, J.R. Kirtley, J.C. Tsang, Phys. Rev. B to be published.
    J.R. Kirtley, S.S. Jha, J.C. Tsang, Solid State Commun. <u>35</u>, 509

(1980).

15. E. Burstein, C.Y. Chen, in [6], p. 346.
16. P.N. Sanda, J.M. Warlaumount, J.E. Demuth, J.C. Tsang, K. Christmann, J.A. Bradley, Phys. Rev. Lett. to be published.
17. R. Dornhaus, R.E. Benner R.K. Chang I. Chabay, Surf. Sci. 101 (1980) to be published.
18. M. Dixit, Phil. Mag. 61 , 1049 (1933).
19. B. Pettinger U. Wenning Proc. Conf. on Vibr. in the Adsorbed layer, Jülich 1978, p. 169.
20. M. Kerker, D.S. Wang, H. Chew, to be published.
21. R.K. Chang, R.E. Benner, R. Dornhaus, K.U. von Raben, B.L. Laube, Sergio Porto Memorial Conf. on Lasers and Applications, Rio de Janeiro/Brazil, 1980 (to be published, Springer Verlag).
22. R.E. Benner, R. Dornhaus, R.K. Chang, B.L. Laube, Surf. Sci. 101 (1980) to be published.
23. R.E. Benner, K.U. von Raben, R. Dornhaus, R.K. Chang, B.L Laube, F.A. Otter, Surf. Sci. to be published.
24. M. Moskowitz, Solid State Commun. 32, 59 (1979) and J. Chem. Phys. 69, 4159 (1978).
25. S.L. McCall, P.M. Platzman, P.A. Wolff Phys. Lett. 77A, 381 (1980).

THE RELEVANCE OF SPATIAL VARIATION IN THE ELECTROMAGNETIC VECTOR

POTENTIAL FOR SURFACE RAMAN SCATTERING

J.K.Sass, H. Neff

Fritz-Haber-Institut der Max-Planck-Gesellschaft
Faradayweg 4-6,
D-1000 Berlin 33

S. Holloway

Institut für Theoretische Physik der Freien Universität
Berlin Arnimallee,3,
D-1000 Berlin 33, West Germany

M. Moskovits

Lash Miller Chemical Laboratories and Erindale College
University of Toronto,
Toronto, Ontario, M5S 1A1 Canada

ABSTRACT

The influence of the non-local optical properties of metal
surfaces on the Raman scattering cross-section is discussed. The
importance of considering the quadrupole term in the induced dipole
moment is demonstrated. A new enhancement mechanism, based on the
influence of the charge fluctuation in the chemisorption bond on
the local phonon field, is proposed.

The application of Raman spectroscopy to surface studies hinges upon
the occurrence of a strong enhancement of the scattering cross-section,
to provide sufficient sensitivity. Numerous experimental observations of enhancement factors of up to 10^6 - fold have been reported in the literature [1]. Compared, however, to the other vibrational techniques described in this volume the applicability of
surface enhanced Raman scattering (SERS) for a particular adsorption
system can still not be established a priori.

This restriction is brought about by our limited understanding of
the physical processes governing the enhancement mechanism.   A
recent review by Furtak and Reyes [2] gives an excellent synopsis
of the wide variety of proposed theoretical models.   In this short
communication, we aim at drawing the reader's attention to the
hitherto neglected influence of the strong spatial variations in
the photon fields near metal surfaces.

     In order to obtain a reasonable description of the electroma-
gnetic vector potential immediately adjacent to the metal surface
it is essential to account for the spilling-out of the electron
charge density from the positive ion core background.   Schematically, Fig. la shows the charge profile of a jellium surface [3] .   The
interaction of electromagnetic radiation with this interface has
been calculated by Feibelman [4] and Fig. lb provides a qualitative
illustration on the essential results.

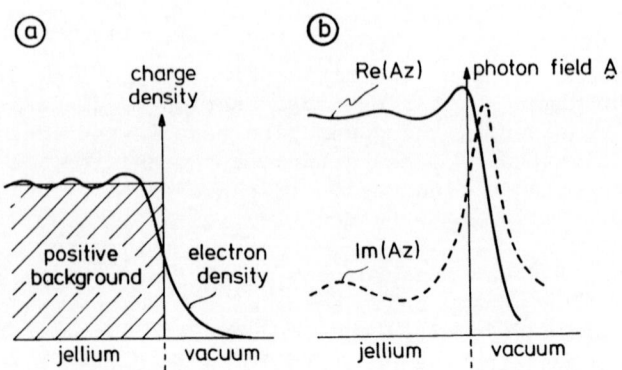

Fig. 1. : Schematic diagram of a) the electron density
          at a jellium surface and b) the real and
          imaginary parts of the normal component of
          the photon field A after Lang [3] and Feibelman [4]

For the purposes of this present discussion, it is most important to observe the dramatic changes of the vector potential $\underset{\sim}{A}$ in that very region where an adparticle would be located.

This importance of including longitudinal waves when describing the optical properties of metal surfaces was first discussed by Sauter and Forstmann [5]. It was later shown by Kliewer [6] that although these effects play a minor role in the total absorption, in those experiments which sample preferentially the surface region they are expected to dominate. Experimentally both photo-emission[7,8] and electroreflectance[9], due to their extreme surface sensitivity, have essentially validified these theoretical predictions. A major result obtained from these studies has been the strong deviation from classical optics which occurs for excitation energies below the bulk plasma frequency $h\omega_p$. In Fig. 2. a family of curves is presented showing that in this particular energy range a strong surface resonance in the photon field appears[8]. In a study of the Al(001) surface, it has been demonstrated that the intensity of the photoemitted electrons effectively scales with the size of this surface peak [8]. From this convincing evidence it may be safely concluded that also in SERS a local description of the metal optics is insufficient.

The cross-section for Raman scattering is determined by the dipole moment induced in the adsorbate by the incident radiation. Writing this as a multipole series[10] we obtain

$$\mu_\alpha = \alpha_{\alpha\beta}E_\beta + \frac{1}{3}A_{\alpha\beta\gamma}\frac{\partial F_\beta}{\partial \gamma} + \text{h.0. terms} \qquad (1)$$

where the Greek subscripts denote Cartesian components and when repeated, indicate a summation. Usually, in Raman studies of gaseous and condensed phases only the first term in eq. (1) needs to be considered since it may be shown that the relative order of magnitude of the quadrupole to dipole terms scales as molecular dimension/light wave length[11]. Near a metal surface, however, the situation is radically changed. The photon field (cf. Fig. 2) experiences dramatic modifications within a very short distance giving rise to field gradients of such magnitude as to make the quadrupole terms in eq. (1) equally important[11].

Clearly the importance of this new scattering mechanism has previously been overlooked. Perhaps the most important aspect that its inclusion brings, is the augmented set of selection rules brought about by the components of the quadrupole tensor. In a recent study of benzene adsorption on an evaporated silver film

the assignment of the observed vibrational bands was discussed on
the basis of both a lowering of the symmetry due to the adsorption
site ($D_{6h} \rightarrow C_{3v}$) and these new selection rules[12].   On a more
general note, this result demonstrates that the symmetry properties
of transition matrix elements may be changed equally well by the
operator as well as the eigenfunctions.

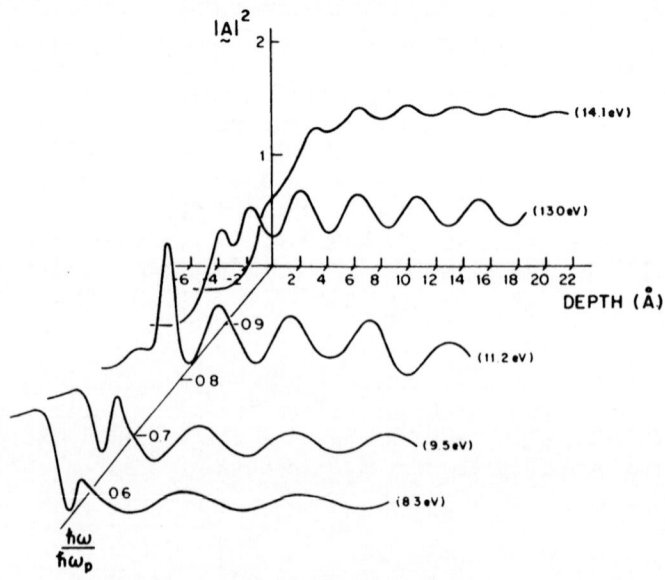

Fig. 2. : Self consistently determined electromagnetic
          fields for jellium (r_s =2).   The positive
          background occupies the half space z>0. Note
          the growth of a strong surface feature as the
          excitation energy decreases away from the
          plasma frequency [8].

        Up until this point, the spatial variation of $\underset{\sim}{A}$ has only been
considered for a smooth, featureless metal surface, i.e. A(r) = A(z).
This picture will be significantly modified in the presence of
adsorbed particles.   Although calculations of $\underset{\sim}{A}$ in the presence
of an adsorbate have not yet been performed we believe it to be
instructive to speculate on the basis of the results obtained by
Lang and Williams[13] for the electron density around an adsorbed
atom.   Fig. 3a shows the electron density contours about a Li
atom on a jellium substrate in the equilibrium position.

Displacements in the position of the adatom as would typically
occur during a vibrational period give rise to a rearrangement
of the charge.   In Fig. 3b is our impression of the changes resulting from an exaggerated extension of the Li-surface distance.
The conclusion to be drawn from Fig. 3. is that the vibration of
the adatom gives rise to a charge fluctuation through the chemisorption bond.   Phenomena such as this have previously been discussed
by McCall and Platzmann[14] and by Persson and Persson[15].

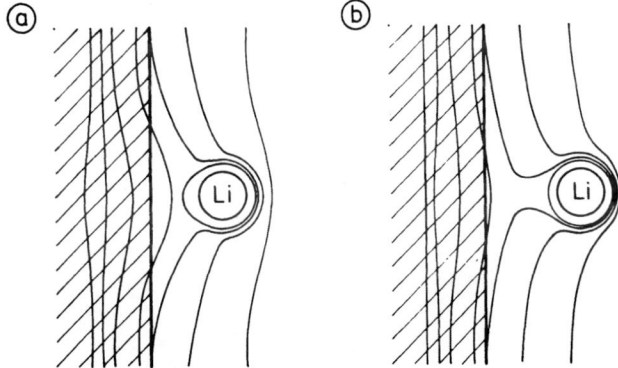

Fig. 3. : Charge density contours for a Li atom adsorbed
          on a jellium surface. a) the equilibrium
          position[13], b) at a larger distance as would
          occur in a vibration.   See text for details.

     Given this picture, the question arises as to the local field
effects resulting from this dynamic charge transfer process.   In
Fig. 4. is presented a somewhat naive attempt at extrapolating the
results obtained by Feibelman[4,8] for the adsorbate-free jellium
surface to this more complex situation.   Looking in a plane through

the center of the adatom we expect the photon field to be modulated,
due to the charge density fluctuation, with the frequency of the
vibration.   This modulation will, in turn, be fed-back in phase
to the molecule suggesting the possibility for a new enhancement
mechanism previously undiscussed in the literature.

Even though Fig. 4 again is only one-dimensional it is quite
apparent from Fig. 3 that the presence of any irregularity at the
surface, albeit an adsorbed molecule or a surface defect[1] , can
give rise to large field gradients in directions parallel to the
surface.   This could explain the apparent experimental observation
that p- and s-polarized light couple with comparable efficiency to
the adsorbate.

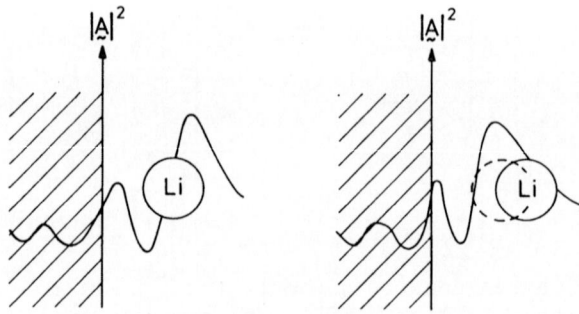

Fig. 4. : Qualitative picture of the influence of the
          charge density fluctuations in the chemisorp-
          tion bond on the local photon field as seen
          by the adsorbed atom (cf. Figs. 1 and 3).

ACKNOWLEDGEMENTS

The authors should like to thank P.J. Feibelman and S. Liu
for valuable discussions.

REFERENCES

1. See for example : A. Otto, in this volume and references therein.
2. T.E. Furtak and J. Reyes, Surface Sci. 93 (1980) 351.
3. N.D. Lang, in Solid State Physics, ed. F. Seitz, D. Turnbull and H. Ehrenreich (Academic, New York, 1973) Vol. 28, p. 225.
4. P.J. Feibelman, Phys. Rev. B12 (1975) 1319.
5. F. Sauter, Z. Physik 203 (1967) 488; F. Forstman, Z. Physik 203 (1967) 495.
6. K.L. Kliewer, Phys. Rev. Lett. 33 (1974) 900.
7. J.K. Sass, H. Laucht and K.L. Kliewer, Phys. Rev. Lett. 35 (1975) 1461.
8. H.J. Levinson, E.W. Plummer and P. Feibelman, Phys. Rev. Lett. 43 (1979) 952 and J. Vac. Sci. Technol. 17 (1980) 216.
9. R. Kötz, D.M. Kolb and F. Forstmann, Surface Sci. 91 (1980) 489.
10. A.D. Buckingham, in Adv. in Chem. Phys. 12 (1976) 107, ed. J.O. Hirschfelder.
11. J.K. Sass, H. Neff, S. Holloway and M. Moskovits, submitted to J. Phys. Chem.
12. M. Moskovits and D.P. DiLella, to be published.
13. N.D. Lang and A.R. Williams, Phys. Rev. B18 (1978) 616.
14. S.L. McCall and P.M. Platzman, Phys. Rev. B22 (1980) 1660.
15. M. Persson and B.N.J. Persson, in this volume.

ADSORPTION OF MARKED ACETIC ACID ON ALUMINA STUDIED BY IETS

S. de Cheveigné, S. Gauthier, J. Klein, A. Léger

Groupe de Physique des Solide de l'Ecole Normale
Supérieure, Université Paris VII   Tour 23
2, Place Jussieu ‑ 75221 Paris Cédex 05 ‑ France

ABSTRACT

The modes of simple carboxylic acids, when adsorbed on alumina are not all undisputably identified. The spectra of marked acetic acid $CH_3C^{18}O^{18}OH$ and CD_3COOD adsorbed on alumina allow us to clarify the assignments.

Although the adsorption of formic acid on alumina[1] to give a formate ion was the first case in which Inelastic Electron Tunneling Spectroscopy (IETS) was shown to allow the study of surface reactions, there is still some doubt as to the assignment of its modes. The same is true of other carboxylic acids[2,3]. This paper discusses the problem and attempts to clarify the situation by the study of isotopic shifts in the case of acetic acid.

IETS uses metal‑insulator‑metal junctions (typically Al‑Aloxide‑Pb) where the insulating layer is roughly 20 Å thick[1]. The first electrode is deposited by evaporation, then oxidized by discharge in a plasma of oxygen. The oxide is then exposed to a vapour of the product to be studied. Finally, after evacuation, the counter electrode is deposited. When a bias is applied to the junction, conduction electrons cross the insulating barrier, exciting (among others) vibrational modes of the adsorbed molecules. The trace of this interaction appears clearly on the second derivative of the electric characteristic I(V), as a spectrum analogous to an infrared or Raman spectrum. The spectra must be taken at T = 4.2 K (or lower) to avoid thermal broadening.

The main characteristics of this spectroscopy are its sensitivity[4] (to a fraction of a monolayer), its resolution (of the order of 5 cm^{-1}) and its range in frequency (200 to 5000 cm^{-1} for vibrational spectroscopy). It is particularly well adapted to surface studies since it is sensitive only to the very thin layer (20 - 30 Å) in which the wavefunction of the tunneling electron is evanescent.

Formic and acetic acid have been well studied by IETS[1,2,5]. There is agreement on the main point i.e. that one observes the formation of the corresponding carboxylate ion. This is proven mainly by the disappearance of the ν C=O and ν C-O vibrations and the appearance of an antisymmetric ν_aCOO$^-$ mode. But some discussion remains as to the attribution of the symmetric ν_s COO$^-$ vibration as well as the neighbouring δCH. The "traditional" attribution for formic acid was ν_s COO$^-$ at 172 meV and δCH at 182 meV. It was first questioned by Lewis[5] et al., Walmsley et al.[6] concluded that it should be reversed after a comparison of the spectra of molecules adsorbed on Al$_2$O$_3$ and on MgO. Shklyarevski et al.[7] arrived at the same conclusion by observing that the 172 meV peak does not shift between the physisorbed and the chemisorbed states of formic acid on Al$_2$O$_3$.

On the other hand, spectra of DCOOD, when sufficiently free of traces of HCOOH[8] , show the peak at 172 meV to be nearly independent of the isotope charge, therefore likely to be the ν_s COO$^-$ If we turn to IR spectroscopy of adsorbed formic acid, we find for the formate ion ν_s COO$^-$ at 169.4 meV and δCH at 170,8 meV[9] . When the ion is adsorbed on Cu, the first mode is shown to have a dipole moment perpendicular to the surface[10]. Studies by EELS of HCOOH on Cu(100)[11] find ν_s COO$^-$ at 168 meV, and do not observe δCH, which has a dipole moment parallel to the metal surface. The disagreement with IR or EELS is not very surprising as both the ν_s COO$^-$ and the ν_aCOO$^-$ involve the bonds to the alumina surface and are sure to be quite sensitive to differences between samples (as suggested also by the width of the ν_aCOO$^-$. But shifts in the position of the ν_s COO$^-$ are such that the situation remains unclear.

The spectra of acetic acid give rise to the same type of discussion. It has a mode at 175 meV (1410 cm^{-1}) and one at 182 meV (1470 cm^{-1}) that are traditionally attributed in that order to the ν_s COO$^-$ and a δ_a CH$_3$ [1,3]. As in the case of formic acid Lewis et al.[5] and Shklyarevski et al[7] propose the contrary. Infrared spectroscopy of CH$_3$COOH on alumina upholds the "traditional" interpretation[12] with δCH at 182.3 meV and ν_s COO$^-$ at 176.5 cm^{-1}. On EELS spectra of acetic acid on Cu (100)[13] ν_s COO$^-$ is at 178 meV, midway between the two values. Finally, published spectra of fluorated or deuterated acetic acid[3] contain sufficient quantities of the normal acid to make the interpretation of the spectra difficult.

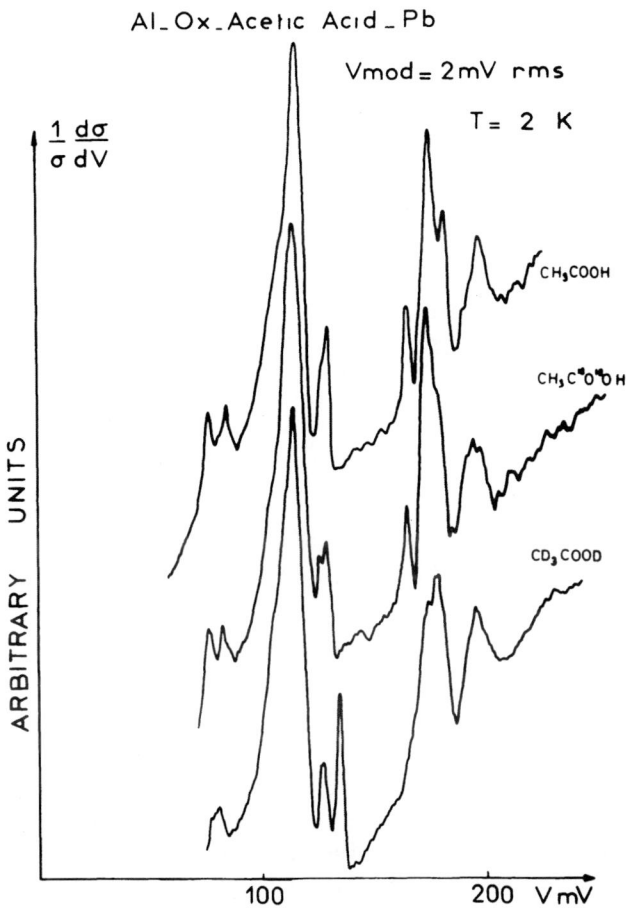

Fig. 1 - IET spectra of acetic acid, ^{18}O-marked acetic acid and deuterated acetic acid.

To try to clear up the situation we drew the spectra of acetic acid $-^{18}O_2$, as well as that of acetic acid-d_4 (Figure 1). When the oxygen mass changes, the peak at 182 meV shifts down to about 173 meV, displacing the top of the peak originally at 175 meV. That is a shift of about 5 %. Other modes are affected by the mass change : the ν_a COO$^-$ mode shifted from 198 to 195 meV (- 1.5 %) and the δOCO scissor mode shifted from 84 to 81 meV (- 3.5 %). These shifts are close to those predicted by a simple mechanical model of a non linear triatomic molecule[14] representing COO$^-$, supposed coupled rigidly to the rest of the molecule and uncoupled to the substrate. The angle of the molecule was taken to be 110°, the elongation force constant $a_{11} = 11.3 \ 10^5$ dynes/cm and the deformation force constant $a_{33} = 3.75 \ 10^5$ dynes/cm. The shifts predicted were $- 5.9$ % for ν_s COO^-, $- 2.7$ % for $\nu_a \ COO^-$ and $- 2.7$ % for δCOO^-. We believe that the shift observed for the peak at 182 meV firmly establishes it as the $\nu_s \ COO^-$.

Most modes involving CH are unaffected : the νCH_3 near 370 meV and $\delta_s \ CH_3$ at 166 meV as well as the modes at 57 and 76 meV. Although classified as COO^- rocking modes, these last two in fact involve movements of the CH_3 radical with respect to the COO^- radical. Since the COO^- is bonded, the mode will be more sensitive to mass changes in the CH_3 , than in the COO^-. Surprisingly, the split between the CH_3 rocks at 130 meV is increased. There is also a slight down-wards in the Al-O mode at 117 meV which shows that the surface to adsorbate $Al-{}^{18}O$ bonds furnish a good part of the intensity of the peak.

The spectrum of the deuterated acid, already discussed in literature[3], is not as clear as could be hoped, due to the presence of residual CH modes. Some cases are simple : νCH_3 shifts as expected by a ratio $\simeq \sqrt{2}$.

Bending modes of CH can be expected to shift by a factor of the order of 1.3 (calculated for a linear triatomic molecule[14] ∞-C-D or ∞-C-H. The result depends very little on the choice of the mass of the left hand atom). The δCH_3 rock at 130 meV moves to form a shoulder at 105 meV. The COO^- rocking modes, which, as we pointed out, involve the movement of CH_3 , are also shifted down : 76 meV to 65 meV and 58 meV to 50 meV. The δCOO^- is also slightly affected : 84 meV to 82 meV. But we do not find the $\nu_s \ COO^-$ isolated as hoped. Part of the peak observed at 180 meV may be due to the excitation of new combination modes : i.e. $180 = 115 + 65$ meV.

In conclusion, we have shown conclusively that the correct attribution for the $\nu_s \ COO^-$ and δCH_3 modes of acetic acid are 182 and 175 meV respectively. This is a crucial point in developping a quantitative theory of Inelastic Electron Tunneling Spectroscopy and in determining eventual selection rules.

REFERENCES

1. J. Klein, A. Léger, Phys. Rev. B 7, 2336 (1973).
2. J.T. Hall, P.K. Hansma, Surf. Sci. 76, 61 (1978).
3. N.M.D. Brown, R.B. Floyd, D.G. Walmsley, J. Chem. Soc. Faraday Trans. II 75, 17, 32 and 361 (1979).
4. P.K. Hansma, Physics Reports 30C, 147 (1977).
5. B.F. Lewis, M. Mosesman, W.H. Weinberg, Surf. Sci. 41, 142 (1974).
6. D.G. Walmsley, W.J. Nelson, N.M.D. Brown, R.B. Floyd.
7. O.I. Shklyarevski, A.A. Lysykh, I.K. Yanson, Sov. J. Low Temp.

TABLE 1. IETS MODES OF ACETIC ACIDS ON ALUMINA

CH$_3$COO$^-$ meV	cm^{-1}	CH$_3$C^{18}O^{18}O$^-$ meV	cm^{-1}	Assignment
58	468	58	468	COO$^-$ rock in plane
76	613	75	605	COO$^-$ rock out of plane
84	677	81	653	δ COO$^-$
116	935	114	919	ν C-C + ν Al-O
128	1032	126	1016	CH$_3$ rock
130	1048	130	1048	CH$_3$ rock
166	1339	166	1339	CH$_3$ scissor sym.
175	1411	174	1403	{ CH$_3$ scissor as.
182	1468			{ ν$_s$ COO$^-$
198	1597	195	1573	ν$_a$ COO$^-$
361	2911	362	2919	ν$_s$ CH$_3$
367.5	2969	368	2968	ν$_a$ CH$_3$
371	2992	372	3000	ν$_a$ CH$_3$

CD$_3$COO$^-$ meV	cm^{-1}	Assignment
50	403	COO$^-$ rock in plane
65	524	COO$^-$ rock out of plane
83	670	δ COO$^-$
105.4	847	CD$_3$ rock
116	935	νC-C + νAl-O
127	1024	CD$_3$ scissor sym.
135	1089	CD$_3$ scissor as.
179	1444	ν$_s$ COO$^-$ (see text)
197	1589	ν$_a$ COO$^-$
260	2097	ν$_s$ CD$_3$
275	2218	ν$_a$ CD$_3$

Phys. $\underline{2}$, 328 (1976).

8. J. Lambe, R.C. Jaklevic, Phys. Rev. $\underline{165}$, 821 (1968).
9. K. Ito, H.J. Bernstein, Can. J. Chem. $\underline{34}$, 120 (1956).
10. M. Ito, W. Suetaka, J. Phys. Chem. $\underline{79}$, 1190 (1975).
11. B.A. Sexton, Surf. Sci. $\underline{88}$, 319 (1979).
12. A.V. Kislev, A.V. Uvarov, Surf. Sci., $\underline{6}$, 399 (1967).
13. B.A. Sexton, Chem. Phys. Lett. $\underline{65}$, 469 (1979).
14. G. Herzberg, Infrared and Raman Spectra of Polyatomic Molecules, Van Nostrand Reinhold, New York (1945).
15. H. Knözinger, P. Ratnasamy, Catal. Rev. Sci. Eng. $\underline{17}$, 31 (1978).
16. W.M. Bowser, W.H. Weinberg, Surf. Sci. $\underline{64}$, 377 (1977).

THE ADSORPTION OF UNSATURATED HYDROCARBONS ON ALUMINIUM OXIDE AND

MAGNESIUM OXIDE : AN INELASTIC ELECTRON TUNNELLING SPECTROSCOPY STUDY

W.J. Nelson and D.G. Walmsley

School of Physical Sciences, New University of Ulster

Coleraine, N. Ireland BT52 1SA

ABSTRACT

The tunnel spectra of three unsaturated hydrocarbons (1-heptyne, 3-hexyne and cyclohexene) adsorbed on aluminium oxide and on magnesium oxide are compared. Intensity variations and frequency shifts of the vibrational modes are interpreted in terms of the likely chemisorptive reactions and the molecular orientations on the surfaces.

Of the various techniques used for the study of "Vibrations at Surfaces" Inelastic Electron Tunnelling Spectroscopy (IETS)[1-4] is among the most recent and least exploited. Although more than 20 laboratories have reported results with the technique these have generally been in Institutions with a background of specialist low temperature physics expertise. Happily there is recent evidence that the method is receiving attention from a wider range of physicists and chemists with specific problems to solve.

IETS has high sensitivity : most spectra are taken from a monolayer of adsorbate on an area of ~ 1 mm^2. The resolution can in principle, be made arbitrarily good by reducing the 5.4 kT thermal smearing through lowering the temperature, T, of the sample under study. Operation at 4.2 K in a liquid helium storage dewar is often adequate though useful improvement can be found in the 1 to 2 K temperature range. A particularly attractive feature of IETS is its sensitivity to modes involving the vibration of hydrogen atoms : CH modes can be seen very clearly. Disadvantages

of IETS are that as yet it is severely limited in the range of
adsorbent surfaces that can be studied and it does not allow
reactions conveniently to be monitored in real time.

Here we report the chemisorption of three unsaturated hydrocarbons on aluminium oxide and magnesium oxide. Previously we have
discussed chemisorption of these and other unsaturated species on
aluminium oxide alone[5]. The availability of data from two oxides
for comparison offers a chance to see differences in sorptive bonding and to test previous reaction assignments. This has already
been harnessed in the case of carboxylic acids where downward shifts
in the symmetric carboxylate modes were observed[6] in adsorbates on
magnesium oxide as compared with the sane modes in the adsorbates
on aluminium oxide.

First consider the spectra of 1-heptyne adsorbed on magnesium
oxide and aluminium oxide as shown in Fig 1a and 1b respectively.
In both cases the oxides were grown on the parent metal in an oxygen
dc glow discharge. From previous work[4] we know that the oxides have
hydroxyl species on their surfaces. The spectrum of the adsorbate on
aluminium oxide shows the $\nu(C{\equiv}C)$ mode with moderate intensity at 2106
cm^{-1} (see Table 1). The acetylenic $\nu(C-H)$ band is also clearly distinguishable at 3300 cm^{-1}; its position is close to that in the free
alkynes and it is narrow. Thus we deduce that the acetylenic proton
is not involved in the surface adsorption but the site of reaction
must be the triple bond since the saturated heptane does nor chemisorb under the conditions of these experiments. On bulk aluminas adsorbed 1-alkynes undergo chemisorption by loss of the acetylenic
proton[7-10] as well as self-reduction and polymerisation[8-10]. The $\nu(C{\equiv}C)$
band is less intense than might be expected from the infrared and
Raman spectra of the free species. This is consistent with the triple
bond being held at an angle, θ, well away from the normal to the
oxide surface. (According to the accepted theory tunnelling intensities are thought to vary as $\cos^2\theta$ [11]). The other bands in the spectrum appear with the expected intensities and in the same positions
as in the free species. One mild exception is a low intensity band
close to 1600 cm^{-1} which matches the $\nu(C=C)$ band in the tunnel spectra of adsorbed alkenes[5]. Apparently some sites on the plasma-grown
oxide can cause reduction, possibly by protonation of the triple-
bond π-electron system of the alkyne. This proposition is borne out
by the reduced intensity of the surface oxide $\nu(O-H)$ band.

On magnesium oxide the triple bond of 1-heptyne is almost
entirely reduced to a double bond. This is seen from the very low
intensity of the $\nu(C{\equiv}C)$ band at 2099 cm^{-1} while the broad $\nu(C=C)$
band at 1555 cm^{-1} is correspondingly stronger than on aluminium
oxide. Its lower energy implies stronger binding to the oxide
surface. Also, the band assigned as $\nu({\equiv}C-H)$ has almost completely
disappeared. We can not say whether the acetylenic proton takes
part in the surface adsorption onto magnesium : the $\nu(=C-H)$ modes

Table 1 – IET[a] , Data of 1-heptyne adsorbed on aluminium and magnesium oxide.

Magnesium oxide host Band position in cm⁻¹ (Relative intensity, I_{Mg})	Aluminium oxide host Band position in cm⁻¹ (Relative intensity, I_{Al})	Relative Intensity Ratio I_{Mg}/I_{Al}	Approximate Mode Description
~ 3660[b] (-)	~ 3640 (-)		ν(O-H)
3280[b] (0.02)	3300 (0.25)	0.08	ν(≡C-H)
2931[c] (1.00)	2936 (sh)	-	
2911[c] (1.21)		-	ν(C-H)
2876[c] (1.66)	2894 (3.00)	0.55	
2845[c] (1.36)	2860 (2.39)	0.57	
2099 (0.09)	2106 (0.69)	0.13	ν(C≡C)
1555[b] (0.24)	1610[b] (0.16)	1.5	ν(C=C)
1432 (1.00)	1441 (1.00)	1.00	CH₂ scissoring
1363 (0.88)	1370 (0.69)	1.27	CH₃ deformation
1293 (0.36)	1296 (0.39)	0.92	CH₂ wag and twist
1111 (0.47)	1120 (0.27)	1.74	
1059 (0.59)	1061 (0.43)	1.37	ν(C-C)| CH₂ rock
	940 (-)		ν(Al-C)
659[d] (0.47)	654 (0.49)	0.96	C≡C-H def
420 (0.56)	-		MgO phonon

a. corrected data; b. broad mode; c. better resolved on Mg; d. superimposed on MgO phonon.

Table 2 – IET. Data of 3-hexyne adsorbed on magnesium oxide and aluminium oxide.

Magnesium oxide host Band position in cm⁻¹ (Relative intensity I_{Mg})		Aluminium oxide host Band position in cm⁻¹ (Relative intensity I_{Al})		Relative intensity Ratio I_{Mg}/I_{Al}	Approximate Mode
~ 3670[a]	(−)	~ 3640	(−)	−	ν(O–H)
2961	(1.24)	2956	(3.03)	0.41	ν(C–H)
2915	(1.24)	2915	(2.74)	0.50	
2865	(0.95)	2870	(2.64)	0.36	
2835	(sh)	2830	(sh)		
2764[a]					
2557	(0.14)	2588	(0.16)	0.87	
2235	(0.13)	2234	(0.44)	0.29	ν(C≡C)
1560[a]	(0.36)	1585	(0.12)	3.00	ν(C=C)
1458	(sh)	1444	(1.00)		
1433	(sh)			1.00	CH₂ scissoring ?
1408	(1.00)				
1363	(0.93)	1368	(0.94)	(0.99)	CH₃ deformation
1307	(sh)				
1297	(0.44)	1315	(0.63)	(0.69)	CH₂ wag and twist
1252	(0.26)	1257	(0.39)	0.66	
1151	(0.21)	1159	(0.36)	0.58	

Table 2 – Continued.

Magnesium oxide host Band position in cm^{-1} (Relative intensity I_{Mg})	Aluminium oxide host Band position in cm^{-1} (Relative intensity I_{Al})	Relative intensity Ratio I_{Mg}/I_{Al}	Approximate Mode
1076 (0.87)	1084 (0.36)	2.41	ν(C-C) | CH$_2$ rock
1065 (sh)	1060 (0.41)		
1005 (0.47)	1020 (0.21)	2.23	ν(Al -O)
	940 (-)		
934 (0.22)	889 (1.05)		
894 (sh)			
876 (0.55)			
773 (0.17)	773 (0.37)	0.46	CH$_2$ rock
650 (0.76)			MgO phonon
425 (0.57)	335 (0.57)		C-C $\equiv$ bend
	294 (-)		Al lattice phonon

a. broad

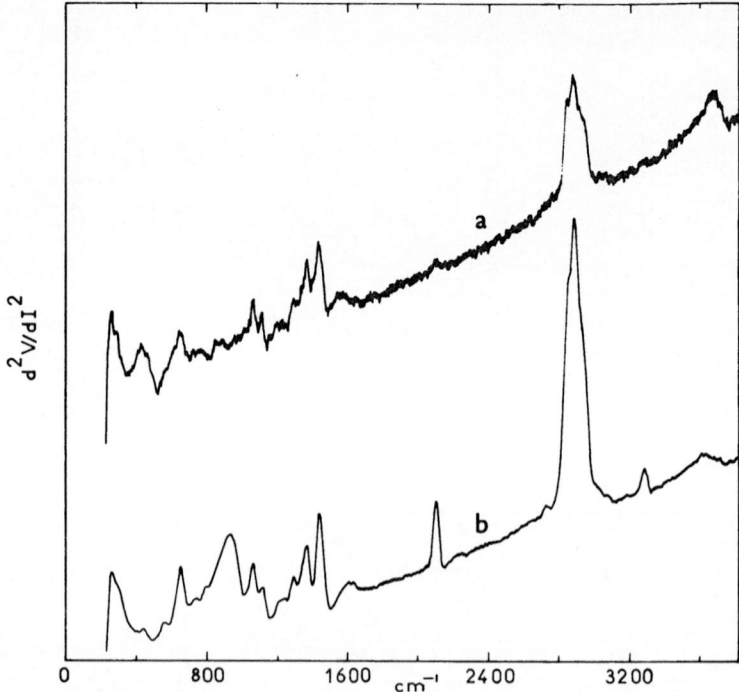

Figure 1 – Tunnel spectrum of 1-heptyne adsorbed on (a) magnesium
oxide, (b) aluminium oxide.

of the terminal bond in adsorbed 1-heptene could not be resolved in
our earlier studies of that system[5]. Some sites on magnesium oxide
are capable of adsorbing 1-heptyne without causing any major perturbation to the triple bond.

The adsorption of 3-hexyne onto magnesium oxide and aluminium
oxide (see Fig. 2 and Table 2) is very similar to that of 1-heptyne.
There is no acetylenic proton in this molecule but the $\nu(C\equiv C)$ bond
is seen quite strongly on aluminium oxide and less so on magnesium
oxide. The $\nu(C=C)$ mode shows a correspondingly greater intensity
on magnesium oxide. The high intensity and good resolution of the
$\nu(C-C)$ modes on magnesium oxide is probably the result of stronger
binding of the C $\equiv$ C bond (note its lower frequency) to the oxide
surface. Magnesium oxide is more basic than aluminium oxide. Also
the magnesium oxide lattice is the more ionic of the two. A
stronger electrostatic binding of the adsorbate to the magnesium
oxide surface is therefore expected. As a result, the orientation
of the $\equiv$C-C bonds may be more nearly perpendicular to the oxide
surface on magnesium oxide and the signal detected will be greater

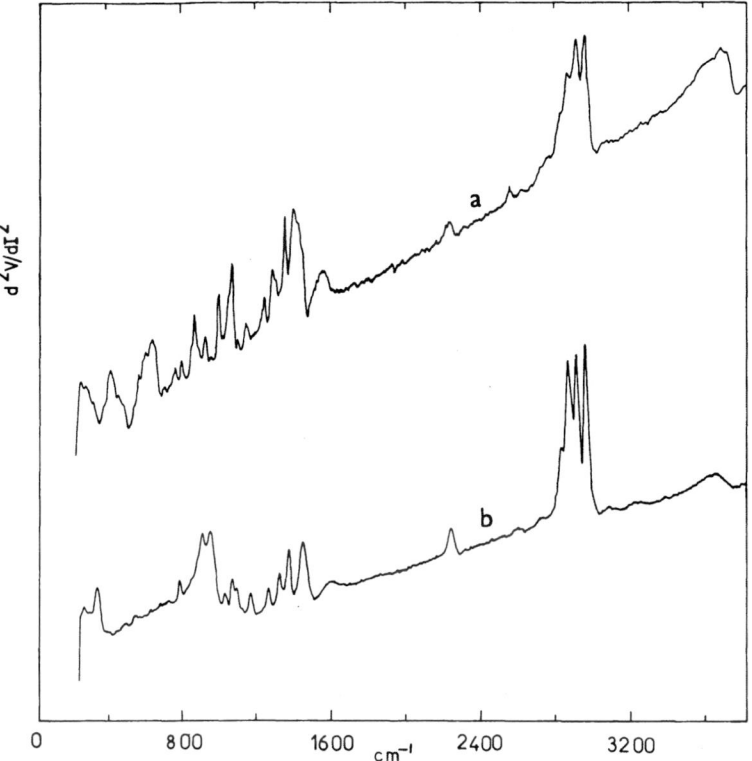

Figure 2 - Tunnel spectrum of 3-hexyne adsorbed on (a) magnesium
          oxide, (b) aluminium oxide.

(as $\cos^2\theta$) in the tunnel spectrum.

     The $\nu(C=C)$ modes of both l-heptyne and 3-hexyne are shifted
to lower energy on magnesium oxide. This too supports the notion
of stronger binding. The fairly large linewidth of this mode on
both surfaces suggests a range of binding strengths at different
sites.

     Binding of the 3-hexyne molecule to the surface at its mid-
point will tend to reduce the coupling of its internal modes and
lead to better resolved lines in the $\nu(C-H)$ region of the spectrum
too; this is observed in both spectra of Fig. 2 as compared with
those of Fig. 1. Our proposed configurations for the adsorbed
l-heptyne are shown in Fig. 4.

     Cyclohexene is a structural isomer of 3-hexyne. When adsorbed

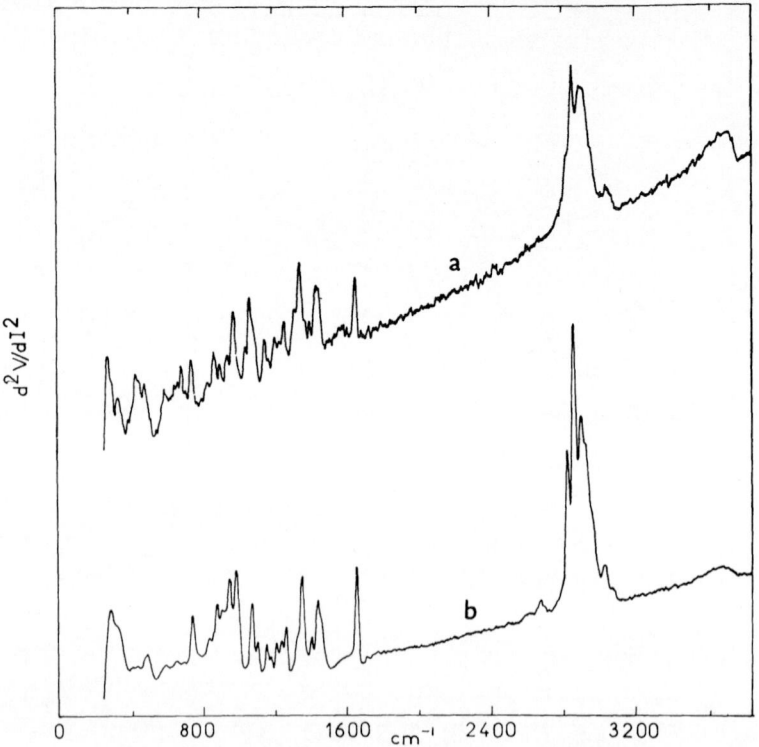

Figure 3 – Tunnel spectrum of cyclohexene adsorbed on (a) magnesium
          oxide, (b) aluminium oxide.

on plasma-grown magnesium oxide and aluminium oxide it yields
the tunnel spectra shown in Fig. 3 (see also Table 3). The ν(C=C)
bond at ~ 1644 cm^{-1} is narrow and moderately strong in both and the
corresponding olefinic ν(C–H) band is resolved at 3013 cm^{-1}
(3022 cm^{-1})on magnesium (aluminium) oxide. By contrast, when
simple alkenes are adsorbed on aluminium oxide[5] the ν(C=C) bands are
relatively weak and broad. Thus we see that the chemisorptive
reaction perturbs the double bond rather less in the cyclic compound. Furthermore, in order to explain the band intensity, the
double bond in cyclohexene may be held more nearly perpendicular
(i.e. at smaller θ) to the oxide surface than shown in Fig. 4.
There is evidence of some perturbed C=C modes in the spectra. On
aluminium oxide the adsorbate gives a low energy tail to the
ν(C=C) mode and on magnesium oxide it displays a separate weak
broad band centred at 1575 cm^{-1}. Both may have their origin in
unusually strong reaction by a small fraction of the adsorbate on
certain surface sites.

Table 3 - IET data of cyclohexene adsorbed on aluminium oxide and magnesium oxide.

Magnesium oxide host. Band position in cm⁻¹ (Relative intensity I_{Mg})	Aluminium oxide host. Band position in cm⁻¹ (Relative intensity I_{Al})	Relative intensity Ratio. I_{Mg}/I_{Al}	Approximate Mode
3640^a	3640 (-)	-	ν(O-H)
3013 (0.32)	3022 (0.4)	0.8	
2941 (sh)	2916 (sh)	-	ν(C-H)
2886 (1.56)	2890 (1.75)	0.89	
	2875 (2.01)	-	
2843 (1.80)	2845 (2.95)	0.61	
2813 (sh)	2815		
2659 (0.14)	2654 (0.19)	0.73	ν(C=C)
1645 (0.70)	1643 (0.99)	0.70	
1575^a (0.12)	1570^a (-)	-	
1434^a (0.64)	1444^a (sh)	-	CH$_2$ scissoring
1388 (0.32)	1424 (0.72)	-	
1358 (sh)	1393 (0.22)	1.45	
1337 (1.00)	1343 (1.00)	-	CH$_2$ wag and twist = C-H in plane deformations
1303 (0.44)	1303 (0.34)	1.00	
1247 (0.42)	1252 (0.47)	1.29	
1219 (0.22)	1227 (0.36)	0.89	
1199 (0.34)	1192 (0.34)	0.61	
1165 (sh)	1156	1.00	
1141 (0.44)	1136 (0.34)	-	
1061 (0.90)	1053 (0.80)	1.29	ν(C-C)| CH$_2$ rock Al-O
970 (0.76)	962^b (0.41)	1.12	
889 {(0.22)	884^b {(0.04)	1.85	skeletal and = C-H out of plane deformations MgO phonon.
319 {(0.10)	819 {(0.05)	5.5 / 2.00	
724 (0.48)	724 (0.47)	1.00	
	673 -		
471 (0.16)	481 (0.22)	0.72	

a. Broad; b. superimposed on Al-O mode.

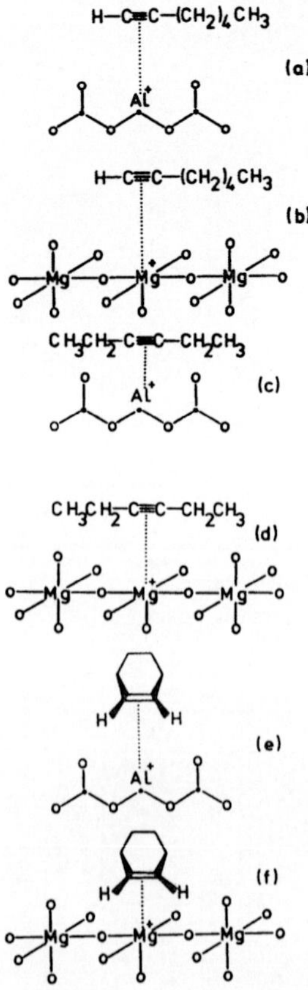

Figure 4 - Proposed configurations for adsorbates :
   (a) 1-heptyne ($HC\equiv CC_5H_{11}$) on aluminium oxide.
   (b) 1-heptyne ($HC\equiv CC_5H_{11}$) on magnesium oxide.
   (c) 3-hexyne ($C_2H_5C\equiv CC_2H_5$) on aluminium oxide.
   (d) 3-hexyne ($C_2H_5C\equiv CC_2H_5$) on magnesium oxide.
   (e) cyclohexene (C_6H_{10}) on aluminium oxide.
   (f) cyclohexene (C_6H_{10}) on magnesium oxide.

        Further studies of unsaturated hydrocarbons on oxides and
oxide-supported metal particles[12],[13] hold out promise of increasing
considerably our knowledge of the chemisorptive and catalytic processes.

ACKNOWLEDGEMENTS .

    We wish to thank Dr. N.M.D. Brown for helpful discussions on
the interpretation of the data. This work was supported by the
Science Research Council, the Department of Education for Northern
Ireland and ICI Corporate Laboratory.

REFERENCES

 1. P.K. Hansma, Phys. Rept. $\underline{30C}$, 145 (1977).
 2. W.H. Weinberg, Ann. Rev. Phys. Chem. $\underline{29}$, 115 (1978).
 3. P.K. Hansma, J. Kirtley, Acc. Chem. Res. $\underline{11}$, 440 (1978).
 4. D.G. Walmsley in Vibrational Properties of Adsorbates (Springer
    Verlag, 1980) to be published.
 5. N.M.D. Brown, W.E. Timms, R.J. Turner, D.G. Walmsley, Journal
    of Catalysis $\underline{64}$, 101 (1980).
 6. D.G. Walmsley, W.J. Nelson, N.M.D. Brown, R.B. Floyd,
    Applications of Surface Science $\underline{5}$, 107 (1980).
 7. J. Saussey, J. Lamotte, J.C. Lavalley, N. Sheppard, J. Chim.
    Phys-chim. Biol., $\underline{72}$, 818 (1975).
 8. M.M. Bhasin, C. Curran, C.S. John, J. Phys. Chem., $\underline{74}$, 3973
    (1970).
 9. S.S. Randava, A. Rehmat, Trans. Faraday Soc., $\underline{66}$, 235 (1970).
10. D.J.C. Yates, P.J. Lucchesi, J. Phys. Chem., $\underline{67}$, 1197 (1963).
11. D.J. Scalapino, S.M. Marcus, Phys. Rev. Lett. $\underline{18}$, 459 (1967).
12. P.K. Hansma, W.C. Kaska, R.M. Laine, J. Amer. Chem. Soc. $\underline{98}$,
    6064 (1976).
13. J. Klein, A. Leger, S. de Cheveigne, C. Guinet, M. Belin,
    D. Defourneau, Surface Science $\underline{82}$, L288 (1979)

VIBRATIONS IN BENZENE AND TOLUENE MONOLAYERS ON THE BASAL PLANE

OF GRAPHITE

R. Stockmeyer and M. Monkenbusch

Institut für Festkörperforschung der Kernforschungsanlage
Jülich GmbH
D-5170 Jülich, FRG

ABSTRACT

The dynamics of an ordered ($\sqrt{7}$ x $\sqrt{7}$) layer of C_6H_6 on microcrystalline graphite (Grafoil, Carbopack) has been studied by neutron spectroscopy. A lattice dynamical calculation on the basis of currently used atom-atom interaction potentials was carried out which explains the experimental data. Similar experiments with $C_6H_5 - CH_3$ in the incommensurate phase showed a strong quasi-elastic scattering intensity which is not present in the commensurate phase but which is also present in the case of a $C_6H_5 - CD_3$ sample. According to distributions of scattered neutrons which have been computed for a 2-dimensional Debye lattice the quasi-elastic scattering intensity is attributed to acoustic modes in the incommensurate adlayer.

INTRODUCTION

Vibrations of rare gas atoms[1] , hydrogen[2] and light hydro-carbon molecules [3,4] have been studied by neutron spectroscopy. With coherently scattering ad-atoms one can in principle observe peaks in the intensity distribution of scattered neutrons at such values of the momentum transfer $\hbar Q$ and the energy transfer $\hbar \omega$ for which single phonon excitations occur. In practice however one can only observe an orientationally averaged inelastic scattering intensity from monolayers on microcrystalline substrates, which yields less direct information on the molecular interactions in question. The incoherent scattering from hydrocarbon molecules gives the opportunity to measure the frequency distribution of the motion of the scattering protons.   Comparing experimental data with theoretical predictions on such frequency distributions one may learn about the molecular interactions.   In the past calculations have been done for an isolated molecule on the graphite surface[5,6,7] only.   Unfortunately this simple situation cannot be studied experimentally.   Besides of technical problems it turned out in the case of C_6D_6 that with decreasing coverage islands of ordered structure are formed.   A homogeneous, low density adlayer can hardly be realised.   This fact indicates the importance of the interactions between the adsorbed molecules.

Benzene layers on graphite have been studied by other methods[8,9,10,11].   A lattice dynamical calculation for this system is not too difficult because the molecule is rather stiff, its form fits to the honeycomb surface structure of the graphite, the C_6H_6 - C_6H_6 interactions can be approximated by a sum of atom-atom potentials[12,13,14] and the same holds for the benzene-graphite interactions[5,6].
The toluene molecule fits less well to the graphite surface potential as indicated by the existence of an incommensurate phase not observed in the case of benzene[15,16].
In section 2. of this paper we present computations on the neutron scattering experiments with benzene and toluene layers on graphite. The experimental data are given in section 3.   Finally in section 4. we discuss what type of calculation should be done to further improve the agreement between theory and experiment.

Table 1.

Intermolecular interaction parameters from [12] for potentials of the form :

$$V_{ij}(r) = A_{ij} \exp{(-B_{ij}r)} - C_{ij}/r^6$$

ij	$A_{ij}[eV]$	$B_{ij}[\text{Å}^{-1}]$	$C_{ij}[\text{Å}^6]$
H...H	149	3.787	1.918
H...C	425	3.669	6.519
C...C	2860	3.611	19.28

Table 2.

Graphite-molecule interaction parameters from [6] for potentials of
the form :

$$V_{ij}(r) = E_{ij} \left\{ (R_{ij}/r)^{12} - 2\,(R_{ij}/r)^6 \right\}$$

ij	$E_{ij}\ [eV]$	$R_{ij}\ [Å]$
$C_M \cdots C_G$	2.956×10^{-3}	3.71
$H_M \cdots C_G$	2.819×10^{-3}	3.21

THEORY

Lattice dynamics of a C_6H_6 monolayer on a graphite [001] surface

The benzene-benzene interactions can approximately be
described by a sum of atom-atom potentials V_{ij} given in table 1.
This approximation is supported by a comparison of theoretical and
experimental data on phonons in bulk benzene[17,18] and justified by
a-priori calculations[19] .   Following[6] we used the atom-atom
potentials of table 2 to represent the interaction between a benzene
molecule with the substrate taking the sum over the carbon atoms
in the two first lattice planes within a distance of 10 Å and
summing over the next 5 carbon layers (Crowell's approximation)[10,31]
we found a minimum in the potential energy of a benzene monolayer
on graphite for the structure shown in Fig. 1.
This structure was confirmed by neutron diffraction data[15] .   The
potential energy then was numerically differentiated at the equilibrium position of the ad-molecules with respect to the center-
of-mass displacements and infinitesimal molecular rotations.
Atomic displacements due to rotation are computed up to the 2nd
order in the infinitesimal rotational coordinates to insure the
correct treatment of all contributions to the 2nd derivatives.
Taking the second derivatives as coupling constants the equations
of motions of the (rigid) C_6H_6 molecule on the (rigid) substrate
were solved in harmonic approximation.   The resulting dispersion
curves of the lattice modes $\omega_\sigma(\vec{q})$; $\sigma = 1, \ldots 6$ are shown in
Fig. 2. for wavevectors $\vec{q}$ along symmetry directions in the surface
Brillouin zone.   At points of high symmetry the center-of-mass and
the rotational motions are decoupled and the corresponding lattice
modes are denoted in Fig. 2. by x,y,z (phonon polarization) and L_x,
L_y, L_z (libron axis) respectively.

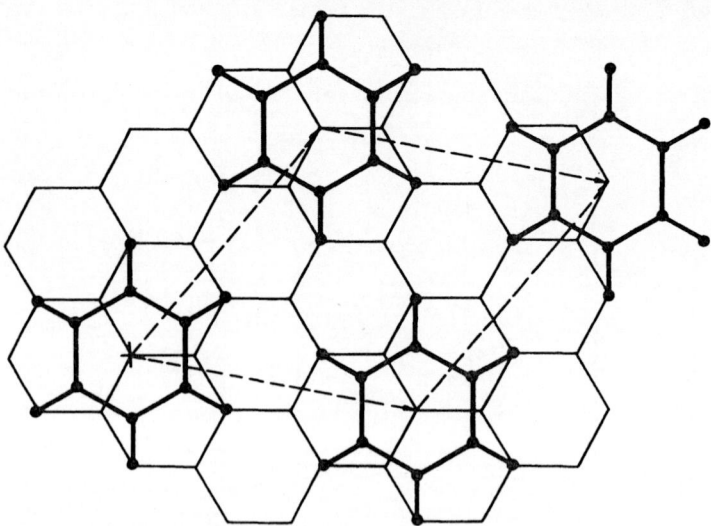

Fig. 1. The minimalisation of the lattice potential energy (sum of
atom-atom potential from Table 1 and Table 2) leads to this
$\sqrt{7}$ x $\sqrt{7}$ structure for a benzene layer on graphite. The C_6H_6
ring lies parallel to the surface at a distance of $Z_0 = 3.2$ Å.
In this structure the surface area per benzene molecule is
36.7 Å^2. (40 Å^2 in a BET monolayer).

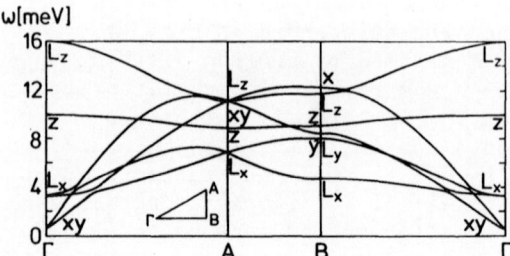

Fig. 2. Dispersion curves of lattice modes in a $\sqrt{7}$ x $\sqrt{7}$ C_6H_6 layer
on a rigid graphite surface with wavevectors along symmetry
directions of the surface Brillouin zone. At points of high
symmetry, Γ, A, B being the corners of an irreducible trian-
gle, the polarisation of center-of-mass vibrations (phonons)
characterized by x, y, z ; librons and the direction of its
axes of rotation are indicated by L_x, L_y, Lz.

The following features of the dispersion curves should be noted for a qualitative understanding of the neutron scattering data (section 3) :

a) Rotations about the surface normal (L_7) give rise to the highest frequencies.

b) The frequencies of phonons with z-polarisation depend weakly on the wavevector thus contributing strongly to the lattice frequency distribution
   $z(\omega) \sim |grad \quad \omega(\vec{q})|^{-1}$ at $\hbar\omega \sim 10$ meV.
   The corresponding proton motion amplitudes are smaller than for a) and c).

c) Long wavelength librons with axis of rotation parallel to the surface occur with small frequencies. Large proton motion amplitudes are to be expected, which may exceed the harmonic region of the atom-atom interaction potentials.

d) Long wavelength phonons with polarization parallel to the surface have a finite frequency, as a displacement of the entire adsorbed layer relative to the substrate requires energy. In consequence there is an energy gap in the lattice frequency distribution; $z(\omega) = 0$ for $\hbar\omega < \omega_c = 0.6$ meV

This energy gap will vanish at a commensurate-to-incommensurate phase transition (toluene), because the "lock-in" of the adlayer into the minima of the surface potential then gets lost.

Neutron scattering cross section
_____

In order to be able to draw conclusions from incoherent, inelastic neutron scattering data, we simulated the experiment by a computation which is based upon the dynamical model presented above. The intensity distribution to be measured from a graphite plus benzene sample as a function of the scattering angle ϑ and the wavelength λ of the scattered neutrons, J_{GB} (λ,ϑ) can be computed as the convolution of the wavelength resolution function of the spectrometer[20,21], $R(\lambda,\vartheta)$, with the "ideal" scattering distribution, S_{GB} (λ,ϑ), which has to be corrected for the detector efficiency and the beam attenuation in the sample by multiplying with a known function F (λ,ϑ)[22] :

$$J_{GB}(\lambda,\vartheta) = \left\{ R(\lambda,\vartheta) \star S_{GB}(\lambda,\vartheta) \right\}_{conv.} \cdot F_{GB}(\lambda,\vartheta) \tag{1}$$

The function $S_{GB}(\lambda, \vartheta)$ contains contributions from different scattering processes :

$$S_{GB}^{01}(\lambda, \vartheta) = \phi_n \cdot \left[S_{GB}^{01}(\lambda, \vartheta) + S_{GB}^{02}(\lambda, \vartheta) \right.$$
$$\left. + S_{GB}^{11}(\lambda, \vartheta) + \ldots \right] \cdot \Delta\lambda \cdot \Delta\Omega \qquad (2)$$

ϕ_n = number of neutrons which have passed the sample
$\Delta\lambda$ = width of the wavelength channel
$\Delta\Omega$ = solid angle covered by the detector at scattering angle ϑ
S_{GB}^{02} = scattering law for two-quanta transitions
S_{GB}^{11} = scattering law for two-fold scattering (elastic plus inelastic)
S_{GB}^{01} = scattering law for one-phonon excitations

For a first comparison, we calculate from our model the dynamic structure function $S_{GB}^{01}(Q, \omega)$ which is related

to $S_{GB}^{01}(\lambda, \vartheta)$ by

$$S_{GB}^{01}(\lambda, \vartheta) = (k/k_o) \cdot S_{GB}^{01}(Q, \omega) \, d\omega/d\lambda$$

$$\vec{Q} = \vec{k} - \vec{k}_o \quad ; \quad k = 2\pi/\lambda \quad ; \quad \omega = (\hbar/2m) \cdot (k^2 - k_o^2) \qquad (3)$$

where $\vec{k}_o$ is the wavevector of the incident neutrons of mass m. In expression (2) the multiple- and the multi- scattering terms S_{GB}^{11} and S_{GB}^{02} respectively are of minor importance and will be discussed later (section 4.).
The main part

$$S_{GB}^{01}(Q, \omega) = S_B^{01}(Q, \omega) + S_G^{01}(Q, \omega) \qquad (4)$$

is a sum of mainly incoherent scattering intensity from the benzene layers, S_{BH} (the scattering cross sections are σ_{inc}^H = 80 barn, σ_{coh}^H = 1.8 barn, σ_{coh}^C = 5.5 barn), and coherent scattering intensity from the graphite, S_G. Interference terms can be neglected. Thus we can interpret the corrected difference intensity

$$J_B(\lambda,\vartheta) = J_{GB}/F_{GB} - J_G/F_G \tag{5}$$

measured with the benzene loaded (J_{GB}) and the degassed graphite sample (J_G) respectively by calculating

$$J_B(\lambda,\vartheta) = \left\{ R(\lambda,\vartheta) \star S_B(\lambda,\vartheta) \right\}_{conv} \tag{6}$$

For this purpose we have calculated on the basis of the lattice dynamics model presented in section 2.1 the incoherent, one-phonon scattering law[25,26]:

$$S_B^{01}(Q,\omega) = \left\langle \frac{\hbar}{2} \sum_{i=1}^{6} \sigma_{inc}^{i} \cdot e^{-2w_i(\vec{Q})} \right.$$

$$\cdot \sum_{\vec{q},\sigma}^{SBZ} \omega^{-1}(\vec{q},\sigma) \cdot \left| \left(\frac{U(\vec{q},\sigma)}{\sqrt{M}} \right) + \vec{\Omega}(\vec{q},\sigma) \times \vec{X}_i \right) \cdot \vec{Q} \right|^2$$

$$\cdot \left. \left[n(\omega) \cdot \delta(\omega + \omega(\vec{q},\sigma)) + (n(\omega) + 1) \, \delta(\omega - \omega(\vec{q},\sigma)) \right] \right\rangle_\Omega \tag{7}$$

The notations used in eq. (7) have the following meaning :
$\langle ....\vec{Q}....\rangle_\Omega$ = average over the orientations of the surface, relative to the direction of momentum transfer $\vec{Q}$ according to the orientational distribution function of the microcrystallites in the Grafoil and Carbopack respectively.

σ_{inc}^{i} = incoherent scattering cross section of proton i in the C_6H_6 molecule.

$W_i(\vec{Q})$ = Debye Waller factor exponent according to the displacement of proton i.

$\omega(\vec{q},\sigma)$ = eigenfrequency of a surface lattice mode with wavevector $\vec{q}$.

σ = 1,2 .....6 accounts for six branches of lattice modes (one molecule per unit cell; 3 center of mass, 3 rotational degrees of freedom)

M = molecular mass

$\vec{X}_i$ = position of proton i in the (rigid) C_6H_6 molecule

$\vec{U}(\vec{q},\sigma)$ = center - of - mass part of the lattice mode at $(\vec{q},\sigma)$

$\Omega_j = \alpha_j/\sqrt{I_j}$ = rotational part of the lattice mode at $(\vec{q},\sigma)$, α_j weighted by the corresponding moment of inertia I_j.

$n(\omega)$ = $(\exp(|\hbar\omega|/kT) - 1)^{-1}$ = (Bose) thermal population factor.

With the scattering law $S_B^{01}(Q,\omega)$ from eq. (7), applying eq.
(3) and eq. (6) we then obtained theoretical predictions for J_B^{01}
(λ,ν) as shown in Fig. 3. and Fig. 4. for Grafoil and Carbopack as
substrate respectively.   In the case of the Grafoil sample the
preferred orientation of the base plane graphite surface parallel to
the neutron scattering plane gives those lattice modes with proton
motions parallel to the surface a larger weight in the angular average of expression (7).   The high frequency peak (L_z modes) is
more intense for Grafoil than for the Carbopack sample.

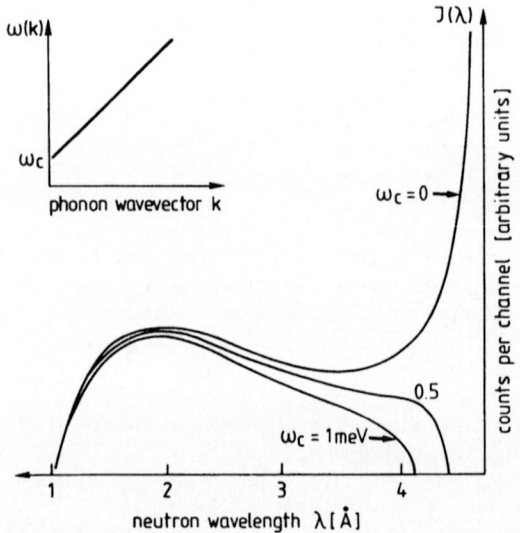

Fig. 3. Incoherent, inelastic neutron scattering from an ordered
        adlayer with linear phonon dispersion curves. Wavelength
        distributions have been computed for the experimental con-
        ditions of the time-of-flight spectrometer SV5-C(FRJ2) at
        a sample temperature T = 100 K ; scattering angle 120°.
        Besides of trivial functions (thermal population factor,
        $\omega \rightarrow \lambda$ scale function) the form of the curves depends on
        the value of the lowest frequency, ω_c, arising from the
        lateral variation of the surface potential.   The gradient
        of the dispersion curve (sound velocity in the assumed
        2-dimensional Debye lattice) determines the inelastic/
        elastic intensity ratio only.

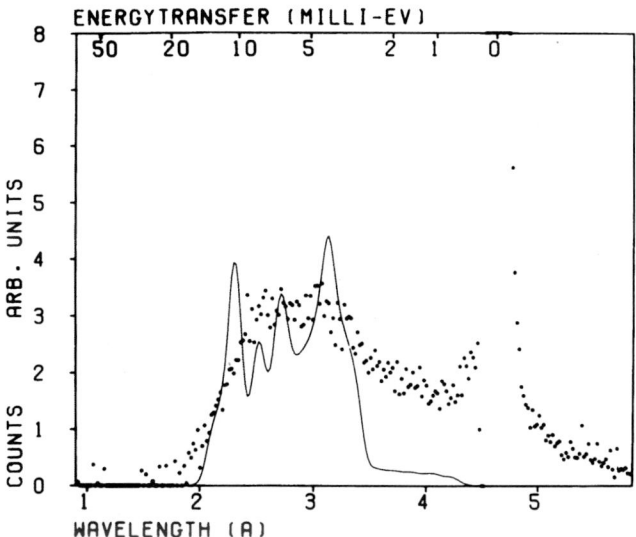

Fig. 4. Wavelength distribution of neutrons scattered from C_6H_6 on
Grafoil (points) ; θ = 0.8 BET monolayer, scattering angle
ϑ = 120°, temperature 100 K. The scattering intensity from
the substrate has been subtracted. The curve is computed
on the basis of the incoherent one-phonon scattering cross
section for the surface lattice dynamics model described
in the text. The λ-resolution of the spectrometer has
been taken account.

Neutron scattering intensity from a 2-dimensional Debye solid

        In order to study the change of the incoherent, inelastic
scattering intensity at the commensurate-to-incommensurate phase
transition of a solid adlayer we have applied a Debye model.
Assuming linear dispersion curves starting at a frequency ω_c
(see Fig. 3.) the lattice frequency distribution is z (ω)$\sim$ (ω- ω_c)
for ω > ω_c and z (ω) = 0 for ω < ω_c.
The wavelength distributions of scattered neutrons to be expected
for the experimental conditions of the spectrometer SV5-C (FRJ2)
have been computed in order to show the near-elastic scattering
intensity connected with z (ω) for $\omega_c \to$ 0.
        The broad peak in the curves of Fig. 3. is due to "trivial"
functions (thermal population factor, $\omega \to \lambda$ scale conversion function). The slope of the assumed linear dispersion curves (sound
velocity or elastic constants of 2-dim. Debye solid) is important
only for the ratio of the elastic to the inelastic incoherent scattering intensity.

EXPERIMENTAL.

Samples and spectrometer

We used a sample of 30 g Carbopack B [27]with a specific
surface area of 100 m^2/g in a sample container of 27 mm diameter
and 120 mm height.   A second sampleholder was filled with 44 g
of Grafoil[28,29] which was oriented with the graphite surface
preferentially parallel to the neutron scattering plane.   A
primary neutron beam attenuation of about 50 - 70 % indicated that
there was strong (002) Bragg reflection ($\vartheta \simeq 91°$).   The Bragg
reflected beam was absorbed to a large extent in horizontal sheets
of Gd, stacked at a distance of 5 mm in the cylindrical sample holder.

Before doing the neutron scattering experiments the samples
were outgassed at 920 K temperature and 10^{-9} Pa pressure.   We
studied C_6H_6, $C_6H_5 - CH_3$ and $C_6H_5 - CD_3$ at coverages of about one
BET monolayer.   Experiments were done at temperatures below 100 K,
where the adsorbates from islands with an ordered structure   on
the graphite surface[15] .   The neutron time-of-flight spectrometer
SC5-C (FRJ2)[20] operated with an ingoing wavelength λ_o = 4.78 Å
(E_o = 3.75 meV) gave a wavelength resolution $\Delta\lambda$ = 0.04 Å nearly
independent of scattered neutron wavelength) which corresponds to
an energy transfer resolution $\Delta\hbar\omega$ = 0.08 meV in the vicinity of
the elastic peak ($\lambda \simeq \lambda_o$).   Distributions of scattered neutrons over
the wavelength 0.5 < λ < 7 Å and the scattering angle 20° < ϑ < 160°
were measured and the intensities from the adsorbate, J_B (λ,ϑ),
were evaluated according to eq. (5) (corrected difference intensity).

Inelastic neutron scattering data

Some of the experimental data from benzene on both substrates
are shown in Fig. 4. and Fig. 5. (points) together with theoretical
curves.   In Fig. 6. (line) data as shown in Fig. 5. have been reduced
in intensity by the thermal population ratio n(ω , T=30 K)/n(ω ,
T= 100 K) and thus compared to data actually taken at 30 K.   The
comparison shows a remarkable disagreement.   Wavelength distributions of neutrons scattered from toluene in the commensurate and in
the incommensurate phase are presented in Fig. 7. For comparison
results from an experiment with adsorbed $C_6H_5-CD_3$ are shown in Fig.8.
In this sample a possible contribution from diffusive modes of the
CH_3 group to the near-elastic incoherent scattering is suppressed
by deuteration.

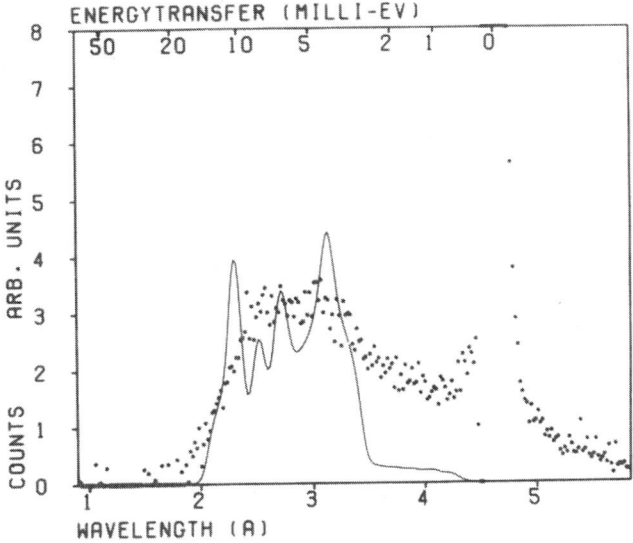

Fig. 5. Wavelength distribution of neutrons scattered from C_6H_6 on
Carbopack B (points) for the same conditions as in Fig. 4,
The experimental and theoretical data curves are different
from those in Fig. 4., because the orientational distribution function of the substrate planes is different.

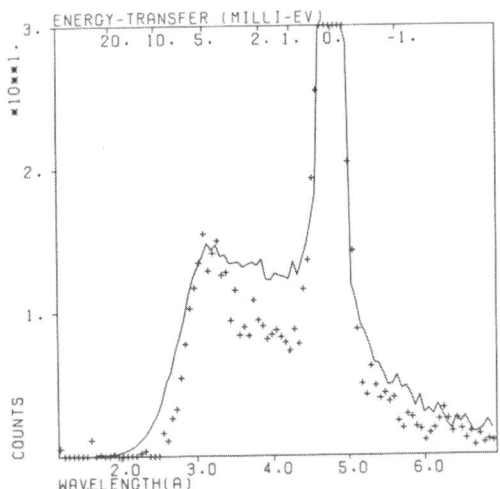

Fig. 6. Wavelength distribution of neutrons scattered from C_6H_6 on
Carbopack B ; $\theta = 0.8$, $\vartheta = 150°$. The data taken at 100 K
and reduced according to the thermal population factor (line)
should coincide with the data taken at 30 K (crosses). The
low lying parts of the phonon dispersion curves connected
with large proton motion amplitudes give rise to anharmonicity effects.

DISCUSSION

Benzene layer lattice dynamics

The lattice dynamical computation for a benzene monolayer
on graphite leads to maxima in the neutron spectra at energies,
where indeed the main part of the inelastic scattering intensity
is found experimentally (Fig. 4,5).   The fine structure of the
computed curves - which include the experimental λ-resolution
already - is not found in the experimental data.   Furthermore
at small energy transfer, $3.5 < \lambda < 4.5$ Å, the measured intensity
is larger than the calculated one.

We first discuss multiple-scattering and multi-phonon transitions (eq. 2) as possible reasons for the differences between
theory and experiment.   We rewrite eq. 5 including the first three
terms of eq. 2. :

$$J_B (\lambda, \omega) = J_B^{01} + J_B^{02} + J_B^{11} \tag{8}$$

The elastic-inelastic double-scattering term J_B^{11} , describing
an elastic scattering process in the graphite followed by a one phonon
deexcitation in the benzene layer (or vice versa), is of the
form

$$J_B^{11}(\lambda, \omega) \sim \int d\omega \int d\delta \ J_B^{01}(\lambda + \delta, \omega) \ W(\omega) \ g(\delta)$$

The average over the scattering angle ω with the weight
function $w(\omega)$ does not disturb the structure of the spectra
with respect to the variable λ , because the incoherent, one-
quantum scattering terms $J_B^{01}(\lambda, \omega)$ contains ω in a slowly variable
intensity factor.   The wavelength dependence is involved because
the neutrons spent a short time between the first and the second
scattering event in the sample, thus the delayed neutrons arriving
at the detector are attributed to a shifted wavelength $\lambda + \delta$.
The probability distribution function $g(\delta)$ however is nonzero
only for values smaller than the resolution width $\Delta\lambda$, which includes
the   effect of sample geometry.   In our calculated curves a possible broadening of inelastic peaks due to multiple scattering
is overestimated because we have taken the elastic peaks from the
experiments without adsorbates as the resolution functions $R(\lambda, \omega)$.
These peaks are slightly asymmetric towards the long wavelength
side because of the double-elastic scattering which is much
stronger than the elastic-inelastic scattering intensity.

The two-phonon term $J_\beta^{02}(\lambda,\nleftarrow)$ in eq. 8. has been computed from the one phonon term J_β^{01} ^{22} . It turned out to contribute a broad distribution, which is numerically important only beyond the benzene lattice frequency cut-off.

Another possible improvement of the presented computation of neutron spectra concerns the dynamical coupling between the ad-sorbate and the substrate (section 2.2). A calculation for a simple model, with parameters adapted to meet benzene on Grafoil in the continuum limit, indicates that effects due to this coupling have a negligible influence on our experimental results.[22]

An important restriction of the computation (presented in section 2) lies in the harmonic approximation. In order to see anharmonic effects experimentally we reduced neutron intensities, which have been measured at 100 K, by the ratio of the thermal population factors, $n(\omega$, T= 30.K)/n(ω , T=100 K) and compared the resulting data with spectra measured at 30 K (Fig. 6.). Apparently at $3.5 < \lambda < 4.5$ Å there is a larger increase of inten-sity with temperature than accounted for by $n(\omega,T)$ in the neutron scattering cross section (eq.7) evaluated in the frame of harmonic lattice dynamics. Experiments at much lower temperature should be made in order to get results compatible with the T = o lattice theory. As to the lack of structure in the ,neutron spectra (Fig. 4 and Fig. 5) the small size of the 2-dimensional crystals (about 15 x15 C_6H_6 molecules) may be most important. The corresponding broadening of the Bragg reflections has been accounted for in the neutron diffraction work[15] by replacing the $\delta(Q - Q_{hk})$ function by a Gaussian. Correspondingly one could approximately describe the finite size effect on the lattice dynamics by a broadening of the dispersion curves (Fig. 2.) thus representing the effect that the eigenfrequencies and eigenvectors of a set of microcrystals are distributed over a band in the (ω,q) plane. (In this case the eigenstates are no longer plane waves with wavevectors $\vec{q}$.)

Toluene

The slight modification of the benzene molecule has a large influence on the monolayer structure[22] and a commensurate - to - incommensurate phase transition occurs at 70 K. temperature. Neutron spectra from a monolayer of $C_6H_5-CH_3$ on Grafoil are shown in Fig. 7 and 8 for sample temperatures of 50 K and 100 K respectively. The apparent change of the spectra is attributed to the change of lat-tice dynamics connected with the phase transition at 70 K. The gene-ral shape of the spectra is in agreement with the theory presented in Fig. 3, assuming that in the incommensurate phase $\omega_c = 0$. An alternative interpretation could invoke an onset of diffusive motions, especially hindered rotations of the CH_3 group.

In order to rule out the interpretation of the near elastic scattering intensity on the basis of assumed CH_3 group diffuse motions we also studied a C_6H_5-CD_3 sample in which the incoherent scattering from the methyl group is reduced by a factor of 20. As Fig. 8. shows, the scattering intensity which we connect with the existence of acoustic phonons in incommensurate adlayers is not removed by deuteration of the methyl group.

SUMMARY

Comparing computed and measured neutron spectra from a benzene monolayer on graphite, we find that the lattice dynamics calculation for an infinite harmonic, plane lattice on a rigid substrate yields maxima in the density of states in a region of frequencies, for which the experimental data also exhibit the main part of the scattering intensity, but the theory is obviously not sufficient for a detailed interpretation of the measured intensities. For small energy transfers, $0 < \hbar\omega < 4$ meV, it has been shown experimentally that anharmonic effects are important.

The change of neutron spectra observed at the commensurate - to - incommensurate phase transition of toluene on graphite is attributed to a vanishing energy gap of the adlayer lattice frequency distribution.

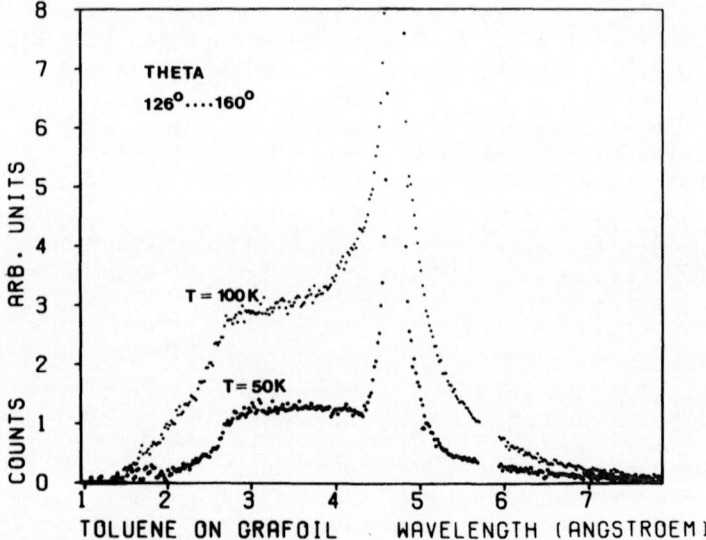

Fig. 7. Wavelength distribution of neutrons scattered from a monolayer of C_6H_5-CH_3 on Grafoil in the commensurate phase (50 K)
and in the incommensurate phase (100 K). Note the increase
of the quasi-elastic scattering intensity, $4 < \lambda < 4.7$ Å,
and compare with Fig. 3.

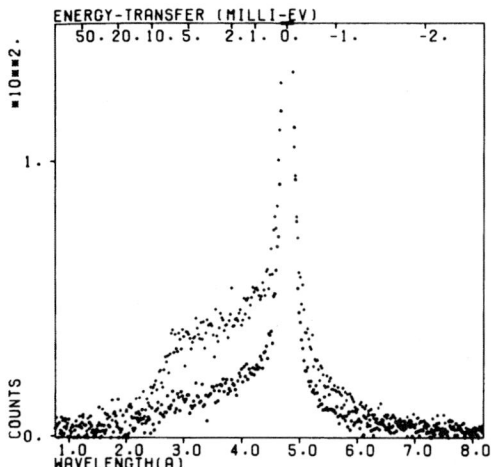

Fig. 8. Wavelength distribution of neutrons scattered from a monolayer of C_6H_5–CD_3 in the incommensurate phase at T = 100 K
for two values of the scattering angle (ϑ = 37° points ;
ϑ = 142° squares). Compared to normal toluene the scattering intensity from the methyl group is reduced by a factor
20 by deuteration. The methyl group mobility does not seem
to be responsible for the quasi-elastic scattering intensity
but in the incommensurate phase of the adlayer low frequency
– long wavelength modes with amplitudes parallel to the
surface exist in spite of the coupling to the substrate.

ACKNOWLEDGMENT

    We gratefully acknowledge stimulating discussions with Prof.
H. Stiller.

REFERENCES

1. H. Taub, K. Caneiro, J.K. Kjems, L. Passell, J.P. Mc Tague,
   Phys. Rev. B 16, 4551 (1977).
2. M. Nielsen, W.D. Ellenson, J.P. Mc Tague in "Neutron Inelastic
   Scattering 1977" Vol. II, p. 433, IAEA Wien, 1978.
3. H. Taub, Proceed. of the Int. Conf. and Symp. on Electron
   Tunneling, Springer Verlag 1977, p. 170.
4. H. Taub, H.R. Danner, Y.P. Sharma, H.L. Mc Murry, R.M. Brugger,
   Phys. Rev. Lett. 39, 215 (1977).
5. C. Pisani, F. Ricca, R. Dovesi, Proceed. 2nd Int. Conf. on
   Solid Surfaces, Jap. J. Appl. Phys. Suppl. 2, pt 2, 269 (1974).

6. L. Battezzati, C. Pisani, F. Ricca, J. Chem. Soc. Faraday
    Trans. II, 71, 1629 (1975).
7. F.Y. Hansen, H. Taub, Phys. Rev. B 19, 6542 (1979).
8. B. Boddenberg, J.A. Moreno, Z. Naturforschung 31a, 854 (1976).
9. S. Ross, J.P. Olivier, "On Physical Adsorption", Wiley Inter-
    science, New York, 1964.
10. W.A. Steele, "The Interaction of Gases with Solid Surfaces",
    Pergamon, Oxford 1974.
11. N.N. Avgul, A.V. Kiselev, Chemistry and Physics of Carbon 6, 1
    (1970).
12. D.E. Williams, J. Chem. Phys. 47, 4680 (1967).
13. D.E. Williams, Acta Crystallogr. A30, 71 (1974).
14. H.A. Govers, Acta Crystallogr. A31, 380 (1975).
15. M. Monkenbusch, R. Stockmeyer in Proceed. of the Int. Conf. on
    "Ordering in Two Dimensions" Lake Geneva (1980), North Holland
    Publ. Comp. Ansterdam.
16. M. Monkenbusch, R. Stockmeyer, to be published in Ber. Bunsen
    Ges. Phys. Chemie (1980).
17. E.L. Bokhenkov, V.G. Fedotov, E.F. Sheka, I. Natkaniec,
    M. Sudnik-Hrynkiewicz, S. Califano, R. Righini, Il Nuovo
    Cimento 44B, 324 (1978).
18. B.M. Powell, G. Dolling, H. Bonadeo, J. Chem. Phys. 69, 2428
    (1978).
19. C. Huiszoon, F. Mulder, Molecular Physics 38, 1497 (1979).
20. R. Stockmeyer, Jül.-Report 1162 (1975).
21. D. Glasenapp, H.J. Stornik, R. Wagner, Jül-Spez - 68 (1980)
    ISSN 0343 - 7639.
22. M. Monkenbusch, to be published, Jül.-Report.
23. G. Armand, P. Masri, L. Dobrzynski, J. Vac. Sci. Technol. 9,
    705 (1971).
24. A.D. Novaco, J.P. Mc Tague, Phys. Rev. B20, 2469 (1979).
25. W. Ludwig, "Festkörperphysik", Akademische Verlagsanstalt
    Wiesbaden 1978.
26. G. Gilat in "Methods of Computational Physics", Vol. 15,
    Academic Press, London 1976.
27. Supelco Inc. Bellefonte, PA Bulletin 738 B.
28. Trade name by Union Carbide for a foil, consisting of partially
    oriented graphite microcrystals.
29. J.G. Dash, "Films on Solid Surfaces", Academic Press, London
    (1975).
30. S.C. Ying, Phys. Rev. B 3, 4160 (1971).
31. A.D. Crowell, J. Chem. Phys. 22, 1397 (1954).

PHOTOACOUSTIC AND PHOTOTHERMAL SURFACE SPECTROSCOPY

S.O. Kanstad, P.E. Nordal

Laser and Applied Optics Laboratory,p.o. Box 303

Blindern, Oslo 3, Norway

ABSTRACT

A specimen subjected to pulsewise illumination, may experience pulsating temperature excursions in its surface layers. The amplitude of those temperature fluctuations can be made to depend on the absorption spectrum of the sample, and may be measured photoacoustically through sound waves generated in an enclosing chamber or in a piezoelectric substrate, or photothermally by observing similar pulsations in thermal reradiation.

We discuss the application of those techniques to the spectroscopic analysis of surfaces and surface species. In particular, the photoacoustic reflection – absorption spectroscopy (PARAS) technique appears to have sensitivities better than 10^{-2} monolayer. Photothermal radiometry (PTR) offers improved flexibility by enabling contact-free measurements to be made. Calculations indicate monolayer sensitivity for PTR, but this has not yet been experimentally verified. On the other hand, PTR provides particular advantages at higher temperatures as will be shown. With PARAS and PTR techniques, observations of surfaces can be made under ambient conditions on naturally occurring samples. Though still under development, therefore, those techniques promise to facilitate real-time observations of surface processes during reaction.

ENERGY TRANSFER PROCESSES IN GAS AND SURFACE REACTIONS

B. Kasemo

Department of Physics, Chalmers University of Technology
S-412 96 Göteborg, Sweden

INTRODUCTION

The transformation of chemical energy in a chemical reaction
takes place via the elementary excitations of the reacting system.
In a gas phase reaction these are the electronic, vibrational,
rotational and translational degrees of freedom modes.   When the
reaction is exothermic the (chemical) reaction energy is channeled
into one or several of these excitations, initially in a non-thermal
manner.   In an endothermic reaction some excitation is necessary
to make the reaction go.   During the reaction the excitation energy
is then converted to chemical energy.

The detailed participation of the excitations in a reaction can
only be revealed by experiments, where the state of excitation of the
reactants and reaction products are carefully controlled.   Such
experiments are the beam and crossed-beam experiments.   Kinetic
data alone have, for such purposes, very limited value, since they
represent averages over the thermal distributions of reactant
excitations[1].
The situation is illustrated in Fig. 1. for a reaction A+BC $\rightleftarrows$ AB+C,
with an activation barrier for the reaction.   The one-dimensional
potential-energy curve can only be specified for given orientations
and directions of the reactants and for specified internal excitations.
Furthermore, for each velocity of the species A, all collisions with
BC molecules having a velocity larger than a critical value (related
but not equal to the activation energy for reaction) may result
in a reaction.   Another velocity of A will give a new set of BC
velocities that may cause reaction.   With internal excitations of
A and/or BC new thresholds will in general be found.   Kinetic

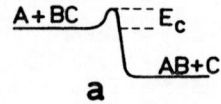

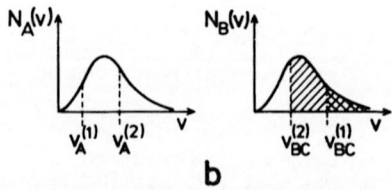

Fig. 1. a) Potential energy along the reaction coordinate for
          an exothermic reaction, A + BC → AB + C with acti-
          vation barrier E_c.
       b) Maxwell-Boltzmann velocity distribution functions
          for the reactants.   Species A, with velocity $v_A^{(n)}$,
          reacts only with species B with velocity $\geq v_{BC}^{(n)}$,
          due to the activation barrier.   Kinetic data
          are obtained as averages over the thermal distribu-
          tions of reactant velocities and internal excitations.

data are the measured, folded result of many such collision events,
averaged over the thermal distribution of velocities and internal
excitations of reactants.   It is obvious that the extraction of
microscopic details from such data is at least difficult, if not impossible.

In a beam experiment under single-collision conditions most of the parameters that are unspecified in kinetic experiments can be controlled.   The velocities, directions, and internal excitations of reactants and products can in principle be measured and varied to give detailed reaction cross sections.

Similar considerations are valid also for the reactions of gas-phase molecules with solid surfaces.   However, in contrast to the gas phase case there are to date very few studies of energy transfer in molecule-solid reactive scattering.   A rapid expansion of the field is foreseen, however.   One modest aim of this paper is to point out that the same kind of powerful information as in gas phase experiments can be obtained from molecule-solid reactive-scattering experiments.   This is done by first describing some of the energy transfer processes observed in reactive gas-phase beam experiments.   Observed energy transfer in recent molecule-solid experiments is then reviewed.   Finally, the similarities and most pronounced differences between the homogenous and heterogenous systems are discussed.

CLASSIFICATION OF SCATTERING

Collisions between two or more molecules are conveniently divided into three different types, namely elastic, inelastic, and reactive scattering, defined in the following way :

Type of scattering	Energy exchange		
Elastic scattering $A+BC \rightarrow A+BC$	translational	$\leftrightarrow$	translational
Inelastic scattering $A^{\star}+BC \rightarrow A +BC^{\star}$	translational + internal $\begin{array}{l}\text{rotational}\\\text{vibrational}\\\text{electronic}\end{array}$	$\leftrightarrow$	translational + internal
Reactive scattering $A+BC \rightarrow AB^{\star}+C$	translational + internal + chemical	$\leftrightarrow$	translational + internal + chemical

*indicates the possibility of internal excitation

In elastic scattering, say between two molecules A and BC, only translational energy is exchanged. Inelastic scattering may involve both translational and internal energy exchange, but the molecules A and BC remain intact. In the reactive scattering, finally, one or several molecular bonds are broken and new ones are formed, so that the products of the reactive scattering are different from the initial colliding species (the reactants). The reactive scattering, which is the only kind dealt with here, involves conversion between translational, internal and chemical energy.

ENERGY TRANSFER IN GAS PHASE REACTIONS

Here we will discuss four examples of energy transfer, two involving transfer of chemical energy to vibrational energy in exothermic reactions, one showing the reverse process in an endothermic reaction, and one illustrating chemical to electronic energy transfer. The examples have been selected to illustrate in a simple way, how internal molecular excitations come into play in chemical reactions.

Chemical energy $\leftrightarrow$ vibrational energy transfer.

The first example is taken from the extensively studied reaction $F+H_2 \rightarrow HF^{\dagger} + H$,[2] which efficiently channels the released chemical energy into vibrational excitations (denoted by $\dagger$) of the product HF molecules. In the $F+H_2$ and $F+D_2$ reactions the vibrational levels with quantum numbers ν_{HF} equal to 1,2 and 3 and ν_{DF} equal to 1, 2, 3 and 4, respectively are populated in a very nonthermal manner, as shown in Fig. 2. It should be noted that a product HF (DF) molecule in the highest populated vibrational state $\nu_{HF}=3$, $\nu_{DF}=4$) corresponds to an almost complete conversion of the reaction energy into vibrational energy. These results were obtained by infrared chemiluminescence, i.e. by measuring the intensities of the radiative decay, $E_{\nu+1} \rightarrow E_{\nu} + h\nu$ of the vibrational excitation[3]. Similar results were obtained in an IR-chemiluminescence study of the $H+Cl_2 \rightarrow HCl^{\dagger}+Cl$ reaction[4]. Here the product HC molecules were observed to have vibrational excitations in the range $\nu \lesssim 6$, the highest number again corresponding to almost 100 per cent conversion of chemical to vibrational energy. The peak in the population occurred at $\nu = 3$.

Application of the "principle of microscopic reversibility" to the exothermic reactions discussed above suggests that endothermic reactions of the same kind should be enhanced by vibrational excitation of one of the reactants. This has been verified by several experiments. One example is the reaction

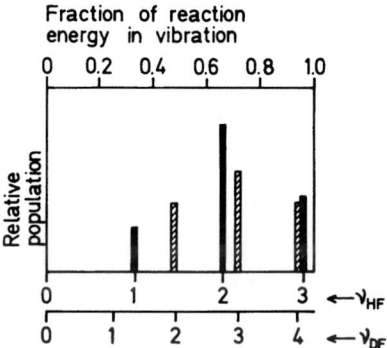

Fig. 2. : The diagram shows the relative populations of the
vibrational levels of HF (DF) produced in the
reaction F + H$_2$ → HF$^+$ + H, and in the corresponding
reaction with D$_2$.   The upper horizontal axes gives
the fraction of the total reaction energy channeled
into vibrational energy (from Ref. 3.).

$$Br + HCl(\nu) \xrightarrow{k_\nu} HBr + Cl$$

which is endothermic by 0.66 eV.   By exciting the reactant
HCl (ν=0) to HCl(ν=2) by a pulsed HCl laser, Arnoldi et al[5] showed
that the reaction rate constant k_ν increases by eleven orders of
magnitude, i.e., $k_2/k_0 \sim 10^{11}$.   The reaction is schematically
illustrated in Fig. 3.

Chemical energy → electronic energy transfer.

A larger number of reactions have been studied, where product
molecules are left in electronically excited states. The radiative
decay of the excitation, chemiluminescence, can be observed spectroscopically and then yield information about the reaction path[6].
The following example of visible chemiluminescence serves the double
purposes of illustrating chemical to electronic energy transfer, and

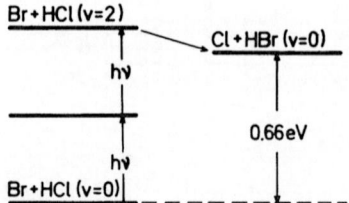

Fig. 3. : Schematic illustration of how vibrational
            excitation of HCl induces the reaction with
            Br (redrawn from Ref. 5.).

of yielding data for comparison with the corresponding molecule-
surface reaction (sec. 4). The example is the class of reactions
$M_2 + XY \rightarrow MX^* + MY$, where M_2 is an alkali-dimer and XY a halogen dimer
(for example Cl_2 or ICl).
This reaction is one of several possible reaction channels, when an
alkali beam reacts with a halogen beam or with a low-pressure
halogen gas. The chemiluminescence spectrum has been identified as
resulting from the radiative decay of the weakly bound, electronically excited alkali halide state (MX^*) to the ground state[7].

The results can be discussed in terms of an "electron jump (harpooning)" mechanism. A somewhat naive but conceptually attractive illustration of the reaction sequence is shown in Fig. 4. When the two dimers come close enough, the valence electron of the alkali-dimer "jumps" to the XY molecule and thanks to the Coulomb attraction, a four-body complex is formed. It eventually separates

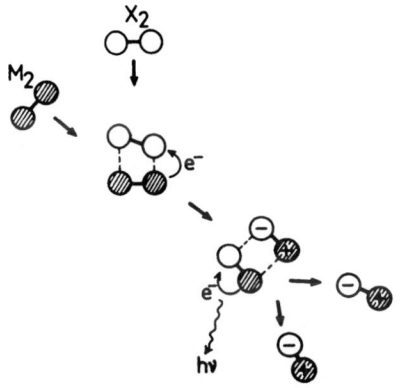

Fig. 4. : Schematic illustration of the four center
reaction $M_2 + X_2 \rightarrow MX + MX^* \rightarrow MX + MX + h\nu$
(M = alkali metal atom, X = halogen atom).

as electronically excited MX^* and ground state MY molecules. It was noted that when X and Y were different, emission takes place from the halide molecule with the least electronegative halogen atom. The photon is emitted when MX^* makes a transition to one of the ground state vibrational levels, or analogously, when the second electron "jumps" from the M valence level to the affinity level of X.

The process is illustrated in Fig. 5. for the case of a two
body collision between, say Na and Cl. The two potential curves
(full lines) represent the ionic and covalent states correlating
with separated ions and ground state atoms, respectively [8]. In
the adiabatic approximation (dashed curves) these states, which are
of the same symmetry, do not cross (symmetry forbidden or avoided
crossing). At finite velocity, however, crossing between the
states may take place [8]. The radiative recombination to a bound
NaCl molecule occurs with a very small probability, $\sim 10^{-5}$ per
collision [9] as estimated from the ratio of the collision time
($\sim 10^{-13}$ s) and the radiative lifetime of the excited state ($\sim 10^{-8}$s).

Another but equivalent way of viewing the curve crossing and
the photon emission is shown in the lower part of Fig. 5., which
illustrates the shift in the Cl affinity level, A, with respect to
the Na valence level, I, as a function of distance, r, between the
atoms. At large internuclear distances A lies above I and the lowest
energy state corresponds to neutral ground state atoms. As r
decreases, A shifts downwards with respect to I. The curve crossing
region at r_c corresponds to the situation where A and I are at the
same energy. As r decreases further A falls below I and the ioni-
cally bound state is the lowest energy state. (This picture is
of course an oversimplification not taking into account, for
example, the mixing of states). The radiative recombination can
now be viewed as an electron transfer from the Na valence level I to
the Cl affinity level when it lies below I, i.e., when $r < r_c$.
In the two body collision this occurs with a very small probability,
as mentioned above. The dominating process is the elastic collision,
which may take place either by [1] an (adiabatic) electron transfer
at r_c, when the atoms approach, followed by the reverse process,
again at r_c, when the atoms separate, or [2] by no electron transfer
at all.

ENERGY TRANSFER IN SURFACE REACTIONS

In this section some of the few published observations of
energy transfer in surface reactions are discussed.
Only such examples are selected, where some microscopic picture
of the reaction dynamics emerges.

Surface chemiluminescence

Several observations have now been made of photon emission
in the reaction between gas phase molecules and metal surfaces
(a rather complete reference list is found in ref. 10). The
example brought up here is the surface chemiluminescence observed
when a solid Na surface reacts with halogen molecules[11]. The

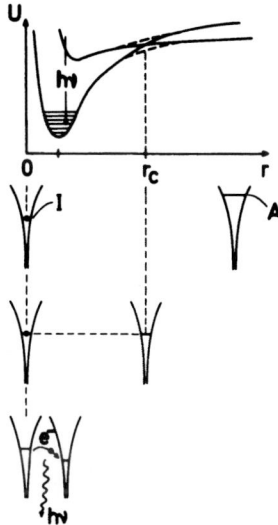

Fig. 5. : a) Ground and excited state potential energy
curves for an alkali-halide dimer. The
curve crossing (solid curves) is an avoided crossing in the adiabatic approximation,
as shown by the dashed curves. The photon
emission is due to the electronic transition from the excited state to the ground
state.

b) Shows an alternative way of viewing the
photon emission and curve crossing. As the
alkali-atom and halogen-atom approach each
other the affinity level of the latter is
shifted downwards relative to the ionization
level of the alkali-atom. In the curve
crossing region they are at the same energy
and at shorter internuclear distances the
halogen affinity level is the lower of the
two levels. An electron jump at $r < r_c$ is
a non-adiabatic process equivalent to the
radiative transition shown in Fig. 5a.

spectra consist of broad emission peaks (FWHM ~ 0.5 eV) in the visible, and the emission probability per reacting molecule was estimated to be in the range $10^{-6} - 10^{-7}$.   Calculated spectra, using a theoretical description of the photon emission[12] adopting a physical picture resembling that for the corresponding gas phase reaction (ref. 7 and sec. 3 above), are in good agreement with the experimental spectra.   The picture of the Na-halogen surface chemiluminescence is visualized in Fig. 6.   Like in the gas phase reaction the first reaction step is assumed to be the electron jump or "harpooning" process, where a Na valence electron is transferred to the Cl_2 affinity level.   As the Cl_2^- molecule comes closer to the surface it eventually dissociates leaving an essentially neutral Cl atom at the surface.   The affinity level of a Cl atom close to a Na surface will fall considerably below the Na Fermi level, as illustrated in Fig. 6b, i.e. a neutral Cl atom configuration represents an excited (hole) state.   The photon emission is attributed to the radiative filling of this hole state in analogy with the corresponding gas phase process.

       The major difference between the gas phase and surface reactions is the presence of the continuum of substrate electron states (both filled and unfilled) in the latter case.   This causes an inherent broadening of the emission band in addition to that caused by transitions taking place at different Cl-surface distances, and by lifetime effects.   The continuum of states open up efficient channels for decay of the adsorbate hole as schematically shown in Fig. 7.[10]   .   The resonance tunneling in a) is most likely the dominating process determining the total lifetime of the hole. Taking the observed band width (0.5 eV) as a measure of the inter-action strength we obtain an estimate of the total lifetime of $\sim 10^{-15}$s.   Assuming the radiative lifetime to be $\sim 10^{-8}$ s as in free molecules one then arrives at a probability for photon emission (Fig. 7c) of $\sim 10^{-7}$ per elementary reaction, in agreement with the experimental findings.   The picture of Fig. 7 is also consistent[10] with the observation of (exo)electron emission[13] simultaneously with the photon emission.   The electron emission can be produced either by the direct Auger decay of the hole state (Fig. 7b).   A prerequi-site is of course that the hole excitation energy is larger than the Na work function.

Infrared surface chemiluminescence

       As in the case of visible chemiluminescence form electronically excited species one might expect that also IR chemiluminescence from vibrationally excited reaction products should yield information about surface reactions.   There are two reasons why IR surface chemiluminescence will be much harder to detect, however.   The

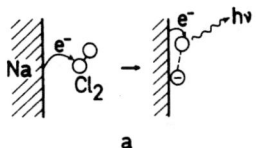

a

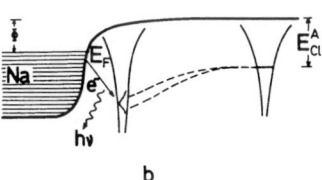

b

Fig. 6. a) Schematic illustration of the surface chemiluminescence observed in the $Na_{solid} + Cl_2$
reaction.

b) Illustrates the shift, broadening, and radiative filling of the affinity level of a Cl
atom approaching a Na surface (in the particular case of Cl, in contrast to the situation
with Br and I the affinity level lies below
E_F at infinite Na - Cl separation).

first and most important one is the usually very long radiative
lifetime $(1 - 10^{-3}$ sec) for vibrational excitations, which makes
radiative decay extremely unlikely at a surface.   On a metal the
total lifetime appears to be of order 10^{-12}s due to interaction
with the metal valence electrons.   On insulators and semiconductors
decay via phonons will probably give a longer, but still very
short total lifetime.   The second reason is that the much less
($\leqslant 10^{-3}$) sensitive detection of IR photons as compared to
visible ones further decreases the possibility to detect IR surface
chemiluminescence.   There is, however, one example where such
emission has been observed [14], namely in the chemical etching of
silicon by XeF_2.   In this reaction XeF_2 reacts with Si to form vola-
tile SiF_4 molecules, some of which radiate at characteristic vibrational frequences (whether the emission really occurs from SiF_4 or
from intermediate reaction complexes SiF_n, $n < 4$ is not clear[15]).
The main reason that the emission could be observed in spite of the
very low emission probability is the large number of elementary
reactions per unit area and time taking place in the chemical
etching reactions.

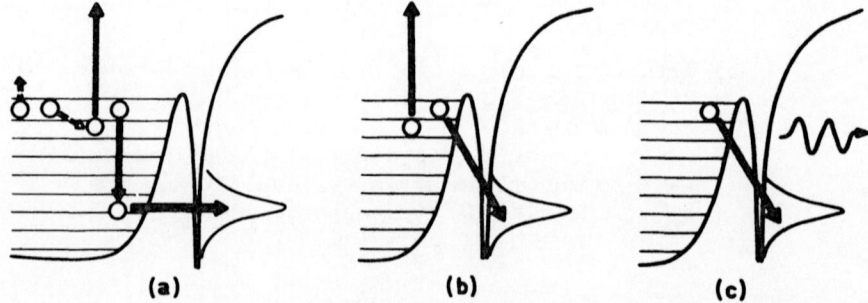

Fig. 7. : Possible decay processes of an excited hole
          state created during the approach of an atom
          to a metal surface.   The three figures illus-
          trate a) resonance tunneling followed by a
          cascade of electron hole pair excitations,
          b) an Auger decay process and c) the radiative
          deexcitation (from Ref. 10.).

Desorption of translationally hot hydrogen molecules

Several observations have been made of desorption of translationally hot H_2 or D_2 molecules from metals (for references see ref. 16). In the most recent work it was demonstrated that D_2 molecules desorbing from sulphur covered Pd (100) belong to either of "two clearly separable groups of molecules, fast and slow ones"[16] and that the relative amount of fast D_2 is strongly dependent on sulphur coverage. It was concluded that an activation barrier for *adsorption* was not sufficient to explain the results. The high translational energy was instead associated with an activation barrier for *absorption*. In the proposed model absorbed hydrogen atoms recombine and desorb without equilibrating in the chemisorbed state.

Desorption of vibrationally excited N_2 from Fe

Another recent publication reports on the first direct observation of vibrationally excited desorption products[17]. The experimental set up consisted of an Fe foil, fed with N_2 on its backside, and an electron-impact-induced fluorescence arrangement to detect vibrationally excited N_2 desorbing from the front side of the Fe foil. N_2, dissociatively adsorbed at the backside diffused as atoms to the front side where association to molecules and subsequent desorption took place. The desorbing N_2 molecules were irradiated by an electron beam. The fluorescence spectra of the produced excited N_2^+ molecules shows that the N_2 molecules desorbing from clean foils have a vibrational "temperature" of 2600 K i.e. much higher than the Fe foil temperature (1150-1325 K). This result has been attributed to the existence of an activation barrier for N recomination and fast desorption without energy relaxation after recombination[17].

As in the case of D_2 desorption from Pd the degree of excitation appears to be sensitive to the amount of sulphur contamination. Increasing sulphur coverage *increased* the fraction of translationally hot D_2 (from Pd) but *decreased* the fraction of vibrationally hot N_2 (from Fe). This observation may be an indication that the sulphur influences the distribution of the excess energy between vibrational and translational energy.

Production of translationally hot CO_2 in the catalytic oxidation of CO

The catalytic oxidation of CO on Pt (111) produces CO_2 molecules with very high (0.6 eV) kinetic energy[18] . This observation has been made in an experiment where a Pt (111) surface with a small oxygen coverage was exposed to a pulsed CO beam. The observed kinetic energy is larger than the estimated available reaction energy for a reaction between an adsorbed O atom and an adsorbed CO molecule. As in the previous two examples the result has been explained in terms of an activation barrier for the reaction. An alternative explanation, which does not appear to be ruled out by the experimental data, is that the reacting CO molecules do not equilibrate on the surface, but performe precursor motion and react as hot surface molecules[19].

COMPARISON BETWEEN SURFACE AND GAS PHASE ENERGY TRANSFER PROCESSES

The possible energy transfer processes in bimolecular reactions is severely restricted by the requirement of energy, momentum, and angular momentum conservation rules. If a third body is participating these rules are relaxed and the reaction cross section is usually increased. The quantization of the internal excitations (rotations, vibrations, electronic excitations) impose additional conditions on the possible energy transfer processes.

At a surface the situation is quite different. The solid can more efficiently supply or pick up momentum via its continuous phonon spectrum. The phonons can also transfer energy, although at rather small amounts per event ($\lesssim 10^{-2}$ eV) if only single phonon processes are considered. On metals the valence band electrons can in principle dissipate any amount of energy, which is a situation drastically different from gas phase reactions. Insulators and semiconductors also have continuous electronic excitation spectra but only above the threshold value given by the band gap.

These inherent differences between gas and surface reaction conditions must be reflected in the behaviour of the reacting systems. There is at present a great lack of both experimental results and theoretical understanding in this area, however. The situation with the "sticking problem" serves as a good illustration.

The theoretical interest in this problem has increased considerably during the last few years, yet such questions as the relative importance of phonon and electronic excitations as energy

dissipation channels in chemisorption, and thus for sticking, on
metals are still unanswered.   The phonon mechanism has been treated
at various levels of sophistication [20a] and the electron hole pair
mechanism has been treated in "friction" models[20b] and "localized
perturbation" models [20c].

The experimental information on the sticking problem is in
most cases restricted to a single number (for a given combination
of solid surface and gas), expressing the probability that a gas
molecule stricking the surface adsorbs (sticks).   The dependance
of the sticking coefficient on velocity and direction is only rarely
given.   With few exceptions it is impossible to deduce from the
available experimental results,whether a non-unity sticking coefficient is due to inefficient energy transfer or to an activation
barrier for adsorption.

A great step forward is taken when the energy loss spectra of
the particles that do not adsorb is measured, may be as a function
of velocity and angle of incidence[21] .   An elegant illustration
of the information that can be obtained from such measurements was
given recently by Hurst et al [22].   They measured the transit time
between the source and the detector for a beam of Xe atoms
interacting with a Pt (111) surface.   From the results the inves-
tigators deduced that the Xe interaction with Pt is predominantly
inelastic.   The data also revealed two reaction channels, one
representing a direct inelastic scattering without trapping at
the surface, and one representing trapping and thermalization of
the Xe atoms followed by thermal desorption.

The differences and similarities between gas and surface reactive scattering discussed above have some important experimental
and theoretical implications.   In both situations much can be
learned about the reaction path by state selection of the incoming
molecules, and by measurements of the angular distribution and
energy state (internal and translational) of the reaction products.

As briefly discussed above the much shorter lifetimes of
internal molecular excitations at surfaces may rule out several of
the powerful techniques used to study gas phase reactions.   Fluorescence and chemiluminescence techniques, for example, may in many
cases not be applicable due to too efficient radiationless deexci-
tations.

The many body character of the molecule-solid scattering is
a severe theoretical complication in comparison with, for example,
a bimolecular reaction.   On the other hand the possibility of careful control of the orientation of adsorbed reactants in gas-surface scattering is a major advantage over the situation in gas phase
scattering.   The manyfold of solid substrates that can be choosen
for gas-surface scattering experiments make a bridging over to the

gas phase scattering case possible.   Scattering of a reactive molecule on a metal surface is as discussed above very different from the corresponding gas phase reaction.   Scattering between a gas molecule and a physisorbed molecule, perhaps on an inert substrate, or between a gas molecule and a molecular solid, however, is much more resembling the gas phase case.

In conclusion I hope to have demonstrated that in studies of reaction dynamics at surfaces much can be learned from the field gas phase reaction dynamics, but also that there are basic differences between these fields, associated with the solid state properties of surfaces.

ACKNOWLEDGEMENT

The author is grateful to B.I. Lundqvist for many valuable comments and discussions during the preparation of this manuscript. The financial support from the Swedish Natural Science Research Council (Grant No. E3106-101) is gratefully acknowledged.

REFERENCES

1. For a discussion of this point see J. Nicholas, "Chemical Kinetics. A modern survey of gas reactions". (Harper and Row Publishers London 1976).
2. For recent results and references see R.F. Heidner III, J.F. Bott, G.E. Gardner, and J.E. Melzer, J. Chem. Phys. 72 (1980) 4815.
3. J.C. Polanyi and K.B. Woodall, J. Chem. Phys. 57 (1972) 1574.
4. J.C. Polanyi in Chemical Kinetics, Physical Chemistry Series One, Vol. 9 p. 135, J.C. Polanyi ed. (Butterworths, London, 1972).
5. D. Arnoldi, K. Kaufman and J. Wolfrum, Phys. Rev. Lett. 34 (1975) 1597.
6. See e.g. M.F. Golde and B.A. Thrush, in : "Advances in Atomic and Molecular Physics", Vol. 11, Eds. D.R. Bates and B. Bederson, (Academic Press, New York 1975) p. 361 and ref. 1 and 4.

7. R.C. Oldenborg, J.L. Gole and R.N. Zare, J. Chem. Phys. 60 (1974) 4032, and W.S. Struwe, J.R. Krenos, D.L. McFadden and D.R. Herschbach, J. Chem. Phys. 62 (1975) 404.
8. For a discussion of these potential curves see T.F. O'Malley, in : "Advances in Atomic and Molecular Physics", Vol. 7, Eds D.R. Bates and I. Esterman (Academic Press, New York, 1971) p. 223.
9. G. Herzberg, "Molecular Spectra and Molecular Structure: I. Spectra of Diatomic Molecules" (Van Nostrand, Princeton, N.J., 1950) 2nd ed pp 400-405.
10. B. Kasemo, E. Törnqvist, J.K. Nörskov and B.I. Lundqvist, Surface Sci. 80 (1979) 179.

11. B. Kasemo and L. Walldén, Surface Sci. 53 (1975) 393.
12. J.K. Nörskov, D.M. Newns and B.I. Lundqvist, Surface Sci 80 (1979) 179.
13. See Refs. 10, and 11, and L. Himmel and P. Kelly, Comments Solid State Phys. 7 (1976) 81.
14. T.J. Chuang, Phys. Rev. Lett., 42 (1979) 815.
15. T.J. Chuang, private communication.
16. G. Comsa, R. David and B.J. Schumacher, Surface Sci. 95 (1980) L210.
17. R.P. Thorman, D. Anderson and S.L. Bernasek, Phys. Rev. Lett. 44 (1980) 243.
18. C.A. Becker, D.J. Auerbach and L. Wharton, to be published.
19. J. Harris, B. Kasemo and E. Törnqvist, submitted to Surface Sci.
20. a) R.M. Logan and R.E. Stickney, J. Chem. Phys. 44 (1966) 195, R.M. Logan and J.C. Keck, J. Chem. Phys. 49 (1968) 860, T.R. Knowles and H. Suhl, Phys. Rev. Lett. 39 (1977) 141, G.P. Brivio and T.B. Grimley, Surface Sci. 89 (1979) 226, E.K. Grimmelmann, J.C. Tully, and M.J. Cardillo, J. Chem. Phys. 72 (1980) 1039,
G. Doyen and T.B. Grimley, Surface Sci. 91 (1980) 51, W. Brenig, Z. Phys. B 36 (1980) 227, R. Sedelmayer and W. Brenig Z. Phys. B. 36 (1980) 245.
20. b) E. Müller-Hartmann, T.V. Ramakrishnan, and G. Toulouse, Solid State Comm. 9 (1971) 99, E.G. d'Agliano, P. Kumar, W. Schaich, and H. Suhl, Phys. Rev. B. 11 (1975) 2122, W. Schaich, Surface Sci. 49 (1975) 221.
20. c) J.K. Nörskov and B.I. Lundqvist, Surface Sci. 89 (1979) 251 R. Brako and D.M. Newns, Solid State Comm. 33 113 , (1980), XX, K. Schönhammer and O. Gunnarsson, Phys. Rev. B 22 (1980), J.W. Gadzuk and H. Metiu, Phys. Rev. B. 22 (1980), XXX, J. Chem. Phys. (to be published), Proc. of ICSS IV, Cannes, Sept. 1980, J. Gadzuk(to be published), B. Gumhalter, Proc. of ICSS IV, Cannes, Sept. 1980 Z. Pensar and M. Sunjic (unpublished).
21. When the interaction is strong i.e. when the adsorption well is >> kT$_{gas}$, the velocity and direction of the molecule at the surface will be determined by the acceleration in the well. In such case the original velocity and direction of motion, far from the surface, is of minor importance.
22. J.E. Hurst, C.A. Becker, J.P. Cowin, K.C. Janda, L. Wharton, and D.J. Auerbach, Phys. Lett. 43 (1979) 1175.

ELECTRON-HOLE PAIRS, MOLECULAR VIBRATIONS, AND RATE PROCESSES AT METAL SURFACES

J.W. Gadzuk and Horia Metiu[*][†]

National Bureau of Standards
Surface Science Division
Washington, D.C. 20234

ABSTRACT

Consequences of the coupling of nuclear motion of an atom or molecule near a metal surface with the electron-hole pairs excitations of the metal are considered from the point of view of vibrational spectroscopy. Special emphasis is placed on the interrelationship between pair excitation, surface localized vibrational structure, and rate processes. Specific realizations discussed here in terms of vibrational spectroscopy include :

a. Pair renormalization of intramolecular vibrational modes.
b. Desorption rates.
c. Pair excitation, trajectory theories, and vibrational modes.
d. Reaction rate theory at surfaces.

INTRODUCTION

In the recent past, considerable theoretical attention has been directed toward the formulation of models which describe the dynamics of various atomic and molecular processes at surfaces. The models we have in mind emphasize the qualitatively unique features of a metallic substrate which make a theory of surface reaction dynamics more than just a simple extension of gas phase molecular reaction theory taken in the large molecule limit[1]. A central rôle

---

* A.P. Sloan Fellow and Camille and Henry Dreyfus Teacher-Scholar

† Permanent address : Department of Chemistry, University of California, Santa Barbara, California 93106

of the substrate, beyond that of providing an infinite momentum
source for elastic scattering of an incident particle beam, has
been that of a heat bath. Both the lattice phonons[2] and the conduction band electron-hole pair excitations[3] can exchange energy
with the incident (reactant) beam in a way which might affect the
outcome of a reactive event. With this in mind, it is desirable
that a theory of surface dynamics, in one way or another, deals
with intrinsically non-adiabatic effects, both vibrational and
electronic. Thus the simple idea of letting particles roam around
on ground state adiabatic potential energy surfaces is not sufficient. A proper theory of surface reaction dynamics must include
provisions for treating on an equal footing :

i)The electronic states of the incident (reactant) and trapped or
   scattered (product) particles.
ii) The states describing the nuclear motion of incident and scattered (continuum) or trapped (quasi-discrete) particles.
iii) The excited electronic or lattice states of the substrate and
   ways for handling energy flow (and transitions) between the
   subspaces of these states.

In this article we discuss possible connections between
reactive surface processes and quantities accessible in vibrational
spectroscopy of adsorbed atoms and molecules such as Franck-Condon
factors, fundamental and overtone frequencies, and linewidths. In
Section II we argue that if the localized vibrational frequency
is substantially larger than the substrate Debye frequency, then
the large (~ 20 cm^{-1}) linewidths observed experimentally are
caused by the fact that the oscillating molecular charge excites
electron-hole pairs in the metal. Thus careful linewidth measurements could yield information pertaining to the coupling of
nuclear motion with substrate pairs which is required input in
theory of reactive surface scattering[4]. As a first example of the
relation between vibrational properties of adsorbed atoms or
molecules and elementary surface processes, we show in Section III
that the vibrational linewidth, due to electron-hole pair excitation, is determined by the same factors which directly influence
desorption rates. Our views of surface reaction dynamics[3f, 4, 5]
involving quasi-adiabatic (diabatic) transitions to bound (chemisorbed) states in which the rates are influenced by pair excitation,
nuclear bound continuum Franck-Condon factors, and vibrational
excitation spectra, are summarized in Section IV. One widely
invoked picture of molecular and surface scattering is based on
semi-classical scattering theory[6] in a limit which we will call
the trajectory approximation. An illustrative classical model
demonstrating the connections between trajectories, vibrational
excitations, and energy transfer to the substrate, as applicable
to problems of adsorption or sticking rates within the context of
our surface reaction theory is presented in Section V. Again,
spectroscopic information may aid in understanding the dynamics

of the rate of population of the surface bound state.

VIBRATIONAL LINE-SHAPES

Recently, polarization modulation infrared reflectivity measurements have produced vibrational line-shapes that are not instrumentally broadened. Furthermore, use of single crystal surfaces and coverages that give good LEED pattern indicate that the broadening may be homogeneous. It has been stated on many occasions that the line-shape is determined by anharmonic interactions between the molecular vibration and substrate phonons. Model calculations do not bear out this assumption. The physical reason for this is that the energy mismatch between the molecular (~ 2000 cm^{-1}) and phonon ($\leqslant 350$ cm^{-1}) frequencies is too great. When the line-shape is computed by perturbation theory in the anharmonic coupling, the virtual processes invoked must conserve energy in order to contribute significantly. Because of the frequency mismatch, it takes roughly at least six phonon excitations to match the energy of one molecular vibration. Such processes can appear only in higher orders of the perturbation theory. However, such terms are substantially reduced since they are multiplied by high powers of small anharmonic constants. As a result the width given by anharmonicity is very small ($\sim$ 2 to 6 cm^{-1}). Note that for the vibrations of the chemisorptive bond, (e.g., carbon-metal stretch of chemisorbed CO) having frequencies of the order 500 cm^{-1}, the anharmonic effects can be substantially larger than for the other modes (e.g., carbon-oxygen stretch)[9].

This indicates that another broadening mechanism must be invoked to explain the large widths observed experimentally[8,10]. For this purpose let us model the chemisorbed diatomic molecule by two charges q_1 and q_2 connected springs, as shown in Figure 1. The oscillating charges located at $\vec{R}_1$ and $\vec{R}_2$ couple to a metallic electron, located at $\vec{r}$, through coulomb forces. We expand this interaction in powers of the nuclear displacements $\delta \vec{R}_i$ ($\vec{R}_i = \vec{R}_i{}^{eq} + \delta \vec{R}_i$), and keep the linear term, which couples the molecular vibration to the electron. Transforming from nuclear displacements to normal modes (of amplitude δQ_α), we can generate a vibration-electron coupling of the form :

$$V_{ev} = \delta Q_\alpha \int g_\alpha^o(\vec{r}) \hat{\rho}(\vec{r}) d\vec{r}, \qquad (1)$$

where $\hat{\rho}(\vec{r})$ is the electron-density operator and $g_\alpha^o(\vec{r})$ is the "bare" electron-phonon coupling. It is important to keep in mind that – in the simplest model – $g_\alpha^o(\vec{r})$ is proportional to the gradient of the coulomb interaction between the charges q_1 and q_2 and the metallic electron[11].

The vibrational line-shape is given by the imaginary part of the phonon Green function [8]:

$$D(\omega) = - i \int_{-\infty}^{+\infty} dt < 0|T\delta Q(t)\delta Q(o)|0 > e^{-i\omega t} \qquad (2)$$

Here $\delta Q(t)$ is the vibrational amplitude in Heisenberg representation, and T is the time ordering operator and $|0>$ is the vibrational ground state.

The procedure used to compute $D(\omega)$ is similar to that followed in studies of electron-phonon coupling in metals[12] with certain modifications caused by the presence of the surface. There are three major physical effects that must be included :

Vertex Renormalization by Screening

The use of the "bare" coulomb interaction between the "effective" molecular charges and the metallic electron at $\vec{r}$ is inadequate. The charge q_1 interacts simultaneously with all the electrons of the metal pushing them out from their unperturbed positions, thus causing the appearance of an induced charge density. This acts so as to diminish the interaction between q_1 and the electron at $\vec{r}$. This is the same type of mechanism as the one leading to the screened asymptotic (large $\vec{r}$) interaction $e^2 r^{-1} e^{-\lambda r}$ between two charges located in a bulk metal.

The effective interaction can be computed by using, for example, the random phase approximation (RPA)[12,13]. The diagrams of Figure 2 illustrate the processes included. In 2a, q_1 and e interact through a "bare" coulomb potential. In 2b, q_1 excites an electron-hole pair at $\vec{r}'$, which recombines at $\vec{r}''$, and interacts with the electron at $\vec{r}$, etc. Summing up all these terms to infinite order we obtain the "renormalized vertex" $g_\alpha(\vec{r})$:

$$g_\alpha(\vec{r}) = g_\alpha^0(\vec{r}) + \iint d\vec{r}' d\vec{r}' \, g^0(\vec{r}') \Pi(\vec{r}',\vec{r}'') \, |\vec{r}''-\vec{r}|^{-1} \qquad (3)$$

where $\Pi(\vec{r}',\vec{r}'')$ is the RPA polarization of the metal. The second term represents the total effect of the polarization of the medium : the action of the molecular charges at $\vec{r}'$ (through g^0) polarizes the medium at $\vec{r}''$; the polarization charge induced at $\vec{r}''$ (given by $\int d\vec{r}' \, g^0(\vec{r}') \, \Pi(\vec{r}',\vec{r}'')$) interacts with the electron at $\vec{r}$ through the coulomb potential $|\vec{r}''-\vec{r}|$. The total effect is obtained by integrating over all the intermediate points $\vec{r}'$ and $\vec{r}''$ in the metal.

The Change of Phonon Frequency and Width

Now that we have a properly screened electron-phonon interaction $g_\alpha(\vec{r})$ we can proceed to compute the influence of the metal electrons on the vibrational line-shape. This is illustrated by

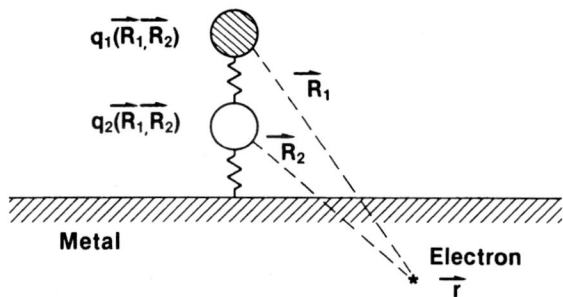

Fig. 1 – A simple model for the interaction of a vibrating diatomic
molecule with the electrons in the metal.

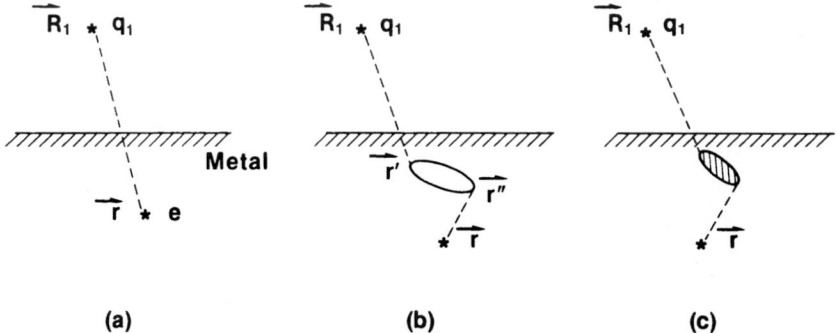

Figure 2 – Diagrams entering into the computation of the screening
of the interaction between an oscillating molecular
charge q_1 and an electron in the metal.

(2a) Interaction through the bare coulomb potential.

(2b) First order diagram indicating the electron–hole
pair excitation.

(2c) The RPA screening, representing the result of the
interactions (26) summed to infinite order.

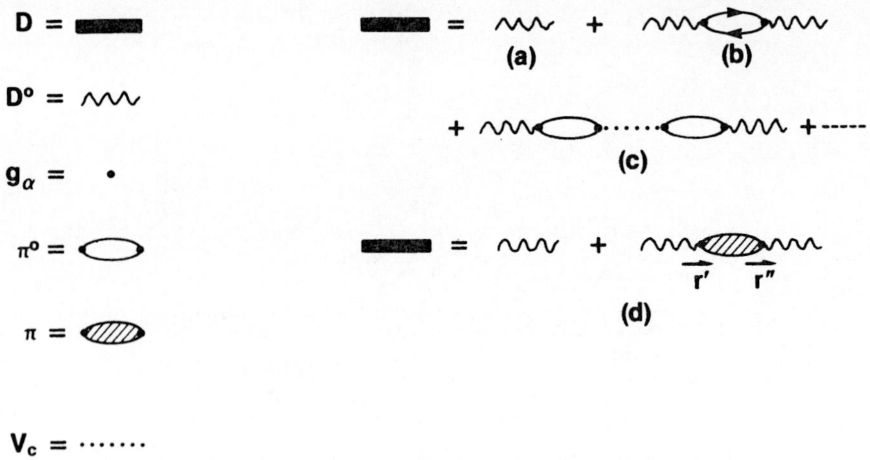

Figure 3 - The diagramatic symbols used in computing the phonon
line-shape. D is the phonon Green function with the
effect of metal electrons included. D^o is the vibrational
Green function in the absence of the interaction with
the electrons. g_α is the renormalized electron-phonon
interaction for the mode α. Π^o is the free electron
polarization. Π is the RPA polarization. V_c is the bare
coulomb potential. Figures 3a, b, c present successive
terms in the perturbation theory. 3d gives the effect
of electron polarization (within RPA) on the vibrational
Green function.

Figure 3 which gives successive terms of perturbation theory. In
zeroth order D equals the phonon Green function in the absence of
the electron-phonon coupling g_α. In first order the phonon may
excite an electron-hole pair which later returns the energy to the
vibration. In second order (Figure 3a) the vibration transfers
energy to an electron-hole pair, which recombines to excite another
electron-hole pair, which returns the energy to the molecule, etc.
Summing to infinite order we obtain Figure 3d. The presence of the
metal electrons has modified the line-shape, adding to it the
second term, representing an event in which the vibration acts on
the electron system at $\vec{r}$ (through the coupling $g_\alpha(\vec{r})$, represented
by a dot) and induces a charge at $\vec{r}'$. This acts back on the
vibration modifying it. This is a "self-energy" contribution in the
sense that the vibration is modified by the polarization induced
by its own action on the electron system.

As a result, the vibrational Green function becomes :

$$D_\alpha(\omega) = \frac{2\Omega_\alpha}{\omega^2 - \Omega_\alpha^2 - 2\Omega_\alpha \Sigma_\alpha(\omega)} \qquad (4)$$

where ω is the phonon-frequency, Ω_α is the normal mode frequency and $\Sigma(\omega)$ is the self-energy :

$$\Sigma_\alpha(\omega) = \int d\vec{r} \, d\vec{r}' \, g_\alpha(\vec{r}) \, \Pi(\vec{r},\vec{r}';\omega) g_\alpha(\vec{r}') \qquad (5)$$

Note that the imaginary part of $D_\alpha(\omega)$ is the line-shape. Since $\Sigma_\alpha(\omega)$ is complex, one can easily show that if $|\Sigma_\alpha(\omega)| << \Omega_\alpha$ or is slowly varying with ω for ω in the neighbourhood of Ω_α, then the line is Lorentzian, $Re\Sigma_\alpha(\Omega_\alpha)$ gives the frequency shift and $Im\Sigma_\alpha(\Omega_\alpha)$ the line-width.

The Metal Polarization

The metal affects the vibration by "screening" the vibration-electron interaction and by adding a self-energy to the vibrational line-shape. Both these effects can be computed if the polarization $\Pi(\vec{r},\vec{r}';\omega)$ is known. The presence of the surface modifies this quantity and makes it rather tedious to compute[14]. One cannot use the bulk value since the effect on a molecule located very close to the surface is mainly determined by the polarization in the interface region. For static calculations one may use density-functional theory[15]. This may give an adequate qualitative result for the frequency shift[16] but it would not give the line-width. The latter is a genuinely dynamic quantity which depends on the electron-hole pair excitations, which in turn depend on the vibrational frequency. In recent work[16] the RPA has been used to compute Π for a jellium model. Numerical applications to the line-width problem are in progress.

It is useful to conclude this discussion by emphasizing that the theory presented here is essentially electrodynamic. The polarization effects would reduce to image charge effects if the phenomelogical Maxwell equations were valid. The self-energy in this case would be proportional to $(\varepsilon(\omega) - 1)(\varepsilon(\omega) + 1)^{-1}$ (from the image formula[17]) where $\varepsilon(\omega)$ is the complex, frequency dependent dielectric constant of the bulk metal. The imaginary part of this quantity gives the line-width. Unfortunately, the wavenumber independent image dipole formula breaks down[14b,18] when charges are very close to the surface. The main reasons for this are as follows : (1) The dielectric response varies continuously through the surface, from the bulk value to that of the vacuum. The phenomenological theory replaces this variation with a discontinuity. Since the molecule is very close to the surface, naive application of this

procedure could yield substantial errors. (2) The field exerted by
the molecule on the metal is spatially inhomogeneous. Therefore
its Fourier transform contains high wave vector components. The
polarization of the metal depends on wave vector (non-locality,
or equivalently, spatial dispersion). If the dielectric constant
measured optically is used in the image formula, spatial dispersion
is implicitly neglected since the photons have practically zero
wave vector. (3) Because of these two factors a microscopic calculation of the frequency dependent Π is required; the use of
phenomenological theory is questionable.

THE RELATION TO THE KINETICS OF DESORPTION

The Master Equation for the Desorption Process

We discuss in this section the rôle of electron-hole pairs in
the desorption process which is a kinetic problem closely connected
to vibrational spectroscopy. Using a one-dimensional model in which
the adsorbed particle is connected to the solid by a Morse potential,
one can derive[19] a Master Equation for the probability $P_n(t)$ of
finding the particle in a state n at time t (see Figure 4) :

$$\delta P_n(t)/\delta t = - \sum_m W_{n \to m} P_n(t) + \sum_m W_{m \to n} P_m(t) \qquad (6)$$

Here $W_{n \to m}$ is the rate of transition from the level n to the level
m. In previous work, this quantity was computed under the assumption
that the vibration stretching the molecule-metal bond is coupled
to lattice phonons. The fact that the oscillations of the molecular
charge with respect to the surface cause electron-hole pair
excitation is generally ignored.

The transition rate of the anharmonic oscillator from a level
E_i to a level E_f, induced by the coupling to the metal electrons,
can be computed from Golden Rule :

$$W_{i \to f} = \frac{2\Pi}{\hbar} \sum_\Lambda \left| <<f \| <\Lambda | [\int d\vec{r} \ g(\vec{r})\rho(\vec{r})] \ \delta Q | G> \| i>> \right|^2 \delta [(\varepsilon_f + \varepsilon_\Lambda - \varepsilon_i - \varepsilon_G)/\hbar]$$

$$(7)$$

Here $\| i>>$ and $\| f >>$ are the anharmonic oscillator states, $|G>$ and
$|\Lambda>$ are the ground and excited states of the metal, and ε_Λ and ε_G
are the corresponding energies. The quantity $g(\vec{r})$ is the renormalized
electron-vibration coupling. Using both well known manipulations
that express a Golden Rule formula in terms of a correlation function,
and also the fluctuation-dissipation theorem, we can rewrite
Equation (7) as :

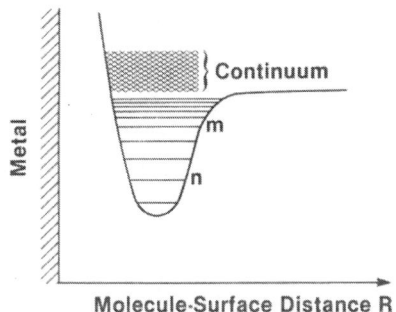

Fig. 4 – Schematic energy diagram used to model the kinetics of
         desorption. To desorb, a molecule must be promoted into a
         continuum state. The rate for this is determined by a
         random walk up and down the energy ladder. The rate of
         going "from one peg to another" is affected by electron-
         hole pair excitations.

$$W_{i \rightarrow f} = -(2/\hbar^2) \, |F_{if}|^2 \int d\vec{r} \; d\vec{r}' \; g(\vec{r}) \; g(\vec{r}') \; \text{Im}\Pi(\vec{r},\vec{r}';\omega_{fi}) \qquad (8)$$

Here F_{if} denotes $\langle\langle f \| \delta Q \| i \rangle\rangle$

   Comparing Equations (8) and (4-5) we see that the transition
rate is proportional to $\text{Im}\Sigma(\omega_{if})$, which is the vibrational line-
width. This quantity represents the metal's ability to receive the
energy $\hbar\omega_i \equiv E_f - E_i$. The factor $|F_{if}|^2$ is proportional to the
rate of the $E_i \rightarrow E_f$ transition for a linearly driven anharmonic
oscillator. The driving force in this case is caused by electron
density fluctuations in the substrate.

The Relationship between the Rate of Thermal Desorption and Vibrational Spectroscopy

There are several ways in which vibrational spectroscopy can provide input into the theory of desorption rate. (1) The obvious contribution would be the determination of the potential binding the molecule to the surface. For example, for a CO molecule we need information concerning the metal-carbon bond. Knowledge of overtone frequencies would allow us to determine the potential. Hopefully, if resonant scattering can be measured, in many cases one will observe several overtones, enough to be able to construct a reasonably accurate potential. Since the desorption theory[19] predicts that the activation energy is smaller than the well depth, it will be of interest to obtain spectroscopically determined values for the well depth, in order to compare them to the desorption activation energy. (2) If the transition frequency is substantially larger than the Debye frequency, and if the oscillator strenght (i.e., the oscillating part of the permanent dipole) is large, then the line-shape is dominated by the electron-vibration coupling. The inverse width of the transition line from i to j is then proportional to the transition rate from j to i. Thus a spectroscopic measurement yields direct information concerning a kinetic quantity.

SURFACE REACTION DYNAMICS

In this section we present some of the salient features of our theory of rate processes at metal surfaces which will be expanded upon in last Section. Since the theory has been discussed in great detail elsewhere[3f,4,5] we mainly quote physical points and end results here.

The basic idea for a reactive surface event goes as follows. Suppose that an atom or molecule with a (thermal) kinetic energy = K_i is incident upon a surface. Furthermore, suppose that due to chemical interactions with the surface, there are two possible electronic states for the coupled system which are labeled 1 and 2. A potential energy curve for nuclear motion can be determined for each electronic state, as shown in Fig. 5. If the incident particle is initially prepared in state 1, nuclear motion will be on curve 1 unless a diabatic transition[20] from $1 \rightarrow 2$ occurs in the vicinity of the curve crossing point $z = R_c$. Let us identify a probability for such a transition as $|\langle \theta_{2 \rightarrow 1} \rangle|^2$ where θ is some complicated operator determined by chemical properties of the incident particle and the substrate. However for the two state system shown in Figure 5, such a transition cannot occur unless the incident energy lines up with a vibrational level of curve 2 and in addition, the overlap integral $= \langle K_i | \lambda \rangle$ (Franck-Condon factor) of the continuum wavefunction $|K_i\rangle$ and the bound vibrational wavefunctions $|\lambda\rangle$ is large.

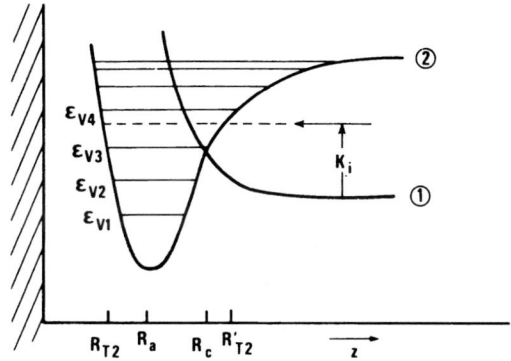

Fig. 5 – Diabatic potential energy curves as a function of Z, the
normal distance from the metal surface, for an incident
particle initially in some electronic state 1 and with
kinetic energy K_i. Curve 1 corresponds to an electronic
state giving rise to a strictly repulsive surface interaction. Curve 2 corresponds to an electronic state which
strongly adsorbs at an equilibrium separation R. The
probability for a diabatic electronic transition from
curve 1 to curve 2 is maximum at the crossing point
$Z = R_c$. The discrete vibrational levels for surface bound
states on curve 2 are labeled ε_{vi}.

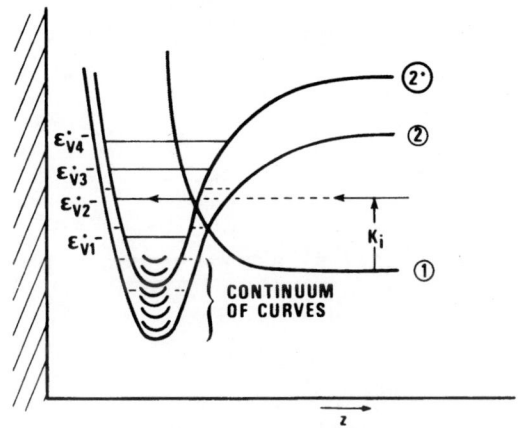

Fig. 6 – Same as Fig. 5, but now the continuum of substrate
electronic excitations gives rise to a continuum of
potential curves, one of which is shown explicitely. In
this case the excited curve 2★ has $\varepsilon^\star_{v2} = K_i$.

Due to the continuum of substrate electron-hole pair excitations, the simple situation depicted in Figure 5 is more accurately represented in Figure 6 where a continuum of curves 2* representing the total electronic system, chemical 1 and 2 states plus substrate electron-hole pairs is shown. Since the 1 → 2 diabatic transition represents a time varying change in electronic charge on the incident particle, this is perceived by the metal electrons as a time dependent field which can excite the substrate, thus providing a mechanism for transferring energy from the incident particle to the substrate. Consequently it is always possible to find some excited potential curves for which $K_i = \varepsilon_\lambda$, thus permitting nuclear motion to continue onto a bound state.

We have shown that the transition probability for a 1 → 2* process is well approximated by a Golden Rule-like expression :

$$P_{2^* \to 1} \approx \frac{1}{\hbar} | < \theta_{2 \leftarrow 1} > |^2 \sum_{\lambda \leqslant K_i} | < K_i | \lambda > |^2 S(K_i - \varepsilon_\lambda) \quad (9)$$

where the sum is over all vibrational states on the pair ground state curve 2 and $S(\omega)$ is the probability for excitation of pair states with energy ω due to the field associated with the 1 → 2 transition or charge rearrangement. Microscopic models depending upon the nature of the 1 → 2 transition must be postulated in order to specify $S(\omega)$, but in general, these models will depend in one way or another on the dynamic structure factor of the surface[3-5]. In any event, to the extent that Equation 9 is a meaningful representation of a surface reaction rate, input from vibrational spectroscopy, in the form of Franck-Condon factors and vibrational levels ε_λ, is required.

OSCILLATORS, TRAJECTORIES, REACTIONS, AND VIBRATIONS

As already noted, a useful approach in the formulation of inelastic reactive scattering is based upon semi-classical scattering theory within the trajectory approximation (TA). The basic elements of the TA are as follows. One starts with a credible zero order potential energy surface which is used to determine the nuclear motion of the incident particle beam. Assuming that the particles remain on this trajectory, the excitation of the other degrees of freedom of the system are calculated as if they were subjected to the prescribed force provided by the particle of the trajectory. Recoil or feedback effects on the trajectory are neglected in this part of the calculation. The probability of excitation calculated in this way is then fed into a complete theory which includes the possibilities for nuclear motion transitions and thus deviations from the initially assumed trajectory. Presumably a satisfactory theory contains some procedures which assure that the trajectory and the outcome of the event are at least self-consistent. In our formal theory of surface reaction dynamics (Section IV), the inter-related rôle of substrate electron-hole pair excitations, surface induced quasi-adiabatic (diabatic) electronic transitions within the reactants, and nuclear motion has been emphasized.

For inelastic surface scattering, whether one is concerned with electronic excitation (electron-hole pairs[21] or plasmons) or phonons within the substrate, most calculable models ultimately reduce to the equivalent of a linearly displaced, forced harmonic oscillator which is exactly soluble[22]. In this section we illustrate, via a simple classical model, the connections between the nuclear motion of an incident beam within the TA, and the degree of substrate excitation resulting from the interaction between the particle and substrate modes. Since these modes are bosons, the essential features of the excitation process are contained in the equations of motion for a single particle couples to a harmonic oscillator. We emphasize that in this section, the harmonic oscillator represents the boson sub-strate excitations and not adparticle vibrational modes.

Forced Oscillator

Suppose that an incident particle of mass M at a position R interacts with a surface through some potential $V(R,r)$ where r is the substrate boson coordinate or oscillator displacement from equilibrium. The equations of motion for the coupled system are :

$$M\ddot{R} = \frac{d}{dR} V(R,r) \tag{10}$$

and

$$m\ddot{r} + kr = \frac{d}{dr} V(R,r) \tag{11}$$

with $k = m\omega_o^2$ the oscillator force constant. Next assume that the interaction is of the form :

$$V(R,r) = V_s(R) + V_d(R,r) \tag{12}$$

where $V_s(R)$ is a static term which provides the momentum transfer necessary to elastically reflect the particle, $V_d(R,r)$ is the dynamic term allowing for energy transfer to the surface excitations, and $\frac{dV_d}{dR} \ll \frac{dV_s}{dR}$ for relevant values of R. Within this scheme, Equation 10 is approximately :

$$M\ddot{R} \approx \frac{d}{dR} V_s(R) \tag{13}$$

which is readily solved for a given $V_s(R)$, yielding $R = R(t)$, the trajectory to be used subsequently. With Equation 12 and this trajectory,

$$\frac{d}{dr} V(R,r) \approx \frac{d}{dr} V_d [R(t),r] \Big|_{r=o} \equiv V_d' [R(t)] \quad . \tag{14}$$

Thus Equation 11 becomes :

$$m\ddot{r} + kr \approx V_d' [R(t)] \tag{15}$$

which is the equation of motion for a driven oscillator where the time dependence of the forcing function enters through the time dependent trajectory. Noting that :

$$r(t) = \frac{1}{2\pi} \int d\omega e^{i\omega t} r(\omega)$$

Equation 15 is solved by Fourier transforming, giving :

$$r(\omega) = \frac{1}{m} \frac{V_d' (\omega)}{\omega_o^2 - \omega^2}$$

and thus :

$$r(t) = \frac{1}{m} \frac{V_d' (\omega_o)}{\omega_o} \sin \omega_o t \tag{16}$$

where :

$$V_d' (\omega) = \int_{-\infty}^{+\infty} dt \, e^{-i\omega t} V_d' [R(t)] \tag{17}$$

for well behaved $V_d'(\omega)$. The energy transferred to the oscillator, as a result of the prescribed driving force, is :

$$\Delta E(\omega_o) = \frac{1}{2} k \, r_{max}^2$$

which with Equation 16 is :

$$\Delta E(\omega_o) = \frac{| V_d'(\omega_o) |^2}{2m} \quad . \tag{18}$$

Equation 18 demonstrates that the propensity for the trajectorized particle to excite the oscillator of frequency ω_o depends upon the

ω_o'th Fourier component of the time dependent interaction $V'_d[R(t)]$. Presumably total excitation for a distribution of uncoupled oscillations (normal modes) of varying frequencies is given by a summation over all modes; that is :

$$\Delta E_{\tau_o t} = \int d\omega \rho(\omega) \quad \Delta E(\omega) \tag{19}$$

where $\rho(\omega)$ is the density of states of oscillator excitations at frequency ω and $\Delta E(\omega)$ is given by Equation 18.

Illustrative Trajectories

Non-reactive event

A typical interaction potential $V_s(R)$ between a particle and a surface is characterized by an attractive potential well of some depth and range determined by the specific chemistry of the systems and by a short range repulsion. For present purposes this is represented by the attractive square well with repulsive hard core shown in Figure 7a. The incident particle, in a continuum state K_i, is elastically scattered and follows the trajectory shown in Figure 7b, according to Equation 13. Assuming a square well of the same range for the dynamic potential $V_d(R,r)$, the time dependent force $V'_d[R(t)]$ for this trajectory is displayed in Figure 7c. The significant dynamical aspect of this force is the time duration over which it acts. This is determined by a combination of the incident particle energy, the well depth and range, and the use of pure classical mechanics for determining the trajectory. Exceptions to the last point will form the basis for relating vibrational spectroscopy to surface reactivity and we will return to this point at great length later.

For the square pulse time dependence shown in Figure 7c, Equation 18 is :

$$V'_d(\omega) = V'_d \int_{-\tau_c}^{\tau_c} dt\ e^{-i\omega t} = 2V'_d\ \frac{\sin \omega \tau_c}{\omega}$$

and hence, from Equation 19,

$$\Delta E(\omega_o) = \frac{2\ V_d'^2\ \sin^2 (\omega_o\ \tau_c)}{m\ \omega_o^2} \tag{20}$$

The intimate connection between the oscillator frequency and the

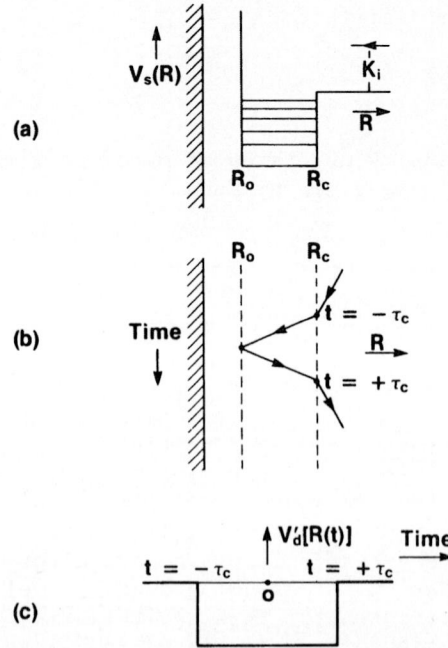

Fig. 7 - (a) Model attractive square well with hard core repulsion
             potential used to illustrate non-reactive scattering
             from surface.

         (b) Elastic trajectory for a particle moving in the
             potential shown in (a).

         (c) Time dependent force on the substrate excitations due
             to particle motion described by the trajectory in (b).

collision time (or length of time the <u>prescribed trajectory</u> puts
the particle in the interaction region) is seen in Equation 20.
Maximum energy transfer occurs when $\tau_c = (2n-1)\pi/2\omega_0$ and no transfer for $\tau_c = n\pi/\omega_0$ where n is an integer. This is due to the fact
that since the first collision transfers energy from the particle
to the oscillator, whereas the second collision transfers energy
from the oscillator back to the particle, the time duration must
be such that an odd number of particle-oscillator collisions occur
if net excitation of the oscillator is to result.

As a specific example of a distribution of oscillators or bosons, consider the electron-hole pair excitations of a Fermi system which are reasonably characterized by a linear density of states :

$$\rho_{e-n}(\omega) \approx \rho_{\epsilon_F}^2 \, \omega \quad (\omega < \omega_c) \tag{21}$$

where ρ_{ϵ_F} is the Fermi level density of states and $\omega_c = \epsilon_c/\hbar$ is a cutoff of order the bandwidth. Equations 20 and 21, inserted in Equation 19, give

$$\Delta E_{\tau ot} \approx \frac{2v_d'^2 \, \rho_{\epsilon_F}^2}{m} \int_o^{\omega_c} d\omega \, \frac{\sin^2 \omega\tau_c}{\omega} \tag{22}$$

In the short τ_c limit $\sin^2 \omega\tau_c \sim (\omega\tau_c)^2$ and

$$\Delta E_{\tau ot}(\tau_c \ll \omega_c^{-1}) \approx \frac{v_d'^2 \, \rho_{\epsilon_F}^2}{m} \, (\omega_c \tau_c)^2 \tag{23}$$

increasing quadratically with collision time. For large τ_c replace the rapidly oscillating term by its mean value = 1/2 and introduce a low frequency cutoff $\omega_{min} \approx 1/\tau_c$ (as is customary is these types of problems). Then Equation 22 integrates to :

$$\Delta E_{\tau ot}(\tau_c \gg \omega_c^{-1}) \quad \frac{v_d'^2 \, \rho_{\epsilon_F}^2}{m} \, /n(\omega_c \tau_c) \tag{24}$$

From Equations 23 and 24 it is clear that the excitation of the substrate electron-hole pairs increases as the length of the collision time does. Since τ_c is determined by the trajectory we have chose, albeit with some physical justification (Equation 13), it is very important to search out other mechanisms which could qualitatively alter the trajectories if we hope to have the basis for a realistic theory of surface reaction dynamics.

Reactive events

As discussed in Section IV, a reactive event is characterized by a surface-induced electronic transition involving the molecular orbitals of the incident particle. After the quasi-adiabatic transition, nuclear motion (the trajectory) is determined by a new set of potential energy surfaces. A significant class of events involve nuclear transitions, both real and virtual, between reactant continuum states and product surface bound states. Due to

pair-excitation intrinsic to the quasi-adiabatic transition, low
lying bound vibrational states of the product potential well are
accessed and the properties of these states in turn dramatically
alter the trajectories. For instance, an energetic situation as
pictured in Figure 5 would be characterized by a basically one
bounce symmetric trajectory of the type just discussed and illustrated in Figure 7b. However if the quasi-adiabatic transition
occurs and it is accompanied by pair excitations so that a nuclear
transition from K_i to some $\varepsilon_v^\star$ vibrational level of curve 2★ is
possible, both from energetic and Franck-Condon considerations, then
the subsequent motion is as shown in Figure 8. The oscillating
structure is the trajectory for a particle in a discrete bound
state of the attractive well of Figure 6. It is this part of the
trajectory whose properties may be inferred from vibrational spectroscopy.

Some general ideas can be noted. First, the trapping of the
particle in the vibrational state greatly enhances τ_c, the
"collision time" for interaction with the pairs. Thus, from
Equations 20 or 24, the energy loss to the electronic excitations
is enhanced which, in a self-consistent manner, increases the
probability for initial excitation of the 2★ electronic state necessary for this trajectory. If (in an oversimplified) harmonic
vibrational well, the characteristic frequency is ω_v, then
$\tau_c \gg 2\pi/\omega_v$ for the discrete structure to be preserved. That is,
on the no-recoil trajectory, several oscillations must be experien-
ced before the particle emerges as a continuum state on curve 1.
Put another way, the "predissociation broadening" must be considerably smaller than the vibrational spacing, if this model is to be
useful. For reasonable curves of the 1 and 2 type, this is usually
so[23].

More importantly, the electron gas "friction"[3c] or self-
energy effects on the properties of (alternatively but equivalently,
the trajectory or the vibrational lineshapes must be considered in
some detail. The Schaich-Suhl[3c] school has elected to formulate the
damping in terms of a friction term added to Equation 13, the nuclear equation of motion; that is, $F_{fric} = M\eta(R)\dot{R}$ where $\eta(R)$ is
a position dependent viscosity felt by the incident particle. If
$\eta(R) = $ constant for $R < R_c, = 0, R > R_c$, then classically the amplitude of the oscillations in the trajectory would be exponentially
damped.

We prefer to view the interaction with the electron gas from
the point of view of a self-energy process. The influences on the
trajectory can be visualized as in a typical mixed representation
diagram shown in Figure 9. The incident particle enters "state" $\varepsilon_{v,n}^\star$
of the attractive well at time $t = -\tau_c$ and oscillates until $t = t_1$
at which point it excites a substrate electron-hole pair and drops
to state $\varepsilon_{v,n-1}^\star$, oscillating with decreased amplitude. At time

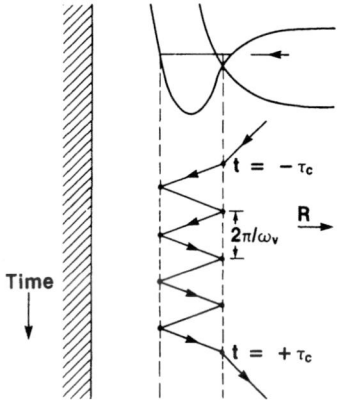

Fig. 8 – Trajectory associated with motion in an attractive well in
         which substrate induced transitions between vibrational
         states are not possible.

$t = t_2$ another pair is excited, etc. Subsequent reabsorption of
this polarization ultimately allows the particle to reemerge on an
elastic trajectory. For real rather than virtual pair emission, the
particle can become trapped. Desorption, as already discussed, is
then a process which is independent of how the particle arrived on
the surface in the first place.

In any event, it should be clear that the properties of the
trajectory are quite altered by the quasi-discrete levels of the
attractive well which in turn are shifted and broadened due to
interactions with the substrate pair continuum, as discussed in
Section II. The energies and linewidths of the "bound states" can

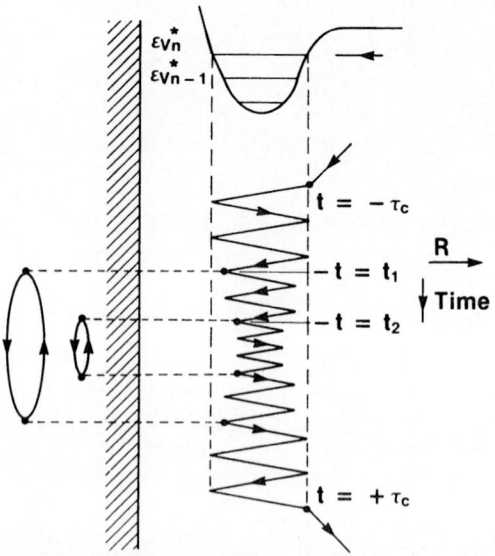

Fig. 9 - Trajectory associated with motion in an attractive well in
which the polarization of the substrate, due to pair excitation, induces transitions between vibrational levels.
Note that at $t = t_1$, t_2, etc. when pairs are excited, the
amplitude of the oscillations in the trajectory decreases
(increases) for pair excitation (absorption) in a manner
which is related to rms displacement of the vibrational
levels.

be determined from vibrational spectroscopy. These quantities
which are directly related to Re $\Sigma_\lambda(\omega)$ and Im $\Sigma_\lambda(\omega)$, the real and
imaginary parts of the localized phonon self-energy respectively,
given by Equation 5, in turn require that the reaction rate expression, Equation 9, be modified to :

$$P_{2\star \leftarrow 1} \simeq \frac{1}{\hbar} |\langle\theta_{2\leftarrow 1}\rangle|^2 \sum_{\lambda \ll K_i} |\langle K_i| \lambda\rangle|^2 \int d\varepsilon \, \rho_\lambda(\varepsilon) \quad S(K_i - \varepsilon) \quad (25)$$

with

$$\rho_\lambda(\varepsilon) = \frac{1}{\pi} \frac{\text{Im } \Sigma_\lambda(\varepsilon)}{[\varepsilon - \varepsilon_\lambda - \text{Re } \Sigma_\lambda(\varepsilon)]^2 + [\text{Im } \Sigma_\lambda(\varepsilon)]^2}$$

From Equation 25, it is clear that vibrational spectroscopy can also provide valuable and necessary information required for understanding reactive rate processes at metal surfaces.

REFERENCES

1. For entries into the literature, see : E.E. Nikitin, "Theory of Elementary Atomic and Molecular Processes in Gases", (Clarenden, Oxford, 1974); M.S. Child, "Molecular Collision Theory" (Academic, London, 1974); "Dynamics of Molecular Collisions, Part A and B", ed. by W.H. Miller (Plenum, N.Y., 1976); "Atom-Molecule Collision Theory : A Guide for the Experimentalist", ed. by R.B. Bernstein (Plenum, N.Y., 1979).

2. For entries into the lattice literature, see : F.O. Goodman and H.Y. Wachman, "Dynamics of Gas-Surface Scattering", (Academic, N.Y., 1976); H. Metiu, J. Chem. Phys. $\underline{67}$, 5456 (1977); M. Shugard, J.C. Tully, and A. Nitzan, J. Chem. Phys. $\underline{66}$, 2534 (1977); J.E. Adams and W.H. Miller, Surf. Sci. $\underline{85}$, 77 (1979); T. Maniv and M.H. Cohen, Phys. Rev. B 19, 4883 (1979); G.S. De, U. Landman, and M. Rasolt, Phys. Rev. B 21, 3256 (1980); G. Doyen and T.B. Grimley, Surf. Sci. $\underline{91}$, 51 (1980); S.A. Adelman, Adv. Chem. Phys. (in press).

3. a) E. Müller-Hartman, T.V. Ramakrishnan and G. Toulouse, Solid State Comm. $\underline{9}$, 99 (1971).
   b) Ch. Steinbruchel and L.D. Schmidt, Phys. Rev. B $\underline{10}$, 4215 (1974).
   c) E.G. d'Agliano, P. Kumar, W. Schaich, and H. Suhl, Phys. Rev. B 11, 2122 (1975); W. Schaich, Surf. Sci. $\underline{49}$, 221 (1975).
   d) J.K. Nørskov and B.I. Lundqvist, Surf. Sci. $\underline{89}$, 251 (1979).
   e) R. Brako and D.M. Newns, Solid State Comm. $\underline{33}$, 713 (1980).
   f) J.W. Gadzuk and H. Metiu, Phys. Rev. B $\underline{22}$, xxx (1980).
   g) K. Schönhammer and O. Gunnarsson, Phys. Rev. B $\underline{22}$, xxx (1980).

4. H. Metiu and J.W. Gadzuk, J. Chem. Phys. (submitted).

5. J.W. Gadzuk and H. Metiu, Proc. of ICSS IV, Cannes, September, 1980.

6. W.H. Miller, Adv. in Chem. Phys. XXV, 69 (1974).

7. a) A.M. Bradshaw and F.M. Hoffmann, Surface Sci. $\underline{72}$, 513 (1978); A.M. Bradshaw, private communication.
   b) W.G. Golden, D.S. Down and J. Overend, J. Phys. Chem. $\underline{82}$, 834 (1978).
   c) K. Horn and J. Pritchard, Surface Sci. $\underline{52}$, 437 (1975).

8. H. Metiu and W.E. Palke, J. Chem. Phys. 69, 2574 (1978); G. Korzeniewski, D. Hone and H. Metiu, Chem. Phys. Lett. 58, 473 (1978); G. Korzeniewski and H. Metiu, J. Chem. Phys. 70, 5174 (1978).

9. S. Andersson, Solid State Comm. 21, 75 (1977); J.W. Gadzuk, Phys. Rev. B 19, 5355 (1979).

10. B.N.J. Persson, J. Phys. C 11, 4251 (1978).

11. For a detailed presentation see : H. Metiu, Handbook of Surfaces and Interfaces, vol. 8, ed. W.H. Weinberg (Garland Press, 1980).

12. See for example : J.R. Schrieffer, Theory of Superconductivity (Benjamin, N.Y., 1964).

13. D. Pines, Elementary Excitations in Solids (W.A. Benjamin, N.Y. 1964).

14. a) D.M. Newns, Phys. Rev. B 1, 3304 (1970); D.E. Beck and V. Celli, Phys. Rev. B 2, 2955 (1970); P.W. Lert and J.H. Weare, J. Chem. Phys. 68; 2221 (1978); T. Maniv and H. Metiu, J. Chem. Phys. 72, 1996 (1980); Phys. Rev. B (in press).
    b) G. Korzeniewski, T. Maniv and H. Metiu, Chem. Phys. Lett. (in press).

15. N.D. Lang and W. Kohn, Phys. Rev. B 7, 3541 (1972); J.A. Appelbaum and D.R. Hamann, Phys. Rev. B 6, 11 2 (1972); E. Zaremba and W. Kohn, Phys. Rev. B 13, 2270 (1976).

16. S. Efrima and H. Metiu, Surf. Sci. 92, 443 (1980).

17. See for example : A. Datta and D.M. Newns, Phys. Lett. 59A, 326 (1976); J.W. Gadzuk, Phys. Rev. B 14, 2267 (1976), Surf. Sci. 67, 77 (1977); R.R. Chance, A. Prock, and R. Silbey, Adv. Chem. Phys. 37, 1 (1978); or S. Efrima and H. Metiu, Surf. Sci. 92, 417 (1980).

18. P.J. Feibelman, Phys. Rev. B (to be published); W.H. Weber and G.W. Ford, Phys. Rev. Lett. (to be published).

19. S. Efrima, K.F. Freed, C. Jedrzejek and H. Metiu, Chem. Phys. Lett. 74, 43 (1980).

20. T.F. O'Malley, Adv. in Atomic and Mol. Phys. 7, 223 (1971); H. Metiu, J. Ross and G.M. Whitesides, Angew. Chem. Intl. Ed. Engl. 18, 377 (1979).

21. K.D. Schotte and U. Schotte, Phys. Rev. 182, 479 (1969); E. Müller-Hartman, T.V. Ramakrishnan, and G. Toulouse, Phys. Rev. B 3, 1102 (1971).

22. R.P. Feynman and A.R. Hibbs, "Quantum Mechanics and Path Integrals" (McGraw-Hill, N.Y., 1965); H. Haken, "Quantum Field Theory of Solids, an Introduction" (North-Holland, Amsterdam, 1976).

23. M.L. Sink and A.D. Bandrauk, Chem. Phys. 33, 205 (1978).

STATIC AND DYNAMIC ASPECTS OF CHEMISORPTION

Bengt I. Lundqvist

Institute of Theoretical Physics
Chalmers University of Technology
S-412 96 Göteborg, Sweden

ABSTRACT

Theoretical results for molecular adsorption on metals are
reviewed. A scenario for the chemisorption of H_2 on Mg(0001) is
presented, as inferred from calculated potential-energy surfaces
and adsorbate-induced electron structure. Various steps of the
chemisorption process, such as activated associative adsorption,
trapping, precursor motion, and activated dissociative adsorption,
are identified. A conceptual framework for the understanding of
their characteristic properties, as well is of the specificity of
adsorption and of the reactivity of metals, as described.
Finally, it is stressed, how the study of vibrations at surfaces can
give valuable information about the chemisorption process.

INTRODUCTION

Chemisorption is the interaction between an adsorbate and a
substrate that results from a sharing of the electrons. A massive
theoretical program is now rapidly developing for the calculation
of the adsorbate-induced electron structure. Several efforts
have also been made from this to calculate the adiabatic potential-
energy curves, and from these to deduce, e.g., adsorbate geometries,
chemisorption energies and vibrational frequencies. Still more
massive is the experimental program to measure such static properties by LEED, UPS, EELS etc. An interplay between theory and
experiment in this area is even starting to be discerned.

Chemisorption also has dynamic aspects, namely when one follows
a molecule from its free state far away from the metal to the

equilibrized state of the whole or fragmented molecule adsorbed on
the surface.   In doing so, one may encounter such concepts as activation barriers for associative adsorption, for migration along the
surface (for the molecule, as well as for possible fragments) and
for dissociation, sticking probabilities, decay of vibrations,
precursor states and active sites.   The theoretical program in this
area is small but growing.   The experimental methods are few, and
the invention of new dynamic experiments should be encouraged, in
particular in view of the great importance of the dynamical processes for surface reactions in general.

Adsorption raises many questions : When a molecule hits a surface, is it adsorbed at all? Is the adsorption atomic or molecular,
i.e., dissociative or associative?   Is the associative adsorption
activated?   Is the dissociative adsorption activated?   Is there
a precursor state, i.e., a molecularly adsorbed state, in which the
molecule is trapped upon hitting the surface and then free to move
along the surface in order to seek out the preferred site of adsorption or dissociation?   Why are adsorption processes so specific,
both with respect to substrate and to individual substrate surfaces?

In the adsorption process, somehow a chemical bond is transformed into a chemisorption bond; where does the chemisorption bond
override the molecular bond?   Trapping of molecules at the surface
means that translational energy is transformed into other degrees
of freedom;   where and how is this energy dissipated?   How does
desorption occur?   How do surface reactions occur, and why are
metals so reactive?   Etc.

Most of these questions lack a definite answer and are likely to
do so for some time.   The field is in rapid development, however,
and certain patterns start to be discerned.   Many of the questions
can be expressed much clearer, if one has a potential-energy (PE)
surface for the molecule-surface interaction to look at, thanks
to the key importance of the PE surface.   The point of this paper
is to briefly review the picture seen from a particular theoretical
perspective, where such a PE surface enters in an important way.
In Section II a scenario for hydrogen molecule interacting with a
Mg(0001) surface is outlined, as it appears from self-consistent
model calculations[1]   Arguments in favour of why this should be
believed are presented in Section III.   Some possible implications
for other surfaces, substrates and adsorbates are drawn in Section
IV.   Section V is dedicated to the central role of the adsorbate-
induced electron structure, and the concluding Section VI is
addressed to some implications on vibrations at surfaces, the
theme of this conference.

A SCENARIO FOR H$_2$ ON Mg(0001)

    In the light of the results of Ref. 1., the following scenario
can be sketched : When a slow hydrogen molecule hits a Mg(0001)
surface (Fig. 1.), the most likely thing to happen to it is that
it bounces off the surface.

    A repulsive potential-energy barrier without any sizable corru-
gation meets it several Ångströms outside the first layer of
magnesium atoms.   For the vaste majority of H$_2$ molecules in a
thermal beam at room temperature the situation is thus pretty close
to that of a He atom .   For hot molecules, actually at most one
molecule per million in a room-temperature beam or molecules in
a superthermal beam, there is a chance to pass the barrier.   In
particular, this is true for the molecules, which happen to hit the
surface on top of a Mg atom (atop position A in Fig. 1) , where
the external potential barrier has a narrow pass (X in Fig. 2[1]).

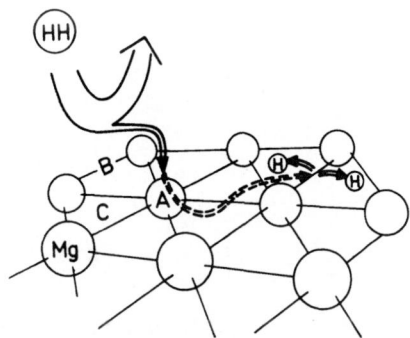

Fig. 1. : Schematic description of the approach of a
          hydrogen molecule to a Mg(0001) surface.
          Atop (A), bridge (B) and center (C) sites
          are indicated.   The dashed curve describes
          a possible path of the molecule in the precur-
          sor state ending with dissociation at a bridge
          site neighbouring to two empty center sites.

Some of the latter molecules penetrate a few Ångströms deeper into
the surface and then bounce off the surface against a steep potential-
energy wall.   The latter is only true for a small fraction of them,
as there are efficient mechanisms to slow the molecule down and, as
a consequence , trap it in the well denoted M in Fig. 2.   A trapped
molecule may perform oscillations in and out from the surface.

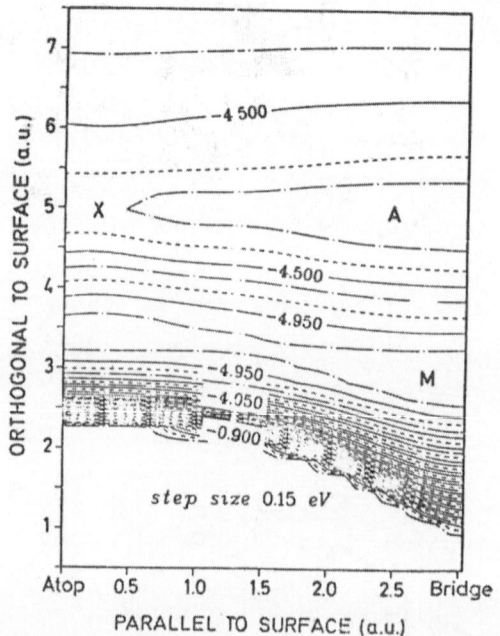

Fig. 2. : Equipotential-energy curves for H_2 parallel
         to a Mg(0001) surface[1] for a fixed H-H
         distance R of 1.5 a.u. The plot is drawn for
         a cut perpendicular to the surface along a row
         of atoms between the atop and bridge sites.
         The coordinate of the figure gives the midpoint
         between the two protons measured relative to
         a Mg atom. The energies (in eV) are relative
         to the free atoms.   The figure shows an acti-
         vation barrier for molecular adsorption (A)
         with a pass (X) over the Mg atoms (atop site).
         A molecular adsorption well (M) with almost
         no structure parallel to the surface is seen.

These vibrations are relatively efficiently damped, however.   At the same time, the adsorbed molecule is almost free to slide along the surface (indicated by the dashed path in Fig. 1.) in the potential well marked M in Fig. 2.   All these properties make the molecularly adsorbed state M, which according to Fig. 3[1] has a bond length only slightly beyond that of the free molecule, a perfect candidate for a precursor state of the kind frequently used in models for kinetic processes at surfaces.[2]

The molecule can dissociate into two hydrogen atoms, adsorbed on the surface. In the atop position this should occur with a relatively high activation barrier (denoted D in Fig. 3.) into bridge (B in Fig. 3.) or preferably center positions for the two H atoms. The dissociation is much more likely to occur, however, if the molecule first slides to a bridge position with two empty neighbouring center positions.   After that, the individual H atoms are almost free to migrate along the surface, however avoiding atop (A) sites and sites earlier occupied by hydrogen atoms.

Although the calculations behind this scenario[1,3-5]do not claim chemical accuracy, a rough indication of the order of magnitude of the various energies that they provide might be in place, in addition to what can be read out of Figs. 2 and 3.   First the energies useful for thermodynamic considerations should be mentioned. Including an estimate of the zero-point vibrational energy, the calculated energy of the molecularly adsorbed state M is -5.0 eV and that of the widely separated H atoms at center positions is -5.2 eV.   This implies that dissociations should be thermodynamically favoured, at least at low temperatures.   These energy values are given with two widely separated hydrogen atoms far outside the metal as an energy reference.   With instead a free molecule (dissociation energy : 4.5. eV) plus a clean metal as a the energy reference, the energy of state M would be -0.5 eV and that of the separated H atoms -0.7 eV, which read literally would imply adsorption and ultimately dissociative adsorption.

The dynamics at the surface is affected by certain energy barriers, for which also numbers can be extracted out of the calculations, again with estimates of the zero-point vibrational motion included.   The "external" barrier towards molecular adsorption in state M should have an activation energy of 0.6 eV, that towards the dissociation of the adsorbed H_2 molecule in the atop position 0.4. eV, and the molecular desorption energy should be 0.9 eV.   The activation energy for migration of the molecule in state M along the surfaces is found small, only 0.1 eV for diffusion from one bridge site to another via an atop position.   One in, e.g., a bridge position, the molecule finds no activation energy for dissociation, the 0.1 eV barrier in the potential-energy curve being overridden by the zero-point vibrations.

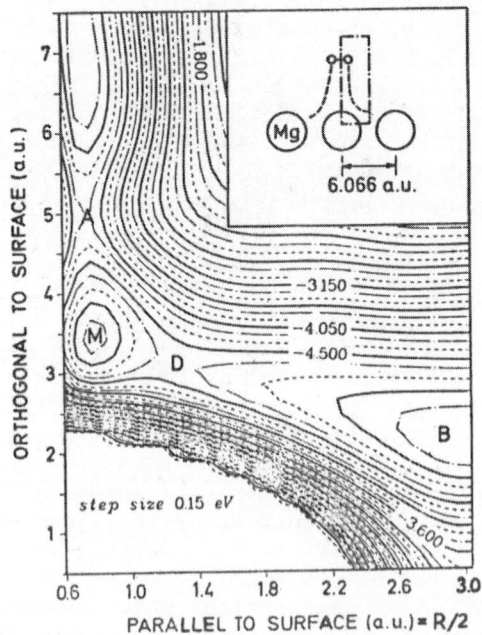

Fig. 3. : Equipotential-energy curves for H_2 dissociating
over the atop site to the bridge sites of a
Mg(0001) surface (see geometry in inset).[1]
Energies (in eV) are relative to the free atoms.
The distance from the surface is measured from
the first atomic layer and the distance parallel to the surface (= half the H-H separation
R) is measured from the atop site towards the
bridge site. As in Fig. 2., the activation
barrier (A) and minimum (M) for molecular
adsorption is seen. Further, an activation
barrier (D) separates the molecular minimum (M)
from the atomic minimum (B), where the H atoms
are dissociated. Note that a molecule that
has surmounted A cannot automatically transfer
its kinetic energy perpendicular to the surface into the H-H vibrations parallel to the
surface. Therefore, the molecule will not
dissociate immediately, even though the barrier
D is of the same size as A.

It should be stressed once again that the numbers given should not be taken literally but only as an indication of the order of magnitude of the various energies involved.

It should be obvious from the above, that a multi-dimensional representation of the potential-energy surfaces reveals a much more elaborate adsorption process than do the "original" one-dimensional Lennard-Jones picture (Fig. 4.).[6]   Furthermore, the calculations support the existence of a <u>chemisorbed</u> molecular precursor state in addition to the physisorbed one in the Lennard-Jones model.

WHY SHOULD THIS SCENARIO BE BELIEVED?

In the previous section some reservations have been expressed about the validity of the scenario described there.   Essentially two reasons will be forwarded in this section in favour of the realism of the presented picture : the reasonableness of the theoretical method used in the calculations, and the "understanding" in terms of electronic structure of what is happening.[7,8]

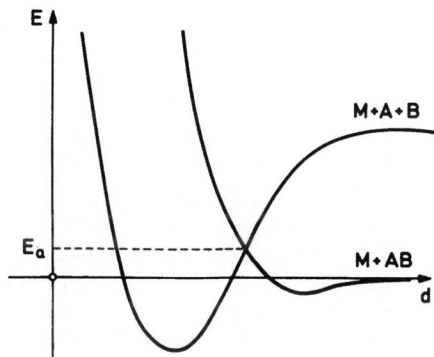

Fig. 4. : The Lennard-Jones potential-energy diagram
         for dissociative adsorption of a molecule AB
         on a metal M.[6]   The molecule-metal potential-
         energy curve (M + AB) is compared with that
         for the interaction of the widely separated
         fragments with the metal (M + A + B). E_a
         denotes the activation energy.

Theoretical methods

A theoretical description of the electronic structure is a
prerequisite for the understanding of the adsorbate-substrate interaction.   The theoretical approach for the description of adsorbate-
induced electron structure is multi-pronged.   Model Hamiltonians,
like the Newns-Anderson one, cluster calculations, layer methods and
embedding schemes all give valuable and often complementary information.[9]

The method behind the potential-energy surfaces of Section II
is a self-consistent calculation within the Kohn-Sham density-functional scheme.[10]   To describe the exchange and correlation effects,
the local-density approximation is applied.[10,11] Behind this scheme
there is a certain amount of understanding of its applicability and
limitations.   Through numerous applications to atoms, molecules and
solids, a relatively clear picture of its quantitative accuracy has
developed.   In applications to the adsorbate-substrate problem one
thus should not expect chemical accuracy but rather an uncertainty
of at most a few tenths of an eV.   For a semiquantitative description of adsorption such precision should be useful.

This of cause requires that the Kohn-Sham equations can be
solved with a sufficient accuracy.   Although the simplicity of
these one-electron equations is the prime attractive feature of
the scheme, the low symmetry of the adsorbate-substrate system
makes the solving far from trivial.   The calculational method
used here is the embedding scheme by Gunnarsson and Hjelmberg.[12]
This scheme keeps the extended nature of the substrate and makes
use of the local nature of the adsorbate-induced changes in, e.g.,
the electron density and potential by expanding the wavefunctions
in a local basis set around the adsorbate.

The fully self-consistent calculations are performed with
the substrate described in the planar uniform-background model,
also called the semi-infinite jellium model.[13] This should be a
very appropriate first approximation for a free-electron metal
like magnesium.   The substrate is characterized by a single
parameter r_s, being the radius in a.u. of a sphere containing one
electron at the bulk electron density.   This model has also been
used by Lang and Williams, with a different method to solve the
Kohn-Sham equations.[14]   The general agreement between the results
is a good support for the embedding scheme and the choice of
embedding region and local basis functions.[15]   The geometry for
the case of H_2 adsorption is shown in Fig. 5.

The effects of the discreteness of the Mg substrate lattice
have been calculated by first-order perturbation theory, with the

result for the hydrogen-plus-jellium problem as a starting point[3].
The effect of the lattice ions on the electrons are described by
Ashcroft pseudo-potentials for each ion.   For a free-electron-
like metal like Mg such a pseudopotential treatment should be
appropriate.   Of course it would be very desirable to go beyond
the first-order perturbation treatment.   In view of the fact
that the lattice corrections are small compared with the relevant
electronic energies, which are of the order of the Fermi energy,
the present treatment should give the major lattice effects,
however.[3]

     The method is thus reasonable on a semi-quantitative level.
It is certainly good on a qualitative level, i.e., when developing
a conceptual picture of adsorption, as described in the next subsection.

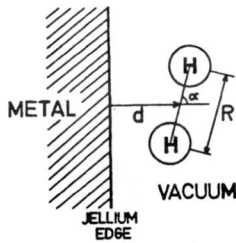

Fig. 5. : The geometry for H_2 adsorption on a metal
          surface in the jellium model.

Understanding chemisorption in terms of electron structure

     In the introduction a great number of questions on adsorption
were raised.   The scenario for H_2 on Mg(0001) might make us rephrase
the questions slightly.   Anyhow, the point of this subsection is
to attempt to answer the questions in terms of the "understanding"
or rather the conceptual picture that develops out of the model
calculations described above.

a) Why does atomic adsorption occur?

Like in all bonding, adsorption occurs, because the new atomic
arrangement can give the whole system a lower total energy.
Ultimately, this lowering can come from the attractive Coulomb
interaction between the electrons and the ionic cores. In hydrogen chemisorption the system benefits from this attraction better
in the adsorption equilibrium configuration than for, e.g., the
atom widely separated from the substrate, through a contraction
of the electronic charge close to the proton.[3] One might think
that this gain in potential energy should be accompanied by a
counteracting increase in kinetic energy, thanks to the increased
gradients of the electronic wave function close to the proton.
Now the important aspect of chemisorption is the overlap between
adsorbate and substrate electron states, and this overlap gives
interference effects that delocalize the electrons. This means
that in the small- and intermediate-overlap configurations increasing kinetic-energy contributions close to the proton can be
balanced by decreasing contributions from the interference or bond
region.[3]

In the strong-overlap configuration, i.e., when the hydrogen
atom is pressed deeper into the metal, the higher electron
density makes the kinetic energy high enough to form a repulsive
wall in the potential-energy surface. Fig. 6. shows, how the cost
in kinetic energy becomes higher, the higher the electron density
of the substrate. In particular for metal substrates with high
electron density, the steep increase (with diminishing adsorbate-
substrate separation d) of the kinetic-energy repulsive wall is
an important factor for the equilibrium position of the adsorbate.
It helps to keep the adsorbed hydrogen off the metal.

For a more detailed localization of the adsorbed atom, e.g.,
for the determination of chemisorption bond lengths, the effects
of the ion-core pseudopotentials have to be considered. It
appears that the hydrogen atom prefers to reside in the regions of
the attractive tails of the pseudopotentials of the ions and
to stay away from the core regions. With an assigned size
of the adsorbed hydrogen atom and with common values of the core
radii, useful rules of thumb for the bondlengths can be given.[3]

b) Why are the molecular forces so much weaker in the surface region?

The dissociation energy of a free H_2 molecule is 4.5 eV, while
that of an adsorbed H_2 molecule is one order of magnitude smaller
or even less. How can this drastic reduction of the intramolecular

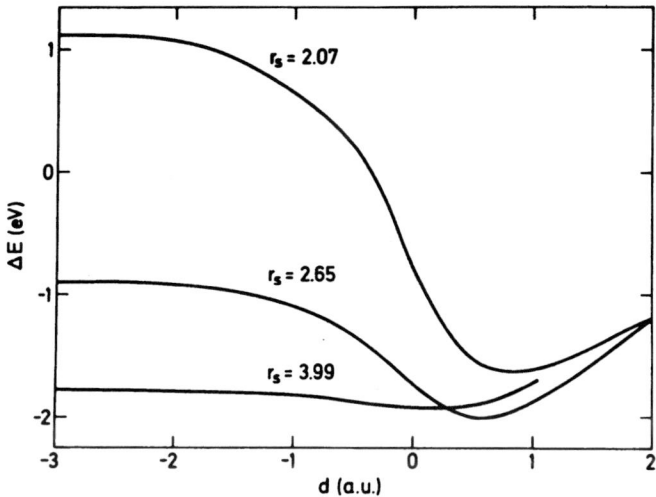

Fig. 6. : Calculated energy curves for H chemisorbed
on three different jellium substrates,
r_s = 2.07 a.u. (Al), r_s = 2.65 a.u. (Mg) and
r_s = 3.99 a.u. (Na).[15] The distance between
the H atom and the jellium edge is denoted
by d.

forces at the surface come about? To see this it is convenient
to separate the total chemisorption energy of H_2 into two parts,
$E_{H_2} = E_{intra} + E_{extra}$,[16] where the extramolecular energy is just
the sum of the chemisorption energies of the two constituting
atoms, as if there were no mutual interaction, $E_{extra}(\vec{R}_1, \vec{R}_2) = E_H(\vec{R}_1) + E_H(\vec{R}_2)$, and E_{intra} the rest.

Both contributions are affected by the overlap between adsorbate and substrate electron states, E_{extra} in the way described in
answering question a). E_{intra} appears to be very much influenced
by the filling of antibonding molecular-orbital resonances, ultimately caused by the overlap.[16] For the case of H_2 standing upright
on some jellium surfaces this is illustrated in Fig. 7.,[4] where
the correlation between E_{intra} and the filling factor q_{AB} is shown.

The parameter q_{AB} is a rough indication of the occupancy of the antibonding $2\overset{\star}{\sigma}$ molecular-orbital resonance, being defined as

$$q_{AB} = \int_{-\infty}^{E_F} dE \; \Delta n_{\Sigma_u} (E),$$

i.e., being an integral over the occupied part of a projection of the projected H_2-induced density of states, with the same symmetry as the free-molecular antibonding state $2\sigma^{\star}$. A prerequisite for the filling of the antibonding resonance is the shifting and broadening of the adsorbate levels due ot the overlap and thus interference between adsorbate and substrate states. In this way electron levels correlating with the antibonding level $2\tilde{\sigma}^{\star}$, unoccupied in the free molecule, end up below the Fermi level of the metal and become occupied. One might anticipate that the occupation of an affinity level like $2\tilde{\sigma}^{\star}$ should carry a huge energy cost, due to the intramolecular Coulomb repulsion. That this is not the case can be blamed on the very efficient screening by the metallic conduction electrons.[7] Therefore, although the overlap affects both E_{intra} and E_{extra} significantly, it affects them in different ways. This in turn means that the intra- and extramolecular energies and their mutual relation vary from one adsorbate-substrate combination to another, a variation that forms the basis for the specificity of the adsorption process.[7,8]

Fig. 7 and 8 illustrate how the filling of the antibonding affinity level weakens the bond and can even make it repulsive.[16] The mechanism for this is analogous to the molecular-orbital explanation of the absence of binding in the He_2 molecule. The intramolecular energy E_{intra} can easily be orders of magnitude less attractive than in the free molecule, which is the case in the region around the equilibrium position for H_2 on for instance Mg. This means that the energy cost for new atomic configurations is small. Implied here is the fact that the ability to fill antibonding affinity levels with electrons is one very important aspect on the reactivity of metals,[7,8] and thus on heterogeneous catalysis. The next step, after this important realization, is then to study the electron structure of different adsorbate-substrate combinations and then understand, why one metal surface is more reactive than another etc.

The Newns-Anderson model[17] has provided us with the central concepts for the description of adsorbate levels. In a simple approximation, the model solution says that to the free-adsorbate level, properly adjusted to the eventual change in occupancy, there should be added a complex energy [17]

$$q(E+i\delta) = \int_{-\infty}^{\infty} \frac{\Delta(E') \; dE'}{E-E'-i\delta} \quad , \tag{1}$$

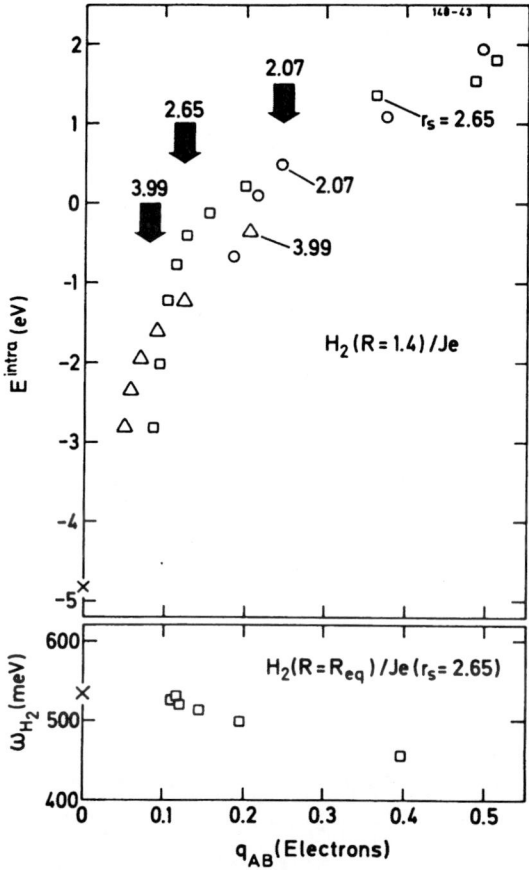

Fig. 7. : Correlations between the number of electrons is the $2\sigma^*$ resonance q_{AB} and the intra-molecular energy, as well as the vibrational frequency ω_{H_2} for a H_2 molecule standing upright on a jellium surface[4]. The square, round and triangular symbols indicate results for r_s = 2.07, 2.65 and 3.99, respectively. The crosses at q_{AB} = 0 give the free H_2 values. The broad vertical arrows show where the total charge transfer across a plane parallel to the surface and through the molecular midpoint is maximal at the indicated r_s -value. The results for E_{intra} are given with a fixed internuclear distance R = 1.4 a.u., while the results for ω_{H_2} are for the equilibrium values R_{eq} at the appropriate d value.

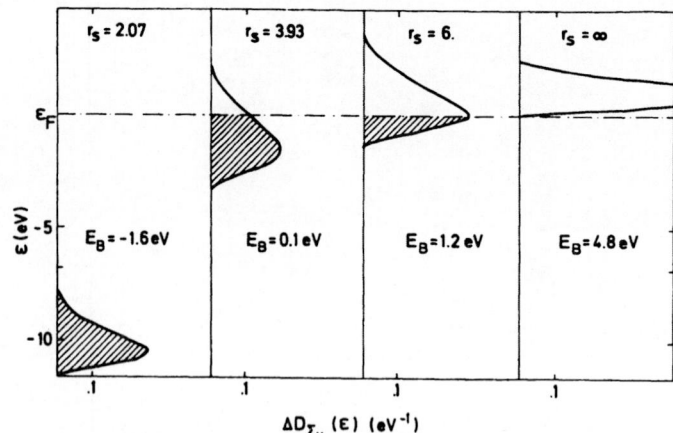

Fig. 8. : Local induced Σ_u density of states for H_2 (R = 1.4 a.u.) in an infinite jellium at different substrate densities, indicated by r_s.[16] The appropriate binding energies E_B are shown.

where the imaginary part gives the width of the adsorbate level,

$$\Delta(E) = \pi \sum_k |V_{ak}|^2 \delta(E-E_k). \qquad (2)$$

For $E = E_a$, the adjusted adsorbate level energy, this expression clearly gives the interpretation of a decay rate of the adsorbate state into substrate states, each with energy E_k and the coupling V_{ak} to the adsorbate state.

While the Newns-Anderson model provides us with concepts, one has to go to first-principle calculations to find, how the adsorbate-induced electron structure varies with, e.g., the distance from the surface. Fig. 9. shows the results from the above-mentioned self-consistent H_2-on-jellium calculations for some features of the adsorbate-induced electron structure.[5] The shift downwards in energy of both the $1\tilde{\sigma}$ and the $2\tilde{\sigma}^*$ levels is seen clearly, as well as the substantial broadening of the adsorbate levels overlapping in energy with the metallic conduction band.

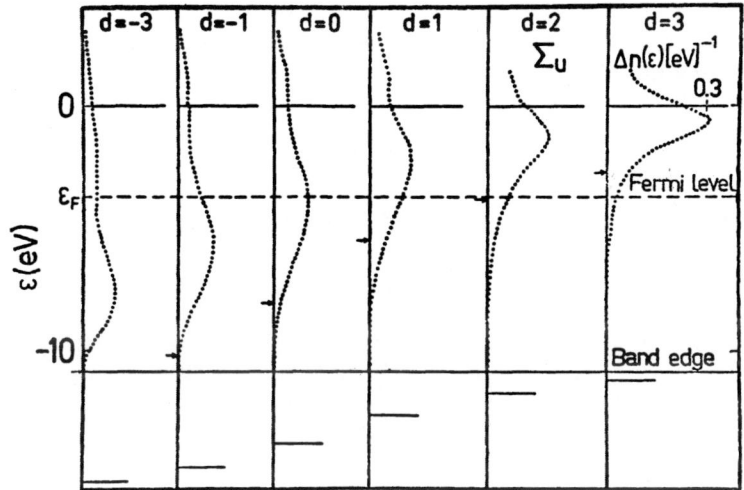

Fig. 9. : H_2-induced density of states $\Delta n(\varepsilon)$ for H_2 lying parallel to a jellium surface ($r_s=2.65$), as a function of the distance d between the jellium edge and the molecular axis.[5] The $2\tilde{\sigma}^*$ molecular-orbital resonance is identified with the Σ_u projection of $\Delta n(\varepsilon)$. The internuclear distance is R = 1.4 a.u. The horizontal lines mark the energies of the localized $1\tilde{\sigma}$ molecular orbital. The arrows denote the values of V°_{eff} (d), the effective electron potential of the clean jellium surface.

A certain correlation between the centroid of the $2\tilde{\sigma}^{\star}$ resonance
and the effective potential $V^{\circ}_{eff}(z)$ felt by the electrons of the
clean substrate can be discerned.   A closer look at such correlations show the rules to be more intricate.[18]   Here it suffices
to state that correlations are generally occuring[14,19] and that
general rules for the variation of adsorbate-induced levels with
the distance from the surface can be formulated.   This means that,
knowing the substrate density of states and the adsorbate orbital
symmetry, one can infer some key features of the adsorbate-induced
density of states.[18]

c) Why is there an activation barrier for molecular adsorption?

In the scenario it was mentioned that most of the H_2 molecules
approaching the surface should bounce off due to a potential barrier.
The reason for this increase in the potential energy of the molecule,
according to Fig. 2. occuring in the region 1 - 3 Å outside the
surface, is basically the same as that for the atom discussed in
subsection a), an increase in kinetic energy of the electrons thanks
to the increased electron density.   The hydrogen molecule should
here resemble the isoelectronic He atom.   While the potential energy of He grows to a high wall, the potential energy of H_2 levels
off to a not-too-high activation barrier, however.   From this
viewpoint, the low barrier should be due to the fact that new levels,
derived from the $2\sigma^{\star}$ level, are introduced.   This makes the Pauli
principle allow an increased occupation of electron states close
to the protons not resulting in a mere increase in kinetic energy
of the electrons.

The barrier can also be viewed in the extra/intra-molecular
picture.[1,4]   As illustrated by Fig. 10,[1]   when the molecule comes
closer to the surface, the atomic chemisorption energy and thus E_{extra}
decrease and get more attractive, while E_{intra} grows.   Whether
there is a barrier or not depends thus very much on the relative
variation of E_{extra} and E_{intra} with the distance to the surface d.
For H_2 outside the Mg(0001) surface, the growth in E_{intra} happens
to be stronger than the decrease in E_{extra} in the outskirts of the
surface region, while the opposite is true further in.   This
variation with d is primarily determined by the overlap between
the adsorbate and substrate states, and as stressed before this
overlap affects the intra- and extramolecular contributions differently.   Already within the jellium model, quite different behaviours of the potential-energy curves can be generated by just
varying the substrate density, the only parameter of the model.[7]
As can be seen from Fig. 2., the barrier lies so far outside
the first atomic layer for H_2 on Mg(0001) that the effect of the

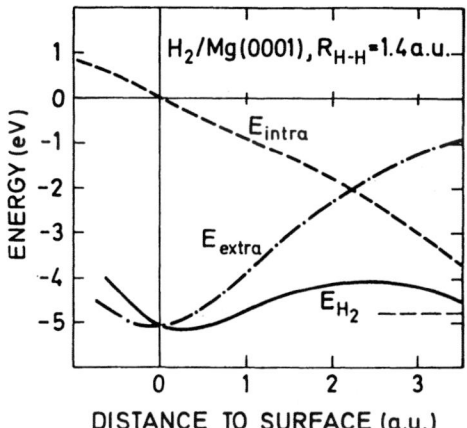

Fig. 10. : Separation of E_{H_2} into E_{extra} and E_{intra}
for a fixed H-H distance of 1.5 a.u. as a
function of distance (relative to the jellium
edge) outside the bridge site of a Mg(0001)
surface.[1] The energy zero is the energy
of a free H atom. The energy of a free H_2
molecule is indicated by a dashed line.

lattice potential on it is small.[7]   On the other hand, it should
be obvious that a stronger pseudopotential or increased overlap
effects, e.g., with the d electrons in transition metals, should
easily reduce or remove the potential barrier.[7]

d) Why is there a local minimum in the molecular potential-energy
   surface?

   Already in the jellium model we can generate quite different
adsorption behaviours, as illustrated in Fig. 11. with three
different substrates, for short called Al, Mg and Na.[4]   With
decreasing substrate density , the molecular potential-energy curve
changes from monotonic (Al), over one with a local minimum (Mg), to
one with an absolute minimum (Na).   The differences are due to

differences in overlap effects.   At high densities repulsive kinetic-
energy effects dominate E_{extra}, and E_{intra} is repulsive, due to
substantial filling of the antibonding resonance, while at low densities the repulsive kinetic-energy contribution can be almost ignored and E_{intra} is relatively strongly attractive.

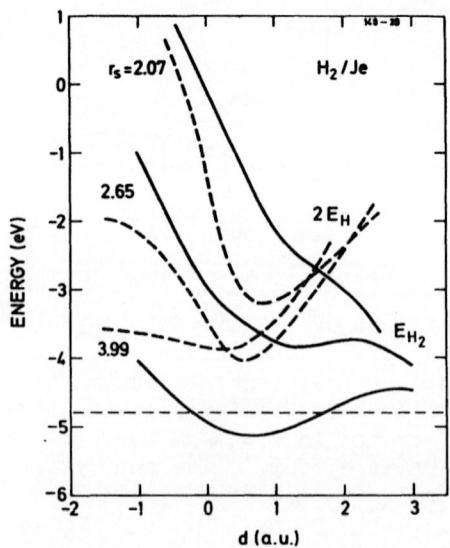

Fig. 11. : Potential-energy curves (E_{H_2}) for H_2 outside
          jellium surfaces with r_s = 2.07 (Al), 2.65
          (Mg) and 3.99 (Na) as a function of d (full
          curve).[4]   As a comparison the dashed curves
          give the potential-energy curves ($2E_H$) of two
          H atoms infinitely apart with equal d.[3]   The
          horizontal, dashed line gives E_{H_2} at d = ∞ ,
          i.e., the free H_2 result.   The internuclear
          distance of H_2 is R = 1.4 a.u.

     In Mg the electron density is obviously favourable enough to
allow the balance between E_{intra} and E_{extra} to result in a local
minimum.   As can be seen by comparing Figs 2. and 11., the pseudopotentials of the lattice give a small but significant contribution
to the chemisorption energy in the molecularly adsorbed state M.

e) Why can the molecule be trapped in state M?

   The molecules passing the barrier A, with activation energy
E_a, can be scattered elastically, inelastically or reactively. If
the excess translational energy $E - E_a$ is lost to other degrees
of freedom of the system, the molecule becomes trapped in the well
M. Such degrees of freedom can be the electron and phonon excitations of the substrate, but also vibrational and rotational modes
of the adsorbate can contribute. The probability that an incoming
molecule is adsorbed, i.e., the initial sticking probability s_o,
thus is the product of the probability for trapping σ , the condensation coefficient, and the Arrhenius factor $\exp(-E_a/kT)$,
$s_o = \sigma\exp(-E_a/kT)$, when a thermal averaging has been made.

   For the dissipation of energy to the substrate both phonon and
electron mechanisms have been proposed in the literature. The
description of sticking is still in too early a stage to allow any
definite conclusions about which mechanism is the most efficient one
in a particular case. Here we will be content with arguing that
both mechanisms should be more efficient in the well M than outside
the barrier A. If the energy transfer is to phonons, the probability
is much larger in the close collisions with the substrate atoms
made possible by the local molecular minimum. Actually, in considerations of the phonon mechanism the long-range attractive part
of the interaction is found to a good approximation to be conservative, so that the gas molecule exchanges energy with the solid only
during the short-range repulsive part of the interaction, beginning
at the bottom of the potential well.[20] The term "collision" is there
referred to the interaction with this repulsive part of the potential, and no doubt the repulsion is stronger at the wall inside
state M than outside the barrier A.

   The energy transfer to electron-hole pair excitations of the
substrate should also occur with a greatly enhanced probability in
the chemisorption well, due to the large rearrangements of the
electronic structure occuring in this region. New adsorbate-
induced levels introduced below the Fermi level during the approach
of the molecule to the surface cause charge transfer between the
substrate and the adsorbate, particularly efficient in metals, where
the conduction electrons have high mobility and lack an energy gap
in their excitation spectrum.[21] Estimates show that for the present
system this mechanism alone would give a condensation probability of
of the order of unity.[21,22] Inherent in such an estimate is an
estimate of the location, where a thermal molecule is loosing all
its translational energy normal to the surface to the excitation of
an intermediate electronic hole state on the adsorbate. This
region stretches from the crest of the barrier and inwards.

This description ties intimately to the picture of adsorbate-induced electron structure that has been described above, e.g., the filling of the antibonding $2\sigma^*$ molecular-orbital resonance on H_2 upon approaching the surface. Downshifting and filling of initially empty adsorbate levels during chemisorption should occur very generally.[7] These are examples, where the adiabatic ground state changes character during the process, analogous to the well known avoided crossings in the gas phase, which happen in a region, where non-adiabatic effects are strong. Recently, Nørskov[22] has developed a simple semi-classical theory for trapping based on this picture. Fig. 12. illustrates schematically the connection between the changes in occupation of the shifted and broadened adsorbate level and the potential-energy curves during one passage in and out of the molecule.[22] While for an infinitely slow molecule only the ground-state adiabatic potential-energy curve is of interest, a moving molecule causes transitions between _diabatic_ potential-energy curves with the adsorbate state $|a>$ filled and empty, respectively. The decay rate of the adsorbate state $1/\tau = 2\Delta/\hbar$, with Δ according to Eq. (2) expressing the dissipation of energy to substrate states, competes with the rate, by which new levels are created below and above the Fermi level due to the motion of the molecule inwards and outwards, respectively. For non-zero τ this leads to the fact that the molecule moves out from the substrate surface on a potential-energy curve that lies above the one for its entering of the surface region. Thus it has lost kinetic (translational) energy during the passage, leaving the substrate with an electron-hole pair excited. With the notations of Fig. 12., the probability for such an energy transfer $\hbar\omega$ can be written

$$P(\omega) = \int_{-\infty}^{E_F} dE_k \int_{E_F}^{\infty} dE_q \ P(E_k \rightarrow E_q) \ \delta(E_q - E_k - \hbar\omega) \qquad (3)$$

where $P(E_k \rightarrow E_q)$ is the probability that an electron-hole pair (q,k) is created during one round trip of the molecule. Assuming, for simplicity, Δ not to vary much and $\dot{E}_a = |dE_a/dt|$ to be constant in the region of interest, one gets[22]

$$P(\omega) = \Delta t^2 \ \omega \ \exp(-\Delta t \omega), \qquad \qquad (4)$$

where $\Delta t = 2\Delta /\dot{E}_a$ is a measure of the time for the resonance to cross the Fermi level.

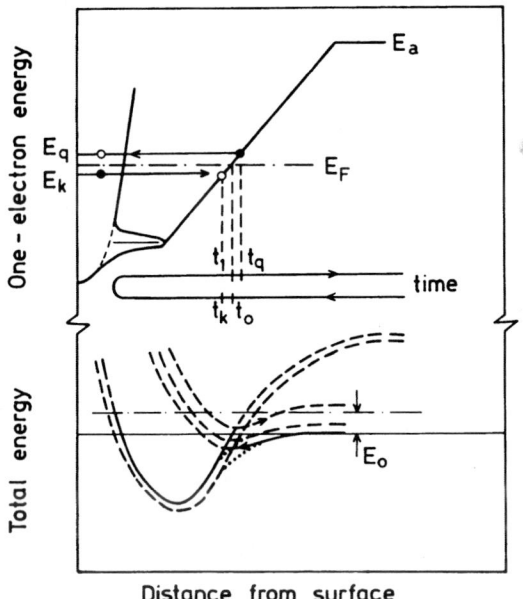

Fig. 12. : Schematic picture of the excitation process[22].
Both a one-electron and a total energy
picture are shown.   The time development is
indicated.   Two sets of diabatic potential
energy curves are shown (dashed curves) with
the adsorbate state |a> filled and empty,
respectively.   The different curves within
each set correspond to various electron-hole
excitations in the substrate.   The adiabatic
curves, where the coupling between the diabatic states is included, is shown (dotted curves).
The trajectory followed by an adsorbate of
energy E_0 is indicated by the full line.   In
this example the adsorbate is trapped.

        This result has been derived using the semi-classical assumption
that only probabilities and not amplitudes are important.   This
is best justified, if the adsorbate resonance crosses the Fermi level in
in a distinct way.   It should thus apply in the limit, where Δt
is small compared to the time $T = t_1 - t_0$ spent near the surface.

Eq. (4) can be regarded as an analog of the famous Landau–Zener
formula for the probability of diabatic crossings in gas–phase
collisions,[23] which also serves the prime purpose to expose the
important concepts and provide a semiquantitative estimate of the
magnitude of the effects.

Considering the fact that the two adsorbate spin states are
degenerate and have energies that cross through the Fermi-level
simultaneously, the probability that the incoming translational
energy E_0 of the adsorbate is dissipated to the substrate electron-
hole pairs during the first round trip is

$$\tilde{P}(E_0) = \int_{E_0}^{\infty} d\omega \int_{0}^{\infty} d\omega_1 \int_{0}^{\infty} d\omega_2 \ P(\omega_1) \ P(\omega_2) \ \delta(\omega_1 + \omega_2 - \omega) \qquad (5)$$

For an H_2 molecule barely having passed the activation barrier
outside the atop position of the Mg(0001) surface, the probability
of loosing at least a thermal energy kT($\sim$0.0025 eV at room temperature) can be estimated to be P(0.0025 eV) = 0.9999.[22]   The low
experimental value for the sticking probability s (10^{-16} or less
for polycrystalline Mg[24]) should accordingly be blamed the activation
energy E_A in $s_0 \cong \tilde{P}(kT) \exp(-E_A/kT)$, the number $E_A = 0.35$ eV
deduced from the measured data being close to the calculated value
0.5 eV.

The limitations of the semi-classical treatment can be assessed to some extent, thanks to the existence of quantum-mechanical
theories of the electronic excitations spectrum, where the
adsorbate is assumed to move on a prescribed trajectory $\vec{R}(t)$.   The
time dependence in the problem is introduced through the R dependence
of the adsorbate-substrate coupling.   Solutions for the weak-[25-28]
and strong-coupling cases[28] have been presented for realistic models
with a boson description of the electron-hole pairs.   If $\vec{R}(t)$ is
a classical trajectory on a diabatic potential-energy surface, these
approaches can be regarded as generalizations of the above simple
theory.

When applied to the strong-coupling limit, the theory of
Schönhammer and Gunnarsson[28] confirms the qualitative features
of the semi-classical theory.   For the case considered in Fig.
12. their formalism gives

$$P(\omega) = P_0 \ \delta(\omega) + (1-P_0) \ \Delta t^2 \ \omega \ \exp(-\Delta t \omega), \qquad (6)$$

which is the result of Eq. (4), renormalized due to the fact that
there is a certain probability P_o that the adsorbate does not
excite any electron-hole pair in the process, given by[26]

$$P_o = \Delta t^2 / (\Delta t^2 + T^2). \qquad (7)$$

Obviously P_o vanishes in the limit, where the level crossing
time Δt is much shorter than the passage time T. This is the limit
in which the semiclassical treatment is expected to hold. With the
parameters associated to Fig. 12., P_o is of the order 0.5 . Consideration of both spins makes the strength of the no-loss peak
$P_o^2 \sim 0.25$, still providing a high excitation probability.

In the limit, where E_a never crosses E_F, i.e., where the
resonance is never more than half filled, the weak-coupling treatment
applies. It implies a P_o differing from unity, even at zero
adsorbate velocity.[26] Thus the adiabatic limit is never reached
completely for reactions proceeding on metals.[27] The typical size
of the excitation energy is small in this limit, though.

As non-adiabatic transitions have been stressed as important
for the trapping, it might be natural to enquire to what extent
such transitions have been observed. It suffices here to just
mention the emission of photons and electrons that has been observed
during adsorption processes.[29] As a metter of fact, chemiluminescence
from halogens becoming adsorbed on a sodium surface can be described
within the above conceptual picture.[30] Both the probability for
light emission and the spectral distribution of the emitted light can
be accounted for. Presently, this is the prime support for the picture.

Although the above results and arguments make it likely that
there are mechanisms that can cause trapping into the molecularly
adsorbed state M, a detailed description of the accomodation of the
molecule into this precursor state is still pending. Of interest
are also questions relating to the energy transfer between internal
excitations of the molecule, rotations and vibrations, and the substrate excitations, the climbing down the ladder of vibrational levels
for the oscillation of the whole molecule relative to the surface,
and the competition between the corresponding rates and the rates
for desorption and for the diffusive and dissociative modes.

f) Why can the molecule migrate along the surface so easily?

The activation barrier for migration along the surface in state
M is very low.   The small corrugation of the molecular PE surface
in Fig. 2. can probably best be understood from Fig. 10.   The
figure illustrates, how the repulsion from E_{intra} prevents the
hydrogen atoms from getting close enough to the substrate atoms to
really feel the difference between the different sites.   As the
ion-core lattice is the source of the corrugation, this makes the
activation energy for molecular diffusion along the surface very
small.

The reason for the low activation barrier for migration of H_2
along the surface is thus the same as the one for the weak corrugation of the scattering potential for, e.g,, He from close-packed
metal surfaces,[31] as documented in diffractive He scattering
experiments.[32]   It is also related to the explanation of the fact
that the Debye-Waller factor for atomic and molecular scattering
is often higher than the bulk Debye-Waller factor.[33]   They
all relate to the fact that the interaction occurs in a region far
outside the first atomic layer.

g) Why is there a weak activation barrier for dissociation?

In subsection b) the weakness of the intramolecular forces
in the surface region was explained in terms of E_{intra} and its
dependence on the occupation of the antibonding $2\sigma^*$ molecular-
orbital resonance.   This modified molecular-orbital picture
also tells us that an increase in the H-H distance R means a
reduced overlap between the atomic orbitals and consequently a
reduced separation between the bonding and antibonding levels,
as illustrated in Fig. 13.[5]   The reduced overlap implies a
decrease in the intramolecular binding energy, enhanced by the
concomitant additional filling of the $2\sigma^*$ resonance.   This
increase in E_{intra} with increasing separation R would lead to an
ever-increasing potential-energy,   had there not been the option
of gaining potential energy by the hydrogen atoms coming into
the regions with stronger attraction from the substrate ions.
As thus the existence and size of a barrier for dissociation
depends on the relative strength of overlap and ion-attraction
effects, one can easily imagine that there can be a rich variation
in behaviour from site to site and from surface to surface.

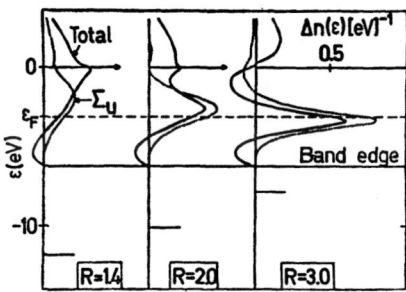

Fig. 13. : H_2-induced electron density of states on a
jellium surface with r_s = 2.65 (Mg), as a
function of internuclear distance R.[5]   The
molecule is placed parallel with the surface
at a distance d = 0.5 a.u.

h) Why does the H atom migrate so easily along the surface?

The model calculations imply that on the Mg(0001) surface the
H atom avoids the atop positions and prefers the bridge and center
positions of the first atomic layer that lies above a center position of the second layer.[3] The activation energy for migration
between bridge and atop positions is only about 0.1 eV.   The migration path is a very broad one, however avoiding the region around
the atop position.[3]

The reason for the low activation energy is similar to that
for H_2 migration, with a repulsive barrier keeping the atom off
the essential range of the ion pseudopotentials. As the atomic
repulsive wall lies a little closer to the first ion layer than
that for H_2, the mechanism is not so efficient in,e.g., the region
close to the atop positions where the pseudopotentials of the ionic

lattice reaches farthest out from the jellium edge.

It should be mentioned that the activation barrier for diffusion inwards the metals is low, too, according to the calculations about 0.4 eV.[3]   This means that adsorption of hydrogen also has to be accounted for in kinetic considerations.

i) Why does H$_2$ desorb?

The potential-energy surfaces discussed above are important also for desorption.   The desorption energy, the energy to lift the H$_2$ molecule from the well M over the barrier X on top of a Mg atom, is, according to the calculation, 0.9 eV with zero-point vibrations accounted for.[1]   So, while the migration of H$_2$ and of H and the recombination of two H atoms occur with ease, there is thus a substantial barrier reducing the desorption rate.   The climbing up the ladder of vibrational levels for the oscillations of the whole molecule towards the substrate is conveniently described by the master equation, which as an essential ingredience has the transition rate between two vibrational levels.[34]   The energy required for the transitions is typically provided by the substrate.   The thermal motion of the substrate atoms, i.e., the phonons, are commonly blamed for this,[35] but in analogy with the sticking problem there are good reasons to expect the thermally excited electron-hole pairs to be at least as important,[36] in particular for a case like H$_2$, with a vibrational frequency above the upper phonon frequency excluding one-photon processes. As yet the relative importance of phonon and electron-hole-pair processes has not been assessed.

WHAT ABOUT OTHER SURFACES, SUBSTRATES AND ADSORBATES?

The scenario has been presented for a particular case, H$_2$ on the dense Mg(0001) surface.   The supporting arguments have been phrased to indicate their expected generality.   This section will briefly comment upon what can be expected for H$_2$ on an open Mg surface, for H$_2$ on transition metals and for other adsorbates.

Hydrogen on an open Mg surface

It has been stressed above that the relative size and variation of the intra- and extramolecular energies E$_{intra}$ and E$_{extra}$, respectively, should vary from one surface to another. To illustrate this, calculations have been performed for an open

Mg surface.   In the absence of any simple hcp surface structure
other than the (0001) surface, the fictitious Mg-fcc(100) surface
has been used for this illustration.   The major differences
from the densely packed (0001) surface are the following :[1] the
external activation barrier for associative adsorption is absent;
the molecular adsorption is more site dependent and stronger in
the bridge position, where the calculated value for the molecular
chemisorption energy is about 1 eV; the barrier for the molecular
migration is substantially higher and the diffusion path more
selective; the barrier for dissociation on the surface is higher,
the calculated value in the bridge position being, about 0.5 eV;
and the barrier for atomic migration is high, one eV or higher.
The rough values given are as read off the potential-energy surface.
Inclusion of the zero-point vibrational energy reduces the stability of the molecule and the activation energy for dissociation by
about 0.2 eV.

The differences between the open and closed-packed surfaces
can directly be understood in terms of differences in E_{intra}
and E_{extra}.   On the fcc(100) surface, the conduction electrons
do not reach as far out from the first atomic layer as on the hcp
(0001) surface.   Compared with Fig. 10., the repulsive wall of
E_{intra} is therefore shifted inwards, relative to E_{extra}, which
is affected more by the local interaction with the metal ions
cores.   From the figure it is clear that this lowers the outer
activation barrier and stabilizes the molecular state.   Similarly,
the other differences get their qualitative explanation.

The above comparison between the hcp(0001) and the "fcc(001)"
surfaces clearly shows the surface specificity of the adsorption
process.

The absence of an outer activation barrier outside the more
open surface is a strong indication of the possibility that adsorption into the precursor state is most efficient at steps on the
(0001) surface, where the structure is more open and thus the
electron density lower.   The molecule can then move to the
terraces and dissociate.   Such an enhanced efficiency for H_2
adsorption at steps has been observed experimentally.[37]

Hydrogen on transition-metal surfaces

With an embedded-cluster model Muscat and Newns[38] have recently helped to bridge the gap between chemisorption on jellium and
chemisorption on transition metals.   Their calculations start
with the problem of H adsorption on a jellium surface.   The
results for the H-induced electron structure of the self-consistent
calculations[15] are fitted to the scattering problem of a single

muffin-tin potential near a finite square barrier for the jellium
surface.   With this parametrized solution to the model problem
the calculations go on by embedding a cluster of transition-metal
muffin-tin potentials, with clusters like TM_3H, TM_7H and $TM_{19}H$.
The last step is not self-consistent however, but based on thought-
fully constructed potentials.[38]

        The following three results should in particular be
mentioned:[38](i) There is a sharpening of the hydrogen- induced den-
sistity of states due to the hybridization of the H resonance with
the d electrons. The bonding-antibonding character of the H-TM
bond is relatively clearly spelled out ; (ii)   The chemisorption
energy, as calculated in a one-electron approximation, shows the
right trend through, e.g., the substrate series Co, Ni and Cu, the
weaker binding on the noble metals being understood in terms of
filling of antibonding states, which are unoccupied on the transition metals; (iii) When applied to a cluster with two hydrogen
atoms, the method predicts the proper "graphite structure" for H
on Ni(111) by finding the interaction energy between two H atoms
in center positions repulsive for nearest neighbours, barely
attractive for next nearest neighbour and definitely attractive for
the third nearest neighbours.

        With the same philosopy, i.e., a hydrogen-induced resonance
hybridizing with the substrate d electrons, the effect of the d
states on the outer potential barrier can be calculated.   A rough
estimate for the extra coupling $V_{d\sigma}$ between the $2\overset{\star}{\sigma}$ resonance
and the appropriate d state can be obtained from the Extended Hüc-
kel Method.   Compared with Mg(0001),copper surfaces should according to such an estimate have a slightly lower activation barrier
towards molecular adsorption and nickel surfaces should have none,
the differences being more pronounced the more open the surface.[7]
In this way the fact that the d band lies higher up in energy in
Ni than in Cu might be the explanation of the observed differences
in activation energy between Cu and Ni.[39]

Other adsorbates

        The key result from the calculations behind Figs. 2 and 3 can
be summed up in the following way :[7] The downshift and filling of
antibonding affinity resonances of molecules on metal surfaces play
an important role for the high reactivity of metals.   Thereby
E_{intra} can become orders of magnitudes less attractive than in
the free molecule, or even repulsive.   This makes the energy cost
for new atomic configurations small.

        Although this result has been derived only for hydrogen
adsorption, it is believed to have general applicability.   It

has support from other kinds of calculations. It suffices here to remind about the fact that the lowering of the C - O stretch mode frequency upon adsorption of CO is commonly interpreted in terms of a partial occupation of the $2\tilde{\pi}^{\star}$ molecular-orbital resonance. Possible consequences for surface reactions[40] and dissociation[7] have been discussed elsewhere.

ADSORBATE-INDUCED ELECTRON STRUCTURE

It should be obvious from the above that the adsorbate-induced electron structure should be the key quantity at focus for both static and dynamic aspects on chemisorption. The ongoing experimental and theoretical effort to study it should be intensified to obtain still more details about it.

VIBRATIONS AT SURFACES

Presently, one of the most sensitive tools for such details is the measurement of vibrations at surfaces, the theme of this conference. So far, the prime quantities obtained from measurements on adsorbates are the frequency ω and width Γ of characteristic adsorbate vibrations in energy-loss and infra-red spectra. The measurement of the frequencies ω is very useful for, e.g., (i) the characterization of potential-energy surfaces $U(\vec{R}_1, \vec{R}_2 \ldots)$, as in the adiabatic and harmonic approximations ω is proportional to the second derivation of U with respect to the appropriate deformation coordinate - it should be obvious from Figs. 2 and 3 that several local minimum and accompanying vibrational frequencies may exist to a particular adsorbate-substrate system - (ii) the identification of adsorption sites, since the potential-energy surface and thus ω depends on the coordination, as illustrated by, e.g., the calculations[3] and measurements[41] of H adsorption, and (iii) the characterization of the adsorbate-induced electron structure, as described in the above examples of reduced stretch frequencies due to increased occupation of molecular antibonding states.

Molecular vibrational levels should get a substantial width Γ upon adsorption. This is due to decay of the vibration into excitations of the substrate, such as phonons, with primarily one-phonon processes, when ω lies within the spectral range of the phonons, but also multi-phonon processes, and electron-hole-pair excitations. The latter process is a truly non-adiabatic effect. The theoretical study of Γ thus parallels that of the trapping probability, concerning decay of vibrational and translational motions, respectively. The width Γ can be related to the fluctuations $(\delta n_a)^2$ in occupation of the relevant adsorbate molecular orbital,[42]

$$\Gamma = 2\pi \ \omega \ (\delta n_a)^2. \tag{8}$$

With a value of the dynamic dipole moment deduced from other experiments, Eq. (8) can give values, for e.g., the lifetime of the C-O stretch vibration for CO on Cu(100)[42], having the same order of magnitude as the experimental number (3×10^{-12}).[43]

Anomalous broadening of vibrational adsorbate levels have in some cases been correlated with reaction activity[44]. Such a connection between the prefactor of an Arrhenius law for the reaction or desorption rate and Γ can be derived from the master equation for the occupation of vibrational levels in the corresponding potential-energy well by mere scaling, provided that the decay goes into one-phonon or electron-hole pair excitations.[45]

As vibrations of adsorbates will be delt with extensively at this Conference, the above short list of connections should suffice here. It is hoped that the perspective presented in this paper will stimulate the study of vibrations at surfaces.

ACKNOWLEDGEMENT

The author is grateful for many stimulating discussions with the many physicists, on whose work this paper has been based, including S. Andersson, O. Gunnarsson, B. Hellsing, H. Hjelmberg, A. Houmøller, P.K. Johansson, B. Kasemo, J.P. Muscat, D.M. Newns, J.K. Nørskov, B. Persson and M. Persson. The support of the Swedish Natural Science Research Council is gratefully acknowledged.

REFERENCES.

1. J.K. Nørskov, A. Houmøller, P.K. Johansson and B.I. Lundqvist, to be published, and A. Houmøller, unpublished report.
2. See, e.g., P. Kisliuk, J. Phys. Chem. Solids 5, 78 (1958), D.A. King and M.G. Wells, Proc. Roy. Soc. London Ser. A 339, 245 (1974), and J. Harris, B. Kasemo and E. Törnqvist, to be published.
3. O. Gunnarsson, H. Hjelmberg and B.I. Lundqvist, Phys. Rev. Lett. 37, 292 (1976); H. Hjelmberg, Surf. Sci. 81, 539 (1979).
4. H. Hjelmberg, B.I. Lundqvist and J.K. Nørskov, Physica Scripta 20, 192 (1979).
5. P.K. Johansson, to be published.
6. J.E. Lennard-Jones, Trans. Faraday Soc. 28, 333 (1932).
7. B.I. Lundqvist, O. Gunnarsson, H. Hjelmberg and J.K. Nørskov, Surf. Sci. 89, 196 (1979).
8. B.I. Lundqvist, P.K. Johansson, A. Houmøller and J.K. Nørskov, in Proceedings of The 1980 Annual Conference of the Condensed

Matter Division of the European Physical Society, Antwerpen
(Ed. J. Devreese), Plenum Publishing Corporation, in print.

9. See, e.g., B.I. Lundqvist, H. Hjelmberg and O. Gunnarsson, Adsorbate induced electronic states, in "Photoemission and the Electronic Properties of Surfaces" (Eds. B. Feuerbacher, B. Fitton and R.F. Willis), ch. 9, p. 227, Wiley, New York (1978), and J.P. Muscat and D.M. Newns, Progr. Surf. Sci. $\underline{9}$, 1 (1978).

10. W. Kohn and L.J. Sham, Phys. Rev. $\underline{140}$, A1133 (1965).

11. See, e.g., L. Hedin and B. I. Lundqvist, J. Phys. C$\underline{4}$, 2064 (1971); O. Gunnarsson and B.I. Lundqvist, Phys. Rev. B$\underline{13}$, 4274 (1976).

12. O. Gunnarsson and H. Hjelmberg, Physica Scripta $\underline{20}$, 192 (1975).

13. This model is put into a perspective in N.D. Lang : Density-functional approach to the electronic structure of metal surfaces and metal-adsorbate systems, in "Theory of the inhomogeneous electron gas" (Eds. S. Lundqvist and N.H. March), Plenum Press, in press.

14. N.D. Lang and A.R. Williams, Phys. Rev. B. $\underline{18}$, 616 (1978) and references therein.

15. H. Hjelmberg, Physica Scripta $\underline{18}$, 481 (1978).

16. B.I. Lundqvist, J.K. Nørskov and H. Hjelmberg, Surf. Sci. $\underline{80}$, 441 (1979).

17. D.M. Newns, Phys. Rev. $\underline{178}$, 1123 (1969).

18. O. Gunnarsson, H. Hjelmberg and J.K. Nørskov, Physica Scripta $\underline{22}$, 165 (1979).

19. H. Hjelmberg, O. Gunnarsson and B.I. Lundqvist, Surf. Sci. $\underline{68}$, 158 (1971).

20. See, e.g., F.O. Goodman, in "Rarefied Gas Dynamics", Proc. 4th Intern. Symp. Toronto, 1964 (Ed. J.H. De Leeuw), Academic Press, New York, 1966, p. 366.

21. J.K. Nørskov and B.I. Lundqvist, Surf. Sci., $\underline{89}$, 251 (1979).

22. J.K. Nørskov, J. Vac. Soc., in press.

23. See, e.g., T.F. O'Malley, Adv. in Atomic and Mol. Phys. $\underline{7}$, 223 (1971).

24. A.L. Reiman, Phil. Mag. $\underline{16}$, 673 (1933); A. Krozer and B. Kasemo, private communication.

25. E. Müller - Hartmann, T.V. Ramakrishnan and G. Toulouse, Phys. Rev. B $\underline{3}$, 1102 (1971).

26. R. Brako and D.M. Newns, Solid State Comm. $\underline{33}$, 713 (1980).

27. J.W. Gadzuk and H. Metiu, Phys. Rev. B, in press.

28. K. Schönhammer and O. Gunnarsson, to be published.

29. See, e.g., B. Kasemo and L. Walldén, Surf. Sci. $\underline{75}$, L379 (1978) and B. Kasemo, E. Törnqvist, J.K. Nørskov and B.I. Lundqvist, Surf. Sci. $\underline{89}$, 554 (1979).

30. J.K. Nørskov, D.M. Newns, and B.I. Lundqvist, Surf. Sci. $\underline{80}$, 179 (1979).

31. N. Esbjerg and J.K. Nørskov, Phys. Rev. Letters, $\underline{45}$, 807 (1980).

32. See, e.g., G. Boato, P. Cantini and R. Tatarek, in Proc. 7th Int. Vac. Congr. and 3rd Int. Conf. Solid Surfaces (Eds. P. Dobrozemsky, F. Rüdemann, F.P. Vlehböck and A. Breth), Vienna, 1977

p. 1377, and J.M. Horne and D.R. Miller, Surf. Sci. <u>66</u>, 365 (1977).

33. P.K. Johansson and B.N.J. Persson, Solid State Commun., in press.

34. Cf. E.E. Nikitin, Theory of elementary atomic and molecular processes in gases, Clarendon Press, Oxford (1974).

35. See, e.g., G.S. De, U. Landman and M. Rasolt, Phys. Rev. B<u>21</u>, 3256 (1980).

36. J.W. Gadzuk and H. Metiu, this volume; B. Hellsing, private communication.

37. R.J. Gole, M. Salmeran and G.A. Somorjai, Phys. Rev. Lett. <u>38</u>, 1027 (1977).

38. J.P. Muscat and D.M. Newns, Phys. Rev. Lett. <u>43</u>, 2025 (1979), Surf. Sci. <u>89</u>, 2821 (1979); Surf. Sci., in print, and to be published.

39. M. Balooch and R.E. Stickney, Surf. Sci. <u>44</u>, 3101 (1974), M. Balooch, M.J. Cardillo, D.R. Miller and R.E. Stickney, Surf. Sci. <u>46</u>, 358 (1974).

40. S. Andersson, B.I. Lundqvist and J.K. Nørskov, Proc. 7th Int. Vac. Congr. & 3rd Int. Conf. on Solid Surfaces (Vienna, 1977) p. 815.

41. See, e.g., S. Andersson, Chem. Phys. Lett. <u>55</u>, 185 (1978).

42. B.N.J. Persson and M. Persson, Surf. Sci. <u>97</u>, 609 (1980); Solid State Commun., in print; M. Persson and B.N.J. Persson, this volume.

43. R. Ryberg, to be published.

44. J. Demuth, H. Ibach and S. Lehwald, Phys. Rev. Lett. <u>40</u>, 1044 (1978).

45. B. Hellsing and M. Persson, private communication.

VIBRATIONAL ACTIVATION AND SURFACE REACTIVITY ; SF_6 INTERACTION

WITH SILICON INDUCED BY INFRARED LASER RADIATION

T.J. Chuang

IBM Research Laboratory
San Jose, California 95193 U.S.A.

ABSTRACT

Infrared-laser-enhanced $Si-SF_6$ interaction has been studied and the surface reaction yields have been determined as a function of the laser wavelength, the laser intensity and the gas pressure in both perpendicular and parallel beam incidences on the solid. The results clearly show that vibrationally excited SF_6 molecules promoted by multiple CO_2 laser photons are very reactive to silicon, particularly when the solid surface is simultaneously exposed to the ir radiation. The laser-induced reaction occurs at both 20°C and −150°C substrate temperatures. The study, therefore, directly illustrates the close correlation between surface reactivity and vibrational activation for the heterogeneous chemical system.

Although it is well recognized that a chemical reaction in the gas phase[1] and in the condensed phase[2] can be induced by the vibrational excitation of the reactants with infrared lasers, the role played by the vibrational activation of a chemical species in a gas-solid system has not been clear[3]. The major difference between a homogeneous chemical system and a gas-solid system arises from the fact that there are electrons in the conduction band and the valence band of the solid. The presence of these electrons can provide additional pathways for energy degradation. Therefore, the chemical dynamics of a heterogeneous system can be quite different from that of a homogeneous medium. In a recent experiment involving SF_6 interaction with silicon, we found that SF_6 molecules could be excited by CO_2 laser pulses to react with silicon causing Si atoms to be chemically removed from the surface[4]. Further study of the $Si-SF_6$ reaction yields as a function of the laser wave-

length, the laser intensity and the gas pressure shows that vibrationally excited SF_6 molecules promoted by CO_2 laser pulses are very
reactive to silicon. As will be reported in this paper, the laser-
enhanced surface reaction occurs at low temperatures as well as at
the room temperature.

SF_6 gas has been chosen for our initial study of the laser-ra-
diation-enhanced surface chemistry because the molecules excited by
multiple photon absorption have been extensively investigated in
recent years[5]. The lifetime[6] and the collisional relaxation rate[7]
of the vibrationally excited states have also been determined.
Such knowledge is of vital importance for our analysis of the complex interactions involved in the heterogeneous chemical processes.
The gas molecule is nonreactive to silicon at room temperature, but
it can etch Si at 1000°C or higher[8].

The experimental apparatus has been described previously[4].
Basically, the surface reaction yield is measured with a Si film on
a quartz-crystal microbalance which has the sensitivity of about
1.9×10^{14} Si atoms/Hz[9]. A Tachisto grating tuned CO_2 laser capable of producing 1.4J pulse energy in the 9-11μm region is used for
excitation. Except in the irradiation of the laser beam parallel
to the solid surfaces, the laser is not focused. In the normal incidence, the unfocused beam of 1.5 cm in diameter covers the entire
solid film. The optical path between the entrance KCl window and
the Si sample is 2 cm, a factor of 3 shorter than the previous arrangement[4]. The experimental procedure involves the cleaning of
the solid surface by Ar^+-ion bombardment, the exposure of SF_6 gas
at a given pressure and the firing of laser pulses at 4 sec. intervals for a desired number of pulses. The net increase in the frequency of the microbalance, directly proportional to the net loss
of Si atoms from the surface, is recorded for analysis.

The surface reaction yields measured in terms of the increase
in the frequency of the Si microbalance (Δf) are found to depend
strongly on the laser excitation wavelength. As shown in figure 1,
the P branch laser lines of the CO_2 (00°1-10°0) transitions can
induce the $Si-SF_6$ reaction for various degrees, but the R branch
lines with the same laser intensities cause no measurable effect.
This wavelength dependence clearly shows that direct excitation of
SF_6 molecules by the laser radiation is a necessary step for the
gas to react with silicon. The relatively broad spectral distribution of the reaction yield further indicates that multiple infrared photons are involved in the vibrational excitation process because the only absorption band of SF_6 in the 920-980 cm^{-1} region
occurs at 948 cm^{-1} for the ν_3 mode with a rather narrow band width
(<10 cm^{-1}). The reaction yield as a function of the laser intensity (I) is shown to have the $I^{3.5 \pm 0.5}$ dependence at 942.4 cm^{-1}, again
suggesting that SF_6 molecules are chemically activated via coherent
multiphoton excitation mechanism.

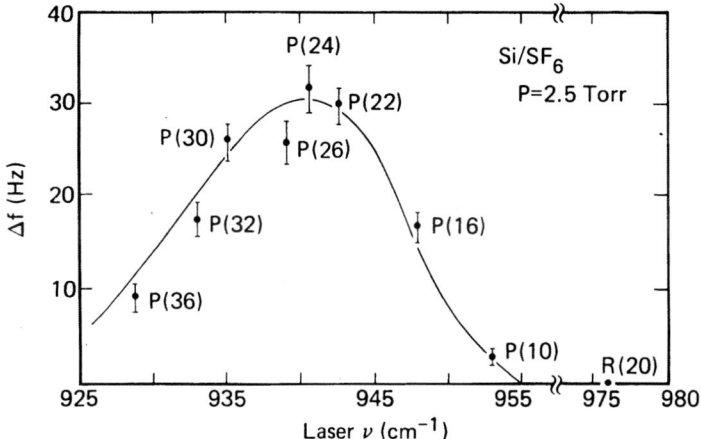

Fig. 1. Si-SF$_6$ surface reaction yield measured with Δf versus the
laser wave number : I = 0.9 J/cm^2 , normal incidence, Si at
20°C, 15 laser pulses for each data point. The CO$_2$ laser
lines in the (00°1-10°0) transitions are also indicated .

The pressure dependence of the Si-SF$_6$ reaction yield at the normal
laser incidence is shown in figure 2. The difference between this
result and the earlier measurement[4] is due to the fact that a much
shorter optical path is used for the present experiment to reduce
the effect of optical attenuation. Under the present condition,
the observed reaction yield increases with the gas pressure, reaches
the maximum at 1.5 Torr and remains practically constant in the
1.5-12.0 Torr region. This behavior is interpreted as follows :
when the pressure increases, the molecular density increases and
more vibrationally excited SF$_6^*$ molecules are produced by the laser
photons. As the pressure increases further, however, molecular collisions become the dominant mechanisms for the SF$_6^*$ relaxation[7].
The effect of collisional deactivation therefore confines the active
SF$_6^*$ molecules that are available to react with Si in the small "interaction region" just above the solid surface. The volume of this
"interaction region" decreases with increasing pressure because of the
reduced mean-free-path for collisional relaxation. Consequently,
the number of SF$_6^*$ molecules that can react with Si remains essentially unchanged in the high pressure region. The effect of collisional deactivation is further demonstrated by introducing Ar buffer
gas into the Si-SF$_6$ system. It is observed that Ar gas can reduce
the surface reaction yield because Ar atoms can deactivate SF$_6^*$ molecules by the mechanism of vibrational-to-translational energy
transfer.

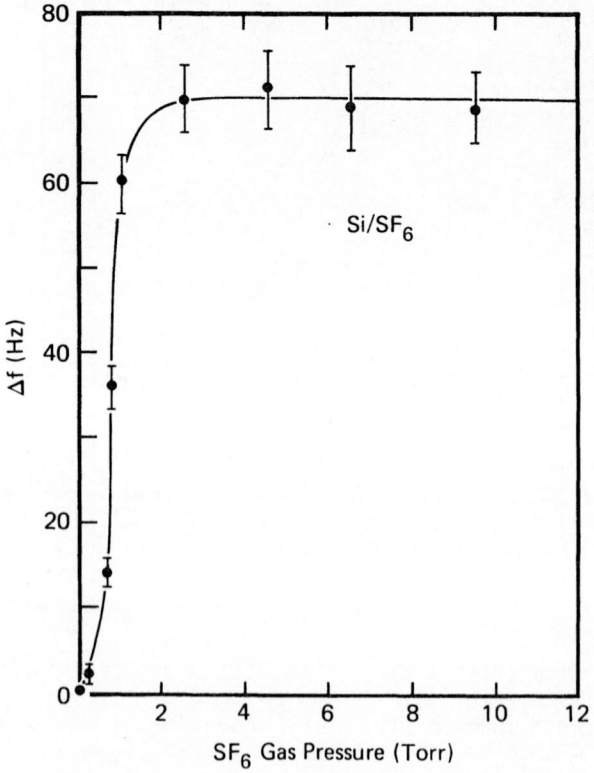

Fig. 2.   Δf versus SF₆ gas pressure : I = 1 J/cm² at 942.4 cm⁻¹,
           normal incidence, Si at 20°C, 30 laser pulses for each
           data point.

When the Si microbalance is positioned parallel to the unfocused laser beam about 1-2 mm above the surface, no major change in
the frequency of the microbalance is detected for up to 200 pulses
at 942.4 cm⁻¹ and 2.0 Torr gas pressure. Apparently, because of
the rapid collisional relaxation rate, most SF₆* molecules are
deactivated before reaching the solid surface. When the laser
beam is focused parallel to the sample surface, however, fluorine
atoms are generated by SF₆ multiple photon dissociation[5],[6].
These thermal F atoms can diffuse by collisions through the gas
phase and still react with the solid as observed from the frequency
increase of the microbalance placed several mm away from the focused light beam. The dynamics of Si-SF₆* interaction is clearly
quite different from that of Si-F reactions.

In a separate experiment for studying the Si-SF$_6$ interaction at low temperatures, a silicon single crystal is installed in an ultra-high vacuum system (base pressure 1×10^{-10} Torr) equipped with a mass spectrometer and an ESCA-Auger spectrometer[10] for analyzing the gaseous products formed in the laser-induced reaction and for surface characterization. The results obtained by electron spectroscopies show that SF$_6$ does not chemisorb on Si at 25°C, but it can be physisorbed on the Si surface at -150°C substrate temperature. When the solid surface covered with adsorbed SF$_6$ is irradiated by the laser pulses with I = 1 J/cm^2 at 942.4 cm^{-1}, a substantial amount of SiF$_4$ is detected by the mass spectrometer in addition to some SF$_6$ desorption. At 976 cm^{-1}, when the laser does not vibrationally excite the adsorbed SF$_6$ molecules, SiF$_4$ formation is not observed. The data indicates that the Si-SF$_6^*$ reaction can happen even at a relatively low temperature. Furthermore, the laser excitation of the Si substrate alone cannot cause the heterogeneous reaction to occur.

REFERENCES

1. J.T. Knudtson and E.M. Eyring, Ann. Rev. Phys. Chem. 25, 255 (1974) ; E. Weitz and G. Flynn, ibid. 25, 275 (1974).
2. M. Poliakoff, B. Davies and A. McNeish, M. Tranquille and J.J. Turner, Ber.Bunsenges. Phys. Chem. 82, 121 (1978) ; E. Catalano and R.E. Barletta, J. Chem. Phys. 66, 4706 (1977).
3. J.T. Yates, Jr., J.J. Zinck, S. Sheard and W.H. Weinberg, J. Chem. Phys. 70, 2266 (1979) ; S.G. Brass, D.A. Reed and G. Ehrlich, ibid. 70, 5244 (1979).
4. T.J. Chuang, J. Chem. Phys. 72, 6303 (1980).
5. P.A. Schulz, Aa. S. Sudbo, D.J. Krajnovich, H.S. Kwok, Y.R. Shen and Y.T. Lee, Ann. Rev. Phys. Chem. 30, 379 (1979).
6. P.A. Schulz, Aa. S. Sudbo, E.R. Grant, Y.R. Shen and Y.T. Lee, J. Chem. Phys. 72, 4985 (1980).
7. I. Burak, P. Houston, D.G. Sutton and J.I. Steinfeld, J. Chem. Phys. 53, 3632 (1970) ; R.D. Bates, Jr., J.T. Knudtson, G.W. Flynn and A.M. Ronn, Chem. Phys. Lett. 8, 103 (1971).
8. L.J. Stinson, J.A. Howard and R.C. Neville, J. Electrochem. Soc. 123, 551 (1976).
9. J.W. Coburn, H.F. Winters and T.J. Chuang, J. App. Phys. 48, 3532 (1977).
10. T.J. Chuang, J. Appl. Phys. 51, 2614 (1980).

INDEX